MODERN ASPECTS OF ELECTROCHEMISTRY

No. 43

Modern Aspects of Electrochemistry

Topics in Number 42 include:

- The electrochemistry and electrocatalysis of Ruthenium in regards to the development of electrodes for Polymer Electrolyte Membrane (PEM) fuel cells
- Breakthroughs in Solid Oxide Fuel Cell (SOFC) anodes and cathodes leading to improved electrocatalysis
- Electrocatalysis of the electrochemical reduction of CO_2 on numerous metals
- The interfacial phenomena of electrodeposition and codeposition, and the need for new theoretical analyses of the electrode-electrolyte interface
- Advantages of scanning tunneling microscopy (STM) in understanding the basics of catalysis, electrocatalysis and electrodeposition
- The role of electrochemistry in emerging technologies including electrodeposition and electroforming at the micro and nano levels, semiconductor and information storage, including magnetic storage devices, and modern medicine

Topics in Number 41 include:

- Solid State Electrochemistry, including the major electrochemical parameters needed for the treatment of electrochemical cells as well as the discussion of electrochemical energy storage and conversion devices such as fuel cells
- Nanoporous carbon and its electrochemical application to electrode materials for super capacitors in relationship to the key role nanoporous carbons have played in the purification of liquids and the storage of energy
- The analysis of variance and covariance in electrochemical science and engineering
- The use of graphs in electrochemical reaction networks, specifically: (1) reaction species graphs, (2) reaction mechanism graphs, and (3) reaction route graphs

MODERN ASPECTS OF ELECTROCHEMISTRY

No. 43

Modeling and Numerical Simulations

Edited by

MORDECHAY SCHLESINGER

University of Windsor
ON, Canada

Mordechay Schlesinger
Department of Physics
University of Windsor
Windsor ON N9B 3P4
Canada
msch@uwindsor.ca

ISBN: 978-0-387-49580-4   e-ISBN: 978-0-387-49582-8
DOI: 10.1007/978-0-387-49582-8

Library of Congress Control Number: 2008936033

© 2009 Springer Science+Business Media, LLC
All rights reserved. This work may not be translated or copied in whole or in part without the written permission of the publisher (Springer Science+Business Media, LLC, 233 Spring Street, New York, NY 10013, USA), except for brief excerpts in connection with reviews or scholarly analysis. Use in connection with any form of information storage and retrieval, electronic adaptation, computer software, or by similar or dissimilar methodology now known or hereafter developed is forbidden.
The use in this publication of trade names, trademarks, service marks, and similar terms, even if they are not identified as such, is not to be taken as an expression of opinion as to whether or not they are subject to proprietary rights.

Printed on acid-free paper

9 8 7 6 5 4 3 2 1

springer.com

Preface

The present volume is devoted to modeling and numerical simulations in electrochemistry. Such a volume cannot be expected to limit itself to the treatment of topics and systems that narrowly defined the field of electrochemistry in the past. On the contrary, the clear demarcation lines between disciplines have become increasingly less pronounced. This positive development, which is considered by many to be the hallmark of the new century, means that related or neighboring fields are in position to cross-fertilize each other as never before, which will benefit everyone. It is in this light that readers of the present volume should view the eight chapters presented here.

Chapter 1 introduces a discussion of the two tools widely considered to be the most important for modeling. These are the finite element and finite difference methods. Chapter 2, by G. Drake, is an in-depth presentation of modeling in atomic physics. It becomes evident that even for as few as three bodies an analytical treatment is not possible and one must resort to a type of modeling. Chapter 3, by Lasia, treats the modeling of impedance of porous electrodes. Chapter 4 is authored by Kottke, Fedorov and Gole. They treat multiscale mass transport in porous silicon gas sensors.

The next three chapters deal with modeling in the area of fuel cells. In Chapter 5, Eikerling and Malek take up the issue of electrochemical materials for PEM fuel cells, while in Chapter 6, Meyers models catalyst structure degradation in PEM fuel cells. Chapter 7, which concludes this group of three chapters, presents a thorough discussion by Weber, Balliet, Gunterman, and Newman of modeling water management in PEM fuel cells. Chapter 8, by Verbrugge, deals with modeling of electrochemical energy storage devices for hybrid electric vehicle applications.

Each chapter is self contained and independent of the other chapters, which means that the chapters do not have to be read in consecutive order or as a continuum. Readers who are familiar with the material in certain chapters may skip those chapters and still derive maximum benefit from the chapters they read.

Finally, thanks are due to each of the fifteen authors who helped make the volume possible.

Windsor, Ontario, Canada   *Mordechay Schlesinger*

Contents

Preface v

List of Contributors ix

1. Mathematical Modeling in Electrochemistry
Mordechay Schlesinger 1

2. High Precision Atomic Theory: Tests of Fundamental Understanding
G.W.F. Drake, Qixue Wu and Zheng Zhong 33

3. Modeling of Impedance of Porous Electrodes
Andrzej Lasia 67

4. Multiscale Mass Transport in Porous Silicon Gas Sensors
Peter A. Kottke, Andrei G. Fedorov and James L. Gole 139

5. Electrochemical Materials for PEM Fuel Cells: Insights from Physical Theory and Simulation
Michael H. Eikerling and Kourosh Malek 169

6. Modeling of Catalyst Structure Degradation in PEM Fuel Cells
Jeremy P. Meyers 249

7. Modeling Water Management in Polymer-Electrolyte Fuel Cells
Adam Z. Weber, Ryan Balliet, Haluna P. Gunterman and John Newman 273

8. Adaptive Characterization and Modeling of Electrochemical Energy Storage Devices for Hybrid Electric Vehicle Applications
Mark W. Verbrugge 417

Index 525

List of Contributors

Ryan Balliet
Department of Chemical Engineering, University of California, 201 Gilman Hall, Berkeley, CA 94720-1462

G.W.F. Drake
Department of Physics, University of Windsor, Windsor, Ontario, Canada N9B 3P4

Michael H. Eikerling
Department of Chemistry, Simon Fraser University, 8888 University Drive, Burnaby, British Columbia, Canada, V5A 1S6

Andrei G. Fedorov
G.W. Woodruff School of Mechanical Engineering, Petit Institute for Bioengineering and Bioscience, Georgia Institute of Technology, 315 Ferst Dr. Atlanta, GA 30332-0405

James L. Gole
G.W. Woodruff School of Mechanical Engineering, School of Physics, Georgia Institute of Technology, 837 State St., Atlanta, GA 30332-0405

Haluna P. Gunterman
Department of Chemical Engineering, University of California, 201 Gilman Hall, Berkeley, CA 94720-1462

Peter A. Kottke
G.W. Woodruff School of Mechanical Engineering, Georgia Institute of Technology, 771 Ferst Dr. NW, Atlanta, GA 30332-0405

Andrzej Lasia
Département de Chimie, Université de Sherbrooke, Québec, Canada J1K 2R1

Kourosh Malek
National Research Council of Canada, Institute for Fuel Cell Innovation, 4250 Westbrook Mall, Vancouver, British Columbia, Canada, V6T 1W5

Jeremy P. Meyers
Department of Mechanical Engineering, The University of Texas at Austin, 1 University Station, Austin, TX 78712

John Newman
Department of Chemical Engineering, University of California, 201 Gilman Hall, Berkeley, CA 94720-1462

Mark W. Verbrugge
General Motors Research and Development, 30500 Mound Road, Warren, MI 48090-9055

Adam Z. Weber
Environmental Energy Technologies Division, Lawrence Berkeley National Laboratory, 1 Cyclotron Rd., Berkeley, CA 94720

Qixue Wu
Department of Physics, University of Windsor, Windsor, Ontario, Canada N9B 3P4

Zheng Zhong
Department of Physics, University of Windsor, Windsor, Ontario, Canada N9B 3P4

1

Mathematical Modeling in Electrochemistry*

Mordechay Schlesinger

Department of Physics, University of Windsor, Windsor ON, Canada N9B 3P4

I. INTRODUCTION

The specific intended theme of the present volume is modeling and numerical simulation in electrochemistry. In general, theoretical analysis and simulation may be performed for a particular phenomenon or a collection of phenomena that occur in a specific system. The objective of the analysis may be to understand and elucidate the phenomena and their impact on system performance and how to design the various components of a system to harvest or to avoid a detrimental influence. Thus, the ultimate objective is to carry out either performance evaluation of an existing system and/or attempt performance prediction of a new design. The aim of this chapter is to introduce and review the nature of mathematical modeling in general and in the context of modern electrochemistry in particular. This chapter attempts to describe how current and emerging trends in hardware and software computer applications and system development are intended to assist practitioners. One rather modern trend is toward the merger of these two disciplines as computer-aided mathematical modeling.

---

* The present chapter is based on Chap. 15 in *Fundamentals of Electrochemical Deposition* 2nd Edition (2006) by M. Paunovic and M. Schlesinger, Reprinted with permission of Wiley.

Most practitioners in the field of electrochemistry resort to the use of commercially available software packages when it comes to modeling. The majority of packages, of which there are a large number, are rather user-friendly and do not require familiarity with their internal content and makeup for their successful application. In most cases, however, such familiarity and understanding may be of invaluable benefit. Understanding the process of modeling should result in acquisition of the system that best fits the setup to be modeled. In the same way, when it comes to possible revisions and other modifications, again, using the software as "black boxes" makes it virtually impossible to obtain optimal results. It is for these reasons that readers who may want to perform modeling will find the present chapter very useful.

As indicated above, there are a large number of modeling packages on the market. Some of those are mentioned below. In the vast majority, differential equations that describe the electrochemical setup are solved using numeric methods. Two of the most common methods are the *finite-difference* method and the *finite-elements* method. These are discussed in some detail in this chapter, including example calculations in Sect. III. We begin with a few general remarks.

The term *modeling* refers to a process of determining an appropriate description of reality that approximates its behavior to some specified degree of accuracy. Models are constructed using well-understood primitive components, or building blocks, defined by their inherent functionality, as well as by their interaction mechanism – typically the manner by which information or data are communicated among them. The activity of producing models promotes a greater understanding of reality by virtue of the need to understand both the primitive components and the linkages between them. Furthermore, to make understanding feasible, models can be adapted to provide analytic and predictive power to researchers, developers and producers alike.

A paper airplane and a plasticine car each represent models that might apply, for example, in aerospace or automotive research and development. That each approximates the reality of actual aircraft and motor vehicles is intuitively obvious, yet it is immediately clear that neither suffices to explain the theory of their operation or to overcome production problems. Paper and plasticine and even fashion modeling provide types of media that appeal to visual appreciation but fail to provide deep understanding of fundamental issues. To overcome this lack, better media that allow for truer representation and insightful analyses are required.

In recent decades, computers have played an increasing role in developing models for research and applied purposes. In particular, the cost of constructing working models, or prototypes, of a research and development (R&D) product such as a plane, car, or even microelectronic component, has grown significantly. Thus, computer systems have been called on to produce simulations of greater accuracy, thereby reflecting reality in ways not achieved previously. The practicalities of modeling in the first decade of the twenty-first century continue to make it a rich and complex activity requiring a broad range of expertise in the use of both tools and fundamental theory.

Mathematical modeling is concerned with describing reality using mathematics (i.e., equations and relationships) and methodologies for solving these. The outcome of a mathematical solution (such as numbers or functions) does not always lend itself to straightforward comprehension or use. Computer modeling, on the other hand, uses mathematics to construct tools to aid in obtaining solutions while providing means to examine, test, and visualize the solution process, hence to simulate reality (i.e., produce a virtual reality). Current and developing generations of computer software and hardware require less fundamental expertise from the users of such systems, striving instead toward more intuitive approaches to productivity and understanding.

By way of illustration, one could consider how the orbits of bodies (planets, moons, etc.) in the solar system are determined. A stick and ball model might suffice to visualize planetary arrangements, but it fails to deal realistically with movement or the mutual (gravitational) interaction between bodies. Using Newton's theories, one can construct an elaborate system of variables and equations whose solution, in terms of various complicated (elliptic) functions and numbers, have little meaning except for the mathematically able. A suitable computer graphics program can transform the mathematical solution into pictures of planets revolving around the sun and incorporate in a straightforward way the effects of one body on the others. This last effect is of fundamental importance when considering the orbit of a human-made satellite traveling through the solar system. What is even more profound, perhaps, is that an end user of such a program can use it as a virtual laboratory to conduct research, even in the absence of mathematical expertise.

II. THE MATHEMATICS OF MODELING

Mathematical modeling is used extensively in electrochemistry, and as new applications arise, techniques of modeling evolve as well. A particular area of interest in electrochemistry is electrostatics. Research in electrostatics is concerned with issues relating to the properties and behaviors of static electric fields about sample element geometries, henceforth referred to as *plate electrode geometries*. These sample geometries contain various charged, uncharged, and neutral components, each of which may consist of a number of possible materials. Specific interest lies in the effects on the electric potential, electric fields, and currents as different geometry properties are varied. For instance, one may ask what would be the consequences of using different materials for different components, in a plating bath, or applying different electric charges to different elements in the system (bath). The configuration or positioning of elements with respect to each other is also relevant. These are representative of the issues that a computer modeling software system for electrostatics should allow a researcher to explore.

In many practical problems the sequence of steps to follow is: (1) geometric and physical properties specification, (2) solution method specification, (3) solution process, and (4) post-processing and analysis. Each of these steps also involves issues of verification as well as of data storage and communication. In Fig. 1 we have represented the previous points as a basic architecture for the design of a computer modeling system. We now describe the details of each of these steps.

1. Geometric and Physical Properties Specification

The first step involves specification of the model properties. Usually, one begins with geometric configuration, i.e., the physical layout of the passive and active elements. This task is facilitated by drawing a facsimile of the system. This process is crude and may give rise to errors in specification, but, after using computer drawing

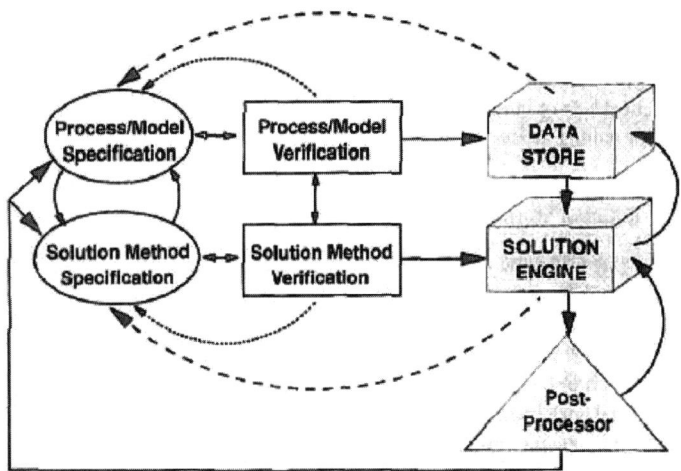

Figure 1. Basic system architecture in the mathematics of the modeling process required to specify, solve, and analyze a physical system. The system permits modeling to proceed linearly or iteratively.

tools that allow for initial rapid sketching, then followed by fine editing, an improved version of the system is readily achieved. Current computer software supports two-dimensional drawing and editing; three-dimensional applications are available for all but the extremely complicated cases.

Although the geometry can be expressed entirely mathematically in terms of formulas and relations, to do so may be awkward for those proficient in mathematics and quite incomprehensible for those who are not. It is often the case that curves and surfaces must be specified functionally, parametrically, or as piecewise sections, all of which add burden and potential error in specification.

Additionally, one must specify the physical properties and individual components of the system. Annotation of the facsimile drawing with digital or analog component properties such as voltage and resistance is accomplished in a straightforward fashion using tabular storage for quick lookup or formula storage that can be interpreted at required points or times for purposes of calculation. For instance, a set of discrete sample values may have been obtained by direct measurement in one case, while in other cases a theoretical formula could be subject to testing for self-consistency using the model.

The process of specification should always be subjected to verification to ensure accuracy and meaning in the data provided. Even without recourse to full-scale calculation of the solution, internal consistency of the geometry can be checked, as can closure of curves or overlap of distinct components, whereas physical properties can be matched, say, with tables of established values representing material properties, or compared against experience accrued by modelers. In Fig. 1 each operational component is connected multiply and reversibly with other components, illustrating the practical side of modeling, where one is often required to repeat steps to correct, clarify, or modify actions taken previously.

2. Solution Method Specification

For the second step one establishes what we term, a *solution method*. The system under consideration may be static, dynamic, or both. Static cases require solving a *boundary value problem*, whereas dynamic cases involve an *initial value problem*. For the illustrative problem, we discuss the solution of a static Laplace (no sources) or Poisson (sources) equation such as (1), here

$$-\nabla \cdot \mathbf{E} = \nabla^2 U(x, y) = \left(\frac{\partial^2}{\partial x^2} + \frac{\partial^2}{\partial y^2} \right) U(x, y) = 4\pi\rho(x, y) \quad (1)$$

for scalar potential $U(x, y)$, electric field vector $\mathbf{E} = \nabla U(x, y)$, and source function $\rho(x, y)$. Our ultimate goal is to obtain solutions of U, hence $\mathbf{E}$, given a specification of the source, $\rho(x, y)$ presumably in the preceding step. As for the source term itself, $\rho(x, y)$, it may also depend on the potential U and even derivatives of same. If these appear to first degree, that is, linear equations in U and derivatives, solutions are generally easier to obtain than with nonlinear forms of (1). For our purposes we will be concerned only with linear cases.

In exceptional circumstances *analytic solutions* exist that are expressible in terms of standard functions. Although numerical methods must be used to obtain solutions in general, it is useful nonetheless to briefly review analytic methods.

Consider (1) with no sources (Laplace) applied to a square plate with U defined everywhere on the boundary. If the problem specification is symmetric under interchange of the x and y directions, the Laplace equation may then be separable with solutions of the general form $U(x, y) = X(x)Y(y)$, i.e.,

$$U(x,y) = A_0xy + B_0x + C_0y + D_0 + \sum_{n=1}^{\infty}(A_n \cos \lambda_n x \cosh \lambda_n y$$
$$+ B_0 \cos \lambda_n x \sinh \lambda_n y + C_n \sinh \lambda_n x \cosh \lambda_n y$$
$$+ D_n \sin \lambda_n x \sinh \lambda_n y) \tag{2}$$

with the various constants $(A_n, B_n, C_n, D_n, \lambda_n,)$ determined by the requirement to fit the boundary conditions. At the corner $x = y = 0$, for example, it follows that $U(0,0) = D_0 + \sum_{n=1}^{\infty} A_n$. "Fitting" the solution to the boundary and determining the (infinite) number of constants is possible only in cases where specific functional or algebraic relationships exist. For example, if the sides of the plate have potentials that vary linearly from $x = 0$ to $x = 1$ at $y = 0$ and 1 (and similarly along y), then we may have

$$U(0 \leq x \leq 1, y = 0)$$
$$= B_0 x + D_0 + \sum_{n=1}^{\infty}(A_n \cos \lambda_n x + C_n \sin \lambda_n x)$$
$$U(0 \leq x \leq 1, y = 1) \tag{3}$$
$$= (A_0 + B_0)x + C_0 + D_0 + \sum_{n=1}^{\infty}[(A_n \cos \lambda_n x$$
$$+ C_n \sin \lambda_n x) \cosh \lambda_n + (B_n \cos \lambda_n + D_n \sin \lambda_n x) \sinh \lambda_n]$$

from which one might deduce that the complete set of A, B, C, D constants $(n > 0)$ are zero; hence, $A_0 = U(1,1) - (U(1,0) - (U(0,1) + U(0,0))$, $B_0 = U(1,0) - (U(0,0)$, $C_0 = U(0,1) - U(0,0)$, and $D_0 = U(0,0)$.

Often, in actual applications, a finite, discrete set of values on the boundary is available through direct measurement. Consequently, therefore, solving for the general solution proves almost impossible. The use of discretely sampled data, however, suggests the need for specialized techniques in order to obtain solutions. At the outset, then, we put aside analytic solutions as being interesting as guides, but rarely useful in practice.

All numerical techniques require application of *sampling theory*. Briefly stated, one chooses a representative sample of points within the region of interest and at each point attempts to calculate iteratively the most accurate solution possible, guided by self-consistency of local solutions with each other and with the boundary conditions specified.

We describe two seemingly contrasting techniques, finite-difference and finite-element methods (1, 2).

Finite-difference methods are based on the specific relationship between the potential at a given sample point and the potentials at nearby, or local, points; the relationship is derived using Taylor type expansion, assuming that the actual potential is continuously differentiable (to at least to second degree).

One category of finite-difference method uses a rectangular grid. In this approach one covers the specified layout with a grid, or mesh, as shown in Fig. 2a. When curvilinear boundaries are involved, it is possible to sample the boundaries with only limited accuracy and then only by using unequal steps in the x and y directions. Using the five-point probe shown in Fig. 2b, at each point one approximates Laplace's equation referring to the four neighboring points above, below, and to either side of the central point. For uni-

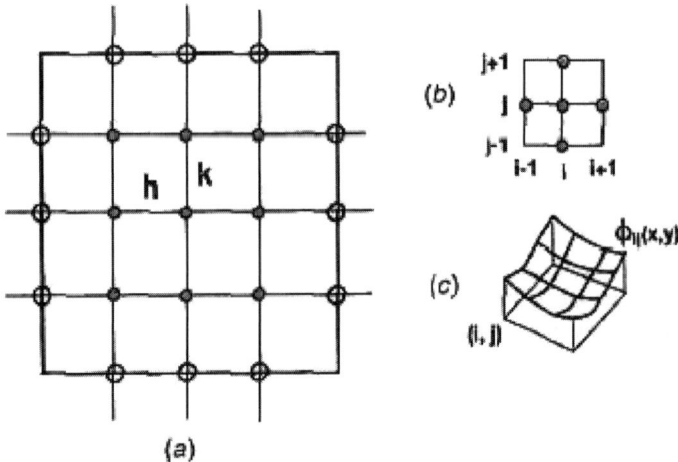

Figure 2. Region of interest for computing potential based on Laplace or Poisson equations, where (**a**) a complete rectangular grid is established to cover the region, which may be adapted to finite-difference techniques using (**b**) a five-point method, or (**c**) a finite-element approach based on sampling functions.

form mesh sizes h and k along the x and y axes, for instance, the second partial derivatives are approximated to second order in the derivatives, about center point potential $U_{i,j} = U(x_i, y_j)$,

$$\frac{\partial^2 U_{i,j}}{\partial x^2} = \frac{1}{h^2}(U_{i+1,j} + U_{i-1,j} - 2U_{i,j}),$$
$$\frac{\partial^2 U_{i,j}}{\partial y^2} = \frac{1}{k^2}(U_{i,j+1} + U_{i,j-1} - 2U_{i,j}). \tag{4}$$

For non-uniform meshes or higher-order derivative approximations, the various U coefficients are more complicated algebraically, but the description that follows is essentially the same for all cases.

Collecting the equations for all sample points (x_i, y_j) into vectors and matrices, one recasts the problem in the form $\mathbf{AU} = \mathbf{b}$, where unknown potentials $U = (U_{I1}....U_{M1}....U_{MN})$ are organized by the N rows and M columns of the rectangular grid, $\mathbf{b}$ is a vector that represents the known values of U on the boundary of the region in question (as well as source information at each point in the region), and $\mathbf{A}$ is an $NM \times NM$ matrix of U-coefficients derived from the approximating equations above. The structure of $\mathbf{A}$ is determined by the U-coefficients; that is, $\mathbf{A}$ may be nonzero in most element positions or quite sparse, as shown in the following matrix for a 4 × 4 mesh:

$$A = \begin{bmatrix} D & B & 0 & 0 \\ B & D & B & 0 \\ 0 & B & D & B \\ 0 & 0 & B & D \end{bmatrix}, D = \begin{bmatrix} 1 & \alpha & 0 & 0 \\ \alpha & 1 & \alpha & 0 \\ 0 & \alpha & 1 & \alpha \\ 0 & 0 & \alpha & 1 \end{bmatrix}, B = \begin{bmatrix} \beta & 0 & 0 & 0 \\ 0 & \beta & 0 & 0 \\ 0 & 0 & \beta & 0 \\ 0 & 0 & 0 & \beta \end{bmatrix}. \tag{5}$$

Matrix $\mathbf{D}$ has 1s along the diagonal, reflecting the use of a normalized discrete Laplace equation with $\alpha = \{-k^2/[2(h^2+k^2)]\}$, and $\mathbf{B}$ is a multiple of the identity matrix with $\beta = \{-h^2/[2(h^2+k^2)]\}$. Matrix $\mathbf{A}$ displays a sparse block structure whose off diagonal coefficients must be less than 1 to converge to a solution.

For the approximation to be valid, the mesh must be sufficiently small; hence, the number of sample points must be rather large. Programming such a mathematical system is straightforward in principle but extremely difficult to compute in practice. Obtaining a solution of the system of simultaneous linear equations is time consuming and in many cases exhibits pathological behavior where the "solution" generated is patently unrealistic. In many cases of interest, the form of matrix **A** can be drastically simplified, however. The five-point grid, for instance, results in the matrix structure shown in (5), which is triblock diagonal: that is, along the diagonal are $M \times M$ blocks that are tridiagonal, and additional diagonal blocks arise on the sub and super-diagonals.

At this point, specification of the finite-difference solution method is complete in that we have chosen to utilize the finite-difference scheme and have specified the mesh properties and the sampling of points required to provide the desired approximation to the derivatives of U. Such systems can be solved efficiently even for N and M large ($>1,000$), although the time required by typical calculations range from several minutes to hours, depending on the type of computers used, which eliminates some from consideration in those time-critical situations. We defer additional discussion of achieving solutions until the next step.

Looking ahead to the issue of solutions, it is important to realize that what is being sought by the solution method is a discrete set of points, $\{x_i, y_j, U(x_i, y_j)\}$ which specify the values of the potentials at the grid locations. To obtain values of the potentials at other points lying between the sampling locations, other techniques can be employed. Straightforward linear interpolation is one such method that is simple to implement and efficient to compute, but it suffers from a lack of sufficient accuracy required in many modeling circumstances. Other forms of interpolation involve use of higher-order polynomials that increase the accuracy of approximation with increased difficulty of implementation and cost of computation. The use of polynomials to evaluate potentials at arbitrary points leads in a natural way to our next method, however.

An alternative approach to finite difference or differencing *involves finite-element methods*. In the former approach one seeks the underlying behavior of U by solving for it numerically at all sample points in the grid. With finite elements one expresses the solution in terms of other functions (whose behavior is well known) appropriately combined to obtain U to desired accuracy throughout the region of interest. The choice of element functions reflects sampling that covers the entire region, notwithstanding curvilinear boundaries, and that are "well behaved" within each element. In many instances the sampling is more flexible and therefore more accurate than with finite difference.

Such methods start by assuming that

$$U(x, y) = \sum_{k=1}^{L} u_k \phi_k(x, y), \tag{6}$$

where the u terms represent a set of unknown "blending" parameters that produce a mixture (linear combination) of L known sampling, or *basis,* functions $\varphi_k(x, y)$. The number of sampling functions is chosen on the basis of experience with similar problems, the existence of symmetries, adaptive analysis, or combinations of these.

One choice of basis function, based on a quadrilateral patch, is illustrated in Fig. 2c. In the figure the element in the ith row and jth column of the mesh is assumed to have a magnitude that varies within the patch; the derivative properties may be important as well. The choice $\varphi_k(x, y)$ is not arbitrary; it is made to reflect certain mathematical qualities derived, perhaps, from prior knowledge of the general behavior of similar systems, as well as properties that simplify the solution process to follow. One immediately practical constraint is that the $\varphi_k(x, y)$ must satisfy the boundary conditions. Another property is that the patches meet smoothly at the intersections; this is usually obtained by continuity of $\varphi_k(x, y)$ to first and second order in the derivatives. It is also convenient in many applications to choose combinations of products of functions separately dependent on x and y reminiscent of the analytic solution (see (2)).

As with finite differences, the finite-element approach can be recast, using vectors and matrices, in the form $\mathbf{Au} = \mathbf{b}$, with $\mathbf{A}$ and $\mathbf{b}$ known and $\mathbf{u}$ to be determined. There are two basic approaches. In the first case, referred to as *collocation,* substitution of (6) in (1) leads to

$$\nabla^2 U(x_p, y_p) = \sum_{k=1}^{L} \nabla^2 \phi_k(x_p, y_p) \mathbf{u}_k = \mathbf{A}_p \cdot \mathbf{u} = \mathbf{b}(x_p, y_p), \tag{7}$$

where (x_p, y_p) refers to the pth sample point [for rectangular grids with constant spacing, M rows and N columns, the quadrilateral element in row r and column c it follows that $p = N(r-1) + c$ and $1 \leq p \leq L = NM$ and $\mathbf{A}_p$ refers to the pth row of the matrix A whose elements are $\mathbf{A}_{p,k} = \nabla^2 \varphi_k(x_p, y_p)$. Thus, L equations are generated as required to determine uniquely the L unknown u_k coefficients.

In the second approach, called the *Galerkin method,* one uses the property that the sampling functions satisfy the boundary conditions to write

$$\int \phi_m(x,y)\nabla^2 U(x,y) dx\, dy$$

$$= \sum_{k=1}^{L} \int \phi_m(x,y)\nabla^2 \phi_k(x,y)\, dx\, dy\, u_k$$

$$= \sum_{k=1}^{L} \int \nabla\phi_m(x,y) \cdot \nabla\phi_k(x,y)\, dx\, dy\, u_k = \mathbf{A} \cdot \mathbf{u}$$

$$= \int b(x,y)\phi_m(x,y)\, dx\, dy = \mathbf{b}. \tag{8}$$

The second step in this equation involves a property called *Green's identity.* Using either method brings one to the point where the solutions of both require the same basic approaches: solving a matrix problem. As in the case of collocation, the L sample points are used to generate the rows of the $\mathbf{A}$ matrix and $\mathbf{b}$ vector whose elements are written

$$\mathbf{A}_{m,k} = \int \nabla\varphi_m(x,y) \cdot \nabla\varphi_k(x,y)\, dx\, dy \quad \text{and}$$
$$\mathbf{b}_m = \int b(x,y)\varphi_m(x,y)\, dx\, dy, \text{respectively.}$$

Solving for $\mathbf{u}$ is greatly facilitated by choosing basis functions whose properties simplify the structure of $\mathbf{A}$, i.e., functions that lead to easily computed methods of matrix inversion. In one such property, called *locality,* each function has a primary value, hence influence, within a given element region and a much smaller, or zero, value, in other regions. For example, if one chooses $\varphi_k(x,y) = a_k xy + b_k x + c_k y + d_k$, which is piecewise linear in both x and y within each patch k, we again have a tri-block diagonal matrix structure, as in (5), and the method is in fact fully equivalent to the five-point finite-difference technique discussed. However, this method cannot be used for the collocation technique, since second derivatives vanish identically and lack continuity at the patch boundaries, and hence will appear bumpy instead of smooth. In many cases of interest, the method involves using *piecewise polynomials* of sufficient order (of at least cubic degree in both x and y) so as to ensure the desired patch edge continuity while allowing for matrix structure that simplifies computation.

In cases where dynamic effects must be considered, the problem is typified by **A** and **b** matrices that are parametrized, say, by time or other quantities. One such example involves modeling corrosion effects where time, acidity, and material thickness and age might be relevant dynamical parameters. In most cases one calculates a sequence of static solutions at time steps t_m, which are then pieced together to fit initial (and final) conditions.

At this point what must be made clearly understood is that although the specific interpretation of the results differs among the methods, the underlying set of analytical tools that one brings to bear on the problem, once it has been transformed into the language of vectors and matrices, is of course, the same.

3. Solution Process

Many methods exist for *solving* the basic form $\mathbf{AU} = \mathbf{b}$ for the potential $\mathbf{U} = \mathbf{A}^{-1}\mathbf{b}$. The methods depend on various features exhibited by the matrices themselves, immediate byproducts of how the problem was set up in the previous stage of specification. A general method, assuming that **A** is nonsingular (determinant is nonzero), is to find the inverse matrix $\mathbf{A}^{-1}$ using techniques such as Gauss elimination. In practice, however, this approach is not computationally workable. Typically, one looks for features of the problem that simplify **A**.

For example, uniform meshes give rise to highly structured and simplified forms of matrices, as in (5), which are amenable to rapid solution techniques but are very sensitive to the size of the mesh: The larger the mesh, the poorer the solution. More complicated meshes and formulations of the approximation scheme used to set up the solution scheme are used more rarely because of difficulties in programming them and their increased cost in time to achieve solution. Similarly, finite-element schemes have varying degrees of success, depending on choice of mesh, sample functions, and so on.

Relaxation methods involve iteratively seeking a convergent solution to the Laplace equation. In the present case, for instance, if we rewrite the coefficient matrix $\mathbf{A} = \mathbf{I} + \mathbf{E}$ where the latter matrix consists of elements that are all "small" compared to 1, the matrix Laplace equation takes the form $\mathbf{U}^{(r+1)} + \mathbf{EU} + \mathbf{b}$. One begins the calculation with values $\mathbf{U}^{(1)} = \mathbf{b}$ [or equivalently $\mathbf{U}^{(0)} = 0$] and iteratively computes successive values $\mathbf{U}^{(r)}$. The calculation terminates when a specified limit of accuracy is achieved. One such measure involves calculating the proportional differences

$$D_{ij}^{(r)} = 2\frac{\left|U_{i,j}^{(r)} - U_{i,j}^{(r-1)}\right|}{U_{i,j}^{(r)} + U_{i,j}^{(r-1)}} \quad (9)$$

stopping when the average is less than the tolerance. The advantages of such methods are speed of computation and circumvention of much of the need for coefficient matrix storage during the course of the computation, although two complete sets of U vectors must be maintained, the "old" one and the updated solution.

Two approaches that rely on the availability of suitable hardware involve pipelining or parallelizing the equations and solution method. In the first technique the calculation is broken into independent stages, each of which performs a complete set of operations on an input data set, producing a processed data set as output. Specialized hardware accepts values and processes them to each subsequent stage of calculation. As each stage is finished, another data set is input immediately, thus ensuring that the pipeline is always full and busy. In parallel computations the strategy is expanded even further to encompass a family of separate processors that accept independent sets of data and simultaneously apply the same or different processes.

4. Postprocessing and Analysis

The last stage of mathematical modeling consists of *interpretation of the results.* Here lies its greatest weakness, at least prior to the introduction of computer systems. To visualize the behavior of many thousands of data points, it is essential to resort to machine-assisted methods (3). In principle, however, the complete database constituting the numerical solution contains all information required to deduce other quantitative results.

Figure 3 outlines some of the many possible choices of postprocessing options of interest to modeling. Questions regarding, say, optimal cost or operating temperature range are expected to produce simple numerical outputs or, perhaps, simple graphical images such as pie charts or histograms. On the other hand, analysis may require intermediate stages at which the general solution properties must be studied to provide clues as to how to proceed to later stages of development.

Another significant issue in analysis of results lies in altering the parameters of the initial specification. Alteration might be achieved in some small way, such as slight geometric deformation or subtle variation (perturbation) of one or more physical properties (e.g., voltage or conductivity), as shown in Fig. 4, which illustrates a slight rotation of the triangular element to produce a *local parametric solution.* It might also be specified as part of a *global parametric solution* to the original problem. In the first case, by restriction of the primary region, within which the change is expected to manifest itself, it is possible to compute minor modifications to the initial solution in rapid fashion, indicated by the darker, solid contours in Fig. 4. Global solutions require substantially more time, but once calculated, they can be used to perform detailed multivariable (dimensional) analysis, particularly cost/benefit analyses and the like.

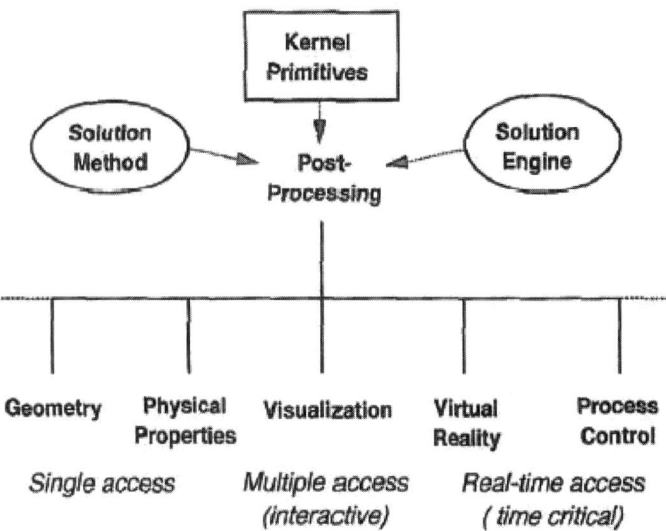

Figure 3. Once a solution has been obtained, the choice of post processing options may be quite varied, with each option requiring different means of accessing the solution data, ranging from one-time access through interactive manipulation of the solution to real-time critical applications, in latter cases requiring re-computation or updating of the solution data.

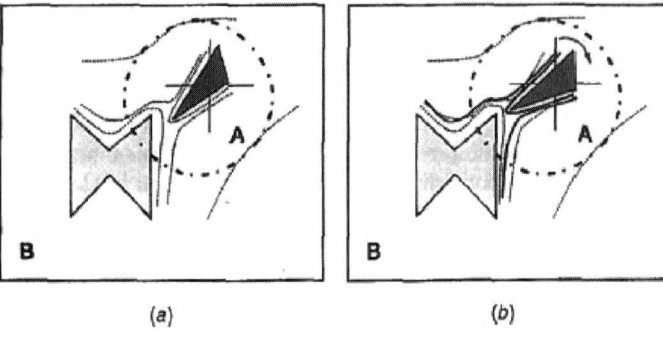

Figure 4. Deformation of a solution state to obtain rapid update may be obtained quickly by restricting computational effort to the region of greatest change.

III. EXAMPLE CALCULATIONS

We begin with the application of the *Finite-Difference* method for solving a linear boundary value problem as follows:

Given the second derivative of the function $y(x)$ in the interval $x = 0$ to $x = 4$ as

$$y''(x) = y(x) + x(x - 4), 0 \leq x \leq 4,$$

the boundary conditions, i.e., the values of $y(x)$ on the two 'edges', are given as

$$y(0) = y(4) = 0.$$

We determine the functions values at $x = 1, 2, 3$ in steps, as follows:

a. Replace the derivative with finite difference approximation.
b. Write down three (ODE) ordinary differential equations.
c. Obtain a system of linear equations.
d. Solve the equations for the (approximate) values of x_i, where $I = 1, 2, 3$.

First, we divide the interval of interest into $n = 4$ subintervals separated by grid points x_0, x_1, x_2, x_3, and x_4.

Mathematical Modeling in Electrochemistry

Next, we calculate the value of $y(x)$ on these grid points and denote them as y_i, $i = 1, \ldots, 4$. From the boundary condition you have $y_1, = y_4, = 0$.)

Finally we proceed as indicated

a. Replace the derivatives with finite-difference approximations as follows:

$$y''(x_i) \approx (y_{i+1} - 2y_i + y_{i-1})/h^2.$$

Here $h = x_i - x_{i-1}$ is the subinterval length (in our case $h = 1$) Also, as in expression 4, this approximate form for the second-derivative is called the *central-difference approximation*.)

b. The approximated ODE at the three inner points, $i = 1 - 3$ are represented as

$$y_2 - 2y_1 + 0 = y_1 + 1 \cdot (1 - 4),$$
$$y_3 - 2y_2 + y_1 = y_2 + 2 \cdot (2 - 4),$$
$$0 - 2y_3 + y_2 = y_3 + 3 \cdot (3 - 4).$$

c. The following system of linear equations, which may be cast in tridiagonal matrix form, yield the required values:

$$-3y_1 + y_2 = -3,$$
$$y_1 - 3y_2 + y_3 = -4,$$
$$y_2 - 3y_3 = -3.$$

d. The system of the linear equations is solved to obtain the values

$$y_1 = 1.857, y_2 = 2.571, y_3 = 1.857.$$

Note: To obtain higher-accuracy solutions, we may divide the interval into more subintervals.

Increase n from 4 as in the previous question, to 8 and decrease h from 1 to 1/2. One will generally obtain a tridiagonal system of $(n - 1)$ linear equations, up to 7 in the present case.

For large n, solving the system of equations with paper and pencil becomes impractical and it is necessary to find algorithms suitable for computation by computers.

We note that the accurate values are

$$y_1 = 1.834, y_2 = 2.532, y_3 = 1.834.$$

Next, we present two worked examples demonstrating application of the *finite element* method described earlier. In the first example, we demonstrate a familiar simple electrostatic problem in which a metal sphere is held at a potential of 1 V and is enclosed in a cubical grounded metal box (Fig. 5). The task at hand is to find the potential distribution in volume V between the sphere and the box.

As is well known from elementary elecrostatics, the equation governing the electrostatic potential, Φ, in the volume V *is* the Laplace equation

$$\nabla^2 \Phi = 0. \tag{10}$$

Another (integral) way of stating this is that for all weighting functions w we may write

$$\int_v w \nabla^2 \Phi \, dV = 0. \tag{11}$$

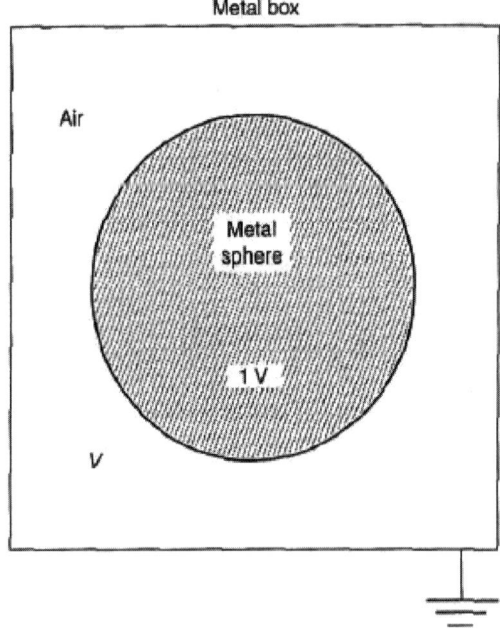

Figure 5. An electrostatic problem.

Mathematical Modeling in Electrochemistry

This form is referred to as the *weighted-residual formulation of* the problem. What is required to be determined is the potential Φ, which solves this equation while it assumes the value 1 on the surface of the sphere and the value 0 on the inner surface of the box. (One may constrain w *to* zero on both surfaces, since this has no effect on the equivalence of the two expressions.) Equation (11) may be used as the launching pad of a finite-element analysis. However, as it stands, it contains second-order derivatives, and it is convenient to reduce these to first-order derivatives. This reduction is accomplished by the use of the vector identity known as Green's identity and is akin to integration by parts. The additional term produced gives a surface integral when the divergence theorem is applied to it and the surface integral vanishes because w is zero on the surfaces. The weighted-residual formulation thus becomes

$$\int_v \nabla w \cdot \nabla \Phi \, dV = 0. \tag{12}$$

To solve this expression numerically, which is, after all, what the finite-element method is all about, the volume V is subdivided into a number of finite elements. For example, tetrahedral shaped elements might be used, as shown in Fig. 6.

Of course, having flat faces, these elements cannot represent the curved surface on the sphere exactly. Using an adequately large

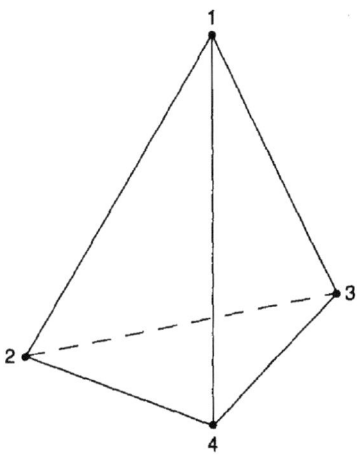

Figure 6. Tetrahedral element.

number of such elements, a sufficiently accurate polyhedral approximation is possible. Inside each tetrahedron the unknown potential is approximated using a polynomial that is first order in the space coordinates (x, y, z). The polynomial can be so written that the four unknown coefficients are the values of the potential at the four nodes (vertices) of the tetrahedron, Φ_1, Φ_2, Φ_3, and Φ_4,

$$\Phi = \Phi_1 \alpha_1 + \Phi_2 \alpha_2 + \Phi_3 \alpha_3 + \Phi_4 \alpha_4. \tag{13}$$

The quantities α, are polynomials whose dependence on x, y, z and on the size and shape of the tetrahedron is known explicitly. The polynomial α, assumes the value 1 at node i, vanishes at the other three nodes, and varies linearly in between. The reader should note that (13) thus provides a complete and continuous representation for the potential within the element while using physically meaningful quantities (the values at the four nodes) as unknown parameters. It also permits potential continuity to be imposed in a simple way between elements. For two elements with a common edge or face to agree in their values of Φ at their common nodes requires only that Φ be continuous everywhere over the edge or face. In actual practice this continuity is achieved by assigning a *global numbering* $1, \ldots, N$, to all the nodes of the finite-element mesh, such that each node has a single global number even if a given node is shared by several tetrahedral. The unknowns in the problem are then the globally numbered values of potential at the nodes $\Phi_i : I = 1, \ldots, N$. Indeed, one may think of the polynomials α_i in the same global way: α_i is a function that is 1 at global node i vanishes at the other $N - 1$ nodes, and is linear in each tetrahedron. The piecewise linear potential in V is then

$$\Phi(x, y, z) = \sum_{i=1}^{N} \Phi_i \alpha_i(x, y, z). \tag{14}$$

The α_i are known as *basis* or *trial functions*. In point of fact, not all N of the Φ are unknown. For a node i that lies on a boundary where the potential is constrained to be 0 or 1, Φ is set to that value from the outset. For simplicity it will be assumed that the first N_f nodes are interior free nodes and the remainder are boundary nodes. The problem therefore has only N_f unknowns. These then are the *degrees of freedom* of the potential function.

Since there are a finite number of degrees of freedom with which to represent the solution, it is not generally possible to satisfy

(12) exactly for all possible weighting functions. It is desirable to have no more than N_f independent weighting functions, so that (12) becomes N_f equations in N_f unknowns. Although it is possible to choose any set of weighting functions, it is convenient to use the basis functions themselves (Galerkin's method). Substituting (14) into (12) with this set of weighting functions yields the following matrix equation:

$$[\mathbf{S}]\{\boldsymbol{\Phi}\} = \{\mathbf{b}\}, \tag{15}$$

where $\{\boldsymbol{\Phi}\}$ is an N_f vector of the unknown potentials, $\{\mathbf{b}\}$ is a known N_f vector that depends on the boundary constraint, and $[\mathbf{S}]$ is a square-symmetric matrix known as the *global* or *stiffness matrix*. The term/entry i, j in $[\mathbf{S}]$ is given by the expression

$$S_{i,j} = \int_v \nabla \alpha_i \cdot \nabla \alpha_i \, \mathrm{d}V. \tag{16}$$

In general, there will be rather a large number of nodes, and consequently, $[\mathbf{S}]$ will be large. Even a modest problem may have $>10^4$ nodes. On the other hand, $[\mathbf{S}]$ is very sparse. The function α_i is nonzero only in those tetrahedra that share node i. If i and j are not nodes of the same tetrahedron, it is evident from (16) that $S_{i,j}$ will be zero. Typically, a row in $[\mathbf{S}]$ may have only 30–50 nonzero entries out of a total of many thousands. It is very important to exploit this sparsity both in constructing $[\mathbf{S}]$ and in solving (15). For example, the assembly of $[\mathbf{S}]$ is done tetrahedron by tetrahedron, not row by row. For each tetrahedron, a 4×4 *local matrix* is calculated from its four basis functions. The 16 entries of this dense local matrix are added to the global matrix in the right places, taking into account the global numbers of the four degrees of freedom. Once the solution $\{\boldsymbol{\Phi}\}$ has been found, the potential at any point can be determined from (14). If an electric field is required, it may be calculated taking the gradient of (14).

As another *example/illustration* of the *finite-element* technique, we proceed as follows. We assume a rectangular mesh of three rows and three columns with uniform step sizes $h = k = 1/5$ along the x and y axes, respectively. Further, assume that potentials on the boundary have values $\{U_{01}, U_{02}, U_{03}, U_{10}, U_{14}, U_{20}, U_{24}, U_{30}, U_{34}, U_{41}, U_{42}, U_{43}\}$, where by $U_{i,j}$ we refer to the top and bottom or left and right rows or columns as $i, j = 0$ and $i, j = 4$, respectively. The five-point sampling Laplace equation has the algebraic form

$$U_{i,j} = \frac{1}{4}(U_{i-1,j-1} + U_{i+1,j-1} + U_{i-1,j+1} + U_{i+1,j+1}), \quad (17)$$

which is just the average of the four neighboring potentials, and the matrix forms

$$\begin{bmatrix} 1 & a & 0 & b & 0 & 0 & 0 & 0 & 0 \\ a & 1 & a & 0 & b & 0 & 0 & 0 & 0 \\ 0 & a & 1 & 0 & 0 & b & 0 & 0 & 0 \\ b & 0 & 0 & 1 & a & 0 & b & 0 & 0 \\ 0 & b & 0 & a & 1 & a & 0 & b & 0 \\ 0 & 0 & b & 0 & a & 1 & 0 & 0 & b \\ 0 & 0 & 0 & b & 0 & 0 & 1 & a & 0 \\ 0 & 0 & 0 & 0 & b & 0 & a & 1 & a \\ 0 & 0 & 0 & 0 & 0 & b & 0 & a & 1 \end{bmatrix} \begin{bmatrix} U_{1,1} \\ U_{2,1} \\ U_{3,1} \\ U_{1,2} \\ U_{2,2} \\ U_{3,2} \\ U_{1,3} \\ U_{2,3} \\ U_{3,3} \end{bmatrix} = \begin{bmatrix} -bU_{1,0} - aU_{0,1} \\ -bU_{2,0} \\ -bU_{3,0} - aU_{4,1} \\ -aU_{0,2} \\ 0 \\ -aU_{4,2} \\ -aU_{0,3} - bU_{1,4} \\ -bU_{2,4} \\ -aU_{4,3} - bU_{3,4} \end{bmatrix}$$

$$= \begin{bmatrix} -b & 0 & 0 & -a & 0 & 0 & 0 & 0 & 0 & 0 & 0 & 0 \\ 0 & -b & 0 & 0 & 0 & 0 & 0 & 0 & 0 & 0 & 0 & 0 \\ 0 & 0 & -b & 0 & -a & 0 & 0 & 0 & 0 & 0 & 0 & 0 \\ 0 & 0 & 0 & 0 & 0 & -a & 0 & 0 & 0 & 0 & 0 & 0 \\ 0 & 0 & 0 & 0 & 0 & 0 & 0 & 0 & 0 & 0 & 0 & 0 \\ 0 & 0 & 0 & 0 & 0 & 0 & -a & 0 & 0 & 0 & 0 & 0 \\ 0 & 0 & 0 & 0 & 0 & 0 & -a & 0 & -b & 0 & 0 & 0 \\ 0 & 0 & 0 & 0 & 0 & 0 & 0 & 0 & 0 & -b & 0 \\ 0 & 0 & 0 & 0 & 0 & 0 & 0 & 0 & -a & 0 & 0 & -b \end{bmatrix} \begin{bmatrix} U_{1,0} \\ U_{2,0} \\ U_{3,0} \\ U_{0,1} \\ U_{4,1} \\ U_{0,2} \\ U_{4,2} \\ U_{0,3} \\ U_{4,3} \\ U_{1,4} \\ U_{2,4} \\ U_{3,4} \end{bmatrix}.$$

(18)

With $a = b = -(1/4)$ (18) has been cast in the form $\mathbf{Au} = \mathbf{b} = \mathbf{BU}$ with $\mathbf{u}$ referring to unknown quantities $U_{i,j}$, at interior points, $\mathbf{b}$ the vector between equal signs that refers to the 12 boundary values of the potentials, also referred to as vector $\mathbf{U}$, and $\mathbf{B}$ the 9×12 coefficient matrix. We state this in this way to emphasize that the unknown potentials are determined using the coefficients and the known potentials and that the matrix structures reflect the differing numbers of knowns and unknowns applying the inverse matrix $\mathbf{A}^{-1}$ to both sides, the explicit solution to this is

$$\begin{bmatrix} U_{1,1} \\ U_{2,1} \\ U_{3,1} \\ U_{1,2} \\ U_{2,2} \\ U_{3,2} \\ U_{1,3} \\ U_{2,3} \\ U_{3,3} \end{bmatrix} = \frac{1}{224} \begin{bmatrix} 67 & 22 & 7 & 67 & 7 & 22 & 6 & 7 & 3 & 7 & 6 & 3 \\ 22 & 74 & 22 & 22 & 22 & 14 & 14 & 6 & 6 & 6 & 10 & 6 \\ 7 & 22 & 67 & 7 & 67 & 6 & 22 & 3 & 7 & 3 & 6 & 7 \\ 22 & 14 & 6 & 22 & 6 & 74 & 10 & 22 & 6 & 22 & 14 & 6 \\ 14 & 28 & 14 & 14 & 14 & 28 & 28 & 14 & 14 & 14 & 28 & 14 \\ 6 & 14 & 22 & 6 & 22 & 10 & 74 & 6 & 22 & 6 & 14 & 22 \\ 7 & 6 & 3 & 7 & 3 & 22 & 6 & 67 & 7 & 67 & 22 & 7 \\ 6 & 10 & 6 & 6 & 6 & 14 & 14 & 22 & 22 & 22 & 74 & 22 \\ 3 & 6 & 7 & 3 & 7 & 6 & 22 & 7 & 67 & 7 & 22 & 67 \end{bmatrix} \begin{bmatrix} U_{1,0} \\ U_{2,0} \\ U_{3,0} \\ U_{0,1} \\ U_{4,1} \\ U_{0,2} \\ U_{4,2} \\ U_{0,3} \\ U_{4,3} \\ U_{1,4} \\ U_{2,4} \\ U_{3,4} \end{bmatrix}$$

(19)

showing how each sample boundary value is probed and participates in the expression of the solution for each interior point in the grid as a weighted average.

Although the general algebraic form is desirable from a computational point of view, it is too difficult to obtain in general and is extremely costly in terms of both computation by algebraic manipulation programs such as Maple (4), used to obtain (18), as well as by storage of the formula expressions. As seen from comparison of the terms in the matrix, the more distant a boundary point is from an interior point, the smaller is its coefficient, consistent with intuition. As the matrix expands to include more sampling points, the coefficients decrease rapidly in value; thus, unless the relative boundary potential values increase in a manner so as to offset the decreasing coefficient values, one can usually ignore the boundary contribution from points beyond some established limit, such as that prescribed by the available floating-point arithmetic hardware.

One may require additional solution points for purposes of presentation or analysis. In many cases, *interpolating polynomials* prove to be useful. Thus, one might approach our example problem using products of quartic equations of the general form $X_j^{(4)}(x) = a_j x^4 + b_j x^3 + c_j x^2 + d_j x + e_j$ and similarly for $Y_i^{(4)}(y)$. An appropriate rationale for this lies in the fact that the quartic has five parameters (order 5, degree 4). By demanding that the curves pass through each of the five sampling points $x = j/4$ ($j = 0, \ldots, 4$) for each of the five families $y = i/4$ ($i = 0, \ldots, 4$), one can always generate unique solutions to the coefficients. Additionally, the coefficients must, of course, be chosen consistent with the boundary conditions and the Laplace equation. Once obtained, the polynomial XY can be used to obtain potential values at arbitrary (x, y), as well as the positions of possible minima and maxima. This approach ultimately breaks down when the number of sample points is larger and the resulting matrix problem becomes intractable. However, this technique suggests that polynomials can be used to fashion approximate solutions that are applicable over broad areas of the interior and boundary regions.

Consistent with the notion of approximating polynomials as in the preceding paragraph, finite-element approaches attempt to simplify the solution process by carefully choosing polynomials so as to minimize the number of coefficients required to determine and simplify the matrix inversion problem. One choice is the Bezier polynomials, defined by

$$B_{t,N}(u) = \frac{N!}{t!(N-t)!} u^t (1-u)^{N-t}; \; B_{t,N}(0) = \delta_{t,0}; \; B_{t,N}(1) = \delta_{t,N}. \tag{20}$$

Curves resulting from the choice of Bezier functions blend the values of the known boundary potentials to produce interior potential values and have the appropriate smoothness properties desired in the final solution. Further, the $B_{t,N}$ have maximum values that distribute evenly through the mesh regions. For instance, for u between 0 and 1/4 in (20) the value of $B_{0,4}$ is greatest, and all other B variables approach minimum values. Thus, $B_{0,4}$ serves to sample that particular range of u values.

Other commonly employed and related sets of approximating polynomials are *Hermite polynomials* and *B splines*. Particularly in the latter case, the functions possess the desired properties of smoothness across patch boundary intersections, strong locality leading to simplification of the A coefficient matrix, and efficiency of computation. In the following discussion the B functions may be viewed, up to specific values, as any of the aforementioned types.

The approximating solution function is

$$u(x, y) = \sum_{i=0}^{N} \sum_{j=0}^{M} U_{j,i} B_{i,N}(x) B_{j,M}(y) \tag{21}$$

which must satisfy the Laplace equation

$$\nabla^2 u(x,y) = 0$$
$$= \sum_{i=0}^{N} \sum_{j=0}^{M} \left[\frac{d^2}{dx^2} B_{i,N}(x) B_{j,M}(y) + B_{i,N}(x) \frac{d^2}{dy^2} B_{j,M}(y) \right] U_{j,i} \tag{22}$$

and the boundary conditions. For the latter we note that the known boundary potentials are sampled at $(0, j/M)$, $(1, j/M)$, $(i/N, 0)$, and $(i/N, 1)$, excluding the corners, and the Bezier functions are either 0 or 1 at the interval endpoints $x, y = 0$ and 1.

In our sample problem there exist nine interior potentials that we must obtain in terms of the 12 known boundary potentials. Thus, we apply (15) at each point $(x, y) = (i/4, j/4)$ with $i, j = 1, \ldots, 3$, to generate the nine equations required to obtain the nine unknown $U_{i,j}$ in terms of the 12 known boundary values. For example, at $(x, y) = (1/4, 1/4)$ one finds that

$$-\frac{243}{32} U_{1,1} - \frac{81}{32} U_{1,2} - \frac{27}{32} U_{1,3} - \frac{81}{32} U_{2,1} - \frac{81}{128} U_{2,2}$$
$$+ \frac{9}{16} U_{2,3} + \frac{27}{32} U_{3,1} + \frac{9}{16} U_{3,2} + \frac{9}{32} U_{3,3}$$
$$= -\left(\frac{3}{512} U_{4,4} + \frac{3}{64} U_{3,4} + \frac{243}{256} U_{0,2} + \frac{135}{512} U_{4,0} + \frac{2187}{512} U_{0,0} \right.$$
$$+ \frac{243}{256} U_{2,0} + \frac{81}{64} U_{0,3} + \frac{9}{32} U_{4,1} + \frac{135}{512} U_{0,4} + \frac{81}{64} U_{3,0}$$
$$\left. + \frac{39}{256} U_{2,4} + \frac{39}{256} U_{4,2} + \frac{3}{64} U_{4,3} + \frac{9}{32} U_{1,4} \right), \tag{23}$$

where the corner potentials are included as well. If these values are available from the specification, they are used directly; otherwise, their values can be approximated consistent with other known values. Equation (23), together with the remaining equations for other interior points, can be cast in matrix form, and from there a solution is deduced.

Equation (23) displays the feature of locality that the blending functions should possess in order to be computationally advantageous: that is, during the process of matrix inversion, one wishes the calculation to proceed quickly. As mentioned earlier, the use of linear approximation functions results in at most five terms on the left side of the equation analogous to (23), yielding a much cruder approximation, but one more easily calculated. The current choice of Bezier functions, on the other hand, is rapidly convergent for methods such as relaxation, possesses excellent continuity properties (the solution is guaranteed to look and behave reasonably) and does not require substantial computation.

A final note is in order. The finite-difference and finite-element techniques are entirely equivalent from a mathematical point of view. What is different about these are the conceptualization of the problem and the resulting computational techniques to be employed. One method is not better than the other, although in particular circumstances one may clearly be superior. The point is that a modeler and modeling systems should account for both methods, as well as others not mentioned here.

IV. ADVANCED CONCEPTS

(Phase field model of an electrochemical system)

This relatively new method (10, 11), employs a so-called phase field variable. That variable, a function of position and time, describes whether the material in question is in one phase or another, i.e., the electrode or electrolyte. The behavior of this variable is governed by a partial differential equation (PDE) that is itself coupled to the relevant transport equations for the material. The interface between the phases is thus described by smooth but highly localized changes of this variable. This approach avoids the mathematically difficult problem of applying boundary conditions at an interface whose location is part of the unknown solution.

The phase field method is powerful because it easily treats complex interface shapes and topology changes. One long term goal of the approach is to treat the complex geometry, including void formation, which occurs during plating in vias and trenches for on-chip metallization.

Early models of the electrochemical interface, focused on the distribution of charges in the electrolyte. These models assume that the charges have a Boltzmann distribution and are subject to Poisson's equation. More recently, density functional models have been applied to the equilibrium electrochemical interface.

These atomic scale models describe the electrolyte with distribution functions, which have maxima at the positions of the atoms and take the electrode to be a hard, idealized surface. The equilibrium distribution of electrons and ions are computed and, finally, kinetic modeling is performed. Phase field models may be viewed as a mean field approximation of atomic scale density functional theories, and the two methods often make similar predictions.

The phase field method has also been used for the subject of solidification. The approach is motivated by the mathematical analogy between the governing equations of solidification dynamics and electroplating dynamics. For example, the solid–melt interface is analogous to the electrode–electrolyte interface. The various overpotentials of electrochemistry have analogies with the supercoolings of alloy solidification: diffusional–constitutional!, curvature, and interface attachment. Dendrites can form during solidification and during electroplating. It is not surprising, however, that one finds significant differences between the two systems. The crucial presence of charged species in electrochemistry leads to rich interactions between concentration, electrostatic potential, and phase stability.

V. COMPUTER ENGINEERING ASPECTS

The steps described earlier have been implemented, to varying degrees, in some commercial packages: Maple (Maplesoft Waterloo, Ontario, Canada), Mathcad (MathSoft, Cambridge, MA), and Maxwell (Ansoft Corp., Pittsburgh, PA), to name but a very few. In recent years two software packages seem also to have become rather widespread: Matlab (the MathWorks Co. Natick Massachusetts) and Abaqus (Abaqus, UK Ltd. Warrington UK).

A vital aspect of constructing modeling systems involves the availability of the correct software packages and tools to solve problems as they arise. In modern electrochemistry it is clear that the speed at which new problems arise may never be addressed by waiting for the next version of a favorite modeling package, not to mention issues such as cost and availability. To address the problems of new system development and construction, a new solution strategy must be developed. Figure 1 illustrates the basic system design and interaction of internal components for the modeling approach discussed in Sect. II.

Since modeling is such a dynamic and largely unpredictable type of activity, it is also the case that the data flow through such a system cannot be considered using a linear model. Out of necessity, the modeler makes certain progress, modifies specifications, performs trial calculations, and requires many alternative views of the same data. Data storage and solution generation are therefore fundamental to the system, with other components placed at a higher level in the system hierarchy.

Communication between logical modules must also be handled efficiently and may be analyzed in terms of data flow as well. An important point in this regard is that in designing the software, data must not become trapped so as to lead to deadlock the entire system, which can result from a failure to deliver data or because the module cannot transfer control to another module. The system design must take into account a consistent model for the file system, data exchange, intermodule communication protocols, data security, and finally, performance and reliability. The modern discipline of software engineering (5) has developed to the extent that design and implementation of such complex software is approaching the same level of dependability required for construction, say, of a major building, although the latter is considered a much easier problem. However, modeling the modeler itself is accomplished using established and testable methods.

Developing the ability to communicate data from one application software to another is also of paramount importance for current application requirements. For example, it may be possible to calculate a certain quantity using only particular software, which is then found lacking in terms of how to process it further. The need to exchange data between applications leads to the need to develop inter-application data exchange protocols. Typically, this is currently done at the text level, which introduces the possibility of error in the output–input transformation, as well as the time to perform the conversion and transfer.

An integral part of modern design approaches involves group activities or conferencing. Although many aspects of a design may be worked on by individuals, the total product will evolve as the result of integrated efforts of a team, possibly separated internationally. The possibility that team members share work and results is an increasing likelihood that must be reflected in modeling software and the supporting hardware systems; current networking capabilities have solved many of these problems, but more work is required, particularly on incorporating conferencing within the design system model.

Parts of this strategy have already been mapped out through the introduction of software products such as Java, an *object-oriented programming language* (6) suitable for immediate integration with Internet tools for distributed application across wide-area networks. In this approach, solutions to specialized problems are made available to users in the form of *applets,* or program modules, which can easily be interfaced with other modules to form higher-order programs. In short a program can be constructed from pieces gathered from around the world; the pieces can be reused, updated, modified, and so on, to fit the needs of the specialist. The delivery system for such approaches involves the notion of clients and servers, the *client–server model,* illustrated in Fig. 7. This model has been used for many years as the basis for operating system software and is undergoing extension and enhancement to meet the needs of modern distributed computing environments.

In a similar vein, users who otherwise might not be able to afford large, complex software packages (costing thousands of dollars, perhaps) may be able to buy time on commercial servers, thereby gaining access to powerful tools while saving precious research and development funds for other uses.

Use of the Internet as a medium for text and graphical image communication and commercial transactions is well established.

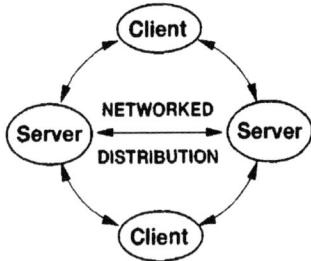

Figure 7. Many-client many-server model for distribution on a network. Both client–server and server–server interaction is permitted, particularly in cases where a service spawns a succession of service calls.

However, use of the Internet as an agent in modeling is not widespread or highly developed at this time. Work (7) has demonstrated the utility of using Java applets to perform fast Fourier transformation (FFT) by downloading appropriate software to a client machine to do the task and to perform submission of data via electronic templates supplied from a server to the client, transmitted back to the server to perform calculations, and subsequent delivery to the client of graphical images, all in the form of applets (8). With such systems it is expected that client accessibility to high-performance software can be increased substantially while reducing the cost proportionately using, say, time of usage billing. Also, distribution from controlled servers implies that software can be maintained and features added more conveniently, ensuring constant state-of-the-art technology to customers.

To develop a distributed modeling system, it is necessary to identify the primitive components essential to the fundamental operation of the system and its various appended and programmable functions, a process referred to as *kernel identification*. Studying such systems provides much of the basis for acquiring insight into what structure and function a kernel should possess (9).

VI. CONCLUSIONS

With the continuing advancement of powerful computer hardware and software systems, the nature of modeling is evolving quickly. Workers at all levels of research and development are increasingly involved with initiation, development, understanding, testing, and production of new products. In general, virtual prototypes are preferred over physical prototypes, primarily because of relative time and cost of production. It is worthwhile to train students and practitioners in the use of certain basic criteria to assist in making decisions on whether to commit resources to the modeling task. Of particular importance in this regard is an appreciation of the role of mathematics in the entire modeling process.

In this chapter we have provided an overview of mathematical modeling from inception of design through specification of solution method, production of solution, and analysis of results. Additionally, we have provided a framework for including computers, particularly current and emerging application software, as vital agents in the modeling process. With regard to both software and hardware developments, the Internet presents a great challenge and opportunity for modeling. Programming languages such as Java have emerged as suitable tools that ensure software reliability and reuse and that permit modeling to occur over a distributed set of computers.

REFERENCES

1. Davis ME (1984) Numerical methods and modeling for chemical engineers, Wiley, New York.
2. Press WH, Teukolsky SA, Vetterling WT, and Flannery BP (1992) Numerical recipes in C: the art of scientific computing, 2nd edn. Cambridge University Press, Cambridge.
3. Foley JD, van Dam A, Feiner SK, and Hughes JF (1994) Computer graphics: principles and practice, 2nd edn. Addison-Wesley, Reading, MA.
4. Abell ML and Braselton JP (1994) Differential equations with maple V, Academic Press, San Diego, CA.
5. Pressman RS (1987) Software engineering: a practitioner's approach, McGraw-Hill, New York.
6. Flanagan D (1996) Java in a nutshell, O'Reilly, Cambridge, MA.
7. Cidambi I (1996) M.Sc. thesis. University of Windsor, Windsor, Ontario, Canada.
8. Martincic F (1997) Honours Bachelor of Computer Science thesis. University of Windsor, Windsor, Ontario, Canada.
9. Marcuzzi E (1997) M.Sc. thesis. University of Windsor, Windsor, Ontario, Canada.
10. Webb JP (1995) Reports Prog. Phys. **58**:1673–1712.
11. Guyer JE, Boettinger WJ, Warren JA, and McFadden GB (2004) Phys. Rev. Rev. E **69**: 021603.

2

High Precision Atomic Theory: Tests of Fundamental Understanding

G.W.F. Drake, Qixue Wu and Zheng Zhong

Department of Physics, University of Windsor, Windsor, Ontario, Canada, N9B 3P4
GDrake@uwindsor.ca

I. INTRODUCTION

The unifying theme of these two volumes is the use of mathematical models to describe various aspects of the physical world around us. In many cases, the system in question is too complicated to be described from first principles, and so various models are introduced, containing perhaps arbitrary parameters whose values can be adjusted to give the best fit to experimental data. In other sufficiently simple cases, an exact analytic solution is possible, at least within the context of various physical approximations. The present chapter will be concerned with systems of intermediate complexity that are too complicated to allow exact analytic solutions, but are simple enough to permit solutions by numerical methods that are essentially exact for all practical purposes.

One such case is the helium atom, consisting of a nucleus and two planetary electrons. This is an example of the three-body problem that has fascinated physicists and mathematicians since the time of Newton. It is analogous to the classical problem of a sun with two planets, or three mutually interacting stars. The simpler problem of a sun with one planet can of course be solved exactly by use of

Newtonian mechanics (at least within the simplified context of objects with perfectly spherical mass distributions and purely gravitational interactions), resulting in Kepler's laws of planetary motion. However, in the years following the first publication of Newton's *Principia* in 1687, the really convincing test was not just the successful derivation of Kepler's laws of planetary motion themselves, but also the tiny deviations from perfectly elliptical orbits of the planets due to additional gravitational interactions between them; i.e., to the many-body nature of the problem. In fact, tiny residual deviations between the predicted and observed motion of the planet Uranus led John Couch Adams in England (1843) and Urbain Le Verrier in France (1845) to the prediction of the then unknown planet Neptune. It was subsequently found in 1846 by astronomers within a few degrees of the predicted position (1). The moral of the story is that it is the tiny deviations from Kepler's laws of planetary motion that provided strong confirmation of Newtonian mechanics and the universal law of gravitation for celestial motion, rather than the explanation of Kepler's laws themselves. It is the search for tiny deviations from theoretical predictions that leads to new physics, but of course one must have reliable theoretical predictions to begin with.

1. The Nonrelativistic Hydrogen Atom

We will deal with the more complicated case of helium in the next section. Let us first discuss the simpler case of the hydrogen atom to introduce some basic ideas and concepts. The results for hydrogen will also provide a valuable guide to the construction of approximate variational solutions for the helium wave function in the following subsection.

At the atomic level, the hydrogen atom is the analogue of the sun with just one planet. We will set aside for the moment corrections due to Einstein's theory of relativity, and focus on a nonrelativistic description. The gravitational interaction energy of two planets with masses m_1 and m_2 is

$$V_G(r) = -\frac{Gm_1m_2}{r}, \qquad (1)$$

where $r = |r_1 - r_2|$ is the distance between the centers of the planets, and G is Newton's universal constant of gravitation (the sign is negative because the gravitational force is attractive). For two charges q_1 and q_2, this becomes instead (in electrostatic units)

$$V(r) = \frac{q_1 q_2}{r}. \tag{2}$$

For the case of a hydrogen atom, we can take $q_1 = -e$ to be the charge of the electron, and $q_2 = Ze$ to be the charge of the nucleus to obtain

$$V(r) = -\frac{Ze^2}{r} \tag{3}$$

for the instantaneous Coulomb interaction potential.

In classical mechanics, the next step is to find the kinetic energy T and add this to the potential energy V to find the total energy $E = T + V$ for the system, which is a constant of the motion. For ordinary conservative systems, this is the same as the Hamiltonian H, and so we can simply write $H = T + V$ with T given in terms of the canonical momenta p_1 and p_2 by

$$T = \frac{p_1^2}{2m_1} + \frac{p_2^2}{2m_2} \tag{4}$$

and so the total Hamiltonian becomes

$$\begin{aligned} H &\equiv T + V \\ &= \frac{p_1^2}{2m_1} + \frac{p_2^2}{2m_2} - \frac{Ze^2}{r}. \end{aligned} \tag{5}$$

This is an example of a two-body central force problem. It is well known from classical mechanics that an important simplification can be achieved by transforming to center-of-mass plus relative coordinates defined by

$$R = \frac{m_1 r_1 + m_2 r_2}{m_1 + m_2}, \quad \text{and } r = r_1 - r_2. \tag{6}$$

A straight-forward substitution then yields the Hamiltonian

$$H = \frac{P^2}{2M} + \frac{p^2}{2\mu} - \frac{Ze^2}{r}, \tag{7}$$

where $M = m_1 + m_2$ is the total mass and $\mu = m_1 m_2 / M$ is the reduced mass. Since the potential depends only on r, R becomes an ignorable coordinate. Without loss of generality we can set $P = 0$ to obtain

$$H = \frac{p^2}{2\mu} - \frac{Ze^2}{r} \qquad (8)$$

and the two-body problem reduces to an equivalent one-body problem with reduced mass μ.

Of course, unlike planets, atoms are inherently quantum mechanical objects where the wave nature of matter is important, as determined by Planck's constant $\hbar$. In this case Newtonian mechanics must be replaced by wave mechanics. The key change is to make the operator substitution

$$p \to \frac{\hbar}{i} \nabla \qquad (9)$$

in H to obtain the second-order differential equation

$$H\psi(r) \equiv \left(-\frac{\hbar^2}{2\mu}\nabla^2 - \frac{Ze^2}{r}\right)\psi(r) = E\psi(r) \qquad (10)$$

known as Schrödinger's equation. Then, instead of having discrete planetary orbits determined by Newton's laws of motion, the Schrödinger equation determines the probability amplitude $\psi(r)$ such that $|\psi(r)|^2 dr$ is the probability of finding the electron in the volume element dr. For bound states, the boundary condition that $\psi(r) \to 0$ as $r \to \infty$ makes this an eigenvalue problem for the discrete eigenvalues E corresponding to the discrete energy levels of a hydrogen atom. The result is the famous Rydberg formula

$$E_n = \frac{\mu Z^2 e^4}{\hbar^2} \left(\frac{-1}{2n^2}\right) \qquad (11)$$

for the energy levels of a hydrogen atom or hydrogenic ion with nuclear charge Ze and principal quantum number n. Since the system is rotationally invariant, one expects the magnitude of the orbital angular momentum l and the component m in the z-direction to be constants of the motion that can be simultaneously quantized. The energies are independent of both l and m. The fact the E_n does not depend on m (i.e., degeneracy with respect to m) is a consequence of the rotational invariance of the potential $V(r)$, but degeneracy with respect to l is a special property of the $1/r$ Coulomb potential analogous to the existence of closed elliptical orbits in the classical case.

The eigenfunctions going with these eigenvalues E_n have the form (2)

$$\psi_{nlm}(\mathbf{r}) \propto R_{nl}\left(\frac{Zr}{a_\mu}\right) \exp\left(-\frac{Zr}{na_\mu}\right) Y_l^m(\theta, \phi), \qquad (12)$$

where $Y_l^m(\theta, \phi)$ is a spherical harmonic, θ and ϕ are the polar angles of the electron, and $R_{nl}(\rho)$ is a radial function given by

$$R_{nl}(\rho) = \rho^l \,_1F_1(-(n-l-1); 2l+2; 2\rho), \qquad (13)$$

where $_1F_1(a; b; z)$ is a confluent hypergeometric function that terminates after $n-l$ terms to give a finite polynomial of order $n-l-1$. This functional form of powers of r times exponentials is important because it will provide a guide in constructing approximate solutions for the helium atom in Sect. II.1.

2. Atomic Units

Concerning the units of energy and distance, the factor $\mu e^4/\hbar^2$ in (11) has a special significance. It can be written in the form

$$\frac{\mu e^4}{\hbar^2} = \frac{e^2}{a_\mu}, \qquad (14)$$

where $a_\mu = \hbar^2/\mu e^2$ is the (reduced mass) Bohr radius. In this form, it is manifestly obvious that the quantity e^2/a_μ has the dimensions of energy. In the limit where the nuclear mass is taken to be infinitely heavy, then $\mu \to m_e$ the electron mass, and $a_\mu \to a_0$ the Bohr radius. For purposes of performing calculations, it is convenient to work in atomic units in which $e = \hbar = m_e = 1$. The atomic unit of energy (the Hartree) is then (3, 4)

$$\frac{e^2}{a_0} = 2R_\infty = \begin{cases} 2.194\,746\,313\,705(15) \times 10^7 \text{ m}^{-1}, \\ 6.579\,683\,920\,722(44) \times 10^{15} \text{ Hz}, \\ 27.211\,383\,86(68) \text{ eV}, \end{cases} \qquad (15)$$

where R_∞ is the Rydberg unit of energy for infinite nuclear mass, and the Bohr radius is $a_0 = 0.529\,177\,208\,59(36) \times 10^{-10}$ m.

For example, from (11), the energy for the ground state of hydrogen (i.e., $Z = 1$ and $n = 1$) is $E_1 = -0.5(\mu/m_e)$, where μ/m_e is the correction factor for finite nuclear mass (called the normal isotope shift). For hydrogen, its numerical value is

$$\frac{\mu}{m_e} = \frac{1}{1 + 1/1836.152\,672\,47(80)}, \tag{16}$$

where $1/1836.152\,672\,47(80)$ is the value of the electron to proton mass ratio. To convert to laboratory units, simply multiply E_1 in atomic units by the appropriate conversion factor from (15).

II. THE NONRELATIVISTIC HELIUM ATOM

As pointed out in Sect. I, helium is an example of a three-body problem. Despite the best attempts of mathematicians since the time of Newton, there exist no exact analytic solutions to the three-body problem, either in classical mechanics or in quantum mechanics. The best that one can hope for is to construct a sequence of approximate numerical solutions that converge in the limit to the exact answer. This section begins with a discussion of the problem to be solved, and then discusses the high-precision variational methods that have been developed to generate solutions that are essentially exact for all practical purposes.

In the center-of-mass frame, the full Schrödinger equation for the helium atom is

$$\left[-\frac{\hbar^2}{2\mu} \left(\nabla_1^2 + \nabla_2^2 + 2\frac{\mu}{M} \nabla_1 \cdot \nabla_2 \right) - \frac{Ze^2}{r_1} - \frac{Ze^2}{r_2} + \frac{e^2}{r_{12}} \right] \Psi(\mathbf{r}_1, \mathbf{r}_2)$$
$$= E \Psi(\mathbf{r}_1, \mathbf{r}_2), \tag{17}$$

where r_1 and r_2 are the position vectors relative to the nucleus (see Fig. 1), and r_{12} is the distance between the two electrons. The term involving $-2(\mu/M)\nabla_1 \cdot \nabla_2$ is called the mass polarization term. Physically, it accounts for the recoil of the nucleus in the center-of-mass frame. This term goes to zero and can be neglected in the limit where the nuclear mass $M \to \infty$, but it is essential to take

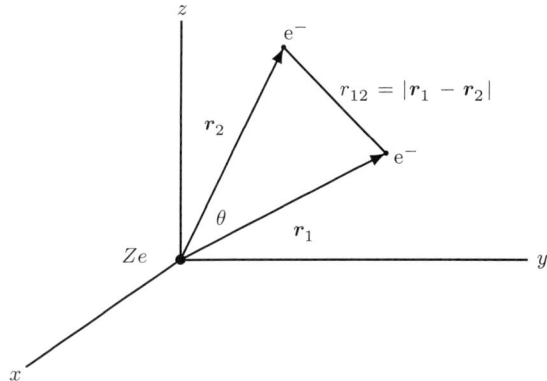

Figure 1. Coordinate system for a helium atom with the nucleus at the origin.

it fully into account for high precision work. It makes an important contribution to the isotope shift.

It is the electron–electron repulsion term e^2/r_{12} in (17) that causes all the difficulty. Without this term, the Schrödinger equation would be separable, and the exact eigenfunctions would be simple products of the hydrogenic functions shown in (12). However, as it stands, exact solutions do not exist in terms of known functions. Over the years, many methods of a more generalized nature have been developed to find approximate solutions for atoms more complicated than hydrogen, such as the Hartree–Fock approximation, density functional methods, configuration interaction methods, and many-body perturbation theory. However, none of these methods is capable of yielding results that are even close to full spectroscopic accuracy. For example, the ionization energy of the 1s2s $^3S_1$ state of helium is known experimentally to an accuracy of five parts in 10^{11}, which means that the total energy must be known to three parts in 10^{12}. In comparison, the accuracy of the Hartree–Fock method is not better than 1%, and the accuracy of even the most sophisticated configuration interaction calculations is not better than one part in 10^7. This is still many orders of magnitude short of the accuracy that is needed.

1. Correlated Variational Basis Sets for Helium

In this section, we discuss the specialized methods that have been developed to achieve a spectroscopic level accuracy of parts in 10^{12} or better for sufficiently simple systems such as helium. Of course corrections for relativistic and quantum electrodynamic effects will ultimately have to be included, but the starting point for the calculation is to find accurate solutions to the non relativistic Schrödinger equation. The effects of finite nuclear mass will also play a key role in studies of the isotope shift, but we begin with the case of infinite nuclear mass, in which case the nucleus can be taken as a fixed point of reference. The Schrödinger equation is then given by

$$\left(-\frac{1}{2}\nabla_1^2 - \frac{1}{2}\nabla_2^2 - \frac{Z}{r_1} - \frac{Z}{r_2} + \frac{1}{r_{12}}\right) \Psi(\mathbf{r}_1, \mathbf{r}_2) = E \Psi(\mathbf{r}_1, \mathbf{r}_2). \tag{18}$$

As already explained, the usual methods of theoretical atomic physics do not yield sufficient accuracy. As long ago as 1929, Hylleraas (5) suggested using the Rayleigh–Ritz variational method to find approximate solutions. The origins of the method go back to the classical mechanics of vibrating systems in the nineteenth century. The basic idea is to construct an approximate trial solution that we have reason to believe resembles the exact solution in its general form, but with adjustable parameters that can be varied to optimize some quantity such as the energy. For example, we already know that the exact solutions for the one-electron hydrogen atom have the form of a sum of powers of r times exponentials, as shown in (12). Using this as a guide, Hylleraas suggested writing the trial solution for helium in the form of a product of one-electron hydrogenic solutions, which makes physical sense, but this by itself cannot go beyond the Hartree–Fock approximation in accuracy because it lacks the correlation energy (as further discussed below). To get around this roadblock, Hylleraas tried extending the variational basis set of functions to create an explicitly correlated trial wave function of the form

$$\Psi(\mathbf{r}_1, \mathbf{r}_2) = \sum_{i,j,k} a_{ijk} r_1^i r_2^j r_{12}^k e^{-\alpha r_1 - \beta r_2} \mathcal{Y}_{l_1 l_2 L}^M(\hat{\mathbf{r}}_1, \hat{\mathbf{r}}_2), \tag{19}$$

where $r_{12} = |\mathbf{r}_1 - \mathbf{r}_2|$ is the interelectronic separation (see Fig. 1), and $\mathcal{Y}_{l_1 l_2 L}^M(\mathbf{r}_1, \mathbf{r}_2)$ is an angular function discussed further below (see (23)). It is the inclusion of these r_{12} terms that makes the basis set correlated, and the wave function nonseparable. The coefficients a_{ijk} are called linear variational parameters, and α and β are nonlinear variational coefficients that set the distance scale for the wave function. The usual strategy is to include all powers such that $i + j + k \leq \Omega$ (a so-called Pekeris shell), where Ω is an integer. If all terms are included within the Pekeris shell, then there are $N = (\Omega+1)(\Omega+2)(\Omega+3)/6$ distinct terms. The basis set therefore grows very quickly approximately in proportion to Ω^3 if all terms are included. For example, Table 1 lists explicitly the powers, i, j, k for the ten basis set members for the case $\Omega = 2$.

For convenience, let $p = 1, \ldots, N$ be a running index such that each p stands for a particular triplet of integer powers $\{i, j, k\}$ appearing in (19). Each term in the summation in (19) then corresponds to a member of a basis set of functions defined by

$$\chi_p(\alpha, \beta) = r_1^i r_2^j r_{12}^k e^{-\alpha r_1 - \beta r_2} \mathcal{Y}_{l_1 l_2 L}^M (\hat{\mathbf{r}}_1, \hat{\mathbf{r}}_2). \tag{20}$$

The central idea of the variational method is to regard the function $\Psi(\mathbf{r}_1, \mathbf{r}_2)$ as a trial solution Ψ_{tr} to the Schrödinger equation (18) with adjustable linear parameters $a_{ijk} \equiv a_p$ and nonlinear parameters α and β. There is then a very simple criterion for choosing the optimum value of these parameters. If one calculates the expectation value of the Hamiltonian H with respect to Ψ_{tr} according to

Table 1.
Example of the 10 basis set members for the case $\Omega = 2$

p	Ω	i	j	k
1	0	0	0	0
2	1	1	0	0
3	1	0	1	1
4	1	0	0	1
5	2	2	0	0
6	2	0	2	0
7	2	0	0	2
8	2	1	1	0
9	2	1	0	1
10	2	0	1	1

$$E_{tr} = \frac{\langle \Psi_{tr}|H|\Psi_{tr}\rangle}{\langle \Psi_{tr}|\Psi_{tr}\rangle} \qquad (21)$$

then E_{tr} is an upper bound to the exact ground state energy E_0, provided that the eigenvalue spectrum is bounded from below. The strategy is then to vary all the parameters in Ψ_{tr} so as to make the trial energy E_{tr} as low as possible, knowing that it can never fall through the exact E_0. This variational minimization of E_{tr} corresponds to solving the system of equations $\partial E_{tr}/\partial a_p = 0$ with respect to all the linear parameters a_p, and also $\partial E_{tr}/\partial \alpha = 0$ and $\partial E_{tr}/\partial \beta = 0$ for the nonlinear parameters. The former minimization gives rise to a system of N linear homogeneous algebraic equations to determine the a_p whose solution is equivalent to solving the generalized N-dimensional eigenvalue problem

$$\mathsf{H}\mathsf{a} = \lambda \mathsf{O}\mathsf{a} \qquad (22)$$

in the basis set of nonorthogonal functions χ_p. Then H is the Hamiltonian matrix in this basis set, O is the overlap matrix, a is the column vector of coefficients a_p for one of the N eigenvectors, and λ is the corresponding eigenvalue. Then by the Rayleigh-Schrödinger variational theorem, the smallest eigenvalue λ_{min} is an upper bound to the true ground state E_0; i.e., $\lambda_{min} \geq E_0$.

The variational optimization with respect to the nonlinear parameters α and β is more difficult because the equations $\partial E_{tr}/\partial \alpha = 0$ and $\partial E_{tr}/\partial \beta = 0$ are transcendental equations that cannot be solved algebraically. Instead, some sort of iterative method must be used to locate the energy minimum on the two-dimensional energy surface as α and β are varied. This problem is further discussed in Sect. III.

This much is standard text book material, and the proof can be found in any book on quantum mechanics. What is not so well known is the Hylleraas–Undheim–McDonald theorem (6, 7), which states that the variational bound property applies not just the lowest eigenvalue λ_{min}, but also to all the higher-lying variational eigenvalues of (22). To be more explicit, if the N variational eigenvalues are ordered from smallest to largest according to $\lambda_q, q = 0, 1 \ldots, N-1$, then $\lambda_0 \geq E_0, \lambda_1 \geq E_1$, and so on. As shown in Fig. 2, all the eigenvalues move inexorably downward as the number of terms N in the basis set is progressively increased, and so approach the exact eigenvalues from above. This is a consequence of the matrix interleaving theorem, which states that as a matrix is

High Precision Atomic Theory: Tests of Fundamental Understanding 43

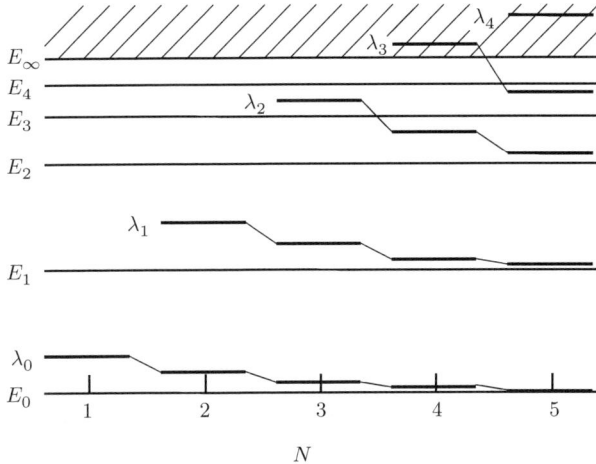

Figure 2. Diagram illustrating the Hylleraas–Undheim–MacDonald theorem. The $\lambda_q, q = 0, \ldots, N-1$ are the variational eigenvalues for an N-dimensional basis set, and the E_i are the exact eigenvalues of H. The highest λ_q lie in the continuous spectrum of H.

progressively enlarged by a process of bordering (i.e., adding an extra row and column), then the old eigenvalues interleave the new.

Even if the eigenvalues λ_p converge to a definite result $\bar{\lambda}_p$ as $N \to \infty$, it is still possible that a finite gap $E_p - \bar{\lambda}_p$ remains in the limit. For example, if no correlation terms r_{12}^k are included in the basis set (which is virtually the same as the Hartree–Fock approximation), then the difference $E_p - \bar{\lambda}_p$ tends to a finite limit called the correlation energy. The condition for the $\bar{\lambda}_p$ to converge to the exact eigenvalues with no finite gap remaining is that the basis set must be complete in the sense that it spans Hilbert space (or more precisely a first Sobolev space (8); i.e., a Hilbert space of functions with square integrable first derivatives). The inclusion of powers of r_{12}, and especially the odd powers, makes the basis set rapidly convergent as Ω increases. As shown by Klahn and Bingel, the basis set is provably complete in the limit $\Omega \to \infty$ (9).

For states of higher angular momentum L, the quantity $\mathcal{Y}_{l_1 l_2 L}^M(\mathbf{r}_1, \mathbf{r}_2)$ in (19) denotes a vector-coupled product of solid spherical harmonics, and the basis set includes a summation over all possible integer values of l_1 and l_2 (with l_2 constrained to be

$l_2 = L - l_1$) such that $l_1 \leq L/2$. The explicit definition in terms of vector coupling coefficients and ordinary spherical harmonics is

$$\mathcal{Y}^M_{l_1 l_2 L}(\mathbf{r}_1, \mathbf{r}_2) = r_1^{l_1} r_2^{l_2} \sum_{m_1, m_2} \langle l_1 l_2 m_1 m_2 | L M \rangle Y^{m_1}_{l_1}(\hat{\mathbf{r}}_1) Y^{m_2}_{l_2}(\hat{\mathbf{r}}_2). \quad (23)$$

Note that the solid spherical harmonics are defined for convenience to include an extra factor of r^l so that the wave function has the correct behaviour at the origin. For S-states, the only contribution is from $l_1 = l_2 = L = 0$, and so $\mathcal{Y}^M_{l_1 l_2 L}(\mathbf{r}_1, \mathbf{r}_2)$ is just a constant $1/(4\pi)$. For P-states, it reduces to just a simple factor of $\sqrt{3} r_2 \cos\theta_2/(4\pi)$ for the case $l_1 = 0, l_2 = L = 1, M = 0$.

Finally, the trial wave function can be explicitly symmetrized or antisymmetrized with respect to the interchange of the electron coordinates $\mathbf{r}_1$ and $\mathbf{r}_2$ to form a triplet state or a singlet state respectively according to

$$\Psi_\pm(\mathbf{r}_1, \mathbf{r}_2) = \frac{1}{\sqrt{2}} [\Psi(\mathbf{r}_1, \mathbf{r}_2) \pm \Psi(\mathbf{r}_2, \mathbf{r}_1)]. \quad (24)$$

2. Early Calculations for Helium

Early calculations done by Hylleraas (5) and others played an important role in the history of quantum mechanics. After all, the old Bohr–Sommerfeld quantum theory worked perfectly well for hydrogen, and so helium was the first real test case for the Schrödinger equation more complicated than hydrogen. Hand calculations with small basis sets containing just a few powers of r_{12} easily recovered nearly all the correlation energy (see Bethe and Salpeter (2) for a review). These results demonstrated the great efficiency of Hylleraas-type basis sets in describing electron correlation, and especially the odd powers of r_{12}.

To emphasize the point about the importance of the odd powers of r_{12}, Table 2 shows the effect of including as many powers of r_1 and r_2 as are needed for complete convergence to the figures quoted, but only one or two powers of r_{12} as indicated. The first line with no powers of r_{12} corresponds to the S-wave limit of a configuration-interaction (CI) calculation, which is essentially the same thing as the Hartree–Fock approximation. The error of 0.672 eV is the correlation energy. The next two lines correspond to the P-wave and

Table 2.
Energies for the ground state of helium obtained with various limited powers of r_{12} in the basis set, but sufficiently many powers of r_1 and r_2 for convergence to the figures quoted

r_{12} terms	Energy (a.u.)	Error (eV)
No r_{12}	−2.879 029	0.672
r_{12}^2	−2.900 503	0.087 6
r_{12}^2, r_{12}^4	−2.902 752	0.026 4
r_{12}	−2.903 496	0.006 20
r_{12}, r_{12}^3	−2.903 700	0.000 65
All r_{12}	−2.903 724	0.000 00

D-wave limits of a CI calculation. The connection with CI calculations can be easily seen by writing the factor of r_{12}^2 in the form

$$r_{12}^2 \equiv |\mathbf{r}_1 - \mathbf{r}_2|^2 = r_1^2 + r_2^2 - 2r_1 r_2 \cos\theta_{12}, \qquad (25)$$

where θ_{12} is the angle between the vectors $\mathbf{r}_1$ and $\mathbf{r}_2$, and then using the spherical harmonic addition theorem to write

$$\cos\theta_{12} = \sqrt{\frac{4\pi}{3}} \sum_{m=-1}^{1} (-1)^m Y_1^{-m}(\hat{\mathbf{r}}_1) Y_1^m(\hat{\mathbf{r}}_2) \qquad (26)$$

which has the form of two P-waves coupled to form an S-state. It can be seen from the first three rows of the table that convergence is quite slow for these CI-type basis sets, and many more partial waves would have to be included to achieve high precision. The fourth line shows the dramatic improvement obtained by including just a single odd power of r_{12}, and the fifth line shows that the first two odd powers reduce the error by another factor of 10 to only 0.000 65 eV. Even though the basis set is in principle complete with just the even powers of r_{12} (9), it is clear that it is the odd powers that are much more efficient in capturing the correlation energy, and hence a Hylleraas basis set is much more rapidly convergent than a CI calculation.

A long sequence of Hylleraas-type calculations with basis sets of increasing size and sophistication culminated with the work of Pekeris and coworkers in the 1960s (see Accad et al. (10)). They showed that nonrelativistic energies accurate to a few parts in 10^9 could be obtained by this method, at least for the low-lying states of helium and He-like ions. However, these calculations also revealed two serious numerical problems. First, it is difficult to improve upon this accuracy of a few parts in 10^9 without using extremely large basis sets where roundoff error and numerical linear dependence become a problem. Second, as is typical of variational calculations, the accuracy is best for the lowest state of each symmetry, but rapidly deteriorates with increasing n.

3. Recent Advances

Over the past 15 years, both of the above limitations on accuracy have been resolved by "doubling" the basis set so that each combination of powers i, j, k is included twice with different exponential scale factors (11–13). Explicitly, each basis function $\chi_{ijk}(\alpha, \beta)$ defined by (20) is replaced by

$$\bar{\chi}_{ijk} = a_A \chi_{ijk}(\alpha_A, \beta_A) + a_B \chi_{ijk}(\alpha_B, \beta_B), \qquad (27)$$

where a_A and a_B are independent variational parameters, and (α_A, β_A), (α_B, β_B) are two sets of exponential scale factors that are common to all the basis set members. A complete optimization with respect to all the exponential scale factors leads to a natural partition of the basis set into two distinct sectors with different distance scales: an asymptotic sector (A) appropriate to the long-range asymptotic behavior of the wave function, and a short-range sector (B) appropriate to the complex correlated motion of the electrons near the nucleus. The greater flexibility in the available distance scales allows a much better physical description of the atomic wave function, especially for the higher-lying Rydberg states where two sets of distance scales are clearly important. However, the multiple distance scales also greatly improve the accuracy for the low-lying states. With care, the basis set size can be reduced by omitting some of the powers i, j, k from one of the two sectors (see (11) for further details).

The advantage gained by doubling the basis set only becomes apparent for sufficiently large Ω. For small basis sets (i.e., values of $\Omega < 4$), there is no significant improvement in either accuracy or numerical stability. In fact, the numerical stability can get worse if the optimum values of the αs and βs are not sufficiently well separated (see Fig. 3 for an example). Ω must exceed a critical value for the optimized basis set to separate into distinct regions.

Recent work has shown that, for Ω sufficiently large, it becomes advantageous to introduce a third partition to give a "triple" basis set (14), in the sense that the numerical stability is better, and the energy is lower for a given total number of terms in the basis set. The optimized values of the αs and βs are illustrated in Fig. 3. However, again, no advantage is gained if Ω is too small because, as can be seen from the figure, the optimized values of the αs and βs are not sufficiently well separated. This illustrates the general principle that lessons apparently learned by experimenting with small basis sets often do not apply for large basis sets.

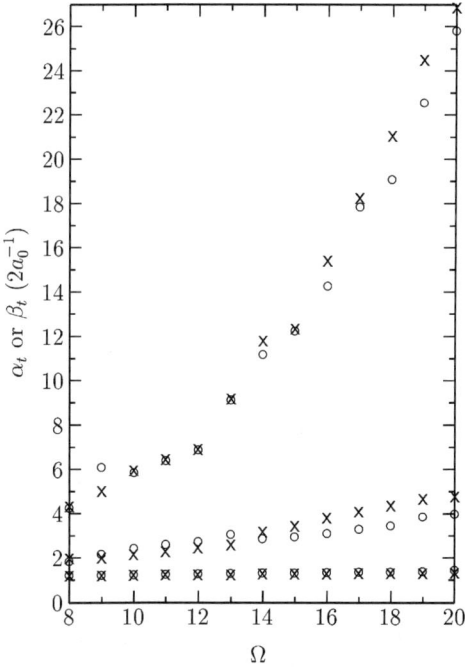

Figure 3. Variation of the exponential scale factors with basis set size for the helium $1s^2\,^1$S state. The three pairs of curves are for a triple basis set with ∘ for α and × for β in each of the three sectors. $\Omega = i + j + k$ is the sum of powers in each sector.

The problem of locating the energy minimum on the multidimensional energy surface is facilitated by the fact that the derivatives $\partial E/\partial \alpha_t$ and $\partial E/\partial \alpha_t$, where t runs over the partitions, can be calculated analytically from the formulas

$$\frac{\partial E}{\partial \alpha_t} = -2\langle \Psi_{\text{tr}}|H - E|r_1\Psi(\mathbf{r}_1,\mathbf{r}_2;\alpha_t) \pm r_2\Psi(\mathbf{r}_2,\mathbf{r}_1;\alpha_t)\rangle, \quad (28)$$

$$\frac{\partial E}{\partial \beta_t} = -2\langle \Psi_{\text{tr}}|H - E|r_2\Psi(\mathbf{r}_1,\mathbf{r}_2;\alpha_t) \pm r_1\Psi(\mathbf{r}_2,\mathbf{r}_1;\alpha_t)\rangle \quad (29)$$

assuming that $\langle \Psi_{\text{tr}}|\Psi_{\text{tr}}\rangle = 1$, and $\Psi(\mathbf{r}_1,\mathbf{r}_2;\alpha_t)$ denotes the terms in Ψ_{tr} that depend explicitly on a particular α_t, and similarly for β_t.

As a final subtlety, the screened hydrogenic wave function $\psi_{1s}^Z(\mathbf{r}_1)\psi_{nL}^{Z-1}(\mathbf{r}_2)\pm$ exchange is included as an additional independent member of the basis set. This corresponds to the physical picture of an inner 1s electron experiencing the full nuclear charge Z, while the outer nL electron sees the screened nuclear nuclear charge $Z-1$. Without this term, rather large basis sets are required just to recover the screened hydrogenic energy

$$E_{\text{SH}} = -\frac{Z^2}{2} - \frac{(Z-1)^2}{2n^2} \quad (30)$$

for Rydberg states, with the inclusion of this term, and a sum over S sectors with different distance scales, the complete trial function becomes

$$\Psi(\mathbf{r}_1,\mathbf{r}_2) = a_{\text{SH}}\psi_{1s}^Z(\mathbf{r}_1)\psi_{nL}^{Z-1}(\mathbf{r}_2) + \sum_{t=1}^{S}\sum_{l_1=0}^{[L/2]}\sum_{i,j,k}^{i+j+k\leq\Omega} c_{ijk,t}^{(l_1)} r_1^i r_2^j r_{12}^k$$
$$\times \exp(-\alpha_t^{(l_1)} r_1 - \beta_t^{(l_1)} r_2)\mathcal{Y}_{l_1,l_2,L}^M(\hat{\mathbf{r}}_1,\hat{\mathbf{r}}_2)$$
$$\pm \text{exchange term}. \quad (31)$$

As an example, Table 3 shows a convergence study for the very well studied case of the ground state of helium (14). The quantity R in the last column is the ratio of successive differences between the energies. A constant or slowly changing value of R indicates smooth convergence, and allows a reliable extrapolation to $\Omega \to \infty$. The results clearly indicate that convergence to 20 or more figures can be readily obtained, using conventional quadruple precision (32 decimal digit) arithmetic in FORTRAN. The very large calculation by

Table 3.
Convergence study for the ground state of helium (infinite nuclear mass case (14). N is the number of terms in the 'triple' basis set

Ω	N	$E(\Omega)$	$R(\Omega)$
8	269	−2.903724377029560058400	
9	347	−2.903724377033543320480	
10	443	−2.903724377034047783838	7.90
11	549	−2.903724377034104634696	8.87
12	676	−2.903724377034116928328	4.62
13	814	−2.903724377034119224401	5.35
14	976	−2.903724377034119539797	7.28
15	1150	−2.903724377034119585888	6.84
16	1351	−2.903724377034119596137	4.50
17	1565	−2.903724377034119597856	5.96
18	1809	−2.903724377034119598206	4.90
19	2067	−2.903724377034119598286	4.44
20	2358	−2.903724377034119598305	4.02
Extrapolation	∞	−2.903724377034119598311(1)	
Korobov (15)	5200	−2.903724377034119598311587	
Korobov extrap.	∞	−2.903724377034119598311594(4)	
Schwartz (16)	10259	−2.903724377034119598311159245194404400	
Schwartz extrap.	∞	−2.903724377034119598311159245194404446	
Sims and Hagstrom (24)		−2.903724377034119598299	
Goldman (17)	8066	−2.90372437703411959382	
Bürgers et al. (18)	24 497	−2.903724377034119589(5)	
Baker et al. (19)	476	−2.9037243770341184	

Schwartz (16), using 104-digit arithmetic, provides a benchmark for comparison. The advantage of the basis sets used here is that they provide results of sufficient accuracy while remaining reasonably compact and numerically stable.

The problem of calculating integrals over the correlated basis functions and evaluating matrix elements of the Hamiltonian for states of arbitrary angular momentum is discussed in detail by Drake (20), and a useful compilation of formulas is given there.

4. Other Computational Methods

Although not yet at the same level of accuracy as variational methods, certain nonvariational methods, such as finite element methods (21), solutions to the Faddeev equations (22), and the correlated-function hyperspherical-harmonic method (23), have their own advantages of flexibility and/or generality. A characteristic feature of these methods is that they provide direct numerical solutions to the three-body problem which in principle converge pointwise to the exact solution, rather than depending upon a globally optimized solution. Recent work by Sims and Hagstrom (24) involves a combined CI-Hylleraas approach, and offers the possibility of extending the method to more complex atomic system. Other methods particularly suited to doubly-excited states are discussed in (25) and earlier references therein.

5. Variational Basis Sets for Lithium

The same variational techniques can be applied to lithium and other three-electron atomic systems. In this case, the terms in the Hylleraas correlated basis set have the form

$$r_1^{j_1} r_2^{j_2} r_3^{j_3} r_{12}^{j_{12}} r_{23}^{j_{23}} r_{31}^{j_{31}} e^{-\alpha r_1 - \beta r_2 - \gamma r_3} \mathcal{Y}_{(\ell_1 \ell_2)\ell_{12},\ell_3}^{LM}(\mathbf{r}_1, \mathbf{r}_2, \mathbf{r}_3)\chi_1, \tag{32}$$

where $\mathcal{Y}_{(\ell_1 \ell_2)\ell_{12},\ell_3}^{LM}$ is again a vector-coupled product of spherical harmonics, and χ_1 is a spin function with spin angular momentum 1/2. As for helium, the usual strategy is to include all terms from (32) such that

$$j_1 + j_2 + j_3 + j_{12} + j_{23} + j_{31} \leq \Omega, \tag{33}$$

and study the eigenvalue convergence as Ω is progressively increased. The lithium problem is much more difficult than helium both because the integrals over fully correlated wave functions are more difficult, and because the basis set grows much more rapidly with increasing Ω. Nevertheless, there has been important progress in recent years (26–31), and results of spectroscopic accuracy can be obtained for the low-lying states.

III. SMALL CORRECTIONS

The high precision variational methods of the preceding section provide essentially exact solutions to the nonrelativistic Schrödinger equation for all practical purposes, but there are several small corrections that must be included as perturbations before a meaningful comparison with experiment can be made. Table 4 summarizes

Table 4.
Contributions to the energy and their orders of magnitude in terms of Z, $\mu/M = 1.370\,745\,624 \times 10^{-4}$ (for ^{4}He), and $\alpha^2 = 0.532\,513\,5450 \times 10^{-4}$

Contribution	Magnitude
Nonrelativistic energy	Z^2
Mass polarization	$Z^2 \mu/M$
Second-order mass polarization	$Z^2 (\mu/M)^2$
Relativistic corrections	$Z^4 \alpha^2$
Relativistic recoil	$Z^4 \alpha^2 \mu/M$
Anomalous magnetic moment	$Z^4 \alpha^3$
Hyperfine structure	$Z^3 gI \mu_0^2$
Lamb shift	$Z^4 \alpha^3 \ln \alpha + \ldots$
Radiative recoil	$Z^4 \alpha^3 (\ln \alpha) \mu/M$
Higher order QED	$Z^5 \alpha^4 + \ldots$
Finite nuclear size	$Z^4 \langle \bar{r}_c/a_0 \rangle^2$

the contributions, expressed as a double expansion in powers of $\alpha \simeq 1/137.036$ and the electron reduced mass ratio $\mu/M \simeq 10^{-4}$. All the terms down to and including the lowest-order Lamb shift (QED) terms of order α^3 Ry can be calculated to very high precision. Until recently, the next-to-lowest order QED terms of order α^4 Ry represented the dominant source of uncertainty. However, in an important sequence of papers, the calculation of these terms has recently been completed by Pachucki (32, 33). The comparison between theory and experiment is sensitive to these terms.

In addition to testing QED, a further application of these high precision results is to use the comparison with experiment for the isotope shift to determine the nuclear charge radius r_c. For this purpose the QED terms independent of μ/M cancel out, and so it is only the radiative recoil terms of order $\alpha^4 \mu/M \simeq 10^{-12}$Ry($\sim$10 kHz) that contribute to the uncertainty. Since this is much less than the finite nuclear size correction of about 1 MHz, the comparison between theory and experiment provides a means to determine the nuclear charge radius.

The objective of this section is to give a brief account of the corrections for finite nuclear mass, relativistic effects, and quantum electrodynamic effects, together with references to the literature for further reading.

1. Mass Polarization

For high precision calculations, and especially for the isotope shift, it is necessary to include also the motion of the nucleus in the center-of-mass (CM) frame. A transformation to CM plus relative coordinates yields the additional $-(\mu/M)\nabla_1 \cdot \nabla_2$ mass polarization term in the modified Hamiltonian

$$H = -\frac{1}{2}\nabla_1^2 - \frac{1}{2}\nabla_2^2 - \frac{Z}{r_1} - \frac{Z}{r_2} + \frac{1}{r_{12}} - \frac{\mu}{M}\nabla_1 \cdot \nabla_2 \qquad (34)$$

in reduced mass atomic units $e^2/a_\mu = 2R_M$, where $a_\mu = (m/\mu)a_0$ is the reduced mass Bohr radius, and $\mu = mM/(m+M)$ is the electron reduced mass, M is the nuclear mass, and $a_0 = \hbar^2/me^2$ is the Bohr radius. The eigenvalues of this equation are therefore also in units of $2R_M = 2(\mu/m)R_\infty$, and so must be multiplied by μ/m to convert to ordinary atomic units for infinite nuclear mass. This common scaling of all the eigenvalues with μ/m is called the *normal* isotope shift.

The contribution from the mass polarization term is different for every state, and so it is called the *specific* isotope shift. It can be treated either by including it as a perturbation (up to second-order), or by including it explicitly in the Hamiltonian. The latter procedure is simpler and more direct, and the coefficient of the second-order term can still be extracted by differencing (34,35). In the perturbation approach, the perturbation coefficients are

$$\Delta E^{(1)} = \langle \Psi^{(0)}|\nabla_1 \cdot \nabla_2|\Psi^{(0)}\rangle, \qquad (35)$$
$$\Delta E^{(2)} = \langle \Psi^{(1)}|\nabla_1 \cdot \nabla_2|\Psi^{(0)}\rangle, \qquad (36)$$

where $\Psi^{(0)}$ satisfies the zero-order Schrödinger equation for infinite nuclear mass, and $\Psi^{(1)}$ satisfies the first-order perturbation equation

$$\left(H^{(0)} - E^{(0)}\right)\Psi^{(1)} + \nabla_1 \cdot \nabla_2 \Psi^{(0)} = \Delta E^{(1)} \Psi^{(0)}. \qquad (37)$$

The total energy for finite nuclear mass up to second order is then

$$E_M^{\text{tot}} = \left[E^{(0)} - \frac{\mu}{M}\Delta E^{(1)} + \left(\frac{\mu}{M}\right)^2 \Delta E^{(2)} + \cdots\right]\frac{\mu}{m_e} \qquad (38)$$

in units of $2R_\infty$. The next term is of order $(\mu/M)^3 \simeq 10^{-12}$, and so is negligibly small.

In the differencing approach, one simply includes the mass polarization term explicitly in the Hamiltonian for some particular value of the nuclear mass M, and finds directly the eigenvalue E_M^{tot}. The disadvantage is that the entire problem has to be solved over again for each different isotope. Instead, one can estimate the the coefficient $\Delta E_M^{(2)}$ to sufficient accuracy by neglecting higher order terms and solving (38) for $\Delta E_M^{(2)}$ to obtain

$$\Delta E^{(2)} \simeq \frac{(m_e/\mu)E_M^{(\text{tot})} - E^{(0)} + (\mu/M)\Delta E^{(1)}}{(\mu/M)^2}. \qquad (39)$$

In summary, just two calculations are needed—one for infinite nuclear mass to determine $E^{(0)}$ and one for any finite mass M to determine $E_M^{(\text{tot})}$. By differencing, $\Delta E^{(2)}$ can then be estimated to sufficient accuracy from (39).

In a completely different approach, a general method for the decomposition of the Schrödinger equation for finite nuclear mass was developed many years ago by Bhatia and Temkin (36), and the effects of mass polarization studied by Bhatia and Drachman (37) for a range of values of μ/M. These authors have also extended the calculation of the second-order mass polarization term for several low-lying states to the He-like ions (38).

2. Relativistic Corrections

This section briefly summarizes the lowest-order relativistic corrections of order α^2 Ry, and the relativistic recoil corrections of order $\alpha^2 \mu/M$ Ry. The well-known terms in the Breit interaction (2) (including for convenience the anomalous magnetic moment terms of order α^3 Ry) give rise to the first-order perturbation correction

$$\Delta E_{\text{rel}} = \langle \Psi_J | H_{\text{rel}} | \Psi_J \rangle, \qquad (40)$$

where Ψ_J is a nonrelativistic wave function for total angular momentum $\mathbf{J} = \mathbf{L} + \mathbf{S}$ (including mass polarization corrections) and H_{rel} is defined by (in atomic units)

$$H_{\text{rel}} = \left(\frac{\mu}{m_e}\right)^4 B_1 + \left(\frac{\mu}{m_e}\right)^3 \Bigg[B_2 + B_4 + B_{\text{so}} + B_{\text{soo}} + B_{\text{ss}}$$
$$+ \frac{m_e}{M}(\tilde{\Delta}_2 + \tilde{\Delta}_{\text{so}}) + \gamma \left(2B_{\text{so}} + \frac{4}{3}B_{\text{soo}} + \frac{2}{3}B_{3e}^{(1)} + 2B_5\right)$$
$$+ \gamma \frac{m_e}{M}\tilde{\Delta}_{\text{so}} \Bigg] \tag{41}$$

with $\gamma = \alpha/(2\pi)$. The factors of $(\mu/m_e)^4 = (1 - \mu/M)^4$ and $(\mu/m_e)^3 = (1 - \mu/M)^3$ arise from the mass scaling of each term in the Breit interaction, while the terms $\tilde{\Delta}_2$ and $\tilde{\Delta}_{\text{so}}$ are dynamical corrections arising from the transformation of the Breit interaction to CM plus relative coordinates (40). These latter terms are often not included in atomic structure calculations, but they make an important contribution to the isotope shift. The explicit expressions for the spin-independent operators are

$$B_1 = \frac{\alpha^2}{8}(p_1^4 + p_2^4), \tag{42}$$

$$B_2 = -\frac{\alpha^2}{2}\left(\frac{1}{r_{12}}\boldsymbol{p}_1 \cdot \boldsymbol{p}_2 + \frac{1}{r_{12}^3}\boldsymbol{r}_{12} \cdot (\boldsymbol{r}_{12} \cdot \boldsymbol{p}_1)\boldsymbol{p}_2\right), \tag{43}$$

$$B_4 = \alpha^2 \pi \left(\frac{Z}{2}\delta(\boldsymbol{r}_1) + \frac{Z}{2}\delta(\boldsymbol{r}_2) - \delta(\boldsymbol{r}_{12})\right) \tag{44}$$

and the spin-dependent terms are

$$B_{\text{so}} = \frac{Z\alpha^2}{4}\left[\frac{1}{r_1^3}(\boldsymbol{r}_1 \times \boldsymbol{p}_1) \cdot \sigma_1 + \frac{1}{r_2^3}(\boldsymbol{r}_2 \times \boldsymbol{p}_2) \cdot \sigma_2\right], \tag{45}$$

$$B_{\text{soo}} = \frac{\alpha^2}{4}\left[\frac{1}{r_{12}^3}\boldsymbol{r}_{12} \times \boldsymbol{p}_2 \cdot (2\sigma_1 + \sigma_2) - \frac{1}{r_{12}^3}\boldsymbol{r}_{12} \times \boldsymbol{p}_1 \cdot (2\sigma_2 + \sigma_1)\right], \tag{46}$$

$$B_{\text{ss}} = \frac{\alpha^2}{4}\left[-\frac{8}{3}\pi\delta(\boldsymbol{r}_{12}) + \frac{1}{r_{12}^3}\sigma_1 \cdot \sigma_2 - \frac{3}{r_{12}^3}(\sigma_1 \cdot \boldsymbol{r}_{12})(\sigma_2 \cdot \boldsymbol{r}_{12})\right]. \tag{47}$$

Finally, the relativistic recoil terms are (39)

$$\tilde{\Delta}_2 = -\frac{Z\alpha^2}{2} \left\{ \frac{1}{r_1}(\boldsymbol{p}_1 + \boldsymbol{p}_2) \cdot \boldsymbol{p}_1 + \frac{1}{r_1^3}\boldsymbol{r}_1 \cdot [\boldsymbol{r}_1 \cdot (\boldsymbol{p}_1 + \boldsymbol{p}_2)]\boldsymbol{p}_1 \right.$$
$$\left. + \frac{1}{r_2}(\boldsymbol{p}_1 + \boldsymbol{p}_2) \cdot \boldsymbol{p}_2 + \frac{1}{r_2^3}\boldsymbol{r}_2 \cdot [\boldsymbol{r}_2 \cdot (\boldsymbol{p}_1 + \boldsymbol{p}_2)]\boldsymbol{p}_2 \right\}, \quad (48)$$

$$\tilde{\Delta}_{\text{so}} = \frac{Z\alpha^2}{2} \left(\frac{1}{r_1^3}\boldsymbol{r}_1 \times \boldsymbol{p}_2 \cdot \boldsymbol{\sigma}_1 + \frac{1}{r_2^3}\boldsymbol{r}_2 \times \boldsymbol{p}_1 \cdot \boldsymbol{\sigma}_2 \right). \quad (49)$$

It is then a relatively straight forward matter to calculate accurate expectation values for these operators.

The above contributions provide a complete account of the relativistic finite nuclear mass and recoil terms of order $\alpha^2 \mu/M$ Ry. These terms are particularly difficult to treat in a relativistic formulation based on the Dirac equation, as discussed by Sapirstein and Cheng (40). For states of high angular momentum, it becomes possible to replace these elaborate calculations based on large variational wave functions with simple analytic expressions derived by the asymptotic expansion method, as discussed in (35).

3. Quantum Electrodynamic Corrections

The QED corrections are generally taken to mean corrections coming from the new physics that lies beyond the level of the Dirac equation together with the Breit interaction. In lowest order, they introduce corrections that are of the order α^3 Ry, and come from such effects as electron self-energy (see Fig. 4), vacuum polarization, and the anomalous magnetic moment of the electron. For one-electron atoms and ions, the theory is very well established, as recently reviewed by Eides et al. (41).

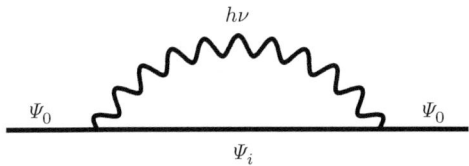

Figure 4. Feynman diagram for the electron self energy.

For low-Z atoms and ions, the QED corrections are conveniently expressed in terms of a double expansion involving powers of αZ and α/π relative to the lowest-order one-loop term of order $mc^2\alpha^5 Z^4$, or $\alpha^3 Z^4$ Ry. The αZ terms represent relativistic corrections to the lowest-order term, and the α/π terms represent higher-order multi-loop corrections. We will begin with a discussion of the lowest-order term, and then include the higher order corrections.

The analysis of Kabir and Salpeter (42), and of Araki (43) and Sucher (44) shows that the lowest order terms can be expressed as the sum of two parts. The first, denoted by $\Delta E_{L,1}$, is simply related to the corresponding hydrogenic Lamb shift and comes from the electron–nucleus interaction. For S-states, one simply replaces the hydrogenic multiplying factor of $(Z^3/\pi n^3)\delta_{l,0}$ for the electron density at the nucleus (summed over the electrons) by the correct matrix element $\langle \delta(\mathbf{r}_1) + \delta(\mathbf{r}_2) \rangle$ for the state in question, and replaces the hydrogenic Bethe logarithm $\beta(nl)$ by the corresponding two-electron value. The second part, denoted by $\Delta E_{L,2}$ comes from QED corrections to the electron–electron interaction.

The starting point for the above is therefore the hydrogenic Lamb shift. Its expansion in powers of αZ and α/π has the form (41,45)

$$\Delta E_L = \Delta E_{L,1} + \Delta E_{L,2} + \Delta E_M + \Delta E_D. \tag{50}$$

We now discuss each of these terms separately. The electron–nucleus term is (41) (in reduced mass units of μc^2)

$$\Delta E_{L,1} = Z\alpha^5 \mu c^2 \{\ln(Z\alpha)^{-2}\mathcal{A}_{5,1} + \mathcal{A}_{5,0} + Z\alpha\mathcal{A}_{6,0}$$
$$+ (Z\alpha)^2[\ln^2(Z\alpha)^{-2}\mathcal{A}_{7,2} + \ln(Z\alpha)^{-2}\mathcal{A}_{7,1} + \mathcal{A}_{7,0}]$$
$$+ (\alpha/\pi)[\mathcal{B}_{6,0} + Z\alpha\mathcal{B}_{7,0}] + O(Z\alpha)^3 + O(\alpha/\pi)^2\} \tag{51}$$

including electron self-energy (see Fig. 4) and vacuum polarization contributions, but omitting the spin-dependent anomalous magnetic moment corrections to the Breit interaction since these terms are counted separately in the evaluation of the Breit interaction itself. The various coefficients up to $\mathcal{A}_{7,2}$ are then

$$\mathcal{A}_{5,1} = (4/3)\langle\sum_i \delta(\mathbf{r}_i)\rangle, \tag{52}$$
$$\mathcal{A}_{5,0} = (4/3)[19/30 - \ln(k_0)]\langle\sum_i \delta(\mathbf{r}_i)\rangle, \tag{53}$$

$$\mathcal{A}_{6,0} = \pi(427/96 - 2\ln 2)\langle\sum_i \delta(\mathbf{r}_i)\rangle, \tag{54}$$

$$\mathcal{A}_{7,2} = -\langle\sum_i \delta(\mathbf{r}_i)\rangle, \tag{55}$$

$$\mathcal{A}_{7,1} \simeq 5.29\langle\sum_i \delta(\mathbf{r}_i)\rangle, \tag{56}$$

$$\mathcal{A}_{7,0} \simeq -30.9\langle\sum_i \delta(\mathbf{r}_i)\rangle, \tag{57}$$

$$\mathcal{B}_{6,0} = \pi^2\left[-\frac{9\zeta(3)}{4\pi^2} - \frac{2,179}{648\pi^2} + \frac{\ln(2)}{2} - \frac{10}{27}\right]\langle\sum_i \delta(\mathbf{r}_i)\rangle, \tag{58}$$

$$\mathcal{B}_{7,0} \simeq -21.55685\langle\sum_i \delta(\mathbf{r}_i)\rangle, \tag{59}$$

where $\ln(k_0)$ is the Bethe logarithm for the particular state in question, with k_0 in units of $Z^2 R_M$, and $R_M = (1-\mu/M)R_\infty$ is the finite mass Rydberg. The subscripts p and q in $\mathcal{A}_{p,q}$ denote the powers of α and $\ln(\alpha)$ respectively.

In the hydrogenic case, the matrix elements of the δ-function are simply $\langle\delta(\mathbf{r})\rangle = Z^3 \delta_{l,0}/(\pi n^3)$ for a state with principal quantum number n and angular momentum l. The above formulas then reproduce the standard expression for the hydrogenic Lamb shift (41). For an N-electron atom, the above coincides with the lowest-order terms discussed explicitly by Kabir and Salpeter (42), together with higher order extensions given by the terms $\mathcal{A}_{6,0}$, $\mathcal{A}_{7,2}$ and $\mathcal{B}_{6,0}$ that are simply proportional to the electron density at the nucleus. However, the coefficients multiplying the δ-function in $\mathcal{A}_{7,1}$ and $\mathcal{A}_{7,0}$ are themselves weakly state-dependent in the hydrogenic case (48) (denoted A_{61} and A_{60} in this reference), and so (56) and (57) are at best approximations to the correct multi-electron expression.

The electron-electron QED terms are similarly given by

$$\Delta E_{\text{L},2} = \alpha^5 \mu c^2[\ln(\alpha)\mathcal{C}_{5,1} + \mathcal{C}_{5,0} + \alpha\ln(\alpha)\mathcal{C}_{6,1} + \alpha\mathcal{C}_{6,0} + O(\alpha^2)], \tag{60}$$

where

$$\mathcal{C}_{5,1} = (14/3)\sum_{i>j}\langle\delta(\mathbf{r}_{ij})\rangle, \tag{61}$$

$$\mathcal{C}_{5,0} = (164/15)\sum_{i>j}\langle\delta(\mathbf{r}_{ij})\rangle - \tfrac{14}{3}\alpha^3 Q, \tag{62}$$

$$\mathcal{C}_{6,1} = -\sum_{i>j}\langle\delta(\mathbf{r}_{ij})\rangle \tag{63}$$

$$\mathcal{C}_{6,0} = \pi\left[\frac{27\zeta(3)}{2\pi^2} + \frac{763}{54\pi^2} - 13\ln(2) + \frac{3,529}{216}\right]\sum_{i>j}\langle\delta(\mathbf{r}_{ij})\rangle \tag{64}$$

with

$$Q = \frac{1}{4\pi}\sum_{i>j}\lim_{\epsilon\to 0}\langle r_{ij}^{-3}(\epsilon) + 4\pi(\gamma + \ln\epsilon)\delta(\mathbf{r}_{ij})\rangle, \tag{65}$$

γ is Euler's constant, and ϵ is the radius of a sphere centered about $r_{ij} = 0$ that is excluded from the range of integration.

The term ΔE_M in (50) is a finite mass correction given to sufficient accuracy for our purposes by

$$\Delta E_M = \left(\frac{-2\mu}{M}\right)\left\{\Delta E_{L,1} + \Delta E_{L,2} - \frac{1}{2}Z\alpha^5 mc^2 \left\langle\sum_i \delta(\mathbf{r}_i)\right\rangle\right\}$$
$$+ Z^2\alpha^5 mc^2 \frac{\mu}{M}\left\{\left[\frac{1}{4}\ln(Z\alpha)^{-2} - \frac{8}{3}\ln(k_0) + \frac{62}{9}\right]\right.$$
$$\left. \times \left\langle\sum_i \delta(\mathbf{r}_i)\right\rangle - \frac{14}{3}Q_1\right\}, \quad (66)$$

where, in parallel with (61),

$$Q_1 = \frac{1}{4\pi}\sum_i \lim_{\epsilon \to 0}\langle r_i^{-3}(\epsilon) + 4\pi(\gamma + \ln\epsilon)\delta(\mathbf{r}_i)\rangle. \quad (67)$$

(The additional term $-\frac{1}{2}Z\alpha^5 mc^2\langle\sum_i \delta(\mathbf{r}_i)\rangle$ in the first line of (66) accounts for the additional mass scaling of the Bethe logarithm in $\Delta E_{L,1}$.)

The remaining term to be discussed is ΔE_D. In the hydrogenic case, this term corresponds to the Coulomb–Dirac energies of order $\alpha^6 mc^2$, and they are known in closed analytic form. However, in the many-electron case, they must be calculated as the nonrelativistic expectation value of set of highly singular operators and second-order contributions from the Breit interaction. The spin-dependent part has been known for many years from the work of Douglas and Kroll (47), but the spin-independent part has only recently been worked out by Korobov and Yelkhovsky (48) for the ground state of helium, and by Pachucki for the low-lying S- and P-states (32, 33). The last reference expresses the final result in a particularly useful form for computational work. Pachucki's results for ΔE_D are quoted in the tables to follow, and the reader is referred to Pachucki's papers for further information on this important advance in the field.

Also included in the tables is the correction ΔE_{nuc} for the finite size of the nucleus. It is an approximation to assume that the nucleus is a point-like source of Coulomb field. If the nuclear charge is distributed over a finite volume of rms radius r_c, then in lowest order the energy correction is (49)

$$\Delta E_{\text{nuc}} = \frac{2\pi Z e^2 r_c^2}{3} \left\langle \sum_i \delta^3(\mathbf{r}_i) \right\rangle. \qquad (68)$$

With sufficient accuracy, this energy shift provides a method of measuring the nuclear charge radius r_c from the isotope shift, as recently reviewed by Drake et al. (50) for certain 'halo' nuclei such as ^{6}He and ^{11}Li that are difficult to measure by any other means.

IV. RESULTS FOR THE ENERGY LEVELS OF HELIUM

The complete singly-excited spectrum of helium is now very well understood, and detailed tabulations of high-precision ionization energies are available for all states up to $n = 10$ and $L = 7$ (20). However, it is the lowest-lying S- and P-states that are of the greatest interest. For example, the QED contributions to the ionization energy decrease in proportion to $1/n^3$, and so they are largest for $n = 1$ and $n = 2$.

On the other hand, the QED corrections become so small for the Rydberg states with high n, such as the $1snd\,^{1,3}$D states, that these states can be taken as known points of reference for the determination of experimental ionization energies from measured transition frequencies. For example, the most accurately known ionization energy of 1 152 842 742.97(6) MHz for the $1s2s\,^3$S$_1$ state comes from adding the measured $2\,^3$S$_1 - 3\,^3$D$_1$ transition frequency of 786 823 850.002(56) MHz (53) to the theoretical ionization energy of 366 018 892.97(1) MHz (20) for the 3^3D$_1$ state.

The various contributions to the energies of the low-lying S-states and P-states are summarized in Tables 5 and 6 respectively, together with a more detailed breakdown of the QED contributions in Table 7. The meaning of the various items is as follows. All entries include the normal reduced mass isotope shift contained in the conversion factor $\mu c^2 \alpha^2 = 2R_M$, where R_M is the reduced mass Rydberg. The entries labelled $\mu c^2 \alpha^2 (\mu/M)_{\text{mp}}$ and $\mu c^2 \alpha^2 (\mu/M)^2_{\text{mp}}$ then represent the first- and second-order mass polarization (mp) corrections to the ionization energy (i.e., the negative ionization potential, relative to He$^+$). The next entries are the lowest-order relativistic corrections of order $\mu c^2 \alpha^4$, and the finite mass corrections of order $\mu c^2 \alpha^4 (\mu/M)$ coming from mass scaling (ms) and mass polarization, including the Stone terms from (48) and (49). ΔE_{nuc} is

Table 5.

Contributions to the (negative) ionization energies of the S-states of helium. The values of the fundamental constants used are $R_\infty = 3\,289\,841\,960.368$ MHz, $\alpha^{-1} = 137.035\,9998$, $\mu/M = 1.370745\,6409 \times 10^{-4}$, $r_c = 1.6730$ fm. Units are MHz

Term	$1\,^1S_0$	$2\,^1S_0$	$2\,^3S_1$
$\mu c^2 \alpha^2$	−5945 405 676.717	−960 331 428.608	−1152 795 881.766
$\mu c^2 \alpha^2 (\mu/M)_{mp}$	143 446.256	8 570.430	6711.192
$\mu c^2 \alpha^2 (\mu/M)_{mp}^2$	−58.146	−16.722	−7.107
$\mu c^2 \alpha^4$	16 901.706	−11 969.811(1)	−57 621.412
$\mu c^2 \alpha^4 (\mu/M)_{ms}$	−154.761	−14.832	−5.642
$\mu c^2 \alpha^4 (\mu/M)_{mp}$	53.370	9.845	2.025
ΔE_{nuc}	29.70(3)	1.995(2)	2.595(3)
ΔE_{QED} (from Table 7)	41 284(36)	2 809.9(1.7)	4 058.8(2.5)
Total (this work)	−5945 234 175(36)	−960 332 037.9(1.7)	−1152 842 741.3(2.5)
Pachucki theory (32, 33)	−5945 204 174(36)	−960 332 038.1(1.9)	−1152 842 741.6(2.5)
Experiment	−5945 204 238(45)[a]	−960 332 041.01(15)[c]	−1152 842 742.97(6)[d]
Experiment	−5945 204 238(45)[b]		

[a] From the $1\,^1S - 2\,^1P$ measurement of Eikema et al. (51)
[b] From the $1\,^1S - 2\,^1S$ measurement of Bergesor et al. (52)
[c] Values determined by combining measured transition frequencies to the n D states with theoretical D-state energies (46)
[d] Values determined by combining measured transition frequency to the $3\,^3D_1$ state (53) with theoretical D-state energies (46)

Table 6.
Contributions to the (negative) ionization energies of the P-states of helium. The values of the fundamental constants used are as in Table 1. Units are MHz

Term	2^1P_1	2^3P_{cm}
$\mu c^2 \alpha^2$	−814 736 669.940	−876 058 183.130
$\mu c^2 \alpha^2 (\mu/M)_{mp}$	41 522.201	−58 230.359
$\mu c^2 \alpha^2 (\mu/M)_{mp}^2$	−20.800	−25.335
$\mu c^2 \alpha^4$	−14 022.122	11 435.310
$\mu c^2 \alpha^4 (\mu/M)_{ms}$	−8.509	4.232
$\mu c^2 \alpha^4 (\mu/M)_{mp}$	3.775	8.384
ΔE_{nuc}	0.064	−0.795(1)
ΔE_{QED} (from Table 4)	48.0(1.0)	−1 255.1(1.0)
Total (this work)	−814 709 147.4(1.0)	−876 106 246.8(1.0)
Pachucki theory (33)	−814 709 145.99(16)	−876 106 246.93(40)
Experiment	−814 709 153.0(3.0)[a]	−876 106 247.35(6)[b]

[a] Value derived by combining the measured $2^1P_1 - 3^1D_2$ transition frequency (54) (including a correction of 0.000 02 cm^{-1} (46) with the calculated energy of the 3^1D_2 state (55)

[b] Value derived by combining 2^3S energy with $2^3S - 2^3P$ measurement (56)

the correction for finite nuclear size. All of the terms up to this point have been well established for many years (46), and are accurate to the figures quoted.

The main quantity of interest in comparison with experiment, and the principal source of uncertainty, is the QED shift ΔE_{QED}. The various contributions are itemized in detail Table 7. All quantities in the leading term of order $\mu c^2 \alpha^5$ are easily evaluated with the exception of the Bethe logarithm. A solution to this problem has recently been developed by Drake and Goldman (57), and Bethe logarithms tabulated for a wide variety of states (see also (58) and (59) for further results and asymptotic expansions). The principal differences from previous work are as follows. The totals of order $\mu c^2 \alpha^5$ are identical to those tabulated previously in (57) and (59), and are close to the ones quoted in the recent papers by Pachucki (32, 33), with exception for the 2^1P_1 state. Here, the difference of 1.44 MHz (59) is large enough to affect the comparison with experiment. The source of the discrepancies between the theoretical values listed here and Pachucki's has been identified as a transcription error, and the present results for the terms of order $\mu c^2 \alpha^5$ should be taken as correct.

Table 7.
Detailed breakdown of the QED contributions of order $\mu c^2 \alpha^5$ and higher

Contribution	1^1S_0	2^1S_0	2^3S_1	2^1P_1	2^3P_{cm}
$\mu c^2 \alpha^5 Z[\ln(Z\alpha)^{-2} \mathcal{A}_{5,1} + \mathcal{A}_{5,0}]$	44708.428	3085.771	4035.790	102.084	−1189.876
$\mu c^2 \alpha^5 [\ln(\alpha) \mathcal{C}_{5,1} + \mathcal{C}_{5,0}]$	−4208.231	−330.373	−36.883	−62.821	−45.502
ΔE_M (finite mass)	−4.689	−0.284	−0.302	−0.027	0.065
$\mu c^2 \alpha^5$ total	40495.508	2755.114	3998.605	39.236	−1235.313
$\mu c^2 \alpha^6 Z^2 \mathcal{A}_{6,0}$	771.109	51.995	67.634	1.663	−20.655
$\mu c^2 \alpha^6 Z \mathcal{B}_{6,0}/\pi$	6.876	0.464	0.603	0.015	−0.184
$\mu c^2 \alpha^6 \ln(\alpha) \mathcal{C}_{6,1}$	30.666	2.494	0.000	0.212	0.000
$\Delta E(2^3P_1 - 2^1P_1$ mixing)				4.755	−4.755
ΔE_D (Pachucki)	52.589	3.34	−1.79	2.18	3.76
$\mu c^2 \alpha^6$ total	861.24	58.29	65.24	8.82	−21.83
$\mu c^2 \alpha^7 Z^3 \ln^2(Z\alpha)^{-2} \mathcal{A}_{7,2}$	−83.628	−5.639	−7.335	−0.180	2.240
$\mu c^2 \alpha^7 Z^3 \ln(Z\alpha)^{-2} \mathcal{A}_{7,1}$	52.289	4.894	5.966	0.170	−1.336
$\mu c^2 \alpha^7 Z^3 \mathcal{A}_{7,0}$	−37.442	−2.525	−3.284	−0.081	1.003
$\mu c^2 \alpha^7 Z^2 \mathcal{B}_{7,0}/\pi$ two-loop binding	−3.948	−0.266	−0.346	−0.009	0.106
$\mu c^2 \alpha^7$ total	−72(36)	−3.5(1.7)	−5.0(2.5)	−0.1(1.0)	2.0(1.0)
Total	41284(36)	2809.9(1.7)	4058.8(2.5)	48.0(1.0)	−1255.1(1.0)

In next higher order, the total contribution of order $\mu c^2 \alpha^5$ is the same as that tabulated by Pachucki (32, 33). The contributions from the terms other than ΔE_D are subtracted from the total in order to display the the new contribution from ΔE_D. It is clear that the ΔE_D term is a relatively small fraction of the total (10–15%) and the dominant contribution comes from the one-loop and two-loop radiative terms contained in $\Delta E_{L,1}$ and $\Delta E_{L,2}$. (The one exception is the 2^1P_1 state where $\Delta E_{L,1}$ happens to be abnormally small.) This provides a justification for approximate calculations in which the ΔE_D term is neglected and the QED shift estimated from $\Delta E_{L,1}$ and $\Delta E_{L,2}$.

Finally, in the absence of more detailed calculations, the terms of order $\mu c^2 \alpha^7$ are estimated from the dominant contributions contained in $\Delta E_{L,1}$. The results are identical to those contained in (57) and (59), but the terms of order $\mu c^2 \alpha^7 Z^3 \ln(Z\alpha)^{-2}$ are somewhat smaller than those of Pachucki (32, 33) because we use ascreened nuclear charge for the outer electron. Otherwise, the contribution from these terms become unreasonably large. This point becomes particularly evident when the contributions are compared with the known spin-dependent terms of the same order (60–62) and with the measured fine-structure splittings.

V. DISCUSSION AND CONCLUSIONS

There are very few problems in physics where exact analytic solutions are possible. Even in the case of the hydrogen atom, the exact analytic solutions to the Schrödinger or Dirac equation that one finds in text books are based on a large number of physical approximations such as a point-like structureless nucleus, the neglect of QED corrections, and the absence of external fields. One can then use the comparison between theory and experiment to look for and examine the physical effects not included in the original theory, and perhaps even discover entirely new physics, such as happened with Adams and Le Verrier leading to the discovery of the planet Neptune.

In most other areas of physics, the many-body nature of the problem precludes a similar kind of program of discovery because the dynamics of the many-body system is too complicated to permit solutions of sufficient accuracy and reliability. There are certainly exceptions, but in most cases a difference between theory and experiment could just as easily be attributed to a defect in the numerical predictions extracted from the underlying theory as to evidence for new physics. This of course is why it is just as important to assign uncertainty estimates to the theoretical predictions as it is to the experimental measurements.

The results summarized in this chapter illustrate that, even though exact analytic solutions are not possible for a three-body problem such as helium, it is still possible to develop specialized numerical methods that are essentially exact for all practical purposes, at least up to the level of the nonrelativistic Schrödinger equation. This then provides a firm platform on which to build the higher order relativistic and QED corrections, and ultimately perform comparisons with high-precision measurements. The existence of the calculations in fact provides a strong motivation to perform the measurements, and so the state-of-the-art for both progresses together. The accuracy of ± 0.06 MHz for the ionization energy of the 1s2s $^3S_1$ state corresponds to a measurement of the QED shift (calculated to be 4058.8 ± 2.5 MHz) to an accuracy of 1.5 parts in 10^5. At this level of accuracy, the results are sensitive to the two-loop binding correction of -0.35 MHz (see Table 7). Of course other aspects of the QED theory must be completed to this same $\alpha^7 mc^2$ level of accuracy for the comparison to be meaningful, but the basic experimental accuracy is in place for the comparison to be made. There is hope that the full two-electron theory can be extended to this level because the spin-dependent parts have already been calculated (60, 62). However, thanks to the recent advances by Pachucki (32, 33) for the term ΔE_D, the atomic theory part of the problem is now complete up to terms of order $\alpha^6 mc^2$.

Results at these levels of accuracy are opening up new research possibilities. For example, the isotope shift provides a unique measurement tool for the determination of the nuclear charge radius for halo nuclei, relative to other known nuclei (30, 50). In addition, as can be seen from the various contributions listed in Table 5, a complete determination of the QED energy shifts to order $mc^2 \alpha^7$ will allow direct measurements of nuclear charge radii from the nuclear volume effect, rather than relative measurements. This will provide an important new technique at the interface between atomic physics and nuclear physics for the determination of nuclear properties in a way that is independent of particular models for nuclear scattering.

ACKNOWLEDGEMENTS

Research support by the Natural Sciences and Engineering Research Council of Canada, and by SHARCNET are gratefully acknowledged.

REFERENCES

1. W. Sheehan, N. Kollerstone, and C.B. Waff, *The Case of the Pilfered Planet*, Scientific American, New York, NY, December (2004).
2. H.A. Bethe and E.E. Salpeter, *Quantum mechanics of one- and two-electron atoms*, (Springer, Berlin Heidelberg New York, 1957).
3. For a complete discussion of units, see W.E. Baylis and G.W.F. Drake, in *Handbook of Atomic, Molecular and Optical Physics*, Edited by G.W.F. Drake (Springer, Berlin Heidelberg New York, 2005), p. 1.
4. The latest values of the fundamental constants can be downloaded from the web site http://physics.nist.gov/cuu/Constants/index.html.
5. E.A. Hylleraas, Z. Phys. **48**, 469 (1928); **54**, 347 (1929). See also *ibid.*, Rev. Mod. Phys. **35**, 421 (1963).
6. E.A. Hylleraas and B. Undheim, Z. Phys. **65**, 759 (1930).
7. J.K.L. MacDonald, Phys. Rev. **43**, 830 (1933).
8. R.N. Hill, Int. J. Quantum Chem. **68**, 357 (1998).
9. B. Klahn and W.A. Bingel, Theor. Chem. Acta **44**, 27 (1977); Int. J. Quantum Chem. **11**, 943 (1978).
10. Y. Accad, C.L. Pekeris, and B. Schiff, Phys. Rev. A **4**, 516 (1971).
11. G.W.F. Drake, in *Long-range Casimir forces: Theory and recent experiments on atomic systems*, Edited by F.S. Levin and D.A. Micha (Plenum, New York, 1993), pp. 107–217.
12. G.W.F. Drake and Z.-C. Yan, Phys. Rev. A **46**, 2378 (1992).
13. G.W.F. Drake, Adv. At. Mol. Opt. Phys. **31**, 1 (1993).
14. G.W.F. Drake, M.M. Cassar, and R.A. Nistor, Phys. Rev. A **65**, 054501 (2002).
15. V.I. Korobov, Phys. Rev. A **66**, 024501 (2002).
16. C. Schwartz, Int. J. Mod. Phys. E – Nucl. Phys. **15**, 877 (2006).
17. S.P. Goldman, Phys. Rev. A **57**, R677 (1998).
18. A. Bürgers, D. Wintgen, and J.-M. Rost, J. Phys. B: At. Mol. Opt. Phys. **28**, 3163 (1995).
19. J.D. Baker, D.E. Freund, R.N. Hill, and J.D. Morgan III, Phys. Rev. A **41**, 1247 (1990).
20. G.W.F. Drake, in *Handbook of Atomic, Molecular and Optical Physics*, Edited by G.W.F. Drake (Springer, Berlin Heidelberg New York, 2005), p. 1355.
21. J. Ackermann and J. Shertzer, Phys. Rev. A **54**, 365 (1996), and earlier references therein.
22. C.-Y. Hu, A.A. Kvitsinsky, and J.S. Cohen, J. Phys. B **28**, 3629 (1995), and earlier references therein.
23. E.Z. Liverts, M.Y. Amusia, R. Krivec, and V.B. Mandelzweig, Phys. Rev. A **73**, 012514 (2006), and earlier references therein.

24. J.S. Sims and S.A. Hagstrom, Int. J. Quantum Chem. **90**, 1600 (2002).
25. S. Chakraborty and Y.K. Ho, Chem. Phys. Lett. **438**, 99 (2007).
26. Z.-C. Yan and G.W.F. Drake, Phys. Rev. Lett. **81**, 774 (1998).
27. Z.-C. Yan and G.W.F. Drake, Phys. Rev. A **66**, 042504 (2002).
28. Z.-C. Yan and G.W.F. Drake, Phys. Rev. Lett. **91**, 113004 (2003).
29. M. Puchalski and K. Pachucki, Phys. Rev. A **73**, 022503 (2006).
30. M. Puchalski, A.M. Moro, and K. Pachucki, Phys. Rev. Lett. **97**, 133001 (2006).
31. K. Pachucki and J. Komasa, J. Chem. Phys. **125**, 204304 (2006).
32. K. Pachucki, Phys. Rev. A **74**, 022512 (2006).
33. K. Pachucki, Phys. Rev. A **74**, 062510 (2006).
34. G.W.F. Drake and Z.-C. Yan, Phys. Rev. A **46**, 2378 (1992).
35. G.W.F. Drake, Adv. At. Mol. Opt. Phys. **31**, 1 (1993).
36. A.K. Bhatia and A. Temkin, Phys. Rev. **137**, 1335 (1965).
37. A.K. Bhatia and R.J. Drachman, Phys. Rev. A **35**, 4051 (1987).
38. A.K. Bhatia and R.J. Drachman, J. Phys. B: At. Mol. Opt. Phys. **36**, 1957 (2003).
39. A.P. Stone, Proc. Phys. Soc. (London) **77**, 786 (1961); **81**, 868 (1963).
40. J. Sapirstein and K.T. Cheng, Can. J. Phys. **85** (2007).
41. M.I. Eides, H. Grotch, and V.A. Shelyuto, Phys. Rep. **342**, 63–261 (2001).
42. P.K. Kabir and E.E. Salpeter, Phys. Rev. **108**, 1256 (1957).
43. H. Araki, Prog. Theor. Phys. **17**, 619 (1957).
44. J. Sucher, Phys. Rev. **109**, 1010 (1958).
45. G.W.F. Drake, Adv. At. Mol. Phys. **18**, 399 (1982).
46. G.W.F. Drake and W.C. Martin, Can. J. Phys. **76**, 679 (1998).
47. M. Douglas and N.M. Kroll, Ann. Phys. (N. Y.), **82**, 89 (1974).
48. V. Korobov and A. Yelkhovsky, Phys. Rev. Lett. **87** 193003 (2001); A. Yelkhovsky, Phys. Rev. A **64**, 062104 (2001).
49. J.D. Morgan III and J.S. Cohen, in *Handbook of Atomic, Molecular and Optical Physics*, Edited by G.W.F. Drake (Springer, Berlin Heidelberg New York, 2005), p. 1355.
50. G.W.F. Drake, Z.-T. Lu, W. Nörtershäuser, and Z.-C. Yan, in *Precision Physics of Simple Atoms and Molecules*, Edited by S. Karshenboim (Springer, Berlin Heidelberg New York, 2007).
51. K.S.F. Fikema, W. Ubachs, W. Vassen, and H. Horgorvorst, Phys. Rev. A **55**, 1866 (1997).
52. S.D. Bergeson, A. Balakrishnan, K.G.H. Baldwin, T.B. Lucatorto, J.P. Marangos, T.J. McIlrath, T.R. O'Brian, S.L. Rolston, C.J. Sansonetti, J. Wen, and N. Westbrook, Phys. Rev. Lett. **80**, 3475 (1998).
53. C. Dorrer, F. Nez, B. de Beauvoir, L. Julien, and F. Biraben, Phys. Rev. Lett. **78**, 3658 (1997).
54. C.J. Sansonetti and W.C. Martin, Phys. Rev. A **29**, 159 (1984).
55. D.C. Morton, Q.X. Wu, and G.W.F. Drake, Can. J. Phys. **84**, 83 (2006), and earlier references therein.
56. P.C. Pastor, G. Giusfredi, P. De Natale, G. Hagel, C. de Mauro, and M. Inguscio, Phys. Rev. Lett. **92**, 023001 (2004).
57. G.W.F. Drake and S.P. Goldman, Can. J. Phys. **77**, 835 (1999).
58. G.W.F. Drake, Phys. Scr. **T95**, 22 (2001).
59. G.W.F. Drake and Z.-C. Yan, Can. J. Phys. **85** (2007).
60. G.W.F. Drake, Can. J. Phys. **80**, 1195 (2002).
61. K. Pachucki and J. Sapirstein, J. Phys. B: At. Mol. Opt. Phys. **35**, 1783 (2002).
62. K. Pachucki, Phys. Rev. Lett. **97**, 013002 (2006).

3

Modeling of Impedance of Porous Electrodes

Andrzej Lasia

*Département de chimie, Université de Sherbrooke, Sherbrooke, QC
Canada J1K 2R1*

I. INTRODUCTION

Porous electrodes are very important in practical applications of electrocatalysis, where an increase in the real surface area leads to an increase in catalytic activity. Porous electrodes are used in gas evolution (water electrolysis, hydrogen and oxygen evolution, chlorine evolution), electrocatalytic hydrogenation or oxidation of organic compounds, in batteries, fuel cells, etc. Good knowledge of the porous electrode theory permits for the construction of the electrodes with optimal utilization of the active electrode material. The porous electrode model was first developed by several authors for dc conditions (1–6) and later applied to the impedance studies.

The electrochemical impedance spectroscopy, EIS, is a method in which small sinusoidal perturbation of the potential or the current is applied and the ac response is measured, from which the complex impedance is obtained at different frequencies (7). It is sensitive to the interfacial processes (e.g., reaction mechanism, adsorption), the surface geometry (planar, spherical, cylindrical, porosity) and the mass transfer (semi-infinite or finite length diffusion, forced mass transfer).

In the previous chapters on impedance a theory of the electrochemical impedance spectroscopy including a general discussion of porous materials (7) and the applications to the reactions involving hydrogen (8) were presented. Understanding of the theoretical basis of the impedance is essential in comprehension of the theory of porous electrodes. In this chapter a simple cylindrical porous electrode model will be presented in the absence and in the presence of the concentration gradient and, later, a distribution of pores and a continuous pore model will be shown. In addition the discussion of the fractal model and the constant phase element were added as they are often related to the electrode roughness.

II. IMPEDANCE OF POROUS ELECTRODES IN THE ABSENCE OF THE ELECTROCHEMICAL REACTION, OHMIC DROP IN THE SOLUTION ONLY

1. Cylindrical Pore

i. Ohmic drop in the solution only

First, let us consider a simple cylindrical ideally polarizable electrode, length l and radius r, in which only the pore walls are electrically conductive and the external surface as well as the bottom disk are nonconducting. The model of such pore is shown in Fig. 1. The

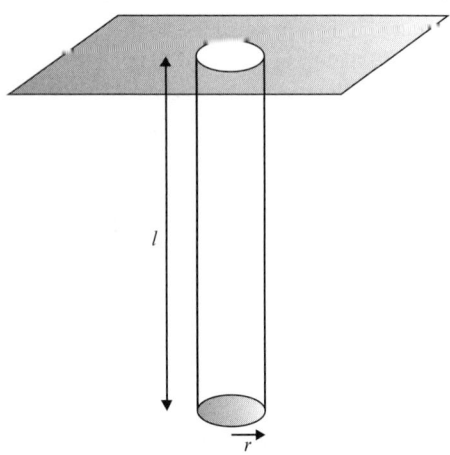

Figure 1. Model of a cylindrical porous electrode. The gray area is nonconductive.

electrically nonconductive surface is shown as gray area. Although there is no dc current flowing in such a system an ac charging current can flow to the pore walls, its density decreasing with the penetration depth due to the ohmic drop in the solution.

An equation for impedance of this electrode was developed by de Levie (9). It is assumed that

(a) The pore is cylindrical
(b) Only the pore side walls are conducting, resistance of the electrode material is zero, $\rho_e = 0$
(c) The pore is filled with the supporting electrolyte characterized by the resistance ρ_s
(d) ac potential gradient exists in the axial direction only, i.e., there is no radial potential gradient

The specific impedance of the pore walls is $\hat{Z}_{el} = 1/j\omega C_{dl}$ (in $\Omega\,\text{cm}^2$), where C_{dl} is the specific capacitance (in F cm^{-2}) of the pore walls, ω is the angular frequency of the ac perturbation and $j = \sqrt{-1}$. The ac potential, $\tilde{E}$, which enters into pore decreases from the initial value at pore orifice, $\tilde{E}_0$, because of the Ohm's law and the ac charging current, $\tilde{I}$, decreases with the pore depth as it flows to the pore walls. The system might be described by a system of two differential equations

$$\frac{d\tilde{I}}{dx} = -\frac{\tilde{E}}{\hat{z}}, \tag{1}$$

$$\frac{d\tilde{E}}{dx} = -r_s\tilde{I}, \tag{2}$$

where $\hat{z}$ is the impedance per pore unit length (in $\Omega\,\text{cm}$), $\hat{z} = \hat{Z}_{el}l/s = \hat{Z}_s l = 1/(j\omega C_{dl}\,2\pi r\,l) = 1/(j\omega C_{dl}\,2\pi r)$, $\hat{Z}_{el}$ is the specific impedance of pore walls, $\hat{Z}_{el} = 1/j\omega C_{dl}$, (in $\Omega\,\text{cm}^2$), $\hat{Z}_s = 1/(j\omega C_{dl}\,2\pi rl)$ is the total impedance of the pore wall (in Ω), s is the total surface area or the pore wall, r_s is the solution resistance in the pore per unit length, $r_s = R_{\Omega,p}/l = \left(\rho_s l/\pi r^2\right)/l = \rho_s/\pi r^2$ (in $\Omega\,\text{cm}^{-1}$) and $R_{\Omega,p}$ is the total resistance of the pore filled with solution, $R_{\Omega,p} = \rho_s l/\pi r^2$, (in Ω). The second derivative of $\tilde{E}$ (see (2)) versus x equals

$$\frac{d^2\tilde{E}}{dx^2} = -r_s\frac{d\tilde{I}}{dx} = \frac{r_s}{\hat{z}}\tilde{E} \tag{3}$$

with the following conditions:

$$x = 0 \quad \tilde{E} = \tilde{E}_0,$$
$$x = l \quad d\tilde{E}/dx = 0.$$

The solution of (3) is

$$\tilde{E} = \tilde{E}_0 \frac{\cosh\left[\sqrt{\frac{r_s}{\hat{z}}}(l-x)\right]}{\cosh\left[\sqrt{\frac{r_s}{\hat{z}}}l\right]}. \tag{4}$$

This equation predicts that the amplitude of the ac signal, which penetrates into the pore, decreases with the distance in pore x. An example of such relation is shown in Fig. 2.

At the top surface of the pore

$$\left.\frac{d\tilde{E}}{dx}\right|_{x=0} = -\tilde{E}_0 \sqrt{\frac{r_s}{\hat{z}}} \tanh\left(\sqrt{\frac{r_s}{\hat{z}}}l\right) = -\tilde{I}_0 r_s \tag{5}$$

and the impedance is defined as a ratio of the phasors (7) of the potential and current

$$\hat{Z}_{\text{pore}} = \frac{\tilde{E}_0}{\tilde{I}_0} = \sqrt{r_s \hat{z}} \coth\left(\sqrt{\frac{r_s}{\hat{z}}}l\right). \tag{6}$$

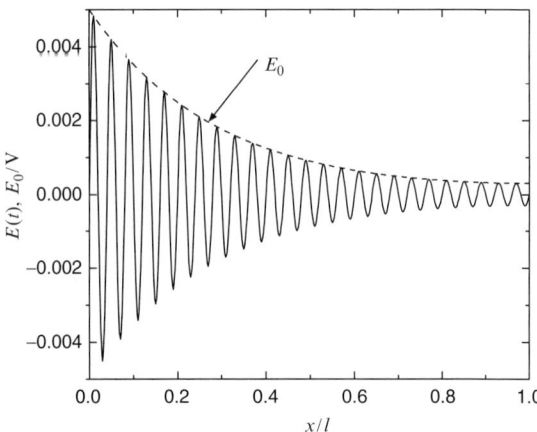

Figure 2. Changes of the ac signal $E(t)$ and its amplitude in the pore according to (4).

This equation might be rearranged into another useful form (10)

$$\hat{Z}_{\text{pore}} = \frac{R_{\Omega,\text{p}}}{\Lambda^{1/2}} \coth\left(\Lambda^{1/2}\right), \quad (7)$$

where $\Lambda = r_s l^2/\hat{z} = \left(2\rho_s l^2/r\right)/\hat{Z}_{\text{el}}$ is the dimensionless admittance of the porous electrode. In the case of n pores and when the solution resistance outside pores is R_{sol} the total impedance is

$$\hat{Z} = R_{\text{sol}} + \frac{\hat{Z}_{\text{pore}}}{n}. \quad (8)$$

Equation (7) predicts a straight line at 45° at high frequencies followed by a vertical capacitive straight line on the complex plane plots, see Fig. 3.

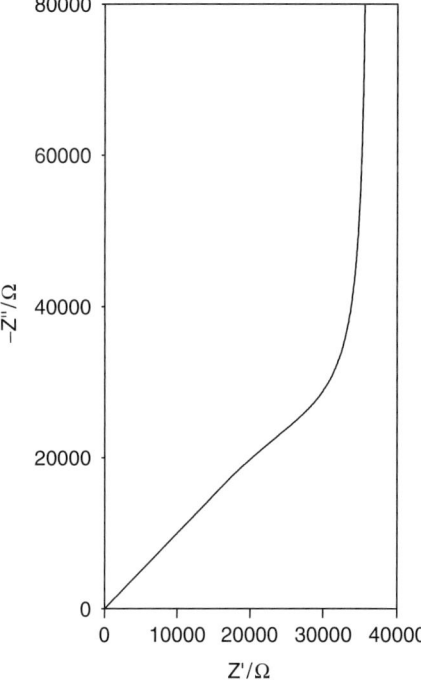

Figure 3. Complex plane plot of the impedance of a porous electrode.

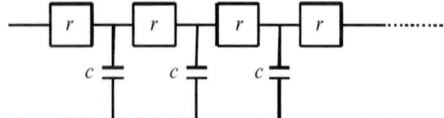

Figure 4. Uniform transmission line representing impedance of a flooded ideally polarized porous electrode, r and c are the solution resistance and double layer capacitance, respectively, of a small element of the pore length.

Equation (7) cannot be represented by a simple connection of the R–L–C elements. It can be represented only by a semi-infinite series, so called transmission line (11, 12), represented in Fig. 4. Of course, it is simpler to represent the impedance by an exact equation than by a semi-infinite series, although both representations are equivalent.

Equation (7) has two limiting forms. At very high frequencies it simplifies to

$$\hat{Z}_{\text{pore}} = \sqrt{r_s \hat{z}} = \frac{R_{\Omega,\text{p}}}{\Lambda^{1/2}} = \frac{\rho_s^{1/2}}{\sqrt{2}\pi r^{3/2}} \sqrt{\hat{Z}_{\text{el}}} = \frac{\rho_s^{1/2}}{2\pi r^{3/2} \omega^{1/2} C_{\text{dl}}^{1/2}} (1-j) \quad (9)$$

taking into account that $1/\sqrt{j} = (1-j)/\sqrt{2}$. Equation (9) indicates that the real and imaginary parts of the impedance are identical which leads to a straight line at 45° on the complex plane plots. It should be stressed that this impedance is independent of the pore length.

At low frequencies $\coth(\Lambda^{1/2})/\Lambda^{1/2} \approx 1/3 + 1/\Lambda$ and (7) becomes

$$\hat{Z}_{\text{pore}} = \frac{R_{\Omega,\text{p}}}{3} + \frac{R_{\Omega,\text{p}}}{\Lambda} = \frac{\rho_s l}{3\pi r^2} + \frac{1}{j\omega C_{\text{dl}}(2\pi r l)} = R_{\text{pore}} + \frac{1}{j\omega C_{\text{pore}}} \quad (10)$$

corresponding to the R_{pore}–C_{pore} connection in series and it displays a straight capacitive line with the resistance $R_{\text{pore}} = R_{\Omega,\text{p}}/3$ and the capacitance equal to the total capacitance of the pore wall (specific capacitance C_{dl} times the surface area $2\pi r l$, $C_{\text{pore}} = C_{\text{dl}} 2\pi r l$).

This behavior may be also explained in terms of the penetration length of the ac signal into pores, $\lambda = l/\sqrt{\Lambda} = \left(r \hat{Z}_{\text{el}}/2\rho_s\right)^{1/2}$.

At high frequencies $\hat{Z}_s \to 0$, $\Lambda \to \infty$, $\lambda \ll l$ and the ac signal cannot penetrate to the bottom of the pore, i.e., the pore behaves as semi-infinite. On the other hand, at low frequencies $\lambda \gg l$ and the ac signal can penetrate to the bottom of the pore and the porous system can be represented by a $R_s - C_{pore}$ connection in series.

In practical porous electrodes the top layer in Fig. 1 is also conductive and contributes to the total electrode impedance (13). In such a case the total impedance consists of the impedance of the porous part, $\hat{Z}_{pore}$, and that of a flat part of the electrode, $\hat{Z}_{flat} = 1/j\omega C_{flat}$. The corresponding electrical equivalent circuit is displayed in Fig. 5.

The complex plane plots of the circuit in Fig. 5 are displayed in Fig. 6. They are different from those in Fig. 3, depending on the ratio of the surface area of the flat surface to that of the pore walls, i.e., on the ratio of two capacitances C_{flat}/C_{pore} assuming the same specific capacitance of the external and pore walls interfaces; at high frequencies there is a curvature of the impedance plot and the slope is higher than 45° while the low frequency the resistance is lower. Analysis of the low frequency impedance indicated that it can be represented by a $R_{pore} - C_{LF}$ connection in series with the capacitance equal to the sum of that of the pore and the flat external surface, $C_{LF} = C_{pore} + C_{flat}$, and the resistance is decreased by the ratio of $\left(1 + C_{pore}/C_{flat}\right)^2$

$$R_{pore} = \frac{R_{\Omega,p}}{3} \frac{1}{\left(1 + \frac{C_{flat}}{C_{pore}}\right)^2} \quad C_{LF} = C_{pore} + C_{flat}. \quad (11)$$

When the external surface area is much smaller than that of the pores the "pure" porous behavior as in Fig. 3 or Fig. 6a is obtained.

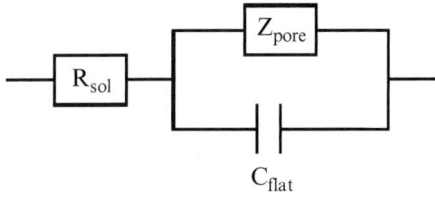

Figure 5. Equivalent electrical circuit of the porous electrode containing external flat surface.

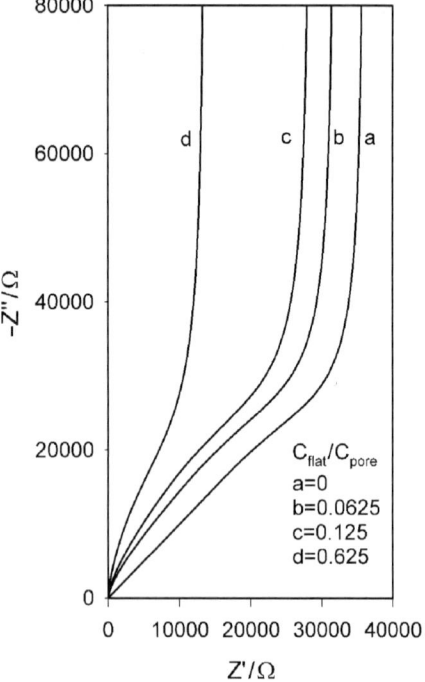

Figure 6. Complex plane plots of the circuit in Fig. 5 for different ratios of C_{flat}/C_{pore}.

Complex nonlinear least-squares approximations of the experimental impedances of different gold porous electrodes permitted to determine the ratio of the flat and porous parts of these electrodes (13).

The case of the ideally polarized porous electrode is quite rarely found in the experimental conditions. It can be observed in the presence of the supporting electrolyte or in the studies of the fuel cells, when nitrogen is used as a feed gas to determine the behavior of the porous layer. However, most studies are carried out in the presence of the red-ox reactions.

2. Other Pore Geometry

Pores of the geometry different from cylindrical were also studied in the literature. de Levie (14) studied the impedance of V-grooved pore. Such pores might be obtained, for example, by scratching the

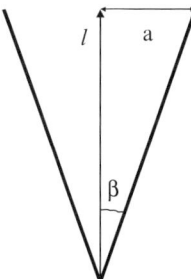

Figure 7. Cross-section of a groove electrode.

electrode surface. A cross-section of such a groove is displayed in Fig. 7.

Assuming such a geometry de Levie developed an analytical expression for the impedance per unit of groove length as:

$$\hat{Z} = \frac{\rho_s}{\tan \beta} \frac{I_0(\lambda)}{\lambda I_1(\lambda)} \tag{12}$$

where β is the angle between groove wall and the normal to the surface, I_0 and I_1 are the modified Bessel functions of zero and first order and:

$$\lambda = 2\sqrt{\frac{\rho_s l}{\hat{Z}_{el} \sin \beta}} \tag{13}$$

l is the groove depth (normal to the surface) and $\hat{Z}_{el}$ is the double-layer impedance per unit of the true surface area. Equation (12) reduces to the impedance of a perfectly flat surface for $\beta = 90°$ and to the impedance of cylindrical porous electrode for $\beta = 0°$. Gunning (15) obtained an exact solution of the de Levie grooved surface not restricted to pseudo-one-dimensional problem in a form of an infinite series. Comparison with the de Levie's equation (12) shows that the deviations arise at higher frequencies or more precisely at high values of the dimensionless parameter $\Omega = \omega C_{dl} a / \sigma_s$, where a is half of the distance of the groove opening, $a = l \tan \beta$ (Fig. 7) and σ_s is the specific conductivity of the solution.

Keiser et al. (16) have studied different pore shapes displayed in Fig. 8. At high frequencies, on the complex plane plots, they display a line at 45° for cylindrical pores, more than 45° for the groove

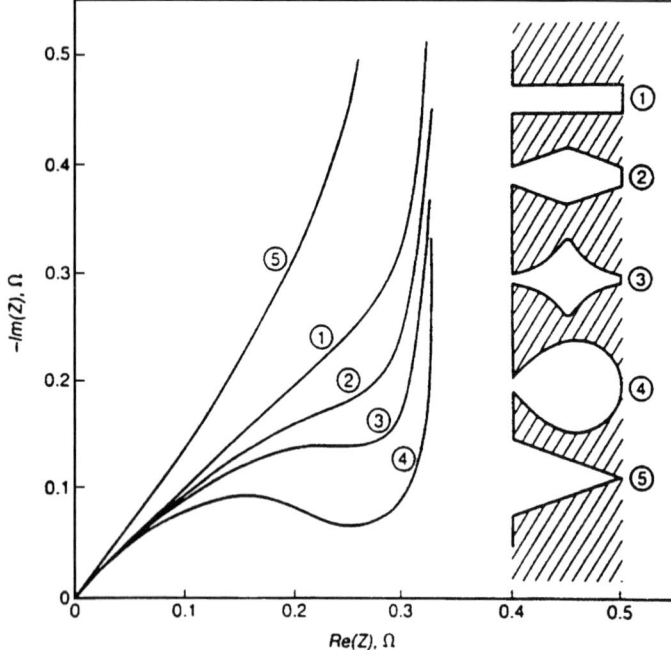

Figure 8. Impedance curves obtained for different pore shapes. Reprinted from (16) with permission from Elsevier.

shape pores or a form resembling a partially or fully developed semicircle. Such a well developed semicircle is clearly visible for pear-shape pores, curve 4.

Similar simulations were later continued for the pear-shape and spherical or bispherical pores presented in Fig. 9 by Hitz and Lasia (17). The obtained impedances are displayed in Fig. 10. It is clear that the shape of the impedance curves depends not only on the pore geometry but also on the size of the opening. In the case of the bispherical pores the effects of each sphere can be observed at different frequencies; at high frequencies the ac signal penetrates only to the first sphere and at lower frequencies it penetrates up to the second (deeper) sphere displaying two overlapping semicircles. Impedance of other arbitrary noncylindrical pores was also simulated using division into small cylinders and matrix calculation (18, 19).

Modeling of Impedance of Porous Electrodes

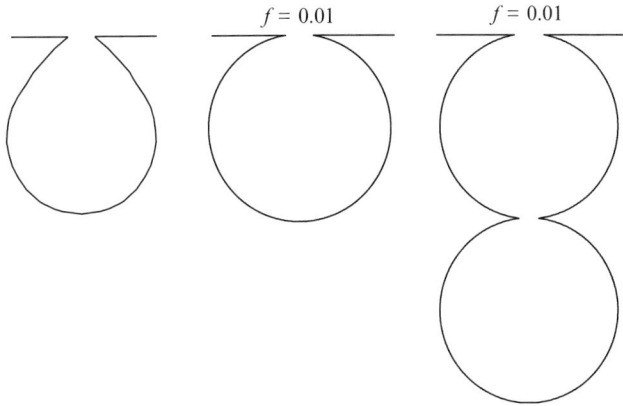

Figure 9. Pore shapes used in the simulation of impedances, f is a fraction of the sphere radius at which the pore was cut-out for the opening. Reprinted from (17) with permission from Elsevier.

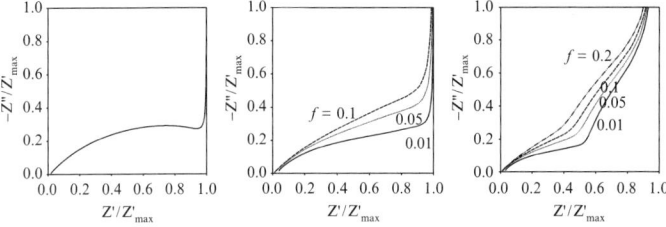

Figure 10. Reduced impedance curves obtained for three shapes displayed in Fig. 9 for different values of the parameter f. Reprinted from (17) with permission from Elsevier.

Formation of the high frequency semicircle related to the pore geometry was observed experimentally (16–21). This indicates that at some electrodes pores of the pear-like shape are present during the continuous gas evolution reaction. An example of such complex plane curves registered during the hydrogen evolution on porous Ni-Al powder electrodes (from which Al was selectively leached out) is displayed in Fig. 11. These impedances were modeled using so called two CPE model consisting of the solution resistance and a connection of two parallel R-CPE elements in series, the high frequency R-CPE element is related to the porosity while the low

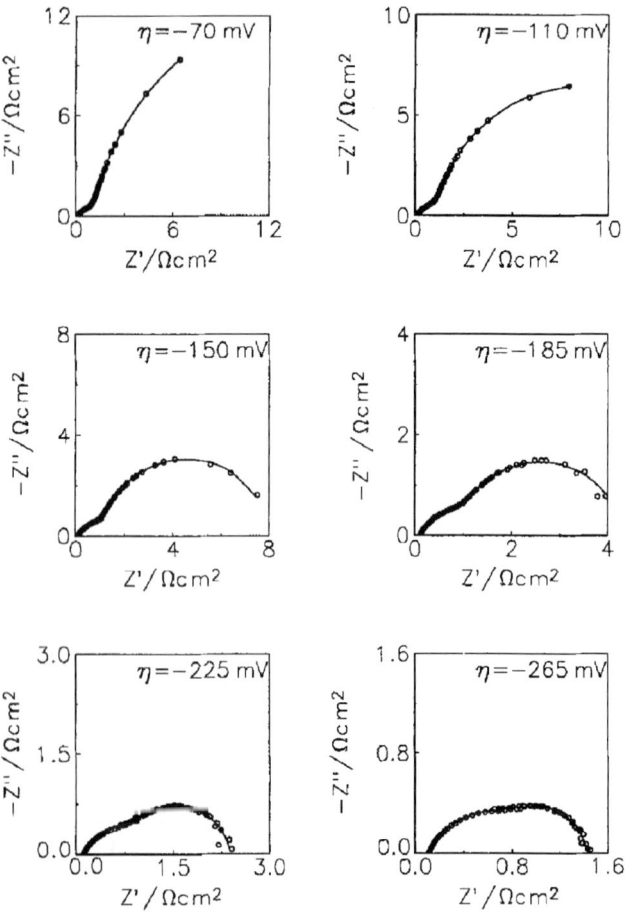

Figure 11. Complex plane plots obtained on the Ni(Al) pressed powder porous electrodes (after leaching out Al) during the hydrogen evolution reaction in 1 M NaOH at different overpotentials. Reproduced with permission from (21). Copyright 1993, The Electrochemical Society.

frequency element to the kinetics of the hydrogen evolution reaction (7, 17–28).

In general, formation of two semicircles on the complex plane plots may be of kinetic origin or the first semicircle might be of geometric and the second of kinetic origin. To distinguish between these models experimental tests were developed (17, 20, 21, 28). If the first semicircles (or deformed semicircle) is of geometric origin, it becomes independent of potential, insensitive to poisons, and depends less on temperature (through the solution resistance).

i. Ohmic drop in the solution and in the electrode material

In certain cases the electrode material is not an ideal conductor and is characterized by the specific electronic resistivity ρ_e. Such effects might be observed for semiconductors, in batteries, oxides, polymers, etc. In this case the pore might be represented as a transmission line in which the ohmic drop appears in two branches. Such circuit is presented in Fig. 12.

Assuming that the electrode impedance does not depend on the position in a pore an analytical equation describing this circuit was developed (24–26)

$$\hat{Z}_{\text{pore}} = \frac{r_s r_e}{r_s + r_e} \left[l + \frac{2\lambda}{\sinh\left(\frac{l}{\lambda}\right)} \right] + \lambda \frac{r_s^2 + r_e^2}{r_s + r_e} \coth\left(\frac{l}{\lambda}\right), \quad (14)$$

where r_s and r_e are the resistances per unit pore length in ($\Omega\,\text{cm}^{-1}$) of the electrolytic solution and electrode material, λ ac signal penetration length defined as

$$\lambda = \sqrt{\frac{\hat{z}}{r_s + r_e}} \quad (15)$$

and $\hat{z} = \hat{Z}_{\text{el}}/2\pi r$ is the electrode interfacial impedance per pore unit length (in $\Omega\,\text{cm}$). Equation (14) might be presented in a form similar to that in (7)

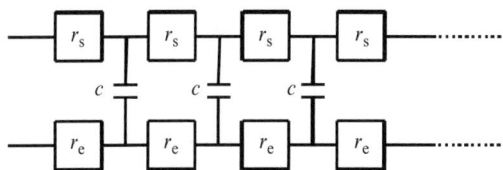

Figure 12. Transmission line for the capacitive porous electrode with the resistance of the solution in pores, r_s, and of the electrode material, r_e.

$$\hat{Z}_{\text{pore}} = \bar{R}_{\Omega,\text{p}} \left[1 + \frac{2}{\Lambda^{1/2} \sinh \Lambda^{1/2}} \right] + \frac{R_{\Omega,\text{p}}}{\Lambda^{1/2}} \coth \Lambda^{1/2}, \quad (16)$$

where

$$\bar{R}_{\Omega,\text{p}} = \frac{r_s r_e}{r_s + r_e} l, \quad R_{\Omega,\text{p}} = \frac{r_s^2 + r_e^2}{r_s + r_e} l, \quad \Lambda = \frac{l^2}{\lambda} = \frac{(l^{2*}(r_s + r_e))}{\hat{z}}. \quad (17)$$

Equation (16) clearly indicates that at very high frequencies pore impedance becomes $\bar{R}_{\Omega,\text{p}}$ instead of zero obtained when the electrode material is well conductive. It is also evident that when $r_e=0$ (16) reduces to (7). The complex plane plots obtained in such cases are shown in Fig. 13 where the impedances of the first, Fig. 13b,

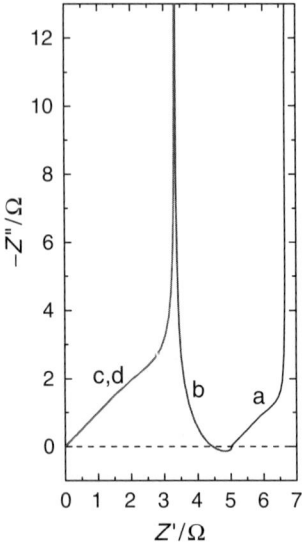

Figure 13. Complex plane plots for the porous electrode according to (14); (**a**) total impedance according to (14), (**b**) impedance of the first term, (**c**) impedance of the second term of (14), (**d**) impedance for $r_e = 0$ (see (7)); parameters used: $r_s = r_e = 200\,\Omega\,\text{cm}^{-1}$, $\hat{z} = 1/j\omega c_{\text{dl}}$, $c_{\text{dl}} = 0.001\,\text{F}\,\text{cm}^{-1}$, $l = 0.05\,\text{cm}$.

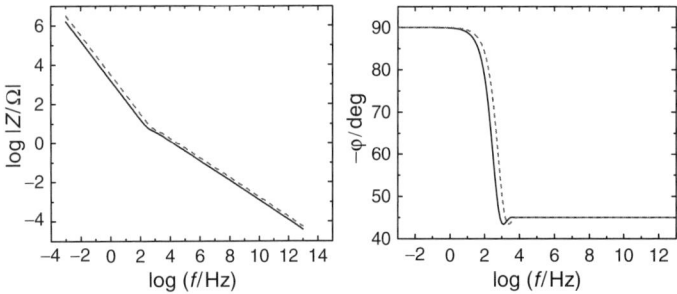

Figure 14. Complex plane and Bode plots of the porous electrode, *continuous line* the second term in (16), *dashed line* (7); parameters as in Fig. 13.

and the second, Fig. 13c, term of (16) are also presented. They are compared with the plot obtained in the case when $r_e = 0$. It is interesting to note that the first term in (14) or (16) represents complex impedance which starts at $\bar{R}_{\Omega,p}$ at high frequencies and initially goes to positives imaginary impedances and, finally, at low frequencies represents a capacitive line. The second term in (16) is formally similar to that of the porous electrode in the presence of ideally conductive electrode, (see (7)) and the complex plane plots look similar, Fig. 13c and d, however, the Bode plots reveal the differences in $\log |Z|$ and the phase angle, Fig. 14.

III. IMPEDANCE OF POROUS ELECTRODES IN THE PRESENCE OF THE ELECTROCHEMICAL REACTION

In general, during the red-ox reaction on porous electrodes a dc potential gradient, dE/dx, and a concentration gradient, dC/dx occur simultaneously. First, two limiting cases in which only one gradient appears practically will be considered and later a general case with the two gradients existing at the same time will be presented. It will be shown that there is a relation between these two gradients.

1. Porous Electrodes in the Absence of dc Current

i. Ohmic drop in the solution only

de Levie (9) has presented a solution for the impedance of porous electrode in the presence of red-ox reaction but in the absence of the dc current that is at the equilibrium potential. In the absence of the dc current there is no dc potential drop in the pore and the faradaic impedance is constant along the pore. This system is described by (1) and (2) in which the electrode impedance consists of the charge transfer resistance, R_{ct}, related to the reaction red-ox and the double layer capacitance and is described as

$$\hat{Z}_{el} = \frac{1}{\frac{1}{R_{ct}} + j\omega C_{dl}}. \tag{18}$$

In this case it was assumed that the ac potential drop occurs because of the solution resistance and the pore walls are a perfect conductor. Although there is no dc potential drop in the pore the oscillating voltage $\tilde{E}$ decreases in the pore because of the solution resistance according to (2) and (4). The impedance of a pore is described by (7), but the electrode impedance is now described by (18). This equation leads to different complex plane plots depending on the ac signal penetration length with respect to the pore length. Three types of plots may be observed (27):

a) Intermediate length pores

In this case a general complex plane plot consists of the straight line at 45° at high frequencies followed by a kinetic semicircle. It is presented in Fig. 15a. It is obtained when the ac signal penetration length is comparable with the pore length, $\lambda \sim l$, and at low frequencies the ac signal penetrates to the bottom of the pore. The high frequency line at 45° is the same as for the ideally polarizable electrode (see (9)). It arises from the coupling of the solution resistance and the double layer capacitance in the pore and does not contain any kinetic information. At low frequencies the impedance becomes real with the value

$$\hat{Z}_{pore}(\omega = 0) = R_p = \sqrt{\frac{\rho_s R_{ct}}{2\pi^2 r^3}} \coth\left(\sqrt{\frac{2\rho_s}{r R_{ct}}} l\right). \tag{19}$$

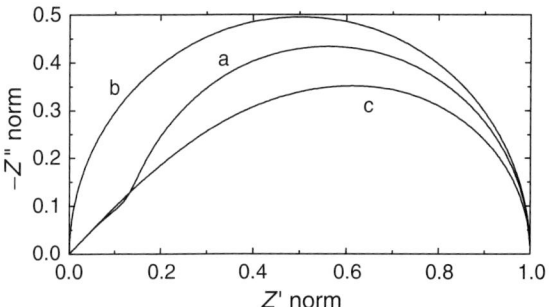

Figure 15. Normalized (i.e., Z/Z'_{max}) complex plane plots for porous electrodes; (**a**) general case, ac signal penetration length comparable with the pore length, $\lambda \sim l$, (**b**) shallow pores, $\lambda \gg l$, (**c**) semi-infinite length pores, $\lambda \ll l$.

b) Shallow pores

When the pores are shallow, i.e., $\lambda \gg l$, the ac signal penetrates to the bottom of the pore and the electrode behaves as a flat one. In this case $\Lambda = (l/\lambda)^2 \to 0$, $\coth(\Lambda^{1/2}) \to \Lambda^{-1/2}$ and (7) becomes

$$\hat{Z}_{pore} = \frac{R_{\Omega,p}}{\Lambda} = \frac{1}{2\pi r l} \hat{Z}_{el}, \tag{20}$$

that is it represents the impedance of a flat electrode with the surface area equal to that of the pore wall $2\pi r l$. In this case a kinetic semicircle is observed on the complex plane plots, Fig. 15b, and no effects of the electrode porosity could be observed.

c) Semi-infinite pores

When the ac signal penetration length is much smaller that the pore length, $\lambda \ll l$, the ac signal cannot reach the bottom of the pore and the pore behaves as semi-infinite. The complex plane plot consists of the high frequency line at 45° followed by a skewed semicircle, Fig. 15c. In this case $\Lambda \to \infty$, $\coth(\Lambda^{1/2}) \to 1$, and the impedance, (7), becomes

$$\hat{Z}_{pore} = \frac{R_{\Omega,p}}{\Lambda^{1/2}} = \left(\frac{\rho}{2n^2\pi^2 r^3}\right)^{1/2} \hat{Z}_{el}^{1/2}. \tag{21}$$

It should be stressed that for semi-infinite pores the charge transfer resistance and the electrode capacitance cannot be extracted from the impedance data. Equation (21), for n pores, can be rearranged (28) to

$$\hat{Z}_{\text{pore}} = \left[\frac{\rho_s}{2n^2\pi^2 r^3}\left(R_{\text{ct}} + \frac{1}{j\omega C_{\text{dl}}}\right)\right]^{1/2} = \left[uR_{\text{ct}} + \frac{u}{j\omega C_{\text{dl}}}\right]^{1/2},$$
(22)

where $u = \rho_s/(2n^2\pi^2 r^3)$ and the only parameters which can be experimentally determined are uR_{ct} and C_{dl}/u. Because the exact geometry and number of pores are not usually known these parameters cannot be determined independently. However, the product of the experimentally accessible parameters uR_{ct} and C_{dl}/u equals to $R_{\text{ct}}C_{\text{dl}}$ and as the double layer capacitance of metallic electrodes is in the range of 20–25 μF cm^{-2} the kinetics of the reaction might be estimated (28).

ii. Ohmic drop in the solution and electrode material

In the case of the electrochemical reaction at the pore surface in the presence of the resistance of the solution and the electrode material the equivalent circuit might be represented as the transmission line depicted in Fig. 16.

The equation describing the impedance of such an electrode is the same as (14) or (16) with exception of the impedance per pore unit length $\hat{z} = 1/[(1/r_{\text{ct}}) + j\omega c_{\text{dl}}] = \hat{Z}_s l$ and the total impedance of the porous electrode, $\hat{Z}_s$, includes the double layer capacitance and the charge transfer resistance. The complex plane plot corre-

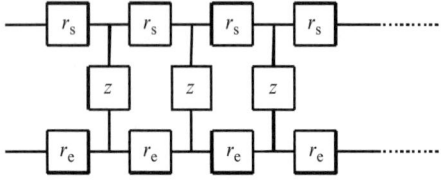

Figure 16. Transmission line representing porous electrode in the presence of the electrode reaction and the resistance of the solution, r_s, and the electrode, r_e.

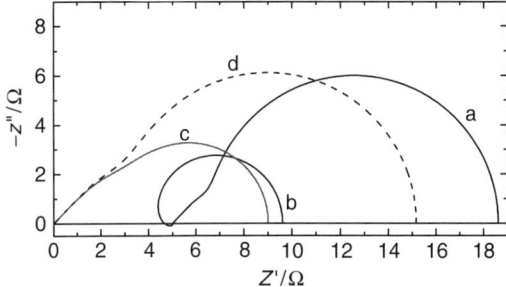

Figure 17. Complex plane plot of the porous electrode in the presence of faradaic reaction with $r_{ct} = 0.6\,\Omega$ cm; (**a**) total impedance (see (16)), (**b**) the first and (**c**) the second term of (16), (**d**) case for $r_e = 0$.

sponding to this model is displayed in Fig. 17. The obtained impedance (a) is a sum of the first (b) and the second (c) term in (16). Because of a peculiar dependence of the first term the total impedance is not simply the impedance of the porous electrode in the case of $r_e = 0$ (d), shifted by the resistance $\bar{R}_{\Omega,p}$.

2. Porous Electrodes in the Presence of dc Current, No Concentration Polarization

i. dc solution

In the presence of a dc current a dc potential gradient develops inside a pore and the charge transfer resistance, which depends on potential, changes with depth in the pore. In this case the value of $\hat{z}$ in (1) depends on x and the equation cannot be directly integrated. In order to solve the problem, first the dc potential as a function of distance from the pore orifice must be estimated (29). The axially flowing dc current, I, which enters the pore, flows towards the walls and its value decreases with the distance. Decrease of this current is proportional to the current flowing to the wall, j, which is defined as a current density

$$\frac{dI}{dx} = -\frac{(2\pi r dx)j}{dx} = -2\pi r j, \qquad (23)$$

where $2\pi r\,dx$ is the surface area of a pore section dx. The current j is described by the Butler–Volmer equation

$$j = j_0 \left[e^{(1-\alpha)nf\eta} - e^{-\alpha nf\eta} \right], \tag{24}$$

where j_0 is the exchange current density, η is the overpotential, n number of electrons exchanged, $f = F/RT$, and α the cathodic transfer coefficient. The Ohm's law in the pore is described by

$$\frac{d\eta}{dx} = -I \left(\frac{\rho_s dx}{\pi r^2} \right) \frac{1}{dx} = -I \left(\frac{\rho_s}{\pi r^2} \right), \tag{25}$$

where $\rho_s dx / \pi r^2$ is the resistance of the section dx of the solution in the pore. Taking the second derivative in (25) and substituting (23) leads to

$$\frac{d^2\eta}{dx^2} = \frac{2\rho_s}{r} j = \frac{2\rho_s j_0}{r} \left[e^{(1-\alpha)nf\eta} - e^{-\alpha nf\eta} \right] \tag{26}$$

which for $\alpha = 0.5$ reduces to

$$\frac{d^2\eta}{dx^2} = \frac{2\rho j_0}{r} \left[e^{0.5nf\eta} - e^{-0.5nf\eta} \right] = \frac{4\rho j_0}{r} \sinh(0.5nf\eta). \tag{27}$$

Equation (26) must be solved taking into account the following conditions:

$$\begin{aligned} x &= 0 \quad \eta = \eta_0, \\ x &= l \quad d\eta/dx = 0, \end{aligned} \tag{28}$$

where η_0 is the value of the overpotential at the pore orifice. This equation was solved in the literature in terms of special functions (3–5, 30) or assuming certain simplifications. (6, 31, 32) The first integration of (26) gives (29)

$$\begin{aligned} \frac{d\eta}{dx} &= -\left\{ \left(\frac{4\rho_s j_0}{r} \right) \left[\frac{\exp[(1-\alpha)f\eta]}{(1-\alpha)f} + \frac{\exp(-\alpha f\eta)}{\alpha f} \right. \right. \\ &\quad \left. \left. - \frac{\exp[(1-\alpha)f\eta_l]}{(1-\alpha)f} - \frac{\exp(-\alpha f\eta_l)}{\alpha f} \right] \right\}^{1/2} \\ &= -\left\{ \left(\frac{4\rho_s j_0}{r} \right) \left[\frac{\exp(b_1\eta)}{b_1} + \frac{\exp(-b_2\eta)}{b_2} - \frac{\exp(b_1\eta_l)}{b_1} \right. \right. \\ &\quad \left. \left. - \frac{\exp(-b_2\eta_l)}{b_2} \right] \right\}^{1/2}, \end{aligned} \tag{29}$$

where $\eta_l = \eta(l)$ is the overpotential at the bottom of the pore and parameters b_i are $(1-\alpha)f$ and αf, respectively. Because the value of η_l is not known a priori (see (29)) cannot be directly integrated. However, assuming semi-infinite length pores, that is $\eta_l = 0$ at $x = l$, (see (29)) reduces to

$$\frac{d\eta}{dx} = -4\sqrt{\frac{\rho_s j_0}{r}} \sinh\left(\frac{b\eta}{2}\right), \quad (30)$$

where $b = 0.5f$. This equation is formally similar to that for the potential distribution in the electrical double layer (33) and has an analytical solution

$$\tanh\left(\frac{b\eta}{4}\right) = \tanh\left(\frac{b\eta_0}{4}\right) \exp\left(-2\sqrt{\frac{\rho_s j_0 b}{r}} x\right). \quad (31)$$

Substitution of (30) into (25) gives the dependence of the total current, I_0, as a function of overpotential

$$I_0 = 4\pi r \sqrt{\frac{r j_0}{\rho_s b}} \sinh\left(\frac{b\eta_0}{2}\right) = 4\pi r \sqrt{\frac{r j_0}{\rho_s b}} \sinh(0.25 f \eta_0). \quad (32)$$

This equation predicts that Tafel slope of $\ln(10)/0.25 f = 0.2366$ V dec^{-1} at $25°$ C which means that it is doubled in comparison with the typical value of 0.1183 V dec^{-1}.

In a general case (29) can be solved numerically by integration and looking for the value of η_l which gives $d\eta/dx = 0$ at $x = l$, analogically to the solution of nonlinear equations (29), from which the dependence of the overpotential and current on position in the pore can be obtained. An example of the dependence of the potential and current on x is displayed in Figs. 18 and 19.

The overpotential decreases with the depth in the pore and the slope $d\eta/dx$ becomes zero at $x = l$. At higher current densities ($j_0 > 10^{-3}$ A cm^{-2}) the overpotential becomes zero at distances $x < l$ and the pore behaves as semi-infinite. The current decreases almost linearly at the lowest exchange current density which is the indication of a negligible potential drop. However, at higher values of j_0 it decreases nonlinearly and drops to zero as the corresponding overpotential drops to zero.

The Tafel curves at porous electrodes corresponding to the same values of the exchange current densities are shown in Fig. 20.

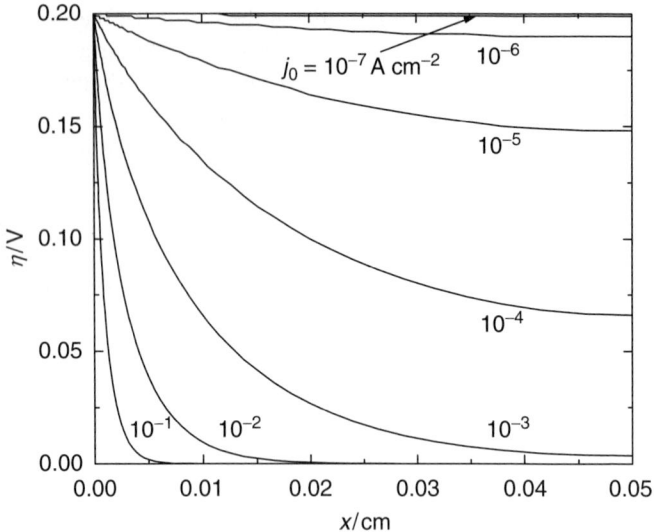

Figure 18. Dependence of the potential inside pore as a function of the distance from the pore orifice for different values of the exchange current density and a constant overpotential; parameters used in simulations: $\eta_0 = 0.2$ V, $\alpha = 0.5$, $l = 0.05$ cm, $r = 10^{-4}$ cm, $\rho_s = 10\,\Omega$ cm, the values of j_0 are displayed in the figure.

It is interesting to notice that for very low exchange current densities and lower overpotentials the Tafel slope is 0.118 V dec^{-1} but it doubles at higher exchange currents and higher overpotentials.

ii. ac solution

Knowing the potential distribution it is possible to calculate the electrode impedance. To do this the pore was divided into a number (1,000–5,000) of small segments Δx and starting from the pore bottom the electrode impedances were added as shown in Fig. 21 using the following equation (see Ref. 29)

$$\hat{Z}_{\text{tot},i} = R\Delta x + 1/\left[1/\hat{Z}_i + 1/\hat{Z}_{i-1}\right]$$
$$= \left(\frac{\rho_s}{\pi r^2}\right)\Delta x + 1/\left[(2\pi r\,\Delta x)/\hat{Z}_{\text{el},i} + 1/\hat{Z}_{i-1}\right], \quad (33)$$

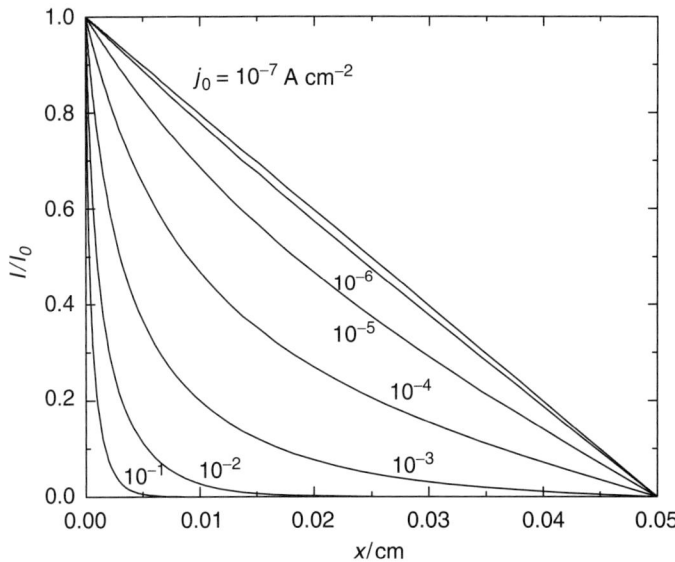

Figure 19. Current distribution expressed as the ratio of the current I at distance x to the total current entering the pore, I_0, as a function of the distance x, at different j_0, other parameters as in Fig. 18.

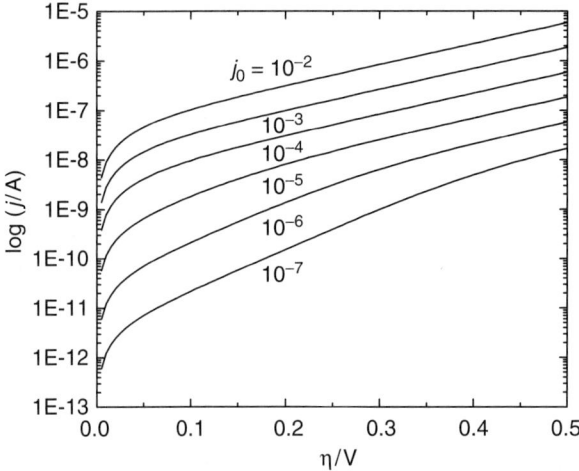

Figure 20. Tafel curves obtained at porous electrodes, parameters as in Fig. 18.

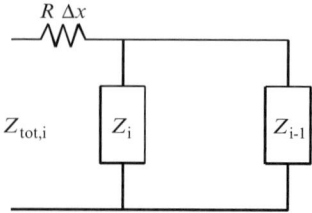

Figure 21. Electrical equivalent circuit used to calculate the electrode impedance.

where $\hat{Z}_{\text{tot},i}$ is the impedance of the pore between x and l, $\hat{Z}_{i-1}$ is the impedance of the pore between $x - dx$ and l, $\hat{Z}_{\text{el},i}$ is the specific impedance of the element dx pore wall at the distance x

$$\hat{Z}_{\text{el},i} = \frac{1}{\frac{1}{\hat{Z}_f} + j\omega C_{\text{dl}}} \tag{34}$$

and

$$\frac{1}{\hat{Z}_f} = \frac{1}{R_{\text{ct}}} = \frac{dj}{d\eta} = j_0 nf \left[\alpha e^{-\alpha nf\eta} + (1-\alpha)e^{(1-\alpha)nf\eta} \right]. \tag{35}$$

The model in Fig. 21 is an extension of the circuit in Fig. 4, where the double layer impedance was replaced by the impedance $\hat{Z}_i = \hat{Z}_{\text{el},i}/2\pi r \Delta x$ (see (34)).

The obtained impedances are displayed in Figs. 22 and 23 and compared with the de Levie's solution which assumes that the faradaic impedance is independent of the distance in the pore. With the increase of the exchange current density, i.e., the electrode activity, and of the overpotential a semicircle corresponding to the penetration of the ac signal to the bottom of the pore changes into a semi-infinite, as predicted in Fig. 15. Similar transitions are observed with increase of the overpotential. It is obvious, that the de Levie's solution underestimates the impedance because, in the real case, the impedance inside the pore increases with x as the overpotential increases. It is interesting to note that for the semi-infinite pore the ratio of the impedances obtained at $\omega \to 0$ using the full model and that of de Levie is equal to $\sqrt{2}$.

de Levie (9) has also noted that the complex plane plots of the skewed impedances obtained for the semi-infinite pore length,

Modeling of Impedance of Porous Electrodes

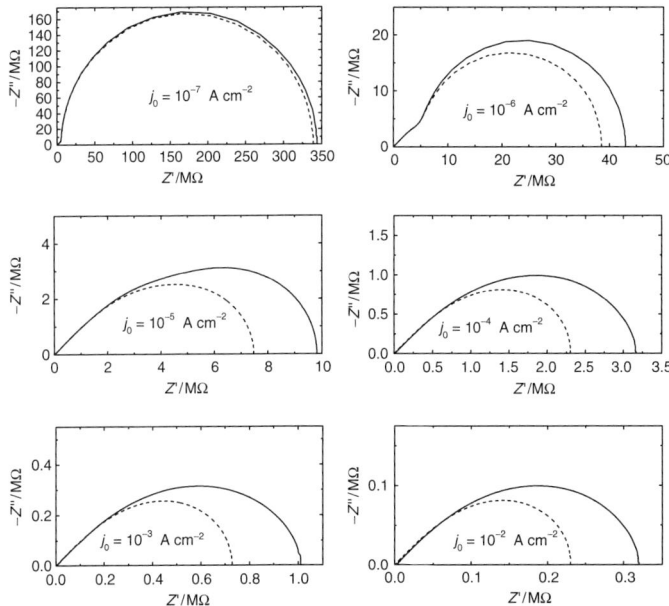

Figure 22. Complex plane plots at a porous electrode in the presence of a redox process at a constant overpotential $\eta_0 = 0.2\,\text{V}$ and different exchange current densities; *continuous line* – simulations (see (33)) *dashed line* – according to de Levie's equation (7); $C_{\text{dl}} = 20\,\mu\text{F cm}^{-2}$, other pore parameters as in Fig. 18.

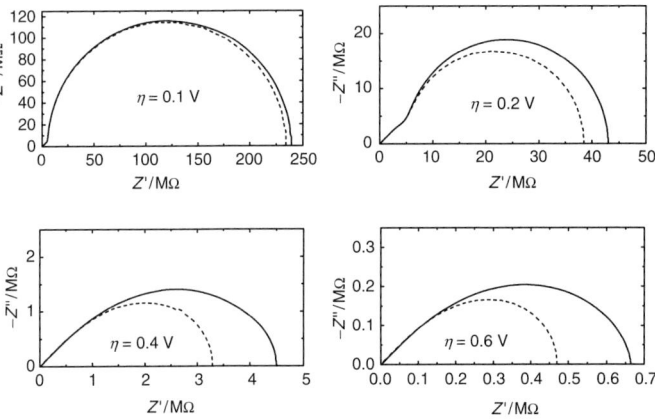

Figure 23. Complex plane plots at a porous electrode at a constant exchange current density $j_0 = 10^{-6}\,\text{A cm}^{-2}$ and different overpotentials; other parameters as in Figs. 18 and 22.

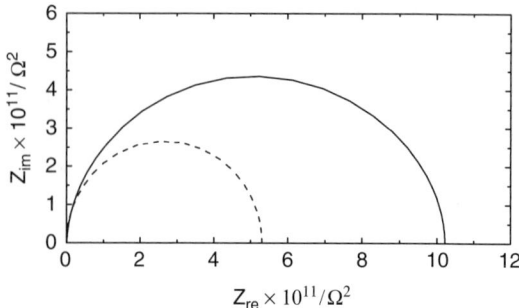

Figure 24. Dependence of squared impedances (see (36)), *dashed line*, and the solution according to Eq. (33), *continuous line*, at semi-infinite lengths porous electrode, $\eta_0 = 0.2\,\text{V}$, $j_0 = 10^{-3}\,\text{A cm}^{-2}$, other pore parameters as in Fig. 22.

(see (21)), may be transferred into ideal semi-circles by the following transformation:

$$Z_{re} = |Z|^2 \cos(2\Phi), \qquad Z_{im} = |Z|^2 \sin(2\Phi),$$
$$\Phi = \arctan\left(-Z''/Z'\right), \qquad |Z| = \sqrt{Z'^2 + Z''^2}. \tag{36}$$

An example of such plot is shown in Fig. 24. For the impedances according to the de Levie's equation it represents a perfect semicircle as a square root of the electrode impedance is originally in (21). However, for the real impedances the semicircle is slightly flattened.

The potential profile in the pores depends on the transfer coefficient through (29) and influences impedances through (35) (see Ref. (29)). Examples of the complex plane plots for three different values of the transfer coefficients are displayed in Fig. 25.

Because the correct numerical solution for the porous electrode in the presence of the potential gradient in pores demands knowledge of the pore parameters and more complex mathematics, in practice, simplified de Levie's equation (7) is used in approximations that is the experimentally measured impedances are fitted to (7). As it has been shown above, for the same pore and kinetic parameters the de Levie's equation underestimates the low frequency impedance by up to 100%, which might be not so important for very porous electrodes characterized by a very large surface area. Moreover, the shape of the real impedance is different from that described by the

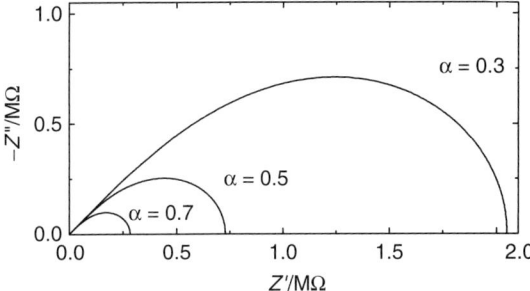

Figure 25. Complex plane plots obtained on a porous electrode at different values of the transfer coefficient, $\eta_0 = 0.2$ V, $j_0 = 10^{-3}$ A cm^{-2}, other parameters as in Fig. 22.

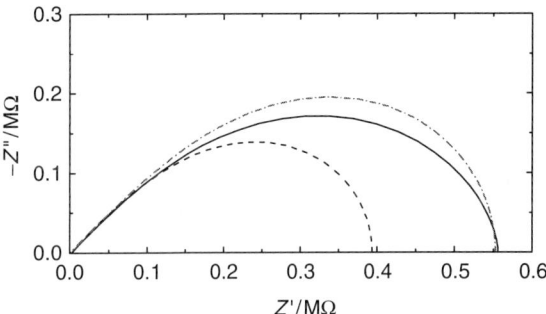

Figure 26. Approximation of the impedance simulated for the cylindrical porous electrode, for $\eta_0 = 0.5$ V, $j_0 = 10^{-5}$ A cm^{-2}, pore parameters as in Fig. 18; *continuous line* approximation by the de Levie's equation (7), *dash-dotted line*, *dashed line* represents impedance simulated for the same kinetic and pore parameters by de Levie's equation.

analytical solution. An example of such an approximation is displayed in Fig. 26.

Comparison of the simulated impedances with the approximation by de Levie's equation reveals that the approximation is not very good and in the medium frequency range the simulated curve lies below that predicted by (7). The assumed value of the charge transfer resistance at $\eta_0 = 0.5$ V is $R_{ct} = 0.306 \, \Omega$ cm^2 and the

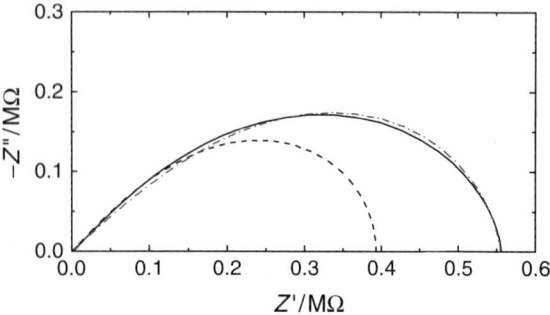

Figure 27. Approximation of the impedance simulated for the cylindrical porous electrode, *continuous line*, with de Levie's equation including the CPE, *dash-dotted line*, parameters as in Fig. 26; *dashed line* represents impedance simulated for the same kinetic and pore parameters by de Levie's equation.

value found from the approximation is $R_{ct} = 0.602$ Ω cm^2, which is practically two times larger, as predicted above. Moreover, the double layer capacitance found is larger, $C_{dl} = 23.3$ μF cm^{-2}, instead of the assumed value of 20 μF cm^{-2}.

In practical applications very often a constant phase element, CPE, is used instead of the pure capacitance, that is the double layer impedance $\hat{Z}_{dl} = 1/(j\omega C_{dl})$ is replaced by $\hat{Z}_{dl} = 1/[(j\omega)^\phi T]$, where $\phi \leq 1$ is a dimensionless parameter and T is a capacitance parameter in μF cm^{-2} s$^{\phi-1}$ (see Chap. 5). Using this model the approximation is much better, Fig. 27, and the small deviations would not be observed in practice because of the experimental errors. In this case the charge transfer resistance is 0.609 Ω cm^2 but the capacitance parameter is much larger, $T = 65.4$ μF cm^{-2} s$^{\phi-1}$ and $\phi = 0.913$. These results indicate that the porous electrode behavior might be mistaken by the CPE which might lead to the overestimation of the double layer capacitance and, in consequence, the electrode real surface area. This problem is especially important when the semi-infinite porous behavior is observed. It has been shown (29) that with the increase of the porous behavior of the electrodes the estimated capacitance parameter increases up to three times when using ϕ as an adjustable parameter while the charge transfer resistance is correctly reproduced.

To estimate the average double layer capacitance, $\bar{C}_{dl}$ in the presence of the CPE behavior Brug et al. (34) proposed an equation:

$$T = \bar{C}_{dl}^{\phi}\left(\frac{1}{R_{sol}} + \frac{1}{R_{ct}}\right)^{1-\phi}. \tag{37}$$

Using this equation to estimate the average double layer capacitance and assuming $1/R_{sol} = 0$ gives $\bar{C}_{dl} = 24.9\,\mu\text{F cm}^{-2}$, which is a value closer to the assumed in simulations.

In summary, presence of the dc current and the potential gradient in pores causes increase in the observed charge transfer resistance as the overpotential decreases and R_{ct} increases in pore, and a certain deformation of the complex plane plots especially visible for semi-infinite pores. The obtained impedance cannot be well approximated by the de Levie's equation and might suggest appearance of the CPE element producing incorrect double layer capacitances. However, for the simplicity de Levie's equation is used in practical applications, although it can overestimate the real surface area by a factor up to three.

3. Porous Electrodes in the Presence of Concentration Gradient, No Potential Gradient

i. dc solution

During the red-ox reaction in a porous electrode the concentrations of the red-ox species change with depth x. This problem was studied by several authors (12, 35–43). A limiting case might be considered in the absence of the dc potential gradient in the pores, $dE/dx = 0$, but in the presence of the concentration gradient, $dC/dx \neq 0$. In this case an analytical solution for the concentration gradient and the impedance (43) might be obtained.

Let us assume the red-ox process

$$O + ne \underset{\overleftarrow{k_b}}{\overset{\overrightarrow{k_f}}{\rightleftarrows}} R, \tag{38}$$

which is described by the current density–potential relation in the pore

$$j(x,t) = nF\left[C_O(x,t)\overrightarrow{k_f} - C_R(x,t)\overleftarrow{k_b}\right], \tag{39}$$

where $C_O(x)$ and $C_R(x)$ are the concentrations of the oxidized and reduced forms in the pore at a distance x from the pore orifice and $\vec{k}_f = k_s \exp\left[-\alpha nf(E - E^0)\right]$ and $\overleftarrow{k}_b = k_s \exp\left[(1-\alpha)nf(E - E^0)\right]$ are potential dependent rate constants, k_s is the standard rate constant, α the cathodic transfer coefficient, E^0 is the standard potential and it was supposed that the potential E is constant in the pore and equals to that at the pore orifice. Let us assume that at the bulk of the solution $C_O = C_O^*$ and $C_R = 0$. Then, (39) may be rearranged to

$$j(x,t) = nFC_O^* \left[a(x,t)\left(\vec{k}_f + \overleftarrow{k}_b\right) - \overleftarrow{k}_b\right], \tag{40}$$

where $a = C_O/C_O^*$ assuming that the diffusion coefficients of both forms are equal and, in consequence, $C_O(x,t) + C_R(x,t) = C_O^*$. Current density flowing to the pore walls between the distance x and $x + dx$ might also be described as

$$j(x) = -\frac{nF}{2\pi r\, dx} \frac{dN_O(x)}{dt}, \tag{41}$$

where N is the number of moles of the oxidized substance being reduced. It can be rearranged to

$$\begin{aligned} j(x,t) &= -\frac{nF}{2\pi r\, dx} \pi r^2 dx \frac{d}{dt}\left(\frac{N_O(x,t)}{\pi r^2\, dx}\right) \\ &= -\frac{nFr}{2} \frac{d\, C_O(x,t)}{dt} = -\frac{nFrC_O^*}{2} \frac{da(x,t)}{dt} \end{aligned} \tag{42}$$

and

$$\frac{da(x,t)}{dt} = -\frac{2}{nFrC_O^*} j(x,t). \tag{43}$$

The total Fick's equation in the pore contains the diffusion term and the loss of the electroactive substance due to electroreduction

$$\frac{\partial a(x,t)}{\partial t} = D\frac{\partial^2 a(x,t)}{\partial x^2} - \frac{2}{nFrC_O^*} j(x,t) \tag{44}$$

or introducing the dimensionless distance $z = x/l$

$$\frac{\partial a(z,t)}{\partial t} = \frac{D}{l^2}\frac{\partial^2 a(z,t)}{\partial z^2} - \frac{2}{nFrC_O^*} j(z,t). \tag{45}$$

To obtain the concentration distribution in pores we are interested in the steady-state solution, $da/dt = 0$

$$\frac{d^2 a(z)}{dz^2} = \frac{2l^2}{nFrDC_O^*} j(z) \qquad \text{with} \quad \begin{cases} a(0) = 1 \\ \frac{da(1)}{dz} = 0 \end{cases}. \qquad (46)$$

There is an analytical solution of (46) is (43)

$$a(z) = \frac{1}{1+P} \left\{ P + \frac{\cosh\left[\sqrt{B}(1-z)\right]}{\cosh\left[\sqrt{B}\right]} \right\}, \qquad (47)$$

where

$$B = \left(\frac{2k_s l^2}{rD}\right)\left(P^{-\alpha} + P^{1-\alpha}\right) = 2\Phi_0^2 \left(P^{-\alpha} + P^{1-\alpha}\right), \qquad (48)$$

$$\Phi_0^2 = \frac{k_s l^2}{rD}, \qquad P = e^{nf(E-E^0)}.$$

The concentration gradient depends on the electrode potential, and parameter B related to the normalized Thiele modulus Φ_0 (44) which characterizes mass transfer in pores. Examples of the concentration gradients obtained at different potentials $E - E^0$ are presented in Fig. 28. As the potential becomes more negative, concentration of the oxidized species decreases. At some potentials part of the pore is not used, concentration in this part of the pore is zero and no further reaction takes place there, for example at $E - E^0 = -0.2$ V only about 30% of the pore is electrochemically active. The current flowing to the pore walls increases with the increase in the negative potential, Fig. 29, however the current divided by the maximal current flowing at the pore orifice $j_{\max} = j(z=0)$ behaves differently; the largest value is observed for $E - E^0 = 0$ and identical values are obtained for the positive and negative potentials, i.e., $j(x)/j_{\max}(E - E^0) = j(x)/j_{\max}[-(E - E^0)]$, displayed in Fig. 29.

Knowing the current flowing at any distance in the pore the total current flowing to the pore walls may be calculated as an integral (43)

$$I = 2\pi rl \int_0^1 j(z)dz = nF(2\pi rl)C_O^* \vec{k}_f \frac{\tanh\sqrt{B}}{\sqrt{B}}. \qquad (49)$$

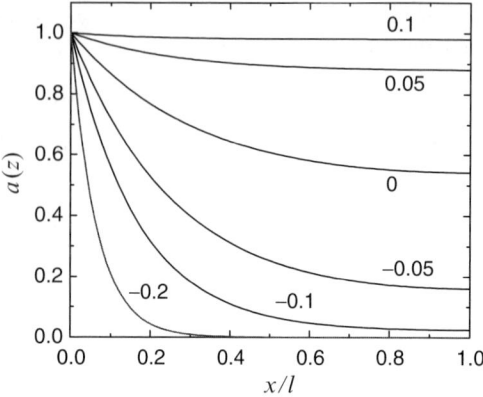

Figure 28. Dimensionless concentration in pores as a function of the distance, $z = x/l$, from the pore orifice at different potentials $E - E^0$ indicated in the graph, in V; parameters used: $l = 0.05$ cm, $r = 10^{-4}$ cm, $k_s = 10^{-6}$ cm s^{-1}, $D_O = 10^{-5}$ cm^2 s^{-1}.

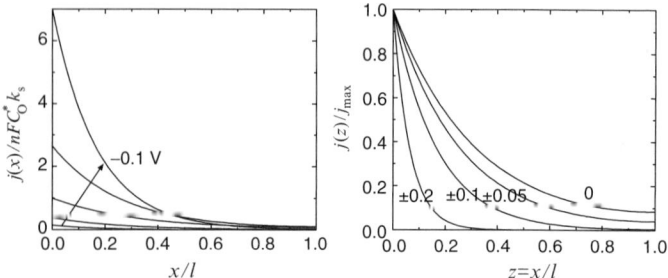

Figure 29. Dependence of the dimensionless current as a function of the distance in the pore for the potentials from 0.1 to -0.1 V, every 0.05 V (an *arrow* shows the direction of the increase in negative potential) and the dependence of the current in the pore divided by the maximal current flowing at the pore orifice.

There are two limiting cases of (49)

(a) For shallow (short and wide) pores, l^2/r is small and $B \ll 1$. In this case $\tanh\sqrt{B}/\sqrt{B} \to 1$ and (49) reduces to

$$I = nF(2\pi rl)C_O^* \vec{k}_f \qquad (50)$$

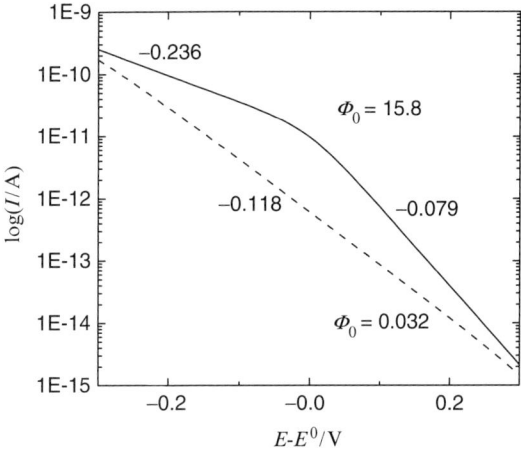

Figure 30. Tafel plots on a porous electrode; *continuous line*: long pores, $l = 0.5$ cm, $\Phi_0 = 15.8$; *dashed line*: shallow pores, $l = 0.001$ cm, $\Phi_0 = 0.032$, other parameters as in Fig. 28.

and a normal Tafelian behavior is observed with the slope $-\ln 10/\alpha n f = -118/n$ mV dec^{-1} for $\alpha = 0.5$ at 25° C.

(b) In the case of long and narrow pores $B \gg 1$ and $\tanh \sqrt{B} \to 1$ and (49) reduces to

$$I = nF(2\pi rl)\, C_O^* \vec{k}_f \frac{1}{\sqrt{B}} = \frac{nF\pi\sqrt{2r^3 k_s D} C_O^*}{\sqrt{P^\alpha + P^{1+\alpha}}} \qquad (51)$$

and two slopes are observed on the Tafel plots, $-\ln 10/\{[(1+\alpha)/2]nf\} = -78.9/n$ mV at more positive potentials and $-\ln 10/[(\alpha/2)nf] = -237/n$ mV dec^{-1} at more negative potentials, assuming $\alpha = 0.5$ at 25° C. The slope at more negative potentials corresponds to the doubling of the typical kinetic Tafel plot observed on flat electrodes. The Tafel plots corresponding to these two cases are displayed in Fig. 30.

ii. ac solution

To obtain the expression for the impedance the expression for current is linearized (7, 41, 43)

$$\Delta j = \left(\frac{\partial j}{\partial E}\right)_a \Delta E + \left(\frac{\partial j}{\partial a}\right)_E \Delta a, \qquad (52)$$

where $\Delta j = \tilde{j}\exp(j\omega t)$, $\Delta E = \tilde{E}\exp(j\omega t)$ and $\Delta a = \tilde{a}\exp(j\omega t)$ and symbols $\tilde{j}, \tilde{E}$, and $\tilde{a}$ are the phasors (7). The faradaic impedance, $\hat{Z}_f$, or admittance, $\hat{Y}_f$ is

$$\frac{1}{\tilde{Z}_f} = \hat{Y}_f = -\frac{\tilde{j}}{\tilde{E}} = -\left(\frac{\partial j}{\partial E}\right) - \left(\frac{\partial j}{\partial a}\right)\frac{\tilde{a}}{\tilde{E}}, \qquad (53)$$

where the negative sign appears because the reduction current was defined as positive. To calculate the impedance it is necessary to determine $\tilde{a}/\tilde{E}$ while the derivatives in parentheses are easily calculated from (40). Therefore, (45) describing diffusion of the red-ox species in the pore must be solved for Δa. Substitution of Δa followed by division by $\exp(j\omega t)$ gives

$$j\omega \tilde{a} = D\frac{d^2\tilde{a}}{dz^2} - \frac{2l^2}{nFrC_O^*}\left[\left(\frac{\partial j}{\partial a}\right)\tilde{a} + \left(\frac{\partial j}{\partial E}\right)\tilde{E}\right] \qquad (54)$$

or expressing the solution in terms of the parameter $\tilde{a}/\tilde{E}$ necessary in (53)

$$\frac{d^2(\tilde{a}/\tilde{E})}{dz^2} = \left(\frac{\tilde{a}}{\tilde{E}}\right)\left[\frac{j\omega l^2}{D} + \frac{2l^2}{nFrDC_O^*}\left(\frac{\partial j}{ja}\right)\right]$$

$$+ \left[\frac{2l^2}{nFrDC_O^*}\left(\frac{\partial j}{\partial E}\right)\right] = \left(\frac{\tilde{a}}{\tilde{E}}\right)\hat{K} + L, \qquad (55)$$

where

$$\hat{K} = \frac{j\omega l^2}{D} + \frac{2l^2}{nFrDC_O^*}\left(\frac{\partial j}{\partial a}\right) = \frac{j\omega l^2}{D} + B \quad L = \frac{2l^2}{nFrDC_O^*}\left(\frac{\partial j}{\partial E}\right) \qquad (56)$$

with the following boundary conditions:

$$\begin{cases} z = 0 & \tilde{a} = 0, \\ z = 1 & d\tilde{a}/dz = 0. \end{cases}$$

The solution of (55) is

$$\frac{\tilde{a}}{\tilde{E}} = \frac{L}{\hat{K}}\left\{-1 + \frac{\cosh\left[\sqrt{\hat{K}}(1-z)\right]}{\cosh\left(\sqrt{\hat{K}}\right)}\right\}. \qquad (57)$$

The total impedance of pore walls consists of the faradaic and the double layer impedances. They are connected in parallel (see (18)) therefore, the faradaic, $\hat{Y}_{f,tot}$, and double layer, $\hat{Y}_{dl}$, admittances must be added in series

$$\hat{Y}_{tot} = \hat{Y}_{f,tot} + \hat{Y}_{dl} = \hat{Y}_{f,tot} + j\omega C_{dl}(2\pi rl) \tag{58}$$

and

$$\hat{Y}_{f,tot} = \int_0^1 \hat{Y}_f \, dz. \tag{59}$$

Substitution of (53) and (57) into (59) and assuming $\alpha = 0.5$ gives an analytical solution

$$\begin{aligned}\hat{Y}_{f,tot} &= u\left(\frac{P}{1+P}\right)\left(1 - \frac{B}{\hat{K}}\right) + \frac{u}{2}\left(\frac{1-P}{1+P}\right)\left(1 - \frac{B}{\hat{K}}\right)\frac{\tanh\sqrt{B}}{\sqrt{B}} \\ &\quad + u\left(\frac{P}{1+P}\right)\frac{B}{\hat{K}}\frac{\tanh\sqrt{\hat{K}}}{\sqrt{\hat{K}}} \\ &\quad + \frac{u}{2}\left(\frac{1-P}{1+P}\right)\frac{B}{\hat{K}}\frac{1}{2\cosh\sqrt{B}\cosh\sqrt{\hat{K}}} \\ &\quad \times \left\{\frac{\sinh\left(\sqrt{\hat{K}} - \sqrt{B}\right)}{\sqrt{\hat{K}} - \sqrt{B}} + \frac{\sinh\left(\sqrt{\hat{K}} + \sqrt{B}\right)}{\sqrt{\hat{K}} + \sqrt{B}}\right\},\end{aligned} \tag{60}$$

where $u = n^2 F f C_O^* \vec{k}_f$ (note a typing error in (14) and (43)). This is a rather complex equation but it simplifies at low and high frequencies. At very low frequencies $\omega \to 0$, $B/\hat{K} \to 1$, $1 - B/\hat{K} \to 0$, and (60) simplifies to

$$\hat{Y}_{f,tot}(\omega \to 0) = \frac{1}{R_p} = u\frac{P}{1+P}\frac{\tanh\sqrt{B}}{\sqrt{B}} + \frac{u}{2}\left(\frac{1-P}{1+P}\right) \\ \times \left[\frac{1}{2\cosh^2\sqrt{B}} + \frac{\tanh\sqrt{B}}{2\sqrt{B}}\right], \tag{61}$$

where R_p is the polarization resistance. At very high frequencies $\omega \to \infty$, $B/\hat{K} \to 0$ and $1 - B/\hat{K} \to 1$ and the admittance becomes

$$\hat{Y}_{f,tot}(\omega \to \infty) = \frac{1}{R_t} = u\frac{P}{1+P} + \frac{u}{2}\left(\frac{1-P}{1+P}\right)\frac{\tanh\sqrt{B}}{\sqrt{B}}. \quad (62)$$

When $\hat{Y}_{f,tot}(\omega \to 0)$ is different from $\hat{Y}_{f,tot}(\omega \to \infty)$ that is at porous electrodes in the presence of the concentration gradient faradaic impedance forms a semicircle on the complex plane plots. These means, that the total impedance exhibits two semicircles for a simple red-ox reaction, which, on a flat electrode, exhibits only one semicircle. The difference between the high and the low frequency admittance equals

$$\begin{aligned}\hat{Y}_{f,tot}(\omega \to \infty) &- \hat{Y}_{f,tot}(\omega \to 0) \\ &= u\frac{P}{1+P}\left(1 - \frac{\tanh\sqrt{B}}{\sqrt{B}}\right) \\ &+ \frac{u}{2}\left(\frac{1-P}{1+P}\right)\frac{\tanh\sqrt{B}}{\sqrt{B}}\left(1 - \frac{\sqrt{B}}{\sinh(2\sqrt{B})}\right).\end{aligned} \quad (63)$$

For flat electrodes $B \to 0$ and (63) becomes

$$\hat{Y}_{f,tot}(\omega \to \infty) - Y_{f,tot}(\omega \to 0) = 0 \quad (64)$$

that is

$$\hat{Y}_{f,tot} = u\frac{P}{1+P} + \frac{u}{2}\left(\frac{1-P}{1+P}\right) = 0.5u. \quad (65)$$

In this case the high and the low frequency impedances are the same, as for flat electrodes, and only one semicircle is observed on the complex plane plots.

For long pores, $B \to \infty$, and at very negative potentials these admittances are

$$\begin{aligned}\hat{Y}_{f,tot}(\omega \to \infty) &= \frac{u}{2}\frac{1}{\sqrt{B}}, \\ Y_{f,tot}(\omega \to 0) &= \frac{u}{4}\frac{1}{\sqrt{B}}\end{aligned} \quad (66)$$

and their ratio becomes constant equal to two

$$\frac{\hat{Y}_{f,tot}(\omega \to \infty)}{\hat{Y}_{f,tot}(\omega \to 0)} = \frac{R_p}{R_t} = 2. \quad (67)$$

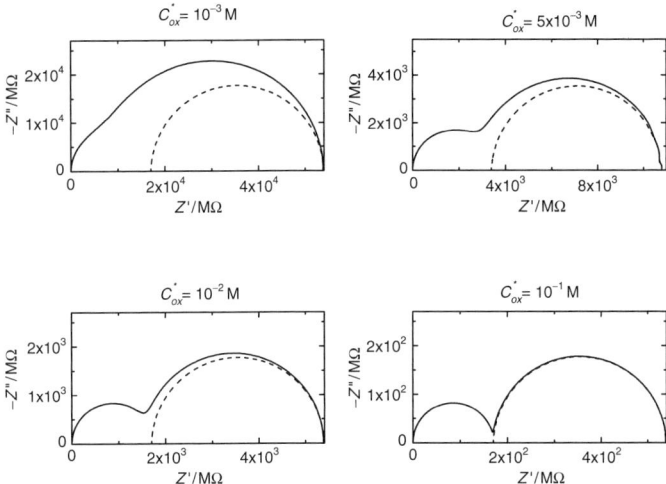

Figure 31. Complex plane plots at a porous electrode at various concentrations of the oxidized species in the absence of the potential gradient in pores; parameters used $E - E^0 = 0$ V, $l = 0.05$ cm, $r = 10^{-4}$ cm, $D = 10^{-5}$ cm^2 s^{-1}, $k_s = 10^{-6}$ cm s^{-1}, $\Phi_0 = 1.58$, concentrations indicated in the graph; total impedance – *continuous line*, faradaic impedances – *dashed line*.

In this case the faradaic impedance is represented as a semicircle starting at $2\sqrt{B}/u$ and ending at $4\sqrt{B}/u$. When combined with the double layer capacitance it displays two semicircles of the same radius on the complex plane plots. Examples of the complex plane plots at different concentrations of the active species and the same potential are displayed in Fig. 31. In this case the ratio of the highest to the lowest faradaic impedance is constant, however, at low concentrations the double layer capacitance is making the first semicircle poorly visible. Two distinct semicircles are well formed at higher concentrations of the electroactive species.

Figure 32 presents the influence of the electrode porosity on the complex plane plots. At very low values of Φ_0 (0.0316) the electrode behaves as flat, the faradaic impedance is represented as one value (point at the graph) and one semicircle is observed on the complex plane plots. With the increase of the Thiele modulus, Φ_0, the faradaic impedance becomes a semi-circle. It growth with Φ_0; for $\Phi_0 = 15.8$ the complex plane plot of the total impedance fol-

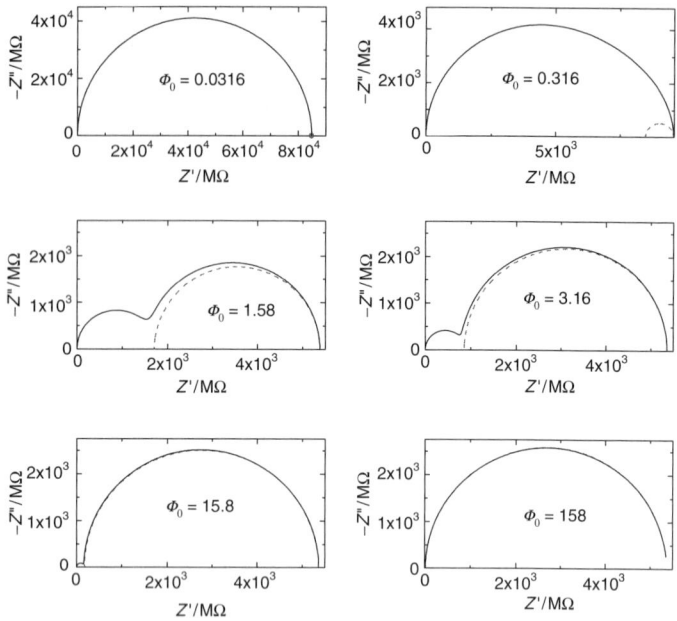

Figure 32. Influence of the electrode porosity on the impedance plots, porosity characterized by different values of the Thiele modulus, Φ_0 (see 48), corresponding to the pore length $l = 0.001, 0.01, 0.05, 0.1, 0.5$, and 5 cm, $r = 10^{-4}$ cm, at a constant potential $E - E^0 = 0$ V; total impedances – *continuous line*, faradaic impedances – *dashed line*.

lows the faradaic impedance with except of a small high frequency semicircle and finally at very porous electrodes, $\Phi_0 = 158$ the total impedance follows the faradaic impedance and the influence of the double layer is invisible.

Figure 33 presents the influence of the electrode potential, $E - E^0$, on the observed complex plane plots. With increase in negative potential the faradaic impedance changes and tends to the condition in (67). The total impedance displays always two semicircles, however, they are separated the best around $E - E^0 = 0$.

In summary, in the case of the concentration gradient in pores and in the absence of the potential gradient, the faradaic impedance becomes a semicircle instead of a constant value. This translates into formation of two semicircles of the total impedance on the complex

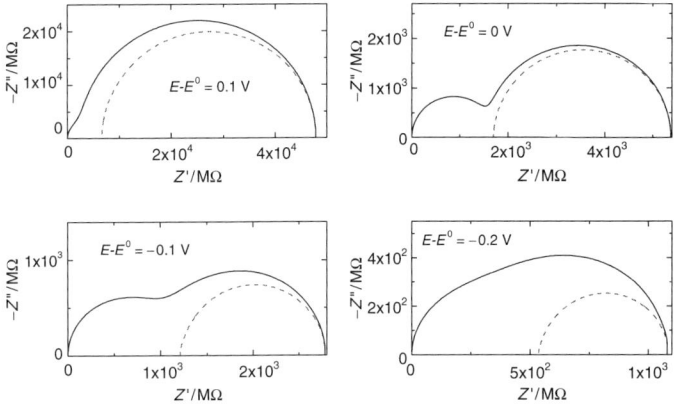

Figure 33. Influence of the electrode potential on the complex plane plots for the porous electrode in the absence of the potential gradient, $\Phi_0 = 1.58$; total impedance – *continuous line*, faradaic impedances – *dashed line*.

plane plots. Formation of two semicircles was observed experimentally on porous gold electrodes (45).

4. General Case: Concentration and Potential Gradient

i. dc solution

In a general case potential and concentration gradients appear in pores simultaneously (35–41, 46). It is assumed that the concentrations of the red-ox species outside the pores are C_O^* and C_R^* and the current might be expressed in terms of the overpotential (47)

$$j(x) = j_0 \left(\frac{C_R(x)}{C_R^*} e^{(1-\alpha)f\eta(x)} - \frac{C_O(x)}{C_O^*} e^{-\alpha f\eta(x)} \right), \quad (68)$$

where both the concentrations and the overpotential depend on the distance from the pore orifice and j_0 is the exchange current density. Assuming, that the diffusion coefficients of both forms are identical one can write

$$C_O^* + C_R^* = C_O(x) + C_R(x). \quad (69)$$

Introducing new dimensionless parameters

$$a(x) = C_O(x)/C_O^* \quad \text{and} \quad m = C_O^*/C_R^*. \quad (70)$$

Equation (68) becomes

$$j(x) = j_0 \left[(m+1) e^{b\eta(x)} - a(x) \left(e^{-b\eta(x)} + m e^{b\eta(x)} \right) \right], \quad (71)$$

where $b = 0.5nf$ for $\alpha = 0.5$. To solve for the concentration and potential gradient (23) and (25) must be solved simultaneously

$$\frac{dI(x)}{dx} = -2\pi r j(x) \quad \text{and} \quad \frac{d\eta(x)}{dx} = -\left(\frac{\rho_s}{\pi r^2}\right) I(x). \quad (72)$$

Taking the second derivative of overpotential and substitution gives (26):

$$\frac{d^2\eta(x)}{dx^2} = \frac{2\rho_s}{r} j(x) \quad (73)$$

or in the dimensionless form

$$\frac{d^2\eta(z)}{dz^2} = \frac{2\rho_s l^2}{r} j(z). \quad (74)$$

This equation must be solved together with (46). They both represent second order differential equations for $\eta(x)$ and $a(x)$ with the following conditions:

$$\begin{aligned} z = 0 & \quad a = 1, \eta = \eta_0, \\ z = 1 & \quad da/dz = d\eta/dz = 0. \end{aligned} \quad (75)$$

Combination of these two equations gives

$$\frac{d^2\eta(z)}{d^2 z} = (nFDC_O^* \rho_s) \frac{d^2 a(z)}{dz^2} = v \frac{d^2 a(z)}{dz^2}, \quad (76)$$

where $v = nFDC_O^* \rho_s$. Equation (76) has an analytical solution

$$\eta_0 - \eta(z) = v \left[1 - a(z) \right]. \quad (77)$$

This means that there is a linear relation between the potential and concentration in the pores. There are two limiting cases:

(a) When $v \ll 1$ V, $\eta_0 - \eta(z) \approx 0$, overpotential is practically constant in the pores and the system is determined by the concentration gradient; this case corresponds to that described in Sect. III.

(b) When $v \gg 1\,\text{V}$, potential gradient in pores is important while the concentration gradient is negligible; this case corresponds to that described in Sect. II.

The value of the parameter v for typical experimental conditions $n = 1$, $D = 10^{-5}\,\text{cm}^2\,\text{s}^{-1}$, $\rho_s = 10\,\Omega\,\text{cm}$, and concentrations, C_O^*, 1 and 10 mM equals $v \sim 10^{-5}$ and 10^{-4} V, respectively. This means, that for the typical concentrations the process is limited by the concentration gradient in pores. Only in the case of very large concentrations of electroactive species or solvent reduction/oxidation (water electrolysis, chlorine evolution in concentrated solutions) the potential gradient may prevail. Substitution of (77) to (46) and (74) gives the following two equations:

$$\frac{d^2 a}{dz^2} = \left(\frac{2j_0 l^2}{nFrDC_O^*}\right) \left[a \left(e^{b(\eta_0 - v) + bva} + m e^{-b(\eta_0 - v) - bva}\right) - (m+1)e^{-b(\eta_0 - v) - bva}\right] \qquad (78)$$

and

$$\frac{d^2 \eta}{dz^2} = \left(\frac{2\rho j_0 l^2}{r}\right)\left[\left(1 - \frac{\eta_0}{v} + \frac{\eta}{v}\right)e^{b\eta} + \left[m\left(1 - \frac{\eta_0}{v}\right) - (m+1) + m\frac{\eta}{v}\right]e^{-b\eta}\right]. \qquad (79)$$

It is sufficient to solve only one of them as the parameters $a(z)$ and $\eta(z)$ are related by (77). The first integration can be carried out analytically but the second one numerically, searching for the values of $a(1)$ or $\eta(1)$ at $z = 1$ (i.e., $x = l$) fulfilling the conditions $da/dz = d\eta/dz = 0$ (41).

Examples of the Tafel plots obtained at porous electrodes are shown in Fig. 34. At lower concentrations of the electroactive species the electroreduction process is controlled by the concentration gradient and the Tafel slope at 25°C is $2\ln(10)/\alpha f = 237\,\text{mV dec}^{-1}$ for $\alpha = 0.5$. At higher concentrations and lower overpotentials the Tafel slope becomes $\ln(10)/\alpha f = 118\,\text{mV dec}^{-1}$ as the concentration gradient is negligible, however, at higher overpotentials and, subsequently, higher currents, the slope increases as concentration and potential gradients appear again.

Examples of the concentration and potential gradients developing in pores are displayed in Fig. 35. At a lower concentration its

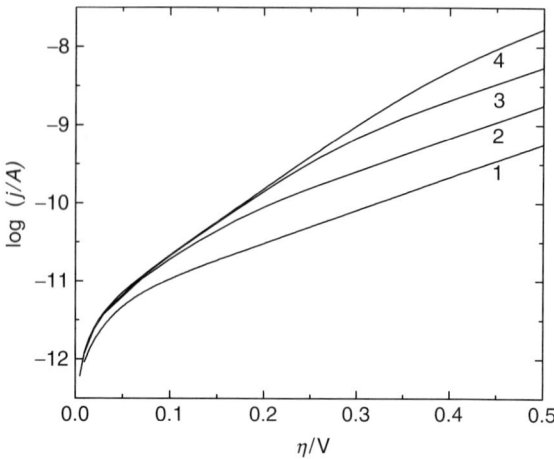

Figure 34. Tafel plots at a porous electrode at different concentrations of the active species. Concentrations: (1) 0.01, (2) 0.1, (3) 1 M, (4) limit at an infinite concentration (no concentration gradient); pore parameters: $l = 0.05$ cm, $r = 10^{-4}$ cm, $\rho_s = 10\,\Omega$ cm, $D = 10^{-5}$ cm^2 s^{-1}, $m = 1$, $j_0 = 10^{-7}$ A cm^{-2}.

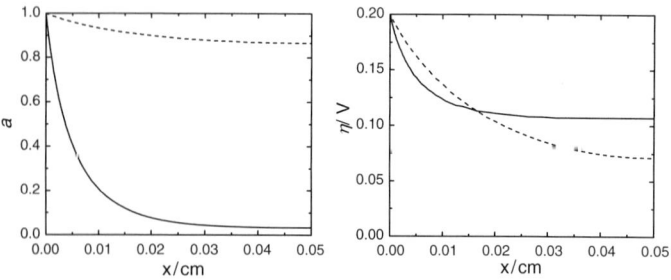

Figure 35. Examples of the concentration and potential gradients developing in pores; *continuous line*: $\eta = 0.2$ V, $j_0 = 9 \times 10^{-4}$ A cm^{-2}, $C_O^* = 0.01$ mol cm^{-3}, $v = 9.65 \times 10^{-2}$ V; *dashed line*: $j_0 = 10^{-4}$ A cm^{-2}, $C_O^* = 0.1$ mol cm^{-3}, $v = 0,965$ V; other parameters as in Fig. 34.

gradient is larger and potential gradient smaller (smaller v) whereas for a larger concentration its gradient is smaller while the whole potential gradient larger (larger v). It should be added that to solve (78) and (79) calculations must be performed with quad precision as certain terms are of similar magnitude and their difference small.

ii. ac solution

To obtain the ac solution one has to solve (52), which leads to the solution (57) for the local impedance in the pore (41). To calculate the total impedance one has to use the concentration and potential distribution in pores and carry out numerical addition of impedances using (33), similarly to the case of the potential gradient only (41). An example of the impedance complex plane plot obtained in the case of the potential and concentration gradients in pores is shown in Fig. 36.

At high frequencies a straight line at 45° is observed indicating that the pore impedance is determined by the potential gradient, analogically to that obtained for the porous electrode in the presence of the potential gradient only, Figs. 22 and 23. At lower frequencies two semicircles related to the concentration gradient appear, as in Figs. 31 and 32.

An influence of the concentration on the impedance spectra is illustrated in Fig. 37. At high frequencies a short straight line at 45° is observed due to the potential gradient. At lower frequencies two semicircles appear, and their diameter decreases with the increase in the concentration. In the case where there is no concentration gradient $\left(C_o^* \to \infty\right)$, impedance due to the potential gradient only is observed, Fig. 37. Differences observed between Figs. 31 and 37 are related to a different definition of the parameter B in both cases; in the former it is independent of concentration (see (48)) because

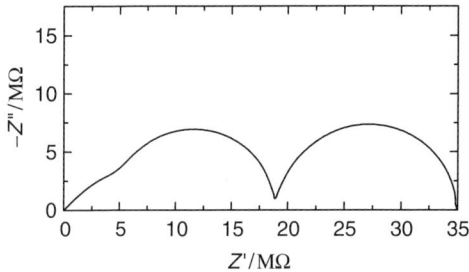

Figure 36. Complex plane plot at a porous electrode in the presence of the potential and concentration gradients; $j_0 = 10^{-5}$ A cm^{-2}, $C_o^* = 1$ M, $\eta = 0.2$ V, other pore parameters as in Fig. 34.

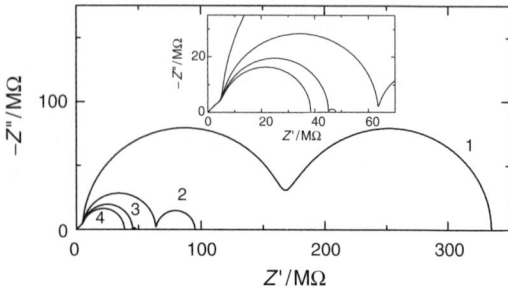

Figure 37. Influence of the concentration of the electroactive species on the impedance spectra. Concentrations: (1) 10^{-4}, (2) 10^{-3}, (3) 10^{-2} mol cm^{-3}, (4) limit at infinite concentration (potential gradient only); $\eta = 0.2$ V, $j_0 = 10^{-6}$ A cm^{-2}, other parameters as in Fig. 34.

it is defined with respect to the standard rate constant while in the latter it is defined as

$$B = \frac{2l^2}{nFrDC_o^*} j_0 \left(P^{-\alpha} + P^{\alpha} \right) \tag{80}$$

and it depends on the concentration because it is defined with respect to the exchange current density. Of course, the exchange current density depends on the concentration which means that in Fig. 37 with the increase in concentration the parameters B and Φ_0 decrease, the standard rate constant decreases but the parameter ν increases. Simulations of the hydrogen pressure buildup in pores were presented in (41).

5. Distribution of Pores

All the simulations presented above were carried out assuming presence of one or multiple pores of the same geometry. However, in real systems there are many different pores of different geometry. The simplest method of taking into account distribution of pores of different sizes is to use the transmission line ladder network, Fig. 4 or Fig. 16, and use different values for the parameters r_i and c_i or interfacial impedances z_i and calculate the total admittance by addition of the admittances of the elements. Such method was used by Macdonald and coworkers (48, 49) and Pyun et al. (50). Although such model might be used in simulation of the impedance spectra assuming changes in parameters with the position in the electrode it is difficult to obtain the pore parameters from the experimental spectra.

Song et al. (51–54) in a series of papers considered distribution of pore parameters for electrodes in the absence of electroactive species. In this case the de Levie's equation (7) was used for the individual pore and in the summation of impedances various distribution functions were used. The penetrability, α, that is the ratio of ac signal penetration depth λ (see Chap. 2.1.(i)) to the pore length was defined as

$$\alpha = \frac{\lambda}{l} = \left(\frac{1}{2l}\sqrt{\frac{r}{C_{dl}\rho_s}}\right)\omega^{-1/2} = \alpha_0 \omega^{-1/2}. \tag{81}$$

where α_0 is called the penetrability coefficient in $s^{-1/2}$ directly proportional to $\sqrt{r}/l$, that is it takes into account distribution of the radius and pore length. The authors assumed different distribution functions of α_0: normal, lognormal, Lorentzian and log Lorentzian. The total impedance is calculated as

$$Z_{tot}^{-1} = \int\limits_{-\infty}^{\infty} Z_p(\alpha_0)^{-1} f(\alpha_0, \mu, \sigma) d\alpha_0, \tag{82}$$

where $Z_p(\alpha_0)$ is the impedance of a pore characterized by the parameter α_0 and $f(\alpha_0, \mu, \sigma)$ is the pores size distribution function (continuous) of the parameter α_0 with the mean value μ and the standard deviation σ. The integration is carried out over the all values of α_0. Figure 38 illustrates the complex plane plots of the ideally polarized electrodes with four different pore size distribution functions. After the high frequency straight line at 45° a low frequency segment deviates from a vertical straight line, predicted for a simple pore; a straight line or a curve are observed at lower frequency range. These plots look similar to the effect of the constant phase element (CPE) (7, 34) especially for the Lorentzian and log-Lorentzian distribution functions. The authors proposed so called electrochemical porosimetry in which they fitted the experimental impedance data to: (a) the discrete sum of the pores with the unknown weighting factors (53)

$$\frac{1}{Z_{tot}} = Y_p \sum_{i=1}^{n} \frac{w_i}{Z_p^*(\alpha_{0,i})}, \tag{83}$$

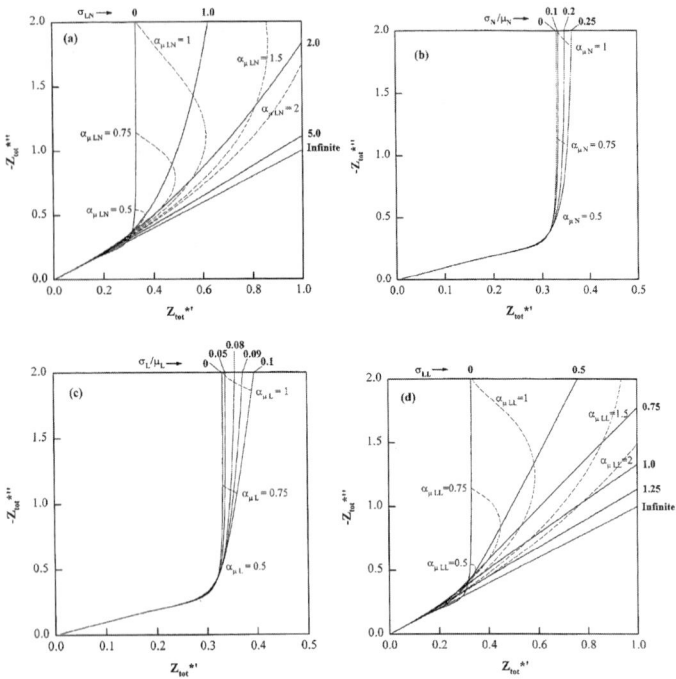

Figure 38. Complex plane plots of distributed porous electrodes; distribution functions: (**a**) log-normal (LN), (**b**) normal (N), (**c**) Lorentzian (L), (**d**) log-Lorentzian (LL); the values of the average value of the distribution parameter α_μ are indicated in the graphs. Reprinted from (52) with permission from Elsevier.

where $Y_p = V_{tot}/l^2 \rho_s$ is the total ionic conductance through pores, V_{tot} is the total pore volume, and w_i is the unknown fraction of pores characterized by the parameter α_0, and (b) by deconvolution of the distribution of the permeability coefficient α_0 using the discrete Fourier transform, DFT. (54) The problem of pore size distribution is described by the Fredholm integral equation:

$$\frac{1}{\hat{Z}_{tot}} = \Psi_t \int \frac{1}{\hat{Z}_p^*(\omega, \alpha_0)} k_v(z) \mathrm{d}z, \qquad (84)$$

where Ψ_t is the total conductance of the electrode, $k_v(z) = (1/\zeta_t)(\mathrm{d}V/\mathrm{d}z)$ is the unknown distribution function, ζ_t is the total pore volume, V is the pore volume, and z is the arbitrary parameter related to the penetrability coefficient α_0. Next, the admittances in

(84) were multiplied by $1/j\omega$ to transform them into capacitances and the DFT deconvolution was carried out to obtain the unknown distribution function. The authors applied their techniques to the determination of the pores size distribution functions of porous carbons (53,54). It should be added that there is a general problem with the determination of the distribution functions. In the first method a very limited number of pore parameters $a_{0,i}$ may be used while the second method leads to the ill-posed problem (55–57) and the determination of the distribution function is a quite difficult process; typically regularization techniques must be used to obtain reasonable results. Although the pore size distribution model leads to a CPE-type impedance behavior at lower frequencies, at the lowest frequencies this model should lead to the purely capacitive behavior (straight line at 90°) which is not usually observed experimentally.

IV. CONTINUOUS POROUS MODEL

In the above chapters impedance of a single pore or a distribution of pores of known geometry was used. However, in real cases of porous electrodes used in e.g., lithium or hydrogen storage batteries, fuel cells, etc., one does not know the exact geometry of the system and the pores do not necessarily resemble those displayed in Fig. 8 but are of a random structure. Real porous electrodes often consist of particles of an electronic conductor, sometimes contain a nonconducting material, and the pores are filled with solution containing electroactive species, for example Li^+ ions in the case of lithium batteries. A continuous theory of porous electrodes in dc conditions was presented by Newman (58). This model is schematically depicted in Fig. 39. The ionic current flowing through the solution, i_2, gradually decreases in the porous electrode while the electronic current, i_1, increases as it is transferred from the solution to the solid matrix, that is at $x = 0$, $I = i_2, i_1 = 0$, $\Phi_2 = 0$ (by definition) and at $x = l$, $I = i_1, i_2 = 0$ while inside the porous electrode $i_1 + i_2 = I$. In this model a continuous variation of the potential Φ_1 in the solid and Φ_2 in the solution is assumed.

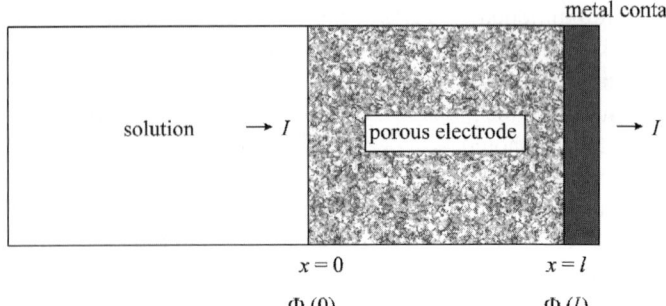

Figure 39. Schematic representation of the continuous porous electrode.

The Ohm's law in the solid matrix and in solution may be written as

$$\frac{\partial \Phi_1}{\partial x} = -\frac{i_1}{\sigma} \quad \frac{\partial \Phi_2}{\partial x} = -\frac{i_2}{\kappa}, \tag{85}$$

where σ and κ are the conductivities of the solid phase and the solution, respectively. The faradaic current which is flowing between the matrix and the solution may be described by the Butler–Volmer equation

$$i_f = ai_0 \left[\exp\left(\alpha_a f \eta_s\right) - \exp\left(-\alpha_c f \eta_s\right)\right] \tag{86}$$

or the current-overpotential equation involving concentrations of the oxidized and reduced species in solution

$$i_f = ai_0 \left[\frac{C_{\text{ox}}}{C_{\text{ox}}^*} \exp\left(\alpha_a f \eta_s\right) - \frac{C_{\text{red}}}{C_{\text{red}}^*} \exp\left(-\alpha_c f \eta_s\right)\right], \tag{87}$$

where a is the specific interfacial area of the pore walls per unit volume of the electrode in cm^{-1}, i_0 is the exchange current density, indices a and c indicate anodic and cathodic processes, respectively, overpotential is $\eta_s = \Phi_1 - \Phi_2 - E_{\text{eq}}$, and E_{eq} is the open circuit (equilibrium) potential. The diffusion equation in pores is given by an analog of (45) (see Refs. 58, 59)

$$\varepsilon \frac{\partial C}{\partial t} = \varepsilon D \frac{\partial^2 C}{\partial x^2} - \frac{a}{F} i_f, \tag{88}$$

where ε is electrode porosity. In the above equations parameters D, κ and σ are the effective values different from the values outside if pores (58, 59)

$$D = \varepsilon^{0.5} D_0; \quad \kappa = \varepsilon^{1.5} \kappa_0; \quad \sigma = (1-\varepsilon)^{1.5} \sigma_0, \quad (89)$$

where index 0 denotes values outside pores.

The impedances might be evaluated by applying small ac perturbation which generates small perturbations of the parameters around their dc values, adding the charging current (60)

$$i_{dl} = C_{dl} \frac{\partial (\Phi_1 - \Phi_2)}{\partial t} \quad (90)$$

and solving the equations numerically. Such calculations were carried out by Meyers et al. (60) for lithium intercalation in the spherical particles assuming lognormal distribution of particle dimensions. The impedance of the electrode is described by

$$\hat{Z} = \frac{\tilde{\Phi}_1(l) - \tilde{\Phi}_2(0)}{\tilde{I}} = \frac{l}{\kappa + \sigma} \left[1 + \frac{2 + \left(\frac{\sigma}{\kappa} + \frac{\kappa}{\sigma}\right) \cosh(\nu)}{\nu \sinh(\nu)} \right], \quad (91)$$

where

$$\frac{l}{\nu} = \left(\frac{\kappa \sigma}{\kappa + \sigma}\right)^{1/2} \overline{(aY)}^{-1/2} \quad (92)$$

and $\overline{aY}$ is the average admittance of all particles in the composite electrode. Examples of the complex plane plots obtained on such electrodes are displayed in Fig. 40. In this figure parameter Ψ was defined as $\Psi = \sqrt{\ln(\bar{r}a/3\varepsilon)}$ and $\bar{r}$ is the mean particle radius. As Ψ approaches zero the distribution function approaches the limit of Dirac's delta function that is all the particles have the same dimension.

It is evident that at high frequencies a skewed semicircle is observed, similar to those presented for a semi-infinite porous electrode, Figs. 22 and 24, while the low frequency part represents the finite-length diffusion, similar to Fig. 6.

This theory was later extended to rechargeable lithium batteries including concentrated solution theory of binary electrolytes and assuming that transference numbers are constant (61). Devan et al. (59) developed an analytical solution for the impedance of

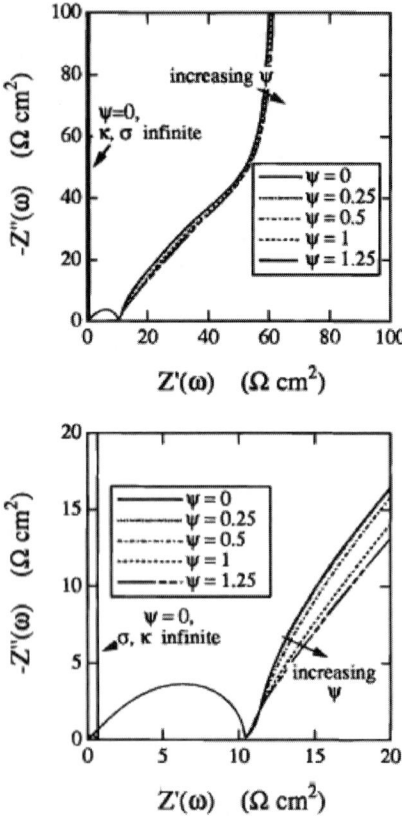

Figure 40. Complex plane plots simulated for porous electrodes with various particle size distribution. The lower is an expanded view at high frequencies. Reproduced with permission from (60). Copyright 2000, The Electrochemical Society.

porous electrode using concentrated solution theory and a linear electrode kinetics that is a linear form of the Butler–Volmer equation: $i_f = i_0 f(\alpha_a + \alpha_c)\eta$. Such a solution (although formally rather complex) permitted study of the influence of different parameters on the impedance plots and the limiting cases of the impedance parameters. An example of the plot is displayed in Fig. 41. It represents two semicircles, the high frequency kinetic skewed arc and the low

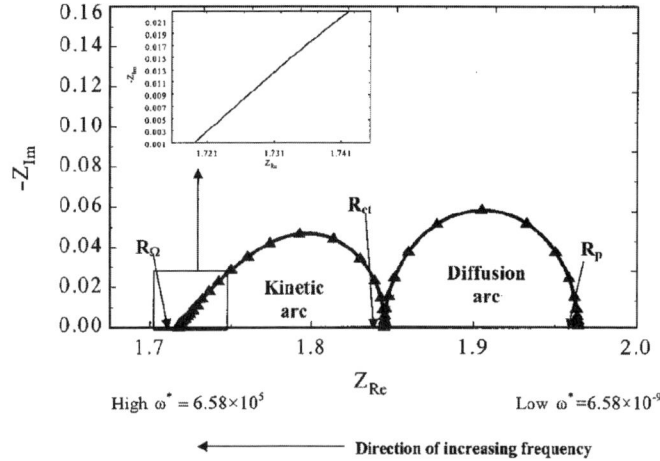

Figure 41. Complex plane plot of the porous electrode, *continuous line*: analytical solution, *triangles*: numerical solution. The *inset* shows the high frequency part of the plot (59). Reproduced with permission from (59). Copyright 2004, The Electrochemical Society.

frequency semicircle due to diffusion in pores. These plots might be compared with the plots for single pore impedance in the absence of the potential drop, Fig. 32, or a general case of a single pore in the presence of the potential and concentration gradient, Fig. 36.

In further studies the continuous porous model was applied to an electrode consisting of distribution of spherical particles of hydrogen absorbing material (62). These authors fitted the model to the experimental impedances of the practical AB_5-type of material. An example of such a fit is displayed in Fig. 42.

Another application was presented by Springer et al. (63) for a polymer fuel cell. The authors also fitted their impedance curves to their model.

It is evident that such a single pore theory cannot describe more complex cases of the real porous electrode. Nevertheless, the numerical simulation demands some additional information about the electrode, particle size distribution, possible heterogeneity of the material, etc., and the electrode studied should be composed of the uniform packing of the particles with their size much smaller than the electrode thickness.

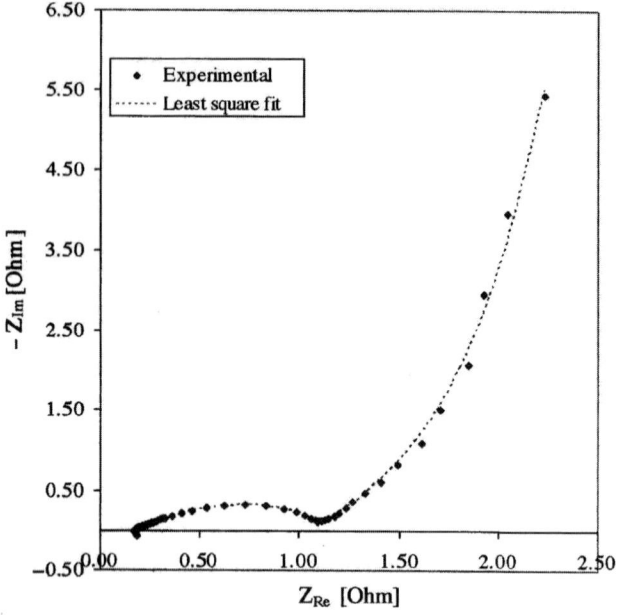

Figure 42. Impedance of an AB$_5$ electrode discharged to 50% and a nonlinear least-squares fit to a model. Reprinted from (62) with permission from Elsevier.

V. FRACTAL MODEL AND THE CPE

1. Fractal Model

As it was mentioned in the preceding chapter the real surfaces are often irregular, consisting of various pores, surface defects, irregularities, etc., and their detailed geometry is usually unknown. In order to describe such surfaces a fractal model was proposed (64). Fractal model describes self-similar surfaces in which further magnification reveals similar structure. Such an approach might also be used to approximate a random roughness of the surface. Fractal model involves a specific scaling. In the example in Fig. 43 (65) simple magnification of the line length l increases its length three times ($a \rightarrow b$) while fractal magnification increases its length to $4l (a \rightarrow c)$. Therefore, the fractal dimension of this line is $D = \ln 4 / \ln 3 = 1.2619$ (65, 66), which is a value larger than one,

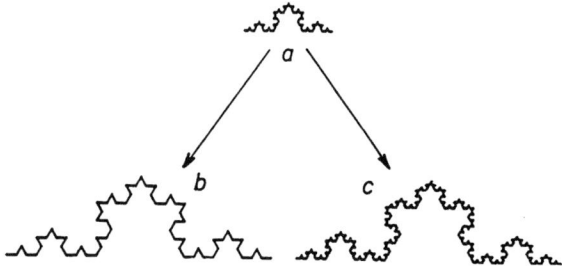

Figure 43. Comparison of a classical (**a**) and fractal (**b**) magnification. Reprinted from (65) with permission from Elsevier.

the value observed at simple magnification. The line consisting of self-similar segments scaled as above is called the von Koch line (67). As the magnification process may be repeated infinite number of times this line is continuous, has an infinite length and is nowhere differentiable. Of course, in practice, such a magnification must be limited by the atomic dimensions and, in practice, used to dimensions between 10 and 0.1 mm (65). Examples of von Koch lines for different fractal dimensions are illustrated in Fig. 44 (66).

In general, the fractal dimension of the von Koch line may be between 1 and 2 and for fractal surfaces the dimensions may be between 2 and 3.

The concept of fractals was introduced to the electrochemical impedance by Le Méhauté and coworkers (68, 69) and later developed by Nyikos and Pajkossy (66, 70–77). They have shown that the fractal geometry of the blocking (i.e., ideally polarizable) interfaces leads to the constant phase element, CPE, behavior

$$\hat{Z}_{CPE} = \frac{1}{T(j\omega)^\phi}, \qquad (93)$$

where T is the parameter related to the electrode capacitance, in F cm^{-2} s$^{\phi-1}$ and ϕ is the constant phase exponent ($0 < \phi < 1$) related to the deviation of the straight capacitive line from 90° by angle $\alpha = 90^0(1 - \phi)$. Because $j^\phi = \cos(\phi\pi/2) - j\sin(\phi\pi/2)$ is a complex number (93) may be expressed as

$$\hat{Z}_{CPE} = \left[\frac{\cos(\phi\pi/2)}{T\omega^\phi}\right] - j\left[\frac{\sin(\phi\pi/2)}{T\omega^\phi}\right], \qquad (94)$$

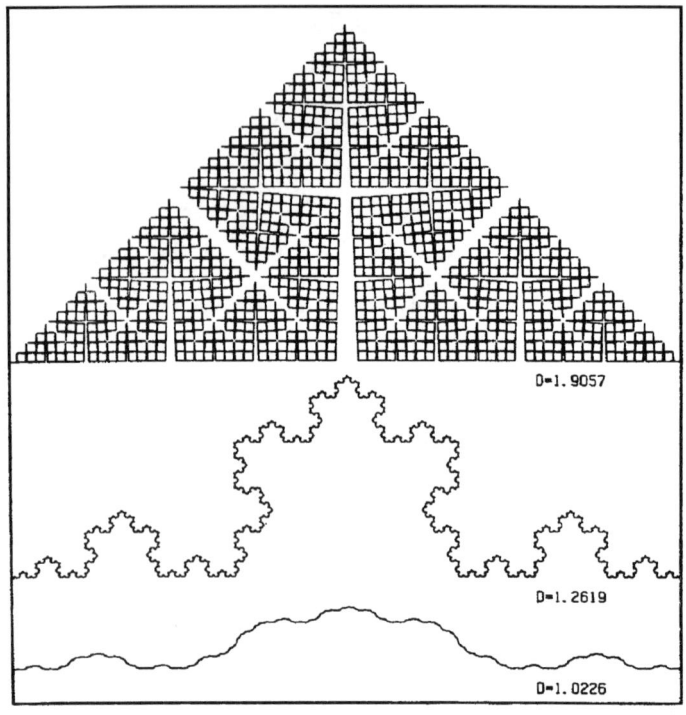

Figure 44. Examples of von Koch lines for different fractal dimensions. Reprinted from (66) with permission from Elsevier.

that is it contains a real and an imaginary parts in the whole frequency range. An example of the complex plane and Bode phase angle plots is presented in Fig. 45. It is obvious that in both cases the phase angle is constant in all the frequency range which gave this factor the name of the constant phase element. Although, in principle, the constant phase exponent may be between $0 < \phi < 1$ in practice it is usually >0.5, typically > 0.8. For the ideal fractal surfaces a simple relation between the coefficient ϕ and the fractal dimension of the surface, D, exists: $\phi = 1/(D-1)$ (66). This would give for $\phi = 0.9$ surface fractal dimension $D = 2.111$.

The fractal theory was subsequently extended to irregular or quasi-random surfaces lacking well defined self-similarity (72, 75, 78, 79). Pajkossy and Nyikos (80) carried out simulations of

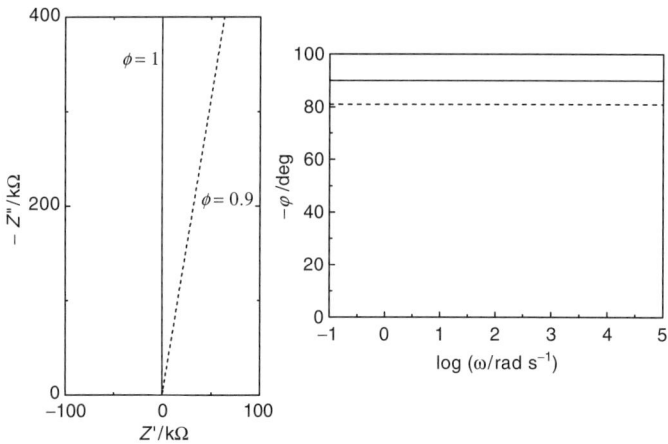

Figure 45. Complex plane and Bode phase angle plots for an ideally polarizable flat electrode, $\phi = 1$, and a fractal electrode, $\phi = 0.9$ and $T = 20\,\mu\text{F cm}^{-2}\,\text{s}^{\phi-1}$.

blocking electrodes with a self-similar spatial capacitance distribution and found that the calculated impedances exhibited the CPE behavior.

It has been found later that although the fractal geometry produces CPE behavior, in practice there is no relation between the CPE exponent and the fractal dimensions (81, 82) although, qualitatively, higher fractal dimensions lead to smaller values of ϕ. This indicates that the impedance technique does not allow for the determination of the surface fractal dimension. Such information can be obtained by the analysis of current-time curves in the presence of diffusion to the surface (72, 73, 83, 84).

The fractal theory has also been applied to the systems with faradaic reaction (65, 72, 73, 85, 86). de Levie (65, 85) has shown that the impedance of a fractal electrode, in the presence of a simple faradaic reaction, is

$$\hat{Z} = R_\text{s} + \frac{1}{b}\left(\frac{1}{\frac{1}{R_\text{ct}} + j\omega C_\text{dl}}\right)^\phi, \quad (95)$$

where the parameters b is given by (66, 85)

$$b = f_\text{g}\rho_\text{s}^{\phi-1}, \quad (96)$$

ρ_s is the solution resistivity and f_g is a geometric factor depending on the fractal surface geometry. It has been stated (85) that this factor is related to the fact that the fractal description ignores details of the surface morphology focusing only on the global response. This means that different surface geometries can have the same fractal dimension but different geometric factors f_g. For flat surfaces $\phi = 1$, $f_g = 1$, $b = 1$ and the second term in (95) reduces to (18). Equation (95) can be used in the absence of a dc current which might introduce additional dc potential drop in the pores/irregularities. Examples of the complex plane plots on fractal electrode in the presence of the faradaic reaction and in the absence of mass transfer effects for different values of the parameter ϕ are displayed in Fig. 46. For $\phi = 1$ that is an ideally flat electrode a semicircle is obtained, it becomes flattened (skewed) for lower values of the parameters ϕ. For $\phi = 0.5$ the shape of the complex plane plot is formally identical, i.e., undistinguishable from the semi-infinite length porous electrode, see Figs. 22 and 23. Equation (95) is formally similar to that found by Davidson and Cole (87) in their dielectric studies. In the fractal model time constants are distributed. The impedance in such a case is expressed as (56, 87–89)

$$\frac{\hat{Z} - Z_\infty}{Z_0 - Z_\infty} = \int_0^\infty \frac{\tau G(\tau)}{1 + j\omega\tau} d\ln\tau \qquad (97)$$

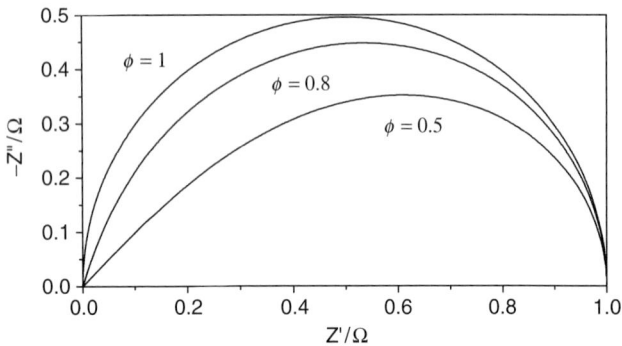

Figure 46. Complex plane plots for the fractal model in the presence of the faradaic reaction (see (95)) for different values of the parameter ϕ.

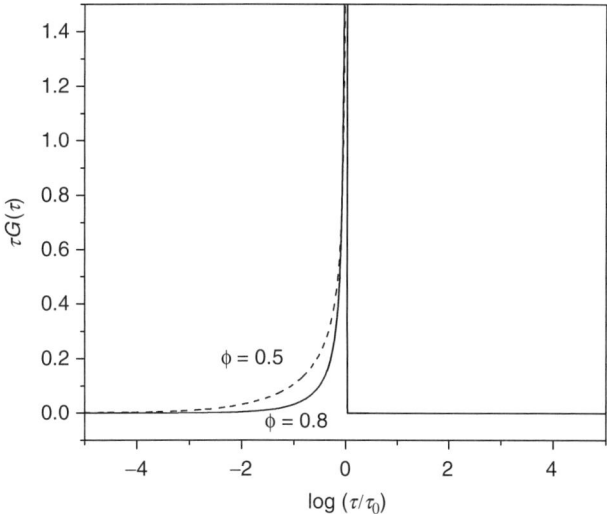

Figure 47. Distribution of the time constants for the fractal model.

while the distribution function $G(\tau)$ is described as

$$\tau G(\tau) = \begin{cases} \frac{\sin(\phi\pi)}{\pi} \left(\frac{\tau}{\tau_0 - \tau}\right)^\phi & \text{for } \tau < \tau_0 \\ 0 & \text{for } \tau > \tau_0. \end{cases} \quad (98)$$

This is an asymmetric distribution function shown in Fig. 47 which causes formation of an asymmetric complex plane plot, Fig. 46. Of course for $\phi = 1$ a Dirac delta function is obtained at $\tau = \tau_0$ and only one time constant is present in the system (no distribution of τ).

Because of the presence of the factor f_g in (95) it is impossible to determine the real value of the charge transfer resistance and double layer capacitance. It is possible to determine the experimental values (note a mistake in (7))

$$C_{\text{dl,exp}} = C_{\text{dl}} b^{1/\phi} \text{ and } R_{\text{ct,exp}} = R_{\text{ct}} b^{-1/\phi}. \quad (99)$$

However, the product of these two values equals to the product of the specific values (per unit of the electrode surface)

$$C_{\text{dl,exp}} R_{\text{ct,exp}} = C_{\text{dl}} R_{\text{ct}} \quad (100)$$

and if the specific double layer capacitance of the electrode material is known, the specific charge transfer resistance can be estimated.

2. CPE Model

A CPE behavior is often found experimentally when working on solid electrodes (90). Such a behavior is observed for the ideally polarized (blocking) electrodes and in the presence of faradaic reactions. The CPE dispersion of blocking electrodes is described by (93). In the presence of faradaic reactions the electrode impedance is described by

$$\hat{Z} = R_s + \frac{1}{\frac{1}{R_{ct}} + (j\omega)^\phi T}. \tag{101}$$

There is an important difference between the CPE (see (101)) and the fractal (see (95)) model. In the former only the double layer is dispersed while in the latter the charge transfer resistance and the double layer capacitance are under exponent and are dispersed. The complex plane plots obtained in the presence of the CPE are displayed in Fig. 48. The presence of the CPE caused rotation of the impedance semicircle center below the real axis.

The CPE behavior follows the Cole and Cole (91) dispersion with the time constant distribution function described as

$$\tau G(\tau) = \frac{1}{2\pi} \frac{\sin[(1-\phi)\pi]}{\cosh[\phi \ln(\tau/\tau_0)] - \cos[(1-\phi)\pi]}. \tag{102}$$

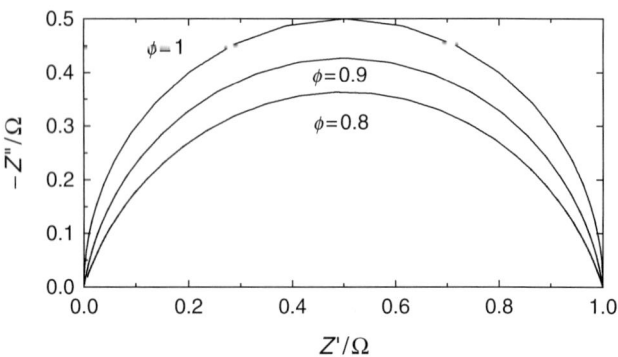

Figure 48. Complex plane plots obtained for the CPE model in the presence of the faradaic reaction without mass transfer effects for various values of the parameter ϕ.

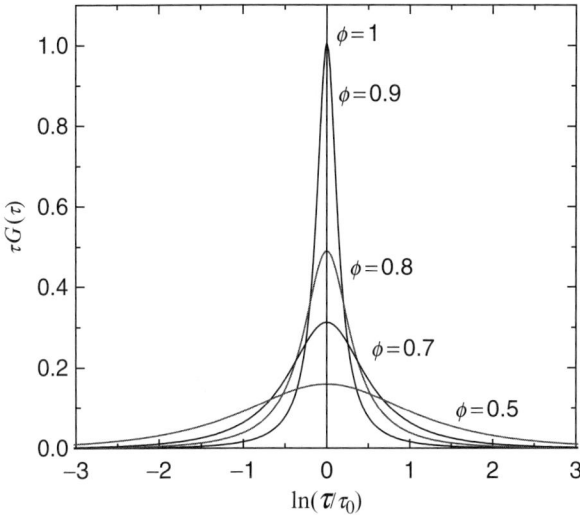

Figure 49. Distribution of time constants in the CPE model (see (102)).

The plot of this function is presented in Fig. 49. It is clear that the distribution functions for the fractal and the CPE models are essentially different.

Analysis of the impedance spectra applying the regularization technique leads to the determination of the distribution functions, although the procedure is numerically difficult (56). It is much simpler to carry out nonlinear least squares fit of the impedances to the equivalent model containing the CPE element.

Brug et al. (34) proposed a simple model for the determination of an average double layer capacitance, $\overline{C}_{dl}$, in the presence of the CPE dispersion. For the blocking electrode the CPE dispersion arises from the dispersion of the time constants, $\tau = R_s C_{dl}$ around an average value $\tau_0 = R_s \bar{C}_{dl}$ and (101) may be rearranged into

$$\hat{Z} = R_s + \frac{1}{(j\omega)^\phi T} = R_s \left[1 + \frac{1}{(j\omega)^\phi (T R_s)} \right] = R_s \left[1 + \frac{1}{(j\omega \tau_0)^\phi} \right] \tag{103}$$

which leads to

$$T = \overline{C}_{dl}^{\phi} R_s^{-(1-\phi)} \tag{104}$$

allowing for the determination of $\bar{C}_{dl}$ from the experimentally determined values of R_s and T. Similarly, in the presence of the faradaic reaction (see (101)) leads to a similar relation (34)

$$T = \bar{C}_{dl}^{\phi} \left[R_s^{-1} + R_{ct}^{-1} \right]^{1-\phi}. \tag{105}$$

There is a lot of confusion in the literature concerning origins of the CPE behavior (82). Initially it was ascribed to the surface microscopic roughness. However, several authors (82, 92, 93) demonstrated experimentally that increase in the microroughness of Pt electrode causes decrease in the deviation from the ideal capacitive behavior, that is the parameter ϕ becomes close to one for the electrochemically roughened surface. Moreover, for very porous gold based electrodes, after initial line at 45° a vertical line showing only small deviation from the ideal behavior is observed on the complex plane plots (13). Simple estimation shows that the microscopic roughness of macroscopically flat electrodes could show the dispersion of the time constants in a very high (kHz to MHz) frequency range, contrary to the experimental observations of dispersion in a very wide frequency range (94). Of course, the fractal model described above also leads to the CPE behavior at blocking electrodes; however, in the presence of faradaic reactions it leads to a skewed semicircles instead of a decrease of the center of the semicircle below the real axis without further deformation observed typically in the experimental conditions (CPE behavior). It has been discussed in the literature that the electrode displaying a CPE behavior cannot be considered purely capacitive (95–98).

In general, one can suggest the following general explanations for the CPE behavior:

1) Physical R_s–C_{dl} dispersion.
2) Kinetic dispersion.

These two phenomena are described in more detail below.

i. Physical R_s–C_{dl} dispersion

This is a simple dispersion of the time constants causing nonuniform current distribution. It can arise from (a) geometric dispersion of solution resistances, that is different resistances at different parts of

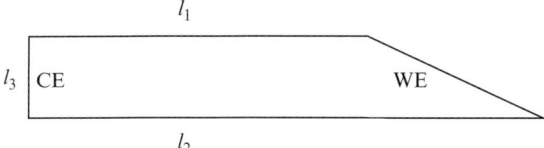

Figure 50. Representation of the Hull cell used in the simulations, *CE* counter electrode, *WE* working electrode, dimensions defined in the figure.

the electrode due to the surface roughness, incorrect positioning of the working electrode (not perpendicular to the auxiliary/reference electrode) causing nonuniform distribution of current at the electrode surface or (b) from dispersion of capacitances of the double layer due to atomic scale surface inhomogeneities (82, 93, 94).

The physical R–C dispersion may be caused by a non parallel setup of the working and counter electrodes (99). Simple simulations of the Hull cell, shown in Fig. 50, are displayed in Fig. 51.

It is evident that when the working and the counter electrodes are not parallel, Fig. 51a–c, there is a distribution of solution resistances and, at a certain frequency range, a slope is observed at the complex plane plots. This dispersion decreases as the two electrodes tend to the parallel position and disappears completely for the parallel position, Fig. 51d.

Similarly, in the presence of the faradaic reaction semicircles distorted at high frequencies are observed on the complex plane plots (99). At parallel electrodes an ideal semicircle with the expected R_{ct} value is observed while at non-parallel electrodes much lower values of the R_{ct} are found, Fig. 52. These distortions are caused by the distribution of the cell resistances.

Jorcin et al. (90) distinguished 2D dispersion, that is dispersion along the electrode plane arising, e.g., from the distribution of capacitances and 3D distribution, that is distribution in the direction normal to the electrode surface arising from the roughness, porosity, varying of the coating properties/composition, etc. It was also found that the electrode displaying a global CPE character may behave as purely capacitive when studied using local impedance method.

Recently, Huang et al. (100–102) studied appearance of the apparent CPE on disk electrodes. It is well known (33) that at disk electrodes the current is distributed non-uniformly in the radial direction

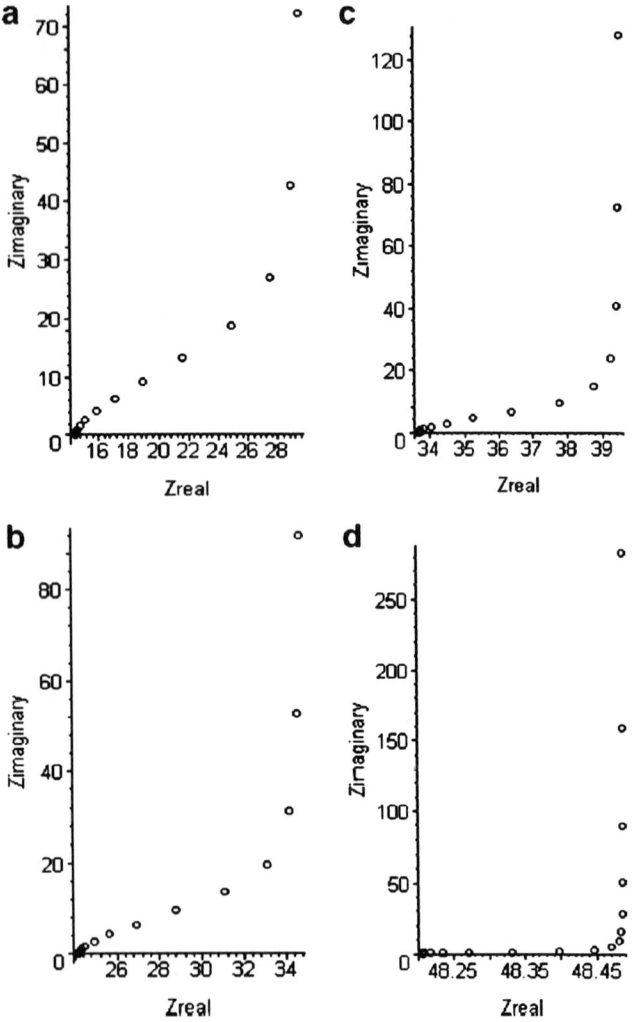

Figure 51. Complex plane plots for the ideally polarizable working electrode in various Hull cells, $\rho_s = 10\,\Omega\,\text{cm}$, $C_{dl} = 20\,\mu\text{F}\,\text{cm}^{-2}$, $l_2 = 5$ cm, $l_3 = 1$ cm, $l_1 =$ (**a**) 1 cm, (**b**) 2 cm, (**c**) 3 cm, (**d**) 5 cm; see Fig. 50 for the definitions of the geometric parameters. Reprinted from (100) with permission from Elsevier.

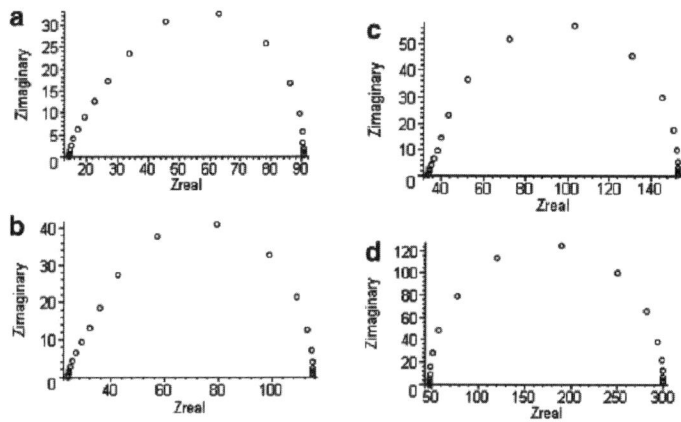

Figure 52. Complex plane plots for the faradaic reaction with $R_{ct} = 250$ Ω cm^2 in various Hull, cell parameters as in Fig. 51. Reprinted from (99) with permission from Elsevier.

(known as the primary (103) and secondary (104) current distribution). In the case of the ideally polarized (blocking) electrode conditions the global impedance displays a CPE like behavior at high frequencies. A complex plane plot of the normalized impedances is shown in Fig. 53. On the other hand, local impedance measured at different parts of the disk, r/r_0, is the largest at the disk center and the smallest at the disk periphery (r_0 is the disk radius). Although the global impedance is purely capacitive the local impedance displays inductive loops at high frequencies, Fig. 54. This behavior is caused by the local ohmic solution impedance which is complex and, at the disk periphery, displays an inductive behavior. Similar CPE-like behavior is observed in the presence of the faradaic reaction. When the solution resistance dominates over the charge transfer resistance the global charge-transfer resistance is determined with large errors.

It is interesting to notice that the average electrode capacitance determined for the disk electrode using formulas developed by Brug et al. (34), (see (104) and (105)) provide the best estimations of the interfacial capacitances (102). The above discussed papers emphasize the geometrical effects on the current distribution at the electrodes leading to a CPE-like behavior.

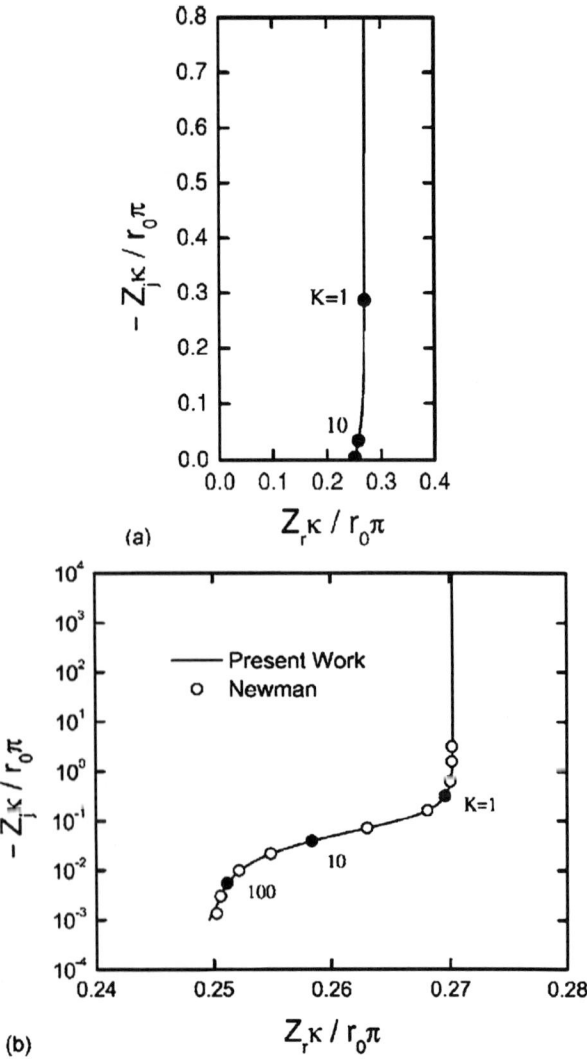

Figure 53. Complex plane plots of the normalized impedances at an ideally polarized disk electrode; κ – solution conductivity, r_0 – disk radius, $K = \omega C_0 r_0/\kappa$ dimensionless frequency, C_0 – the interfacial capacitance. Reproduced with permission from (100). Copyright 2007, The Electrochemical Society.

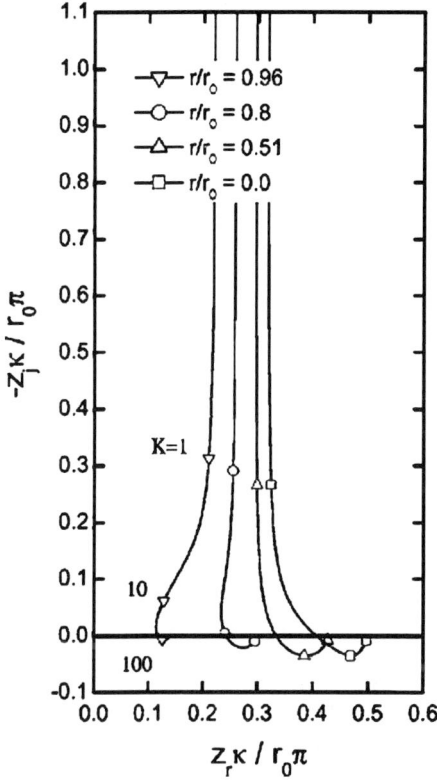

Figure 54. Complex plane plots of the normalized local impedance of the blocking disk electrode. Reproduced with permission from (100). Copyright 2007, The Electrochemical Society.

ii. Kinetic dispersion

This dispersion is related to the slow, i.e., kinetically limited, adsorption of ions (mainly anions) at the electrode surface. It has been found that in very clean solutions at monocrystalline electrodes the CPE parameter ϕ is very close to unity, e.g., at Au(111) in 0.1 M HClO$_4$ it is 0.997 (105) which indicates practically an ideal capacitive behavior. However, in the presence of specifically adsorbed anions this value is decreased. This behavior was explained assuming kinetically and diffusion controlled ionic adsorption (106–113).

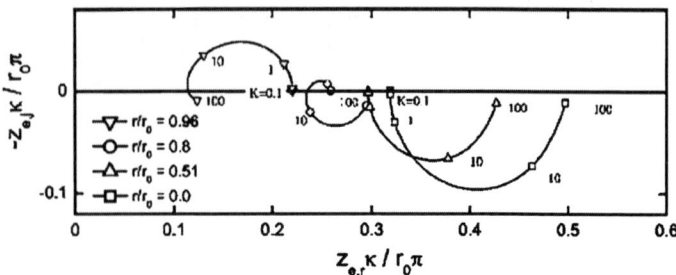

Figure 55. Complex plane plots of the normalized ohmic impedance at different positions at the ideally polarized disk electrode. Reproduced with permission from (100). Copyright 2007, The Electrochemical Society.

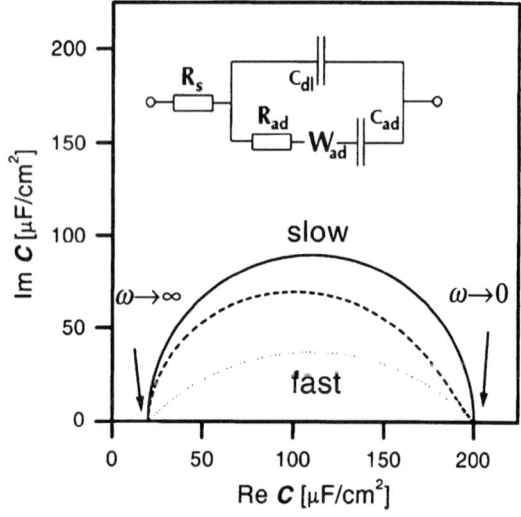

Figure 56. Complex plane plots of the complex capacitance, $\hat{C}$, defined by (109) for limiting cases of slow (*continuous line*), fast (diffusion limited, *dotted line*) and intermediate (*dashed line*) adsorption. Equivalent electrical model for this process is in the inset. Reprinted from (106) with permission from Elsevier.

In this model the faradaic impedance consists of the adsorption impedance, C_{ad}, adsorption resistance, R_{ad}, and mass transfer impedance, $\hat{Z}_{W,ad}$, the corresponding model is displayed in Fig. 56.

The net rate of adsorption reaction, v, is given by (110)

$$v = v_{\text{ad}} - v_{\text{d}} = \frac{d\Gamma}{dt}, \tag{106}$$

where subscripts "ad" and "d" denote adsorption and desorption and Γ is the surface excess of adsorbed ions. The current related to adsorption is expressed as

$$j = \frac{dq^M}{dt} = -\gamma F v, \tag{107}$$

where γ is the electrosorption valency and q^M excess charge density in a metal electrode. Applying the standard linearization procedure the model elements are obtained.

These elements, Fig. 56, are described by the following equations:

$$C_{\text{ad}} = \gamma F \frac{\left(\frac{\partial v}{\partial E}\right)}{\left(\frac{\partial v}{\partial \Gamma}\right)}; \quad R_{\text{ad}} = -\frac{1}{\gamma F \left(\frac{\partial v}{\partial E}\right)};$$

$$\hat{Z}_{\text{W,ad}} = \frac{\sigma_{\text{ad}}}{\sqrt{j\omega}} = -\frac{1}{\gamma F \sqrt{j\omega D}} \frac{\left(\frac{\partial v}{\partial C}\right)}{\left(\frac{\partial v}{\partial E}\right)}, \tag{108}$$

where C is the bulk concentration of adsorbed ions, D diffusion coefficient of ions, and σ_{ad} is a Warburg coefficient. Because the complex plane plots are quite featureless the analysis was carried out by transforming the impedances into complex capacitances

$$\hat{C} = \frac{1}{j\omega \left(\hat{Z}_{\text{tot}} - R_{\text{s}}\right)} = C_{\text{dl}} + \frac{1}{\frac{1}{C_{\text{ad}}} + \sigma_{\text{ad}}\sqrt{j\omega} + R_{\text{ad}}}. \tag{109}$$

Examples of simulated impedances for kinetically slow, fast that is diffusion limited, and intermediate adsorption rates are shown in Fig. 56. For the kinetically controlled reaction a perfect semicircle is observed. It becomes distorted in the presence of mixed kinetic-diffusion control. Such an analysis was applied to study anionic adsorption on several well-defined metallic surfaces (106–113).

VI. CONCLUSIONS

In the present review the geometric effects in the electrochemical impedance were presented. The EIS is a sensitive method that can be used in characterization of various porous electrodes. The simple pore models which are most often used are only an idealization of the real electrodes. In fact, very rarely well defined pores exist and such materials must be prepared in a special way. The real electrodes contain distribution of various pores. Using simple pore models it is possible to achieve good approximations of the experimental impedances obtaining the average values of pore parameters. This approach is most often used in practice as it is simple, however, the parameters obtained might be different from the real ones.

More complex models with the distribution of pores demand good knowledge of the electrode geometry and their distribution. Although the distribution of pores might also be determined experimentally, the mathematical description involves solution of the Fredholm's integral equation, which is an ill-posed problem. The solution using the regularization techniques is difficult to obtain. The models of continuous porous electrodes are more advanced and describe better the real materials assuming that the materials are homogeneous in the axis perpendicular to the electrode surface. Some additional parameters characterizing the electrode materials are necessary to obtain the correct solution. However, in practice, only the numerical solutions are obtained and the mathematical treatment is more complex.

Care must be taken to differentiate between the geometric and kinetic nature of the observed plots and additional experiments at different temperature, potentials and in the presence of poisons might be necessary to distinguish between different models.

REFERENCES

1. A. N. Frumkin, *Zh. Fiz. Khim.* **23** (1949) 1477.
2. O. S. Ksenzhek and V. V. Strender, *Dokl. Akad. Nauk SSSR* **106** (1956) 487.
3. O. S. Ksenzhek, *Russ. J. Phys. Chem.* **36** (1962) 331.
4. A. Winsel, *Z. Elektrochem.* **66** (1962) 287.
5. F. A. Posey, *J. Electrochem. Soc.* **111** (1964) 1173.
6. J. M. Bisang, K. Jüttner, and G. Kreysa, *Electrochim. Acta* **39** (1994) 1297.
7. A. Lasia, in *Modern Aspects of Electrochemistry*, Vol. 32, Ed. by B. E. Conway, J. Bockris, and R. E. White, Kluwer/Plenum, New York, 1999, p. 143.

8. A. Lasia, in *Modern Aspects of Electrochemistry*, Vol. 35, Ed. by B. E. Conway and R. E. White, Kluwer/Plenum, New York, 2002, p. 1.
9. R. de Levie, in *Advances in Electrochemistry and Electrochemical Engineering*, Vol. 6, Ed. by P. Delahay, Interscience, New York, 1967, p. 329.
10. L. M. Gassa, J. R. Vilche, M. Ebert, K. Jüttner, and W. J. Lorenz, *J. Appl. Electrochem.* **20** (1990) 677.
11. R. de Levie, *Electrochim. Acta* **8** (1963) 751.
12. I. D. Reistrick, *Electrochim. Acta* **35** (1990) 1579.
13. R. Jurczakowski, C. Hitz, and A. Lasia, *J. Electroanal. Chem.* **572** (2004) 355.
14. R. de Levie, *Electrochim. Acta* **10** (1965) 113.
15. J. Gunning, *J. Electroanal. Chem.* **392** (1995) 1.
16. H. Keiser, K. D. Beccu, and M. A. Gutjahr, *Electrochim. Acta* **21** (1976) 539.
17. C. Hitz and A. Lasia, *J. Electroanal. Chem* **500** (2001) 213.
18. K. Eloot, F. Debuyck, M. Moors, and A. P. Peteghem, *J. Appl. Electrochem.* **25** (1995) 326.
19. K. Eloot, F. Debuyck, M. Moors, and A. P. Peteghem, *J. Appl. Electrochem.* **25** (1995) 334.
20. L. Chen and A. Lasia, *J. Electrochem. Soc.* **139** (1992) 3214.
21. L. Chen and A. Lasia, *J. Electrochem. Soc.* **140** (1993) 2464.
22. A. Lasia, in *Current Topics in Electrochemistry*, Vol. 3, Research Trends, Trivandrum, India, 1993, p. 239.
23. A. Lasia, *Int. J. Hydrogen Energy* **18** (1993) 557.
24. J. Bisquert, G. Garcia-Belmonte, F. Fabregat-Santiago, and A. Compte, *Electrochem. Commun.* **1** (1999) 429.
25. J. Bisquert, G. Garcia-Belmonte, F. Fabregat-Santiago, N. S. Ferriols, P. Bogdanoff, and E. C. Pereira, *J. Phys. Chem. B* **104** (2000) 2287.
26. J. Bisquert, *Phys. Chem. Chem. Phys.* **2** (2000) 4185.
27. P. Los, A. Lasia, and H. Ménard, *J. Electroanal. Chem.* **360** (1993) 101.
28. L. Birry and A. Lasia, *J. Appl. Electrochem.* **34** (2004) 735.
29. A. Lasia, *J. Electroanal. Chem.* **397** (1995) 27.
30. I. Roušar, K. Micka, and A. Kimla, *Electrochemical Engineering*, Vol. II, Elsevier, Amsterdam, 1986, p. 133.
31. K. Scott, *J. Appl. Electrochem.* **13** (1983) 709.
32. S. I. Marshall, *J. Electrochem. Soc.* **138** (1991) 1040.
33. A. J. Bard and L. R. Faulkner, *Electrochemical Methods. Fundamentals and Applications*, J. Wiley & Sons, Inc., New York, 2001.
34. G. J. Brug, A. L. G. van der Eeden, M. Sluyters-Rehbach, and J. H. Sluyters, *J. Electroanal. Chem.* **176** (1984) 275.
35. J. S. Newman and C. W. Tobias, *J. Electrochem. Soc.* **109** (1962) 1183.
36. L. G. Austin and H. Lerner, *Electrochim. Acta* **9** (1964) 1469.
37. S. K. Rangarajan, *J. Electroanal. Chem.* **22** (1969) 89.
38. K. Scott, *J. Appl. Electrochem.* **13** (1983) 709.
39. M. Keddam, C. Rakomotavo, and H. Takenouti, *J. Appl. Electrochem.* **14** (1984) 437.
40. C. Cachet and R. Wiart, *J. Electroanal. Chem.* **195** (1985) 21.
41. A. Lasia, *J. Electroanal. Chem.* **428** 155 (1997).
42. A. Lasia, *J. Electroanal. Chem.* **454** 115 (1998).
43. A. Lasia, *J. Electroanal. Chem.* **500** (2001) 30.

44. H. Wendt, S. Rausch, and T. Borucinski, in *Advances in Catalysis*, Vol. 40, Academic Press, New York, 1994, p. 87.
45. R. Jurczakowski and A. Lasia, *J. Electroanal. Chem.* **582** (2005) 85.
46. S. Rausch and H. Wendt, *J. Appl. Electrochem.* **22** (1992) 1025.
47. A. J. Bard and L. R. Faulkner, *Electrochemical Methods*, Wiley, New York, 2001, p. 105.
48. D. D. Macdonaldm, M. Urquidi-Macdonald, S. D. Bhaktam, and B. G. Pound, *J. Electrochem. Soc* **138** (1991) 1359.
49. D. D. Macdonald, *Electrochim. Acta* **51** (2006) 1376.
50. S.-I. Pyun, C.-H. Kim, S.-W. Kim, and J.-H. Kim, *J. New. Mater. Electrochem. Syst.* **5** (2002) 289.
51. H. K. Song, Y. H. Jung, K. H. Lee, and L. H. Dao, *Electrochim. Acta* **44** (1999) 3513.
52. H. K. Song, H. Y. Hwang, K. H. Lee, and L. H. Dao, *Electrochim. Acta* **45** (2000) 2241.
53. H. K. Song, J. H. Sung, Y. H. Jung, K. H. Lee, L. H. Dao, M. H. Kim, and H. N. Kim, *J. Electrochem. Soc.* **151** (2004) E102.
54. H. K. Song, J. H. Jang, J. J. Kim, and S. M. Oh, *Electrochem. Commun.* **8** (2006) 1191.
55. L. M. Delves and J. Walsh, *Numerical Solution of Integral Equations*, Clarendon Press, Oxford, 1974.
56. F. Dion and A. Lasia, *J. Electroanal. Chem.* **475** (1999) 28.
57. P. Kowalczyk, S. Savard, and A. Lasia, *J. Electroanal. Chem.* **574** (2004) 41.
58. J. S. Newman, *Electrochemical Systems*, Second Edition, Prentice Hall, Englewood Cliffs, NJ, 1991.
59. S. Devan, V. R. Subramanian, and R. E. White, *J. Electrochem. Soc.* **151** (2004) A905.
60. J. P. Meyers, M. Doyle, R. M. Darling, and J. Newman, *J. Electrochem. Soc.* **147** (2000) 2930.
61. M. Doyle, J. P. Meyers, and J. Newman, *J. Electrochem. Soc.* **147** (2000) 99.
62. A. M. Svensson, L. O. Valøen, and R. Tunold, *Electrochim. Acta* **50** (2005) 2647.
63. T. E. Springer, T. A. Zawodzinski, M. S. Wilson, and S. Gottesfeld, *J. Electrochem. Soc.* **143** (1996) 587.
64. B. B. Mandenbrot, *The Fractal Geometry of the Nature*, Freeman, San Francisco, 1982.
65. R. de Levie, *J. Electroanal. Chem.* **281** (1990) 1.
66. L. Nyikos and T. Pajkossy, *Electrochim. Acta* **30** (1985) 1533.
67. H. von Koch, *Ark. Mat. Astron. Fys.* **1** (1904) 681.
68. A. Le Méhauté and G. Crépy, *Solid State Ionics* **9/10** (1983) 17.
69. A. Le Méhauté, G. Crépy, and A. Hurd, *C. R. Acad. Sci. Paris* **306** (1988) 117.
70. L. Nyikos and T. Pajkossy, *J. Electrochem. Soc.* **133** (1986) 2061.
71. L. Nyikos and T. Pajkossy, *Electrochim. Acta* **31** (1986) 1347.
72. T. Pajkossy and L. Nyikos, *Electrochim. Acta* **34** (1989) 171.
73. T. Pajkossy, *J. Electroanal. Chem.* **300** (1991) 1.
74. L. Nyikos and T. Pajkossy, *Electrochim. Acta* **35** (1990) 1567.
75. A. P. Borossy, L. Nyikos, and T. Pajkossy, *Electrochim. Acta* **36** (1991) 163.
76. A. Sakharova, L. Nyikos, and T. Pajkossy, *Electrochim. Acta* **37** (1992) 973.
77. T. Pajkossy, *Heterogen. Chem. Rev.* **2** (1995) 143.

78. E. Chassaing, R. Sapoval, G. Daccord, and R. Lenormand, *J. Electroanal. Chem.* **279** (1990) 67.
79. M. Filoche and B. Sapoval, *Electrochim. Acta* **46** (2000) 213.
80. T. Pajkossy and L. Nyikos, *J. Electroanal. Chem.* **332** (1992) 55.
81. M. Keddam and H. Takenouti, *Electrochim. Acta* **33** (1988) 445.
82. T. Pajkossy, *J. Electroanal. Chem.* **364** (1994) 111.
83. T. Pajkossy and L. Nyikos, *Electrochim. Acta* **34** (1989) 181.
84. R. de Levie and A. Vogt, *J. Electroanal. Chem.* **278** (1990) 25; **281** (1990) 23.
85. R. de Levie, *J. Electroanal. Chem.* **261** (1989) 1.
86. W. Mulder, *J. Electroanal. Chem.* **326** (1992) 231.
87. D. W. Davidson and R. H. Cole, *J. Chem. Phys.* **19** (1951) 1484.
88. J. R. Macdonald, *Impedance Spectroscopy*, Wiley, New York, 1987.
89. G. P. Lindsay and G. D. Davidson, *J. Chem. Phys.* **73** (1980) 3348.
90. J. B. Jorcin, M. E. Orazem, N. Pebere, and B. Tribollet, *Electrochim. Acta* **51** (2006) 1473.
91. K. S. Cole and R. H. Cole, *J. Chem. Phys.* **9** (1941) 341.
92. A. Sakharova, L. Nyikos, and Y. Pleskov, *Electrochim. Acta* **37** (1992) 973.
93. A. Kerner and T. Pajkossy, *J. Electroanal. Chem.* **448** (1998) 139.
94. Z. Kerner and T. Pajkossy, *Electrochim. Acta* **46** (2000) 207.
95. P. Żółtowski, *J. Electroanal. Chem.* **443** (1998) 149.
96. A. Sadkowski, *J. Electroanal. Chem.* **481** (2000) 222.
97. P. Żółtowski, *J. Electroanal. Chem.* **481** (2000) 230.
98. A. Sadkowski, *J. Electroanal. Chem.* **481** (2000) 232.
99. B. Emmanuel, *J. Electroanal. Chem.* **605** (2007) 89.
100. V. M. W. Huang, V. Vivier, M. E. Orazem, N. Pebere, and B. Tribollet, *J. Electrochem. Soc.* **154** (2007) C81.
101. V. M. W. Huang, V. Vivier, M. E. Orazem, N. Pebere, and B. Tribollet, *J. Electrochem. Soc.* **154** (2007) C89.
102. V. M. W. Huang, V. Vivier, I. Frateur, M. E. Orazem, and B. Tribollet, *J. Electrochem. Soc.* **154** (2007) C99.
103. J. Newman, *J. Electrochem. Soc.* **113** (1966) 501.
104. J. Newman, *J. Electrochem. Soc.* **113** (1966) 1235.
105. Z. Kerner and T. Pajkossy, *Electrochim. Acta* **47** (2002) 2055.
106. T. Pajkossy, T. Wandlowski, and D. M. Kolb, *J. Electroanal. Chem.* **414** (1996) 209.
107. T. Pajkossy, *Solid State Ionics* **94** (1997) 123.
108. T. Pajkossy and D .M. Kolb, *Electrochim. Acta* **46** (2001) 3063.
109. Z. Kerner, T. Pajkossy, L. A. Kibler, and D. M. Kolb, *Electrochem. Commun.* **4** (2002) 787.
110. Z. Kerner and T. Pajkossy, *Electrochim. Acta* **47** (2002) 2055.
111. T. Pajkossy, L. A. Kibler, and D. M. Kolb, *J. Electroanal. Chem.* **582** (2005) 69.
112. T. Pajkossy and D. M. Kolb, *Electrochem. Commun.* **9** (2007) 1171.
113. T. Pajkossy, L. A. Kibler, and D. M. Kolb, *J. Electroanal. Chem.* **600** (2007) 113.

4

Multiscale Mass Transport in Porous Silicon Gas Sensors

Peter A. Kottke, Andrei G. Fedorov and James L. Gole

Georgia Institute of Technology, G. W. Woodruff School of Mechanical Engineering, Atlanta, GA 30332-0405, USA
Petit Institute for Bioengineering and Bioscience, Atlanta, GA 30332-0405, USA

I. INTRODUCTION

Porous silicon (PS) is a material that has garnered considerable research attention over the past 15 years. It is formed by the dissolution of single crystalline silicon. The resulting material's morphology depends upon the silicon doping and the dissolution process. The dissolution process can be varied by changing the applied current and illumination, solvent conditions, and etching time, producing a diverse range of pore diameters (1–12) which can be made to vary from the 1 to 10 nm[2–6] range (nanoporous silicon) to sizes in the 1–3 µm range (9) (microporous silicon). Interestingly, different dissolution processes lead to very different pore sizes. One can fabricate a range of hybrid structures between two limiting well-defined PS morphologies: (1) PS fabricated from aqueous electrolytes which consists of highly nanoporous, structures, and (2) PS fabricated from nonaqueous electrolytes, which is comprised of open and accessible microporous structures with deep, wide, well-ordered channels that display a crystalline Si (100) influenced pyramidal termination.

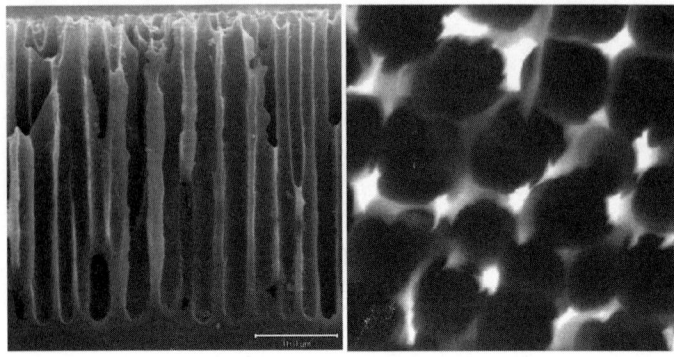

Figure 1. Close-up views of the hybrid porous silicon film: side and top view (from "Nano structures and Porous Silicon: Activity and Phase Transformation in Sensors and Photocatalytic Reactors" by J. L. Gole, S. E. Lewis, A. G. Fedorov, and S. Prokes, SPIE Proceedings Volume 5929, "Physical Chemistry of Interfaces and Nanomaterials IV", 2005).

The ability to control the interplay of these two regimes of porosity provides a means to exploit both the bulk and surface properties of the resulting porous membrane. In fact, the hybrid microporous/nanoporous structure etched into a silicon framework as depicted in Fig. 1, representing an extrapolation of the Probst and Kohl study (10), provides a useful platform for the construction of a conductometric PS-based sensor. All dissolution processes seem to result in mono- or bidisperse pore size distributions (13), with the typical diameters for the two sizes of pores being of the order $\sim 1\,\mu m$ and $< 20\,nm$. In this chapter, the larger ($\sim 1\,\mu m$ pores) will be called micropores, and the smaller ($< 20\,nm$) pores will be called nanopores. This terminology is not universal! For monodisperse pore diameter porous silicon, either micro or nanopores may be present. Because the synthesis conditions that lead to a given morphology have been much perfected, reproducible PS production is now possible, a feature that is necessary for practical utility.

The physico-chemical pheomenon that initially generated intense interest in porous silicon is its visible photoluminescence (PL) at room temperature excited by UV light (14–16), and much early research has focused on the basic physics of and applications for the porous silicon PL. Another important characteristic of porous silicon is its compatibility and ease of integration with silicon based microchip devices. Thus, porous silicon offers the potential for a low cost easily fabricated device that can be combined with silicon based circuits.

Interest in using porous silicon for sensors was motivated by its high accessible surface area (per unit volume), which is a result of its open branching-porous morphology. It is also found that porous silicon's electrical properties (e.g., electrical impedance) change as a result of the introduction and adsorption of different chemicals on the PS surface. These characteristics of porous silicon, such as (i) electrical and luminous property variation in response to the presence of specific chemical species, (ii) the promise of manufacturing simplicity, and (iii) good sensitivity (as a result of a high surface area to volume ratio), make it a potentially useful substrate material for chemical sensing devices.

As is the case for all chemical sensors, the response of porous silicon gas sensors is generated upon interaction of a target chemical (analyte species A) with the chemical recognition interface, either the silicon surface or another material intimately coupled to the silicon. This interaction occurs in nanopores of the PS gas sensors, and is often facilitated by the presence of a metal or metal oxide.

The main purpose of this chapter is to review transport processes underlying the transient response of PS chemical sensors and to provide theoretical tools that will facilitate the formulation of mathematical models of porous silicon gas sensors. The resulting first principles models consist of governing mass transfer equations which can be solved analytically in some cases, or for more complex configurations and with sufficient computing resources, can be solved through numerical simulations. A motivation for formulating such models is that at present both interpretation of results and design of the sensors are in some respects empirical. For example, recent purely experimental investigations (17) of the effect of changing the sensor perimeter while maintaining its area and porosity the same demonstrate the need for improved understanding of mass transport to guide sensor development efforts.

A notable advantage obtained through the use of a first principles model is that calibration becomes a measurement of transport properties and surface kinetics parameters, in addition to the properties inherent to the physics of the relevant sensing mechanism. Because the results of calibration are quantities with specific meanings in a physics-based model, the resulting description of sensor response can be used to predict changes in sensor behavior as experimental conditions or sensor parameters are varied, thus enabling targeted optimization efforts and improved data interpretation.

A secondary purpose of this chapter is to demonstrate routes to further simplification so that *analytical* or at least *semi-analytical* solutions to the mass transport model may be possible. In general, the solution of a mathematical model of the sensor system is too complex and can only be accomplished numerically using, for instance, finite volume, finite element, or finite difference techniques, implemented through either commercially available or user-written computer codes. When using numerical simulation, for any given combination of input parameters and sensed species concentrations, only a single prediction of the resulting sensor output is obtained. The resulting parametric space, even after nondimensionalization, is large, filling this space is time consuming and tedious, and extracting useful insight from the results is difficult. To be useful to practitioners, a theoretical tool must be accurate yet simple enough not to require special skills for implementation, and it must be expressed in terms of meaningful and accessible physical parameters. Analytical solutions are therefore profoundly more useful than numerical simulations, and it is often worth introducing carefully validated simplifications in the model to obtain them. We will demonstrate application of such analytical approaches in this chapter in detail.

1. Mass Transport

When describing behavior of the PS chemical sensor, one is concerned with the mass transport of a representative analyte species, A, which is one of many chemical species within a gaseous mixture. The transport occurs across two or three length scales, e.g., Figs. 2 and 3. There are two relevant length scales if only a single pore size is characteristic of the PS morphology (e.g., purely nano-PS), and three length scales need to be considered if the porous silicon has two pore sizes (e.g., hybrid micro-nano PS, see Figs. 1, 2 and 3). We break our description down into

(1) Macroscale transport, which is analyte mass transport in the bulk gas mixture, to and from the porous silicon, within the sensor flow cell

Multiscale Mass Transport in Porous Silicon Gas Sensors

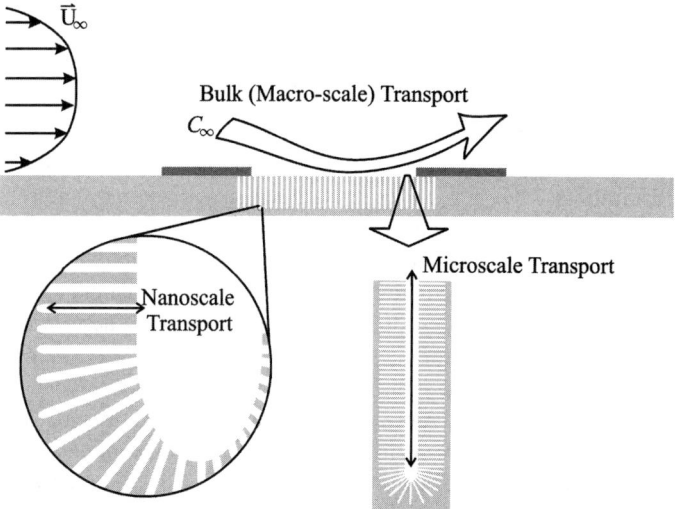

Figure 2. Schematic representation of porous silicon gas sensors highlighting the multiple length scales for transport from the bulk to the sensing surface in the nanopores.

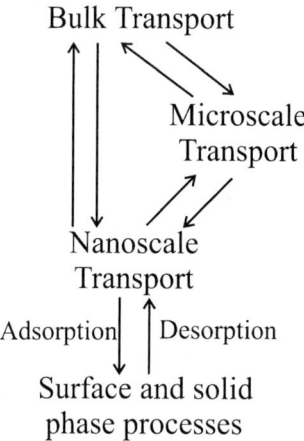

Figure 3. Transport process hierarchy for hybrid (including microporous) and nanoporous PS.

(2) Microscale transport, which is transport through the micropores from the bulk to the nanopores, and
(3) Nanoscale transport, which is transport within the nanopores to the sensing interface of PS

We allow for purely nanoporous silicon. We do not include the possibility of micropores without nanopores, as this morphology is not very applicable for sensors. The impact of the existence of multiple length scales is twofold: the same numerical simulation cannot be uniformly applied across drastically different length scales; and, the dominant physics (and resulting mathematical description) is different for the macro, micro, and nanoscale.

It should be noted that the transport of the gaseous analyte to the porous silicon surface represents only the first stage of the processes that result in transient sensor response. The response may depend also on how the analyte species alters the porous silicon to produce a signal. It is also possible that the interaction of the analyte species with the porous silicon can have feedback into the mass transport. Additionally, other components of the gas mixture might directly or indirectly affect the sensor response. Other species could directly poison (permanently de-activate) the sensing interface or elicit an interfering response from the sensor, i.e., the sensor may not be perfectly selective. Also, other species could directly reduce the sensor response, for instance, by competing with the analyte for vacant surface sites for adsorption. Furthermore, other species could affect the sensor response by indirectly affecting the transport of the analyte species in the gas or on the surface.

It is instructive to keep in mind a very simple description of a typical sensor response, considering the example sensor shown in Fig. 2. The active portion of the sensor is the porous silicon surface. The sensor response is assumed to be related to the amount of analyte species adsorbed on the active surface (typically given in terms of fractional coverage, θ, which is the ratio of occupied adsorption sites to the total number of available sites. This adsorbed amount will change as a result of changes in the bulk concentration of the analyte species. The timescale for the sensor response will depend on the analyte transport from the bulk to the PS surface, and thus an ability to predict it is a major goal of the analysis. Provided changes in the bulk concentration of the analyte are sufficiently slow (when compared to the sensor response timescale), the sensor will eventually give a steady-state signal, which will correspond to equilibrium conditions (not to be confused with sensor saturation). In other words, the steady-state sensor response will correspond to conditions of zero net mass transport to the active surface, and the rates of adsorption and desorption of the analyte species will be balanced.

II. MACROSCALE (BULK) TRANSPORT

Modeling of mass transport at the macroscale is very well developed. Understanding the application of well-demonstrated mass transfer principles to bulk transport in porous silicon sensors is, therefore, mostly an exercise in correctly simplifying the formulation, understanding the limitations of the simplifying assumptions, and correctly linking the macroscale transport to the micro- and nanoscale transport. In any case, idealizations need to be made, and the determination of whether or not they are acceptable requires that we answer two questions: (1) What benefits do the idealizations provide? and, (2) What is the cost of the idealizations? Usually the answer to the first question is *simplicity*: simplicity of formulation; simplicity of solution; and, simplicity of interpretation. The answer to the second question is *accuracy*: quantifiable and reasonably good accuracy for an acceptable model. If the idealization results in a loss of physical realism, the cost is unequivocally unacceptable. In our general formulation of macroscale mass transport we apply idealizations that have been demonstrated to provide considerable benefit with little cost (18–21), as long as certain conditions are met.

1. General Formulation for Macroscale Mass Transport

The starting point for the formulation of a mathematical description of bulk mass transport is a statement of mass conservation under the continuum assumption (22, 23). We will limit ourselves to the consideration of those cases for which there is no homogeneous chemical reaction involving analyte of interest. However, it should be mentioned that chemical sensors can certainly benefit from exploiting bulk reactions cleverly. Such reactions can enhance selectivity, sensitivity, and resolution in some cases, but their utility in porous silicon gas sensors has not been demonstrated (or perhaps has not yet been investigated). Mass conservation for the analyte species is given by

$$\frac{\partial c_A}{\partial t} = -\nabla \cdot \vec{J}_A, \quad (1)$$

where c_A is the mass concentration of species A, t is time, and $\vec{J}_A$ is the local mass flux of species A (a vector). Other than the continuum assumption and the neglect of homogeneous chemical reactions, no idealizations are present in (1).

Next we split the mass flux vector into two components: an advective component, $\vec{J}_{A(\text{adv})}$; and a generalized diffusion component, $\vec{J}_{A(\text{diff})}$. The advective component is simply the flux of the analyte due to bulk motion of the gaseous mixture, and is given by

$$\vec{J}_{A(\text{adv})} = \vec{u} c_A, \quad (2)$$

where $\vec{u}$ is the mass averaged, or bulk, velocity of the gaseous mixture. The generalized diffusion component, under the assumption of sufficiently small gradients, is proportional to the gradient of a generalized chemical potential

$$\vec{J}_{A(\text{diff})} = -\alpha_A \nabla \mu_A, \quad (3)$$

where the proportionality "constant", α_A, and the potential, μ_A, could typically be functions of species concentrations, temperature, pressure, and the strength of applied force fields. The chemical potential in (3) is defined by the conditions needed for equilibrium (24). A complete formulation would next include equations of conservation of energy (typically expressed as a thermal energy balance and relevant for non-isothermal operation) conservation of momentum (required to predict the flow field), and all necessary constitutive relationships. The resulting equations are, in most cases, linked partial differential equations; however, in some cases, such as non-negligible thermal radiation effects, the nature of the equations is more complex, e.g., integro-differential. The partial differential equations describing mass, momentum, and energy conservation require boundary and initial conditions. The resulting system of equations comprises the mathematical model of the physical phenomena. With no further simplification this physical model is complete and accurate; however, it is too complex to be very useful. Fortunately, porous silicon gas sensors can be described accurately after making considerable further idealizations, and solving the momentum and energy equations can usually be avoided. We simplify the model by assuming

(1) Temperature variations are sufficiently small so as to have no effect on species mass fluxes or bulk flow.
(2) Concentration variations do not produce buoyancy effects in the bulk flow.
(3) The chemical potential μ_A can be described using the ideal solution approximation, and all external fields are negligible.
(4) The resulting diffusive mass transport is Fickian (binary), with a constant diffusion coefficient D_A (typically $\sim 10^{-5}$ m^2 s^{-1} for gases at STP) (25).

With assumptions (1–4), the diffusive mass flux of the analyte species, i.e., (3) is given by

$$\vec{J}_{A(\text{diff})} = -D_A \nabla c_A \quad (4)$$

and combining (1), (2) and (4) yields

$$\frac{\partial c_A}{\partial t} = -\nabla \cdot \left(\vec{u} c_A\right) + D_A \nabla^2 c_A. \quad (5)$$

Provided flow velocities are small compared to the speed of sound, the bulk flow is incompressible, i.e., $\nabla \cdot \vec{u} = 0$, allowing further simplification of (5), resulting in a standard transient advection–diffusion equation for the macroscopic mass transport

$$\frac{\partial c_A}{\partial t} = -\vec{u} \cdot \nabla c_A + D_A \nabla^2 c_A. \quad (6)$$

Equation (6) can be generally applied to many problems of single or multi-species gas transport, including chemical sensing applications (20). More general descriptions of multicomponent diffusion in chemical sensors are given by Phillips and Fedorov, who discuss the limits of the binary Fickian diffusion approximation (19). These limits should be checked when there is an interfering species, I, that is a species other than the analyte of interest which is affected by the sensing mechanism, and hence develops appreciable concentration gradients in the sensor. In such situations, the binary diffusion model is not applicable when

(1) The ratio of the interfering species' Maxwell–Stefan diffusivity in the inert gas, D_{IG}, to the analyte species' Maxwell–Stefan diffusivity in the inert gas, D_{AG}, is very large, i.e. $D_{IG}/D_{AG} \gg 1$ (this condition is not likely to be found is PS sensors).
(2) The baseline concentrations of both the analyte and interfering species are at least 100 ppm *and* the concentration of one of the species is two orders of magnitude greater than that of the other.

For more general information the reader is referred to an excellent review of Maxwell–Stefan diffusion by Krishna and Wesselingh (26).

Equation (6) can be further simplified under certain limiting conditions: the bulk concentration may be steady-state (or quasi-steady-state); the effects of bulk flow may be negligible; and, the concentration variations in the direction normal to the sensing element may be negligible. The determination of the appropriateness of any of the resulting simplifications can be accomplished using the method of scale analysis.

2. Scale Analysis for Macroscale Transport

With scales for length, L_s, time, t_s, velocity, u_s, and analyte species concentration, c_s, the governing equation, (see (6)) takes on the nondimensional form

$$\nabla^{2*} c_A^* = \frac{L_s^2}{D_A t_s} \frac{\partial c_A^*}{\partial t^*} + \frac{u_s L_s}{D_A} \vec{u}^* \cdot \nabla^* c_A^*. \tag{7}$$

The asterisks denote dimensionless quantities, i.e., $\nabla^{2*} \equiv L_s^2 \nabla \cdot \nabla$, $c_A^* \equiv c_A/c_s$, $t^* \equiv t/t_s$, $\vec{u}^* \equiv u_s^{-1} \vec{u}$, and $\nabla^* \equiv L_s \nabla$.

The choice of correct scales is important when recognizing possible simplifications to (7). If the scales are chosen correctly, they will form dimensionless groups which multiply terms of $O(1)$ (see Ref. (27)). The choosing of scales and their use in simplifying (7) are discussed next for the conditions leading to a steady-state purely diffusional model for macroscale mass transport (the simplest case, but the easiest to understand, and hence most instructive).

i. Example: Advective timescale, small Peclet number

An example of the simplification of (7) is obtained for the case that the appropriate timescale is an advective timescale

$$t_s = \frac{L_s}{u_s}. \tag{8}$$

Then (7) becomes

$$\nabla^{2*} c_A^* = Pe \left(\frac{\partial c_A^*}{\partial t^*} + \vec{u}^* \cdot \nabla^* c_A^* \right), \tag{9}$$

where Pe is the Peclet number for mass diffusion, $Pe = \frac{u_s L_s}{D_A}$. Thus the quasi-steady-state purely diffusional governing equation

$$\nabla^{2*} c_A^* = 0 \tag{10}$$

is appropriate when the timescale is the advective timescale and the Peclet number is small. In fact, from (9) and (10), it is clear that the solution to the steady-state diffusion problem is the leading order asymptotic solution of (9) for $Pe \to 0$. A difficulty that arises with asymptotic solutions is that their accuracy is, in general, unknown without computation of higher order terms; notationally this concept is expressed using the "little o" (28). The questions that one would hope to have answered prior to using (10) are therefore

(1) What are the appropriate length and velocity scales
(2) Under which conditions is the timescale for the mass transport problem defined by (8) appropriate; and
(3) How small must the Peclet number be to make the leading order solution (i.e., steady-state, pure diffusion) accurate

If the assumption that advection is negligible is correct, then the relevant length scale, L_s, is the diffusional length scale, which will be of the order of a characteristic length for the largest source of perturbation to the concentration. For the porous silicon gas sensor, L_s should be a characteristic dimension of the region within which the analyte species is removed from the bulk. The velocity scale, u_s, should be the largest characteristic velocity in the region of the concentration disturbance.

There are several possible time scales for the problem. For mass diffusion the time scale is $t_s = L_s^2/D_A$. All time varying boundary conditions yield a time scale that depends on the boundary condition and the form of its variation with time. For example, a pulsing gas velocity (29) would have a time scale which is the reciprocal of the pulsing frequency. Particularly important for the problem at hand is the time scale for the perturbation to the concentration at the porous silicon surface. This time scale would be either the time scale for a nanopore or micropore response, t_n and t_μ, respectively. A final time scale is the physical time elapsed following the initial disturbance at t_0, $t_s = t - t_0$. The time scale in (8) is the appropriate time scale when the boundary conditions are not time varying (or are varying slowly) and $L_s^2/D \ll L_s/u_s, t - t_0$, or, in terms of dimensionless parameters

$$Pe \ll 1 \text{ and } Fo_c \gg 1. \tag{11}$$

In addition to the Peclet number, Pe, which was defined previously, the new nondimensional group in (11) is the mass diffusion Fourier number, $Fo_c = D(t - t_0)/L_s^2$.

The answer to the question "How small must Pe be" for the steady-state diffusion formulation to be accurate is, strictly speaking, dependent upon the case at hand. Experience suggests that if $Pe < 0.1$ the approximation can be made with some confidence (18). Previously we remarked that this example is trivial. That is because the true steady-state solution of the problem, (see (10)) with vanishing mass flux boundary conditions on all surfaces is the equilibrium solution, i.e., the concentration has everywhere its bulk value. However, there is the possibility that the micro or nanopore responses are slow, i.e., $L_s^2/D \ll t_\mu, t_n$. Then, the macroscale transport problem could be quasi-steady-state, and (10) would apply with a non trivial result.

ii. Example: Advective timescale, large Peclet number

In the previous section we considered the simplification to the governing equation for transport of the analyte species from the bulk to the porous silicon surface when the relevant time scale is the advective timescale, and the Peclet number is small. Naturally, the question arises, what happens when the Peclet number is large? Physically, a large Peclet number corresponds to instantaneous transport of the analyte species to the sensor surface without any diffusive limitations (zero thickness boundary layer for mass transport). The governing equation,

$$\frac{\partial c_A^*}{\partial t^*} + \vec{u}^* \cdot \nabla^* c_A^* = 0 \qquad (12)$$

is purely advective, so that concentration at the sensor surface is influenced only by changes in the upstream bulk concentration.

iii. Example: Diffusive timescale, no advection

In the absence of a forced flow of the bulk gas, the advective timescale is not defined and clearly does not apply. The relevant timescales therefore become

(1) The time scale for the perturbation to the concentration at the porous silicon surface based on nanopore or micropore response, i.e., t_n and t_μ, respectively.
(2) The observation timescale $t_{\text{obs}} = t - t_0$.
(3) The diffusion time scale, $t_{\text{diff}} = L_s^2 / D_A$.

Possible timescale hierarchies resulting in simplification are

$$\text{Case (a) } t_{\text{obs}} \gg t_{\text{diff}}, \left(t_\mu \text{ or } t_n\right)$$
$$\text{Case (b) } t_{\text{diff}} \gg t_{\text{obs}}, \left(t_\mu \text{ or } t_n\right)$$
$$\text{Case (c) } \left(t_\mu \text{ or } t_n\right) \gg t_{\text{diff}}, t_{\text{obs}}$$

Case (a) implies that steady-state conditions have been achieved, (10) is the appropriate governing equation, and for the porous silicon sensor the equilibrium condition has been achieved. Case (b) corresponds the limit of small Fourier number, $Fo_c = Dt_{\text{obs}}/L_s^2 \ll 1$, where the response is governed by the transient diffusion equation

$$\frac{\partial c_A^*}{\partial t^*} = \nabla^{*2} c_A^*. \qquad (13)$$

Here, time has been made nondimensional with a diffusive time scale, $t^* = tD_A/L_s^2$. Case (c) is the quasi-steady-state case, for which (10) is the governing equation but the boundary conditions vary, and thus the concentration is not at equilibrium yielding transient sensor response on the t_μ or t_n timescale.

III. NANOSCALE TRANSPORT

The active surface for sensing in porous silicon gas sensors is most probably the surface of the nanopore. Also, although this is not the ideal configuration (16, 30–45), a porous silicon gas sensor may be nanoporous, with no micropores (Fig. 3). Therefore we present the description of nanoscale transport first, before that for the microscale. Recall that in our definition the nanopore diameters are ~ 10 nm. We consider morphologies in which the ends of the nanopores are closed, so that there is no net hydrodynamic flow into the pores. Thus advective transport is neglected. The nanopores are presumably long (as compared to pore diameter), with lengths $L_n \sim 1\,\mu$m. We assume that no Kelvin condensation occurs in the nanopores, so the transport is always in gas phase. Describing the transport within the nanopores requires that we first answer two fundamental questions: is the continuum assumption valid, and if so, is the diffusion Fickian?

1. Continuum Assumption for Nanopores

Under the continuum assumption, physical, chemical, and thermodynamic quantities are uniformly distributed within infinitesimally small volumes, thus allowing them to be described by a value at a "point" (46). The alternative is typically to adopt a molecular approach. Here, specific interactions occurring at the atomic level are the starting point from which one builds up to a larger picture (24). The most accepted justification for the adoption of the continuum assumption is that the characteristic system/process length scale is large when compared to the characteristic molecular length scale (46). Alternatively, the continuum assumption is considered acceptable when the characteristic time scale for the described phenomena is large when compared to molecular time scales (47). Thus, it is necessary to identify the process length and time scales, as well as the molecular length and time scales to determine an appropriate formalism to describe transport in nanopores.

The most obvious length scales for transport in nanopores are pore length and radius, L_n and R_n, respectively. For comparison to molecular length scales the most conservative choice for system length scale is the pore radius, $R_n \sim 10$ nm (13). The molecular length scale to which this should be compared is the mean free path between collisions in the gas phase, λ, which is given by (in the limit of ideal gas approximation)

$$\lambda \sim \frac{\mu}{P}\sqrt{\frac{\Re T}{M}}, \qquad (14)$$

where μ is the viscosity, $\Re$ is the universal gas constant, T is the temperature, P is the pressure and M is the mole average molecular weight of the gas mixture (48). Assuming air at standard conditions (with only trace quantities of the analyte), yields $\lambda \sim 50$ nm. The molecular time scale, t_{mol}, is the mean time between collisions, which, for air at equilibrium under standard conditions, is $t_{\text{mol}} \sim 10^{-10}$ s (48). Candidate system time scales include the time scale for mass transport in the gas phase, t_{diff}, the adsorption time scale, t_{ads}, the desorption time scale, t_{des}, and the time scale for the externally imposed concentration variation, t_{force}. As we will see in the next section, we can define an effective diffusion coefficient, $D_{\text{eff}} = D_{\text{eff}} = R_n \sqrt{\Re T/M}$, which, when combined with the pore radius as the system length scale, gives the mass transport time scale,

$$t_{\text{diff}} = \frac{L_s^2}{D_{\text{eff}}} \sim \frac{R_n}{\sqrt{\Re T/M}} \sim 30\,\text{ps}. \qquad (15)$$

Adsorption and desorption time scales vary depending upon substrate surface and analyte species, but they typically fall in the range $t_{\text{ads}} \sim 10^{-6}$ to 10^{-3} s (49). External forcing time scales are typically $t_{\text{force}} \sim 1$ s or longer (29).

Considering the length and time scales above, it is evident that the standard justifications for the use of the continuum assumption are not applicable. The requirement that the characteristic system length scale is large when compared to the characteristic molecular length scale is equivalent to the requirement that the Knudsen number, $Kn \equiv (\mu/PR_n)\sqrt{\Re T/M}$, be small (50). In fact, in PS nanopores, $Kn \approx 5$. Similarly (and not coincidentally) the ratio of the mass transport time scale to the average time between collisions is $t_{\text{diff}}/t_{\text{mol}} \approx 0.3 \sim Kn^{-1}$. Thus, it would appear that the continuum assumption should not be applied and that the governing equation for mass transport should be the more general Boltzmann transport equation (48, 50). However, it has been demonstrated that a Fickian-like continuum model can still be used to describe mass transport in gasses at moderate Knudsen numbers provided the diffusion coefficient is appropriately modified to account for rarefaction effects (50, 51).

2. Diffusion Model for Nanopores

In the discussion on the application of the continuum assumption we introduced the concept of the Knudsen number, Kn, which is the ratio of the average distance traveled by a molecule in the gas phase prior to collision with another gas molecule, to the average distance traveled by a molecule in the gas phase prior to collision with a solid boundary. We found that for nanopores this ratio will typically be greater than unity. Thus, traditional Fickian "bulk" diffusion, in which Brownian motion induced by multiple, frequent intermolecular collisions drives the mass transport, is not a good model. Knudsen diffusion is a description of mass transport based on the concept that a gas molecule that hits a solid surface is instantly adsorbed and then diffusely desorbed (Fig. 4) (25). The resulting Knudsen diffusion coefficient

$$D_{Kn} = R_n \sqrt{\Re T/M} \qquad (16)$$

depends upon the pore radius, R_n, and is independent on the gas pressure (density). As the radius shrinks, more frequent collisions with the wall impede transport, and D_{Kn} becomes smaller.

Adsorbed analyte molecules can also move through a thermally activated "hopping" process termed surface diffusion (26, 51, 52). The driving force is a gradient in surface coverage, which is linked

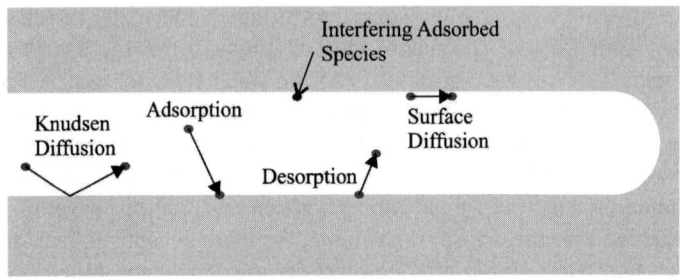

Figure 4. Nanoscale transport processes for PS.

to the concentration in the gas phase by the kinetics for fast (relative to adsorption/desorption) gas phase transport processes, and through the equilibrium relationships for slow processes. A surface diffusion coefficient can be defined, and its typical values for gasses at 300 K are $D_{\text{surf}} \sim 10^{-10}\,\text{m}^2\,\text{s}^{-1}$. Surface diffusion provides a parallel path for mass transport with combined Knudsen and bulk diffusion. Thus, the dominant diffusion mode can be found by first finding the smaller of D_{Kn} and D_A and comparing that value to D_{surf}. For nanopores from 1 to 10 nm in radius, $D_{Kn} \sim 10^{-7}$ to $10^{-6}\,\text{m}^2\,\text{s}^{-1}$, which suggests that Knudsen diffusion is the dominant mass transport mode.

3. Simplified Model for Nanopore Mass Transport

Based on the assumption that advection is negligible within the nanopores, and the recognition that the dominant transport mode is Knudsen diffusion, the nondimensional governing equation for mass transport of analyte species in the nanopore is

$$\frac{\partial c_A^*}{\partial t^*} = \nabla^{*2} c_A^*, \tag{17}$$

where time has been made nondimensional using a diffusive timescale based on the Knudsen diffusion coefficient and the nanopore radius, $t^* = t D_{Kn}/R_n$. In our simplified model, we treat the nanopore as a cylinder, and assume uniform surface properties and a uniform (though perhaps time varying) concentration at the cylinder opening (Fig. 2). The governing equation for the resulting axisymmetric model can be expressed in cylindrical coordinates

$$\frac{\partial c_A^*}{\partial t^*} = \frac{1}{r^*}\frac{\partial}{\partial r}\left(r^* \frac{\partial c_A^*}{\partial r^*}\right) + \frac{\partial^2 c_A^*}{\partial z^{*2}}. \tag{18}$$

In addition to the imposed boundary condition at $z^* = 0$ of a time varying, spatially uniform concentration

$$c_A^*\left(r^*, 0, t^*\right) = f\left(t^*\right) \tag{19}$$

boundary conditions are needed at the side and end walls, as well as at the central axis of the nanopore.

The boundary conditions at the side and end walls require that the rate of Knudsen diffusion to the wall is balanced by the difference between the rate of adsorption, j_a, and desorption, j_d

$$-D_{Kn}\frac{\partial c_A}{\partial r}(R_n, z, t) = j_\text{a} - j_\text{d} \qquad (20)$$

and

$$-D_{Kn}\frac{\partial c_A}{\partial z}(r, L_n, t) = j_\text{a} - j_\text{d}. \qquad (21)$$

Under the assumptions that adsorption sites are all equivalent and uniformly distributed on the surface, that each adsorbed molecule occupies a single site, that there is no interaction between adsorbed molecules, and that there are no interfering species, the rate of adsorption is given by $j_\text{a} = k_\text{a} c_s c_A^* N (1 - \theta)$ where j_a is the local rate of removal of species A from the gas by adsorption at time t^* at axial position z^*, k_a is the rate coefficient for adsorption, N is the number of active sites per unit area of surface, and θ is the fraction of active sites per unit area that are filled with the adsorbed species. With the same assumptions, the rate of desorption, j_d, is given by $j_\text{d} = k_\text{d} N \theta$. This adsorption model gives, at equilibrium, the Langmuir isotherm for fractional coverage: $\theta_\text{eq} = K c_s c_A^* / (1 + K c_s c_A^*)$ where $K = k_\text{a}/k_\text{d}$. Using the selected adsorption model, the boundary conditions at the walls, (see (20) and (21)) become

$$-\frac{\partial c_A^*}{\partial r^*} = \frac{k_\text{a} R_n N}{D_{Kn}} c_A^* [1 - \theta(z^*, t^*)] - \frac{k_\text{d} R_n N}{c_s D_{Kn}} \theta(z^*, t^*) \qquad (22)$$

and

$$-\frac{\partial c_A^*}{\partial z^*} = \frac{k_\text{a} R_n N}{D_{Kn}} c_A^* [1 - \theta(r^*, t^*)] - \frac{k_\text{d} R_n N}{c_s D_{Kn}} \theta(r^*, t^*). \qquad (23)$$

The boundary condition at the centerline $r^* = 0$ is simply a statement of symmetry,

$$\frac{\partial c_A^*}{\partial r^*} = 0. \qquad (24)$$

With a prescribed, axisymmetric initial condition,

$$c_A^*(r^*, z^*, 0) = g(r^*, z^*). \qquad (25)$$

Equations (18), (19), and (22)–(25) when combined with equations describing the change in fractional coverage (see Sect. V) comprise a well posed, first principles based, simplified model of analyte transport-adsorption in a nanopore. They can be modified in a straightforward manner to accommodate other adsorption models (e.g., the Freundlich adsorption model) (53). In keeping with the goal of seeking insight through rational idealization, we will simplify the model even further.

4. Nanopore Transient Response

We are interested in obtaining information about the time scale for the response of the nanopores to changes in concentration at their opening. To quantify this behavior analytically, we consider the case of a nanopore that is initially devoid of the analyte, i.e., $g(r^*, z^*) = 0$ in (25). A step change is applied to the pore opening, and the value to which the concentration is raised is chosen as the concentration scale, so that $f(t^*) = 1$ in (19). At least for short times, while the fractional coverage, $\theta(z^*, t^*)$, is small, only adsorption (no desorption) occurs at an appreciable rate, so that the right hand sides of (22) and (23) become $\frac{k_a R_n N}{D_{Kn}} c_A^*$. Finally, we assume that, due to the very small radius of the nanopore, the concentration variation in the radial direction will be small. Thus, the difference between the concentration at the wall and a cross-sectional area weighted average concentration will be negligible. Integrating (18) with respect to r^*, from $r^* = 0$ to $r^* = 1$ (we have chosen the pore radius as the length scale, and the integration is performed after first multiplying throughout by r^*) we obtain

$$\frac{\partial \bar{c}_A}{\partial t^*} = \frac{\partial^2 \bar{c}_A}{\partial z^{*2}} + \left(\frac{\partial c_A^*}{\partial r^*}\right)_{\text{surface}(r^*=1)}, \tag{26}$$

where $\bar{c}_A \equiv \int_0^1 r^* c_A^* dr^*$ is the aforementioned average concentration. Applying the boundary condition at the wall, i.e., (22) with the RHS modified for small fractional coverage, (26) becomes

$$\frac{\partial \bar{c}_A}{\partial t^*} = \frac{\partial^2 \bar{c}_A}{\partial z^{*2}} - \frac{k_a R_n N}{D_{Kn}} c_A^*(1, z^*, t^*). \tag{27}$$

Having removed the radial coordinate from the governing equation, we now rescale the problem, using the pore length L_n as the length scale, so that $\bar{z} \equiv z/L_n = z^*/\beta$ and $\bar{t} \equiv t D_{Kn}/L_n^2 = t^*/\beta^2$. We also apply our requirement that $\bar{c}_A(z^*, t^*) \approx c_A^*(1, z^*, t^*)$. The resulting rescaled problem is

$$\frac{\partial \bar{c}_A}{\partial \bar{t}} = \frac{\partial^2 \bar{c}_A}{\partial \bar{z}^2} - \alpha^2 \bar{c}_A, \tag{28}$$

where $\alpha^2 = k_a L_n N / D_{Kn}$. The boundary conditions and initial conditions for (28) are

$$\bar{c}_A\left(0, \bar{t}\right) = 1, \tag{29}$$

$$\frac{\partial \bar{c}_A}{\partial \bar{z}}\left(1, \bar{t}\right) = -\alpha^2 \bar{c}_A\left(1, \bar{t}\right) \tag{30}$$

and

$$\bar{c}_A\left(\bar{z}, 0\right) = 0. \tag{31}$$

Considering (28)–(31), we recognize two distinct possibilities for the transient response of the nanopore, an adsorption timescale

$$t_n \sim \frac{L_n}{k_a N} \tag{32}$$

and a diffusion timescale

$$t_n \sim \frac{L_n^2}{D_{Kn}}. \tag{33}$$

We will see that these timescales also appear if we apply a more rigorous analytical solution method. Details of this solution method are now considered.

5. Analytical Solution

Because we are concerned with the initial transient response of the pores, and because it does not affect the main result of our analysis, we replace (30) with a condition of no flux at the pore end (essentially neglecting analyte adsorption at the pore termination point as compared to much larger side walls)

$$\frac{\partial \bar{c}_A}{\partial \bar{z}}\left(1, \bar{t}\right) = 0. \tag{34}$$

Equations (28), (29), (31) and (34) can be solved analytically by the equivalent methods of separation of variables or eigenfunction expansion (54–56).

First we find the solution to the related steady-state problem, i.e.,

$$0 = \frac{d^2 c_{ss}}{d\bar{z}^2} - \alpha^2 c_{ss}, \tag{35}$$

$$c_{ss}(0) = 1, \tag{36}$$

$$\frac{\partial c_{ss}}{\partial \bar{z}}(1) = 0. \tag{37}$$

The solution to (35) and (36) is

$$c_{ss} = \cosh(\alpha \bar{z}) - \tanh(\alpha) \sinh(\alpha \bar{z}). \tag{38}$$

Then, by letting $\bar{c} = c_{ss} + c_H$, we obtain a differential equation for c_H

$$\frac{\partial c_H}{\partial \bar{t}} = \frac{\partial^2 c_H}{\partial \bar{z}^2} - \alpha^2 c_H \tag{39}$$

with homogeneous boundary conditions

$$c_H(\bar{t}, 0) = 0, \tag{40}$$

$$\frac{\partial c_H}{\partial \bar{z}}(\bar{t}, 1) = 0 \tag{41}$$

and an initial condition derived from combining (31) and (38)

$$c_H(0, \bar{z}) = -\cosh(\alpha \bar{z}) + \tanh(\alpha) \sinh(\alpha \bar{z}). \tag{42}$$

To solve the partial differential equation we first solve the related eigenvalue problem

$$X'' - \alpha^2 X = -\mu X, \tag{43}$$

$$X(0) = 0, \tag{44}$$

$$X'(1) = 0 \tag{45}$$

with resulting eigenfunctions given by

$$X_n = \sin(\lambda_n x), \tag{46}$$

$$\lambda_n = \pi \left(n + \tfrac{1}{2}\right) \quad n = 0, 1, 2..\infty, \tag{47}$$

$$\mu_n = \lambda_n + \alpha^2. \tag{48}$$

An orthogonal series expansion solution is found by projecting the initial condition, (see (42)) onto the span of eigenfunction basis $\{\sin(\lambda_n x)\}_{n=0}^{\infty}$:

$$p\{c_H(0, \bar{z})\} = -\sum_{n=0}^{\infty} b_n \sin(\lambda_n \bar{z}) + \tanh(\alpha) \sum_{n=0}^{\infty} d_n \sin(\lambda_n \bar{z}), \tag{49}$$

where p denotes the projection, e.g.,

$$c_H\left(\bar{t}, \bar{z}\right) = p\left\{c_H\left(\bar{t}, \bar{z}\right)\right\} = \sum_{n=0}^{\infty} \omega_n\left(\bar{t}\right) \sin\left(\lambda_n \bar{z}\right). \tag{50}$$

The coefficients in (49) are

$$b_n = \frac{\langle \cosh(\alpha\bar{z}), \sin(\lambda_n\bar{z})\rangle}{\langle \sin(\lambda_n\bar{z}), \sin(\lambda_n\bar{z})\rangle} \tag{51}$$

and

$$d_n = \frac{\langle \sinh(\alpha\bar{z}), \sin(\lambda_n\bar{z})\rangle}{\langle \sin(\lambda_n\bar{z}), \sin(\lambda_n\bar{z})\rangle}. \tag{52}$$

The inner product in (51) and (52) is defined as

$$\langle F, G\rangle = \int_0^1 F G \, d\bar{z}. \tag{53}$$

The orthogonality of $\sin(\lambda_n \bar{z})$ in the range, $0 \leq \bar{z} \leq 1$, is such that

$$\langle X_n, X_m\rangle = \begin{cases} 0 & \text{if } n \neq m \\ \frac{1}{2} & n = m. \end{cases} \tag{54}$$

Upon substitution of (50) into (39), we have

$$\sum_{n=0}^{\infty} \omega_n'(t) X_n = \sum_{n=0}^{\infty} \omega_n(t) \left[X_n'' - \alpha^2 X_n\right] \tag{55}$$

or, using (43)

$$\sum_{n=0}^{\infty} \omega_n'(\bar{t}) X_n = -\sum_{n=0}^{\infty} \omega_n(\bar{t}) \mu_n X_n. \tag{56}$$

Equation (56) is satisfied if it is satisfied for each n, a condition met by the solutions of the ordinary differential equation

$$\omega_n' = -\mu_n \omega_n \tag{57}$$

yielding

$$\omega_n = a_n \exp\left(-\mu_n \bar{t}\right) \tag{58}$$

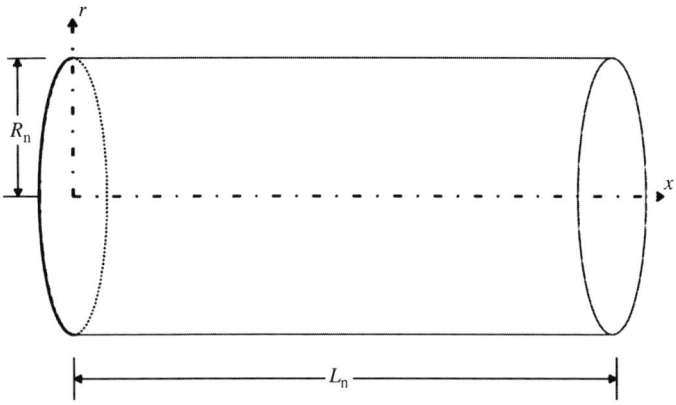

Figure 5. Idealized cylindrical geometry used to analyze transient axisymmetric diffusion in a nanopore.

so that

$$c_H\left(\bar{t}, \bar{z}\right) = \sum_{n=0}^{\infty} a_n \exp\left[-\left(\lambda_n + \alpha^2\right)\bar{t}\right] \sin(\lambda_n \bar{z}). \quad (59)$$

Equation (59) satisfies the governing equation and boundary conditions; therefore, all that remains is to find the coefficients a_n that satisfy the initial condition, (see (49)). The required coefficients, a_n, are given by

$$a_n = -b_n + \tanh(\alpha) d_n. \quad (60)$$

Solutions for large and small α are plotted in Fig. 6. Equation (59) shows that this example problem supports the conclusions given in the previous section regarding time scales. If α^2 is large the decay to the steady-state solution will occur on the adsorption timescale $L_n/k_a N$, while if it is small then the decay will be dominated by the diffusion timescale $t_s/\lambda_0 = L_n^2/2D_{Kn}$ (Fig. 6).

IV. MICROSCALE TRANSPORT

The discussion of microscale transport follows the nanopore discussion for two reasons. First, PS sensors with only nanopores might be used, for which the micropore discussion is irrelevant. Second, when

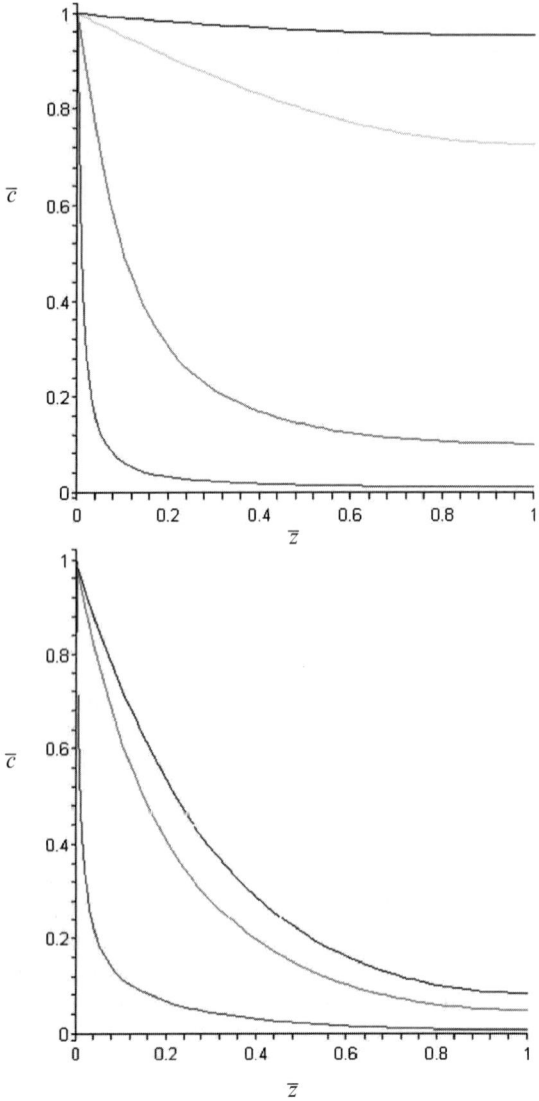

Figure 6. Plots of normalized analyte concentration in the nanopore as a function of position at several scaled times $\bar{t} = 0.01$ (*red*), 0.1 (*green*), 1 (*yellow*), infinity, i.e. steady-state, (*blue*). The *top* and *bottom* plots contrast results for different ratios of the timescales for adsorption and diffusion. In the *top* plot $\alpha = \sqrt{0.1}$, while in the *bottom*, $\alpha = \sqrt{10}$.

micropores (with diameters ~ 1 μm) are present, they serve only to link the bulk to the nanopores, and no new insight is required. Thus, an understanding of the microscale transport is extremely easy to obtain as a special case once the bulk and nanoscale transport are understood.

As was the case for nanopore transport, for micropores with blind ends, there is no net bulk flow through a pore and, therefore, we can assume that advective effects are negligible. The first determination that must be made is the mode of diffusive transport. As was the case for the nanopore, we calculate the Knudsen number to compare the average distance traveled by a molecule in the gas phase prior to collision with another gas molecule, to the average distance traveled by a molecule in the gas phase prior to collision with a solid boundary. For the micropore we find that the Knudsen number is small, $Kn \sim 0.1$, thus classical Fickian diffusion dominates and mass transport can be modeled using the bulk diffusion coefficient.

The only intrinsic timescale for the micropore is the diffusive timescale, i.e.,

$$t_\mu = \frac{L_\mu^2}{D_A}. \tag{61}$$

Thus the governing equation for the analyte concentration in the micropore is

$$\frac{\partial c_A^*}{\partial t^*} = \nabla^{*2} c_A^*, \tag{62}$$

where the nondimensionalization has been accomplished with the diffusive timescale given by (61), with the microprore length $L_\mu \sim 10^{-5}$ m.

V. MULTISCALE FORMULATION

We have so presented an idealized model which allows the determination of relevant timescales of PS gas sensor response. Table 1 provides a summary of the timescales.

The information in Table 1 can be used to provide a quick guide to sensor design improvement. Ideally, sensor response is instantaneous. Actual response occurs on a timescale dictated by the largest timescale in Table 1. Therefore, efforts to improve sensor response

Table 1.
Timescales for PS gas sensor mass transport

Timescale	Bulk	Micropore	Nanopore
Diffusion	L_{diff}^2/D_A	L_μ^2/D_A	L_n^2/D_{Kn}
Advection	L_{adv}/U_∞	N/A	N/A
Adsorption	N/A	N/A	$L_n/k_a N$
Forcing	t_{force}	N/A	N/A

should focus on reducing the largest timescale. Conversely, efforts that reduce timescales other than the largest will have negligible effect.

First we discuss the time scales for bulk transport. In the absence of significant bulk flow ($U_\infty \to 0$) and without forcing, the diffusive timescale is the only timescale, and it is calculated using the characteristic sensor length as the diffusion length. For example, a 1 mm square sensor, i.e., $L_{\text{char}} = 10^{-3}$ m, would have a macroscale diffusive timescale $t_{\text{diff}} \sim L_{\text{char}}^2/D_A \sim 0.1$ s. The diffusion timescale can be made smaller by reducing the diffusion length through advective effects. For large Peclet number $L_{\text{char}} U_\infty/D_A \gg 1$, the diffusion length is the boundary layer thickness. For instance, for laminar flow and large Peclet number in air, $L_{\text{diff}} \sim L_{\text{char}} Pe^{-1/2}$. Using this modified diffusion length to calculate the diffusion timescale one recovers the advective timescale, with $L_{\text{adv}} = L_{\text{char}}$. Thus, ideally one can reduce the macroscale time scale to an arbitrarily low number by increasing the bulk velocity. The penalty one pays is usually in fabrication and operation costs associated with achieving high flow rates.

The "forcing" timescale for the bulk transport is simply the timescale at which changes are applied. If the forcing timescale is smaller than the largest transport timescale, then the sensor response is always transient. With sufficiently accurate modeling, transient response can be exploited to obtain useful information, but for simple sensor operation and greater reliability of results it is typical to desire a steady signal.

The only micropore timescale is the diffusive timescale $t_\mu \sim 10^{-5}$ s. Because the microscale timescale varies with the square of the micropore length, it may seem reasonable to reduce the micropore length to lower the response time. A main purpose of micropores is to enhance sensitivity, by increasing the number of nanopores, thus, decreasing the micropore length will simultaneously reduce sensitivity. This penalty will come with no benefit if other timescales are greater than t_μ, and even if they are not, the loss in sensitivity must be weighed against the reduction in response time when considering making micropores smaller or eliminating them altogether.

In the nanopores, the diffusive timescale is proportional to square the length of the nanopore and inversely proportional to the radius of the nanopore, $L_n^2/R_n\sqrt{\Re T/M} \sim 10^{-8}$ s. Firm values of the adsorption timescale are difficult to obtain, but estimates typically are in the milli- to microsecond range (49, 54).

Based on the timescale comparison, an obvious strategy for sensor design emerges. First, we observe that the limiting timescale is either the bulk diffusion timescale, or, if that timescale is reduced due to advective effects, the adsorption timescale most likely limits the sensor response. A bulk flow of $U_\infty \sim 1$ m/s for the earlier example of a 1 mm^2 sensor would cause a transition in the response from macroscale diffusion limited to nanoscale adsorption limited. Once the sensor response is adsorption limited, mass transport to and from the surface no longer plays a role in sensor response. In fact, the entire multiscale model can be reduced to an ordinary differential equation for fractional surface coverage, in which the fractional coverage is increased due to adsorption and decreased by desorption, with the surface concentration of the analyte species being the bulk concentration, c_∞. For the simple adsorption model described earlier, the resulting model would be

$$\frac{d\theta}{dt} = k_a c_\infty (1-\theta) - k_d \theta \qquad (63)$$

with arbitrary initial condition,

$$\theta(0) = \theta_0. \qquad (64)$$

If the sensor is allowed to equilibrate prior to changes in bulk concentration, i.e., $t_{\text{force}} > t_{\text{ads}}$, then the initial coverage would be the equilibrium coverage based on the previous bulk concentration, c_p, i.e., $\theta_0 = Kc_p/(1+Kc_p)$ where $K = k_a/k_d$.

Equations (63) and (64) can be solved analytically, yielding

$$\theta = \theta_{\text{eq}}\{1 - \exp[-(k_a c_\infty + k_d)t]\} + \theta_0 \exp[-(k_a c_\infty + k_d)t], \qquad (65)$$

where $\theta_{eq} = Kc_\infty/(1 + Kc_\infty)$. Equation (65) describes the sensor response in the limit of infinitely fast analyte transport.

VI. CONCLUSIONS

In this chapter we have presented an idealized formulation of mass transport in porous silicon gas sensors. The results can be directly applied to analysis and design of sensors provided the underlying assumptions are valid, and so long as the interaction of the adsorbed analyte and the substrate that produces a signal is fast. In the event that one wished to introduce additional complexity where we have opted for simplicity, it should be straightforward to add to the presented formulation.

It is not uncommon for experimental results to be related to a simplified formulation. The flaw with many such discussions is that they begin with leaps in formulation requiring oft-unstated assumptions. For instance, assumptions of infinitely fast kinetics are common. Unfortunately, such formulations are doomed to failure. Through careful identification of the possible timescales and comparison of their magnitudes, as presented herein, one can skirt the pitfalls of ad hoc formulation.

For porous silicon gas sensors, the conclusions of the analysis are straightforward. Sensor size and micropore length should be made sufficiently small (within the limits imposed by sensitivity requirements) so that when combined with a moderate bulk flow, if possible, the response is limited by the adsorption timescale. Thus, efforts to improve PS gas sensors should focus on enhancing chemical activity for analyte adsorption of the sensing interface.

REFERENCES

1. M. Hejjo, A. Rifai, M. Christophersen, S. Ottow, J. Carstensen, and H. Föll, *J. Electrochem. Soc.* **147** (2000) 627.
2. A. G. Cullis, L. T. Canham, and P. D. J. Calcott, *J. Appl. Phys.* **82** (1997) 909 and references therein.
3. M. J. J. Theunissen, *J. Electrochem. Soc.* **119** (1972) 351.
4. V. Lehmann and H. Föll, *J. Electrochem. Soc.* **137** (1990) 653.
5. H. Föll, *Appl. Phys. A* **53** (1991) 8.
6. P. C. Searson, J. M. Macaulay, and F. M. Ross, *J. Appl. Phys.* **72** (1992) 253.

7. I. Berbezier and A. Halimasui, *J. Appl. Phys.* **74** (1993) 5421.
8. V. Lehmann, *J. Electrochem. Soc.* **140** (1993) 2836.
9. C. Levy-Clement, A. Lagoubi, and M. Tomkiewicz, *J. Electrochem. Soc.* **141** (1994) 958.
10. E. K. Probst and P. A. Kohl, *J. Electrochem. Soc.* **141** (1994) 1006.
11. V. Lehmann and U. Gösele, *Adv. Mater.* **4** (1992) 114.
12. S. Rönnebeck, S. Ottow, J. Carstensen, and H. Föll, *Electrochem. Solid State Lett.* **2** (1999) 126.
13. G. X. Zhang, in *Modern Aspects of Electrochemistry*, Vol. 39 Ed. by C. G. Vayenas, R. E. White, and M. Gamboa-Adelco, Springer, Berlin, Heidelberg, New York, 2006, p. 65.
14. J. L. Gole, E. C. Egeberg, E. Veje, A. F. d. Silva, I. Pepe, and D. A. Dixon, *J. Phys. Chem. B* **110** (2006) 2064 and references therein.
15. J. L. Gole and S. E. Lewis, in *Nanosilicon*, Ed. by S. Kumar, Elsevier, London, 2008, p. 147.
16. J. L. Gole, S. Lewis, and S. Lee, *Phys. Status Solidi A* **204** (2007) 1417.
17. E. Galeazzo, H. E. M. Peres, G. Santos, N. Peixoto, and F. J. Ramirez-Fernandez, *Sens. Actuators B-Chem.* **93** (2003) 384.
18. P. A. Kottke and A. G. Fedorov, *J. Electroanal. Chem.* **583** (2005) 221.
19. C. Phillips and A. G. Fedorov, *Sens. Actuators B-Chem.* **99** (2004) 273.
20. C. Phillips, M. Jakusch, H. Steiner, B. Mizaikoff, and A. G. Fedorov, *Anal. Chem.* **75** (2003) 1106.
21. P. A. Kottke and A. G. Fedorov, *J. Phys. Chem. B* **109** (2005) 16811.
22. V. G. Levich, *Physicochemical Hydrodynamics*, Prentice-Hall, Englewood Cliffs, NJ, 1962.
23. R. B. Bird, W. E. Stewart, and E. N. Lightfoot, *Transport Phenomena*, Wiley, New York, NY 2002.
24. J. Israelachvili, *Intermolecular and Surface Forces*, Academic Press, San Diego, CA, 1992.
25. A. L. Hines and R. N. Maddox, *Mass Transfer Fundamentals and Applications*, Prentice Hall PTR, Englewood Cliffs, NJ, 1985.
26. R. Krishna and J. A. Wesselingh, *Chem. Eng. Sci.* **52** (1997) 861.
27. A. Bejan, *Convection Heat Transfer*, Wiley, New York, NY, 1995.
28. L. G. Leal, *Laminar Flow and Convective Transport Processes Scaling Principles and Asymptotic Analysis*, Butterworth-Heinemann, Newton, MA, 1992.
29. S. E. Lewis, J. R. DeBoer, and J. L. Gole, *Sens. Actuators B-Chem.* **122** (2007) 20.
30. L. Seals, J. L. Gole, L. A. Tsa, and P. J. Hesketh, *J. Appl. Phys.* **91** (2002) 2519.
31. A. Foucaran, F. Pascal-Delannoy, A. Giani, A. Sackda, P. Combette, and A. Boyer, *Thin Solid Films* **297** (1997) 317.
32. S. E. Lewis, J. R. DeBoer, J. L. Gole, and P. J. Hesketh, *Sens. Actuators B-Chem.* **110** (2005) 54.
33. P. Fürjes, A. Kovács, C. Dücsö, M. Ádám, B. Müller, and U. Mescheder, *Sens. Actuators B-Chem.* **95** (2003) 140.
34. M. Björkqvist, J. Salonen, and E. Laine, *Appl. Surf. Sci.* **222** (2004) 269.
35. M. Björkqvist, J. Salonen, J. Paski, and E. Laine, *Sens. Actuators A-Phys.* **112** (2004) 244.

36. A. Foucaran, B. Sorli, M. Garcia, F. Pascal-Delannoy, A. Giani, and A. Boyer, *Sens. Actuators A-Phys.* **79** (2000) 189.
37. D. G. Yarkin, *Sens. Actuators A-Phys.* **107** (2003) 1.
38. E. J. Connolly, P. J. French, H. T. M. Pham, and P. M. Sarro, in *Sensors*, 2002. Proceedings of IEEE, Vol. 1, 2002, p. 499.
39. M. P. Stewart and J. M. Buriak, *Adv. Mater.* **12** (2000) 859.
40. J. Y. Jin, N. K. Min, C. G. Kang, S. H. Park, and S. I. Hong, *J. Korean Phys. Soc.* **39** (2001) S67.
41. M. Fichera, S. Libertino, and G. D'Arrigo, Vol. 5119 (R.-V. Angel, A. Derek, and C. Ricardo, eds.), SPIE, 2003, p. 149.
42. L. DeStefano, I. Rendina, L. Moretti, A. M. Rossi, A. Lamberti, O. Longo, and P. Arcari, Vol. 5118 (V. Robert, A. Xavier, B. K. Laszlo, and R. Angel, eds.), SPIE, 2003, p. 305.
43. M. Ben Ali, R. Mlika, H. Ben Ouada, R. M'Ghaieth, and H. Maâref, *Sens. Actuators A-Phys.* **74** (1999) 123.
44. S. Zairi, C. Martelet, N. Jaffrezic-Renault, R. M'Gaïeth, H. Maâref, and R. Lamartine, *Thin Solid Films* **383** (2001) 325.
45. S. Zairi, C. Martelet, N. Jaffrezic-Renault, F. Vocanson, R. Lamartine, R. M'Gaïeth, H. Maâref, and M. Gamoudi, *Appl. Phys. A-Mater.* **73** (2001) 585.
46. R. F. Probstein, *Physiochemical Hydrodynamics: An Introduction*, Wiley, New York, NY, 1994.
47. H. Daiguji, P. Yang, A. J. Szeri, and A. Majumdar, *Nano Lett.* **4** (2004) 2315.
48. G. A. Bird, *Molecular Gas Dynamics and the Direct Simulation of Gas Flows*, Clarendon Press, Oxford, 1994.
49. M. K. Gobbert, S. G. Webster, and T. S. Cale, *J. Electrochem. Soc.* **149** (2002) G461.
50. S. Roy, R. Raju, H. F. Chuang, B. A. Cruden, and M. Meyyappan, *J. Appl. Phys.* **93** (2003) 4870.
51. D. N. Jaguste and S. K. Bhatia, *Chem. Eng. Sci.* **50** (1995) 167.
52. R. Aris, *The Mathematical Theory of Diffusion and Reaction in Permeable Catalysts*, Clarendon Press, Oxford, 1975.
53. H. Lu, W. Ma, J. Gao, and J. Li, *Sens. Actuators B-Chem.* **66** (2000) 228.
54. N. Matsunaga, G. Sakai, K. Shimanoe, and N. Yamazoe, *Sens. Actuators B-Chem.* **83** (2002) 216.
55. E. Butkov, *Mathematical Physics*, Addison-Wesley, Reading, MA, 1968.
56. N. Matsunaga, G. Sakai, K. Shimanoe, and N. Yamazoe, *Sens. Actuators B-Chem.* **96** (2003) 226.

5

Electrochemical Materials for PEM Fuel Cells: Insights from Physical Theory and Simulation

Michael H. Eikerling and Kourosh Malek

Department of Chemistry, Simon Fraser University, 8888 University Drive, Burnaby, BC, Canada V5A 1S6
Institute for Fuel Cell Innovation, National Research Council of Canada, 4250 Wesbrook Mall, Vancouver, BC, Canada V6T 1W5

"My advice is to go for the messes – that's where the action is."
(S. Weinberg, Scientist: Four Golden Lessons, Nature 426, 389 (2003).)

Abstract This chapter focuses on the role of physical theory, molecular simulation, and computational electrochemistry for fundamental understanding, diagnostics, and design of Polymer Electrolyte Fuel Cells (PEFCs). Development of stable and inexpensive materials is the most important technological hurdle that PEFC developers are currently facing. A profound insight based on theory and modeling of the pertinent materials will advise us how fuel cell components with optimal specifications could be made and how they can be integrated into operating cells. This chapter highlights major challenges and perspectives in research on electrochemical materials for fuel cells, arising at scales from Ångstrom to meters. Topics include proton conduction, nanoparticle electrocatalysis, self-organization in complex media, effective properties of random heterogeneous media, role of water in various components and at various scales and effectiveness of catalyst utilization.

I. INTRODUCTION

Over the past 20 years, intense efforts in research worldwide have explored fuel cells as the next generation of energy conversion technologies. PEFCs could replace internal combustion engines in vehicles and provide power to a plethora of stationary and portable applications (1, 2).

Nowadays almost all leading car manufacturers are researching and developing Polymer Electrolyte Fuel Cells (PEFCs). In worldwide tests of thousands of fuel cell-driven cars and buses, with a mileage totaling more than 3 million km, fuel cells have been performing very well. Some companies, like General Motors, Daimler-Chrysler, Ford and Toyota, are close to deploying sample series of PEFC driven vehicles. Many electronics giants like Toshiba, Samsung, Motorola, and NEC, as well as smaller startups like MTI Micro Fuel Cells, Medis, Polyfuel, Neah Power, Protonex, Jadoo Power Systems (all United States) and Smart Fuel Cell (Germany) massively pursue the consumer markets for micro fuel cell technology. These fuel cells, based on direct methanol fuel cell (DMFC) technology, could fulfill increased energy density demands of portable laptops, digital cameras, etc. and power hungry electronic gadgets, which are an increasing burden on batteries. Furthermore, larger PEFC systems and solid oxide fuel cells (SOFCs) are expected to play a crucial role in the decentralization of electrical power supplies.

Although various types of fuel cells are under development, which differ in the type of electrolyte, feasible operating conditions, and primary applications (1), we will focus in this chapter on Polymer Electrolyte Fuel Cells (PEFCs), the most versatile and nifty member of the fuel cell family.

Figure 1 shows the principal layout and basic processes in PEFCs under standard operation with hydrogen as a fuel. Anodic oxidation of H_2 produces protons that move through the polymer electrolyte membrane (PEM) to the cathode, where reduction of O_2 produces water. Meanwhile, electrons, produced at the anode, perform work in the external electrical load. The spatial separation of anodic and cathodic reactions, warranted by the proton-conducting PEM, enables the direct conversion of the enthalpy released in the net reaction $H_2 + \frac{1}{2}O_2 \rightarrow H_2O$ into electrical energy, with the only by-products of this process being waste heat and water.

Electrochemical Materials for PEM Fuel Cells

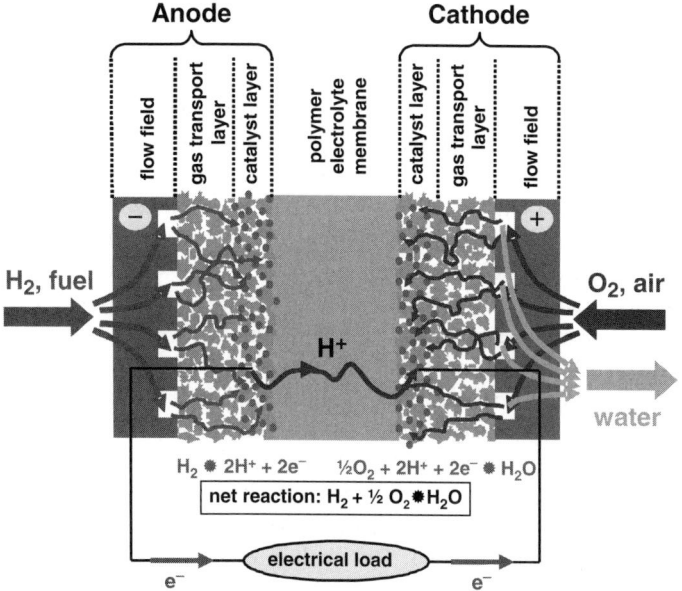

Figure 1. Principal layout of Polymer Electrolyte Fuel Cells under standard operation with hydrogen, depicting 7-layer structure and basic processes.

Unrivalled thermodynamic efficiencies of electrochemical energy conversion, high energy densities, and ideal compatibility with hydrogen render PEFCs a primary solution to the global energy challenge. Further characteristics include fuel flexibility and modular design of fuel cell stacks, which make them, in principle, amenable to varying operating conditions and power requirements of widely different applications.

In spite of these assets and the abundance of promising demonstrations of fuel cell powered vehicles and devices, the success of PEFCs at the commercial stage is far from being guaranteed. New generations of fuel cells have to surpass established energy conversion technologies in power density, operational flexibility, stability, and cost. To give an idea of the magnitude of this challenge: currently used PEFC stacks in prototype cars are about 100 times more expensive than conventional combustion engines ($\sim$\$30 kW^{-1}), albeit distinctly inferior in durability (3).

Ten years ago it was a commonly held view among major stakeholders in industry and academic institutions that critical progress could be achieved with engineering-type optimization on the basis of the existing materials. Nowadays, it is widely recognized that progress in fuel cell technology hinges on breakthroughs rather than incremental changes in the design, fabrication, and implementation of innovative materials. Specific targets for improvement involve (i) increasing the rates of charge transfer and overall current conversion at electrodes, (ii) minimizing voltage losses incurred by the coupled transport of protons, water, electrons, and gaseous reactants in membrane-electrode assemblies, and (iii) establishing uniform distributions of reactants and reaction rates in all components and at all length scales at the cell and stack level. These are formidable multidisciplinary challenges with fundamental physics and chemistry, physical and chemical diagnostics, materials science, and engineering all having contributions to make.

Chemical reactions and thermodynamic considerations of PEFCs are seemingly simple. The challenges for optimizing operation are hidden in the fine microscopic details of structures and processes. The multilayered design of a single operational PEFC is depicted in Fig. 1. The typical 7-layer structure consists of a proton-conducting Polymer Electrolyte Membrane (PEM) sandwiched between anode and cathode. Each electrode compartment consists of an active catalyst layer (CL) that accommodates the finely dispersed catalyst nanoparticles, a gas diffusion layer (GDL), and a flow field (FF) plate. Fuel cell operation entails proton migration and water fluxes in PEM, circulation and electrochemical conversion of electrons, protons, reactant gases, and water in PEM and catalyst layers and vaporization/condensation in pores and channels of CLs, GDLs and FFs. All components in Fig. 1 have to cooperate well in order to optimize the highly non-linear interplay of transport and reaction. It can be easily estimated that this optimization involves more than 50 parameters.

In this situation, strategies that rely predominantly on empirical materials design and engineering approaches to optimization of operation are deemed to failure. Recent years have witnessed ever-increasing efforts towards establishing a theoretical framework and molecular modelling approaches as a basis for fundamental innovation in fuel cell materials and operational optimization (4–14). These theoretical efforts are accompanied by vital progress in diagnostics of structures and fundamental processes (15, 16), using e.g., small angle scattering techniques and quasi-elastic neutron scattering to unravel the complex morphology and processes in PEMs (17, 18), or STM-based methods (19, 20) and advanced spectroscopic techniques in electrocatalysis (21–23), to name only a few. Moreover, new methods have emerged in spatially-resolved diagnostics of operating fuel cells. Impedance spectroscopy adapted to segmented cells and neutron scattering have been specifically developed to unravel relations between transport processes, local reaction rate distributions, and global performance (24–26). Meanwhile, in materials science avenues are being pursued that could eventually replace random composite materials by specifically designed ordered or nanostructured architectures in electrodes and membranes.

The multidisciplinary challenges suggest the merit of concerted efforts. A general strategy towards new materials should be based on understanding the relations between fundamental interactions in many particle systems, mechanisms of spontaneous or directed structural organization, effective properties of heterogeneous media, and performance. The latter should be evaluated locally with high spatial resolution and rated globally in terms of power output, voltage efficiency, catalyst utilization, and water-handling capabilities of the fuel cell.

Our focus in this chapter is on providing an overview of the state of affairs in theory and molecular modelling of materials for PEFCs. As discussed in (27), the theoretical framework fulfils an integrating function in this endeavour, linking the various disciplines in fuel cell research. At the fundamental level, theory helps to unravel complex relations between chemical and morphological structures and properties, bridging scales from molecular to macroscopic resolutions. Understanding these relations could facilitate the design of novel, tailor-made fuel cell materials. In fuel cell diagnostics, theory relates *ex situ* properties of materials to their *in situ* fuel cell performance. It could identify root causes of non-optimal fuel cell operation, which in many cases are not amenable to direct measurements. Theory provides valuable input for cell and systems optimization. Approaches in engineering lead to uncontrollable results, if they are based on oversimplified structural models and unsettled understanding of fundamental physics. For example, it is pointless to study water management in a PEFC without appropriate structural pictures of PEMs, CLs, GDLs, and FFs; all these structural elements have to cooperate well in a properly balanced cell.

It would, however, by far exceed the scope of this chapter to present a complete review of the recent progress in this field. Instead we will give a systematic account of relevant techniques and relate them to major challenges in materials science. We will cover theoretical aspects from fundamental interactions to structure formation to effective physical properties to functioning in an operational cell. Wherever possible we will provide examples of the links between theoretical and modeling studies, materials design, diagnostics of structures, processes and operation, and performance evaluation and optimization. We will limit ourselves to systems containing aqueous based proton conductors and catalyst layers based on nanoparticles of Pt, which operate at $T < 100°C$.

The materials of greatest interest are PEMs and CLs. They fulfill key functions in the cell and at the same time offer to most compelling opportunities for innovation. The membrane provides the basis for electrochemical energy conversion through the spatial separation of anode and cathode compartments. To a large extent the PEM determines architecture and feasible operating conditions of the complete fuel cell. Proton conduction is an ubiquitous fundamental phenomenon that is of interest for a range of systems in soft matter (bio)physics and materials science. In addition to high proton conductivity ($>0.1 \text{ S cm}^{-1}$), PEMs should be impermeable to gases and possess sufficient chemical and mechanical robustness. A sufficient amount of liquid water is needed in pertinent PEMs, like Nafion, Dow and the like, in order to facilitate proton conduction. Migrating protons pull along water molecules in a process known as electro-osmotic drag (28). Due to the humidification requirements and the coupled proton and water fluxes via electro-osmosis, PEMs play a key role for the water balance in the complete cell.

The catalyst layers, the cathode side in particular, are the powerhouses of the cell. It is in them that the full competition of reactant transport, electrochemical conversion, as well as transport and conversion of water unfolds. In conventional PEFCs, CLs are fabricated as random heterogeneous composites. Two significant steps in their development were the advent of highly dispersed Pt nanoparticle catalysts with particle sizes in the range of 1–10 nm, deposited on high surface area carbon, as well as the impregnation with Nafion ionomer. These steps enabled the reduction of catalyst loadings from about 4 mg Pt cm^{-2} (in the 1980s) to about 0.2 mg Pt cm^{-2}.

Separate percolating networks of C/Pt, Nafion-ionomer, and pores provide pathways for electrons, protons, reactants, and water. Electrochemical reactions proceed only at those Pt particles, where all reactants meet. The random three-phase morphology imposes statistical constraints on the utilization of Pt. The relatively high thickness ($\sim$10 μm) leads to non-uniform reaction rate distributions due to impeded transport. Overall, these conditions cause inefficient utilization of Pt and problematic voltage losses, with those due to oxygen reduction in the cathode catalyst layer (CCL) ($\sim$400 mV) diminishing cell efficiency by 30–40%. An increase by a factor 10 in the surface area of Pt would reduce these losses by 60–120 mV. Pt is an expensive and limited resource, however. For a 60 kW fuel cell vehicle, the cost of Pt is over \$2,400 at current loading and market price (3). What's worse, replacing combustion engines in all vehicles in the United States by fuel cell drive systems at no penalty in power would completely exhaust all known reserves of Pt. The single most important hurdle to be cleared on the route to commercialization of PEFC technology is, therefore, to raise the ratio of current conversion rate to Pt loading by a factor of 10–100. In principle, all other components in the fuel cell have to be optimized in order to ensure the most uniform reaction rate distribution in the catalyst layers and, thereby, ensure a high utilization of the Pt-based catalyst.

II. LENGTH SCALES AND RELEVANT APPROACHES IN MATERIALS MODELING

Figure 2 illustrates the length scales from Ångstrom to meters along the route from molecular structure of materials to fuel cell operation. Theoretical studies of molecular mechanisms of proton transport in PEMs require quantum mechanical techniques, i.e., *ab initio* calculations based on density functional theory. The molecular mechanisms determine the conductance of aqueous nano-sized pathways in PEMs. At the same resolution, elementary steps of adsorption, surface diffusion, charge transfer, recombination, desorption proceed on the surfaces of nanoscale catalyst particles. These fundamental processes control the electrocatalytic activity of the catalyst surface. Studies of stable conformations of supported

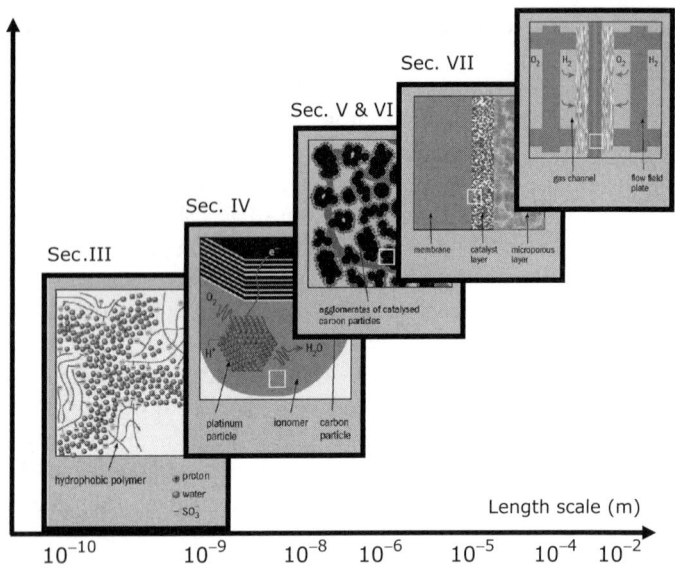

Figure 2. Multi-scale phenomena in PEFC, from fundamental proton transport in PEM (Sect. III), to kinetic mechanisms at nanoparticle electrocatalysts (Sect. IV), to structure formation (Sect. V) and effective properties (Sect. VI) of complex composite materials, to transport, reaction and performance at the macroscopic device level (Sect. VII). (*See Color Plates*).

nanoparticles, as well as of the elementary processes on their surface warrant density functional theory calculations and kinetic modeling of reactivities based on Monte Carlo simulations or Mean Field approximations. At the mesoscopic scale, interactions between molecular components control the processes of structural formation, that lead to random phase-segregated morphologies in membranes and catalyst layers. Such complex processes can be studied by coarse-grained molecular dynamics simulations. Complex morphologies of the emerging media can be related to relevant effective properties that characterize transport and reaction, using concepts from the theory of random heterogeneous media. Finally, conditions for operation at the macroscopic device level can be defined and balance equations for involved species, i.e., electrons, protons, reactant gases and water, can be established on the basis of fundamental conservation laws. Thereby full relations between structure, properties, and performance could be established, which in turn would allow to predict architectures of materials and operating conditions that optimize fuel cell operation.

In the following we give a brief outline of the major topics to be addressed in separate sections of this chapter.

Proton transport (PT) in aqueous media and polymer membranes (Sect. III). *Ab initio* molecular dynamics calculations by Tuckerman and coworkers have unraveled molecular details of the structural diffusion of excess protons in bulk water (29–31). In PEMs, conditions for genuine bulk-water-like PT are, however, hardly ever encountered (2, 32, 33). Similar to proton transport in biophysical systems, rates of PT in PEMs are strongly affected by confinement of water in nanochannels, electrostatic effects at interfaces, and desolvation phenomena (34–40). In the case of cellular membranes and lipid monolayers, systematic experimental studies have exhibited strong dependences of lateral interfacial proton migration on packing density of proton-binding surface groups, length and chemical structure of these groups, and flexibility of acid head groups (36, 41–43). In PEMs the interfaces between charged polymeric sidechains and water account for differences in membrane morphology, stability, state of water, and proton-conductive abilities. It is still unclear whether the observed increase in the activation energy of PT from 0.12 eV at high levels of hydration to >0.35 eV at lowest water uptake of PEMs (44) is due to a change in the molecular mechanism of proton mobility, a morphological transition, or both (45, 46). In archetypical PEMs based on perfluorinated sulfonic acid ionomers, i.e., Nafion and the like, predominant PT along interfaces between polymer and water under conditions of minimal hydration drastically reduces the proton conductivity. In view of membrane design, this raises the general question whether high proton mobility in aqueous-based PEMs is possible under conditions of minimal hydration and at elevated temperature. Obviously the answer could have a tremendous impact on design strategies in membrane research. The focus on unraveling interfacial mechanisms of PT is, thus, motivated by vigorous activities in materials science targeting to demonstrate fuel cell membranes with high proton conductivity under minimal hydration and/or elevated temperatures (47, 48). Another look aside to the plethora of experimental studies on lateral proton transport at biomembranes and Langmuir monolayers provides encouraging insights in this realm.

It was suggested that lateral proton mobility along these interfacial systems could be rather high at a critical packing density of protogenic surface groups – as high as $\frac{1}{2}$ the value of proton mobility in bulk water (36, 42, 43). In spite of their obvious ubiquitous importance, molecular mechanisms of PT at dense interfacial arrays of protogenic surface groups (SGs) are rather unexplored. In Sect. III, we will discuss recent studies on structural correlations and surface-mediated mechanisms of proton transfer in fuel cell membranes. In principle, these mechanisms warrant *ab initio* quantum mechanical simulations, based on density functional theory as pursued by Roudgar et al. (33), Paddison (14), Elliott and Paddison (49), Paddison and Elliott (50) and Eikerling et al. (51). Molecular simulations based on the EVB approach, applied for instance by Peterson and Voth (52), Spohr (13) or Spohr et al. (53), or continuum dielectric approaches, explored in (37, 38) have focused as well on the role of acid-functionalized surface groups on proton conduction mechanisms in water-filled pores of PEMs. However, the latter approaches employ empirical correlations between interfacial structure and mechanisms of proton transport.

Nanoparticle electrocatalysis (Sect. IV). In conventional catalyst layers electrical current is generated at Pt nanoparticles that are randomly dispersed on a carbonaceous substrate (54). All relevant electrochemical reactions in PEFC have been proven to be sensitive to sizes and surface structures of Pt particles (23, 55–57). Obviously, a reduction in particle size improves the surface-to-volume ratio of catalyst atoms. This has a direct impact on catalyst utilization, a key target in fuel cell development, because only surface atoms could be potentially active. Yet, the relation between particle size and electrocatalytic activity is highly non-trivial, since the size of the particles also affects electronic and geometric properties at their surface (23, 58). A better understanding of the relationships between particle size, heterogeneous surface structure and activity, including those effects, could be critical in view of the design of highly performing catalyst layers (59–62). Moreover, there is an often-neglected role of the substrate on the nature and strength of particle-substrate bonding, which is important in view of degradation due to ripening of catalyst particles or loss of active area.

Self-organization in PEMs and CLs at the Mesocopic Scale (Sect. V). Previous topics dealt with modeling and simulation of fuel cell materials and their properties at the molecular or nanoscopic scale. In addition to theoretical and computational tools at the molecular level, meso-scale simulations can describe the morphology of heterogeneous materials and rationalize their effective properties beyond length- and time-scale limitations of atomistic simulations. In Sect. V, we describe how meso-scale computational tools can be used to investigate the self-organization phenomena during fabrication of CLs (63). Such studies will help us to understand structural-related properties for further developing the design of novel materials for PEFC. Improving the fabrication of CLs provides opportunities to improve the Pt catalytic utilization by increasing the interfacial area between Pt particles and ionomer (64–66). This is achieved by mixing ionomer with dispersed Pt/C catalysts in the ink suspension prior to deposition to form a CL. The solubility of the ionomer depends upon the choice of a dispersion medium. This influences the microstructure and pore size distribution of the CL (66). The random morphology of the emerging composite medium in turn has a marked impact on transport and electrochemical properties. In principle, only catalyst particles at Pt/liquid water interfaces are electrochemically active. Electrochemical reactions, thus, occur on the walls of wetted pores inside and between agglomerates as well as at interfaces between Pt and wetted ionomer. Relative contributions of these distinct types of interfaces will depend on the corresponding interfacial areas. Agglomerates are usually referred to as the building blocks of CLs at the mesoscopic scale. They consist of carbon/Pt particles and are vital for the catalyst utilization and the overall voltage efficiency of the layer (64–68). Sec. V presents a recently introduced meso-scale computational method to evaluate key factors during fabrication of CL. These simulations rationalize structural factors such as pore sizes, internal porosity, and wetting properties of internal/external surfaces of such agglomerates. They help elucidating whether or not ionomer is able to penetrate into primary pores inside agglomerates (69). Moreover, dispersion media with distinct dielectric properties can be evaluated in view of capabilities for controlling sizes of carbon/Pt agglomerates, ionomer domains, and the resulting pore network topology. These insights are highly valuable for the structural design of catalyst layers with optimized performance and stability.

Effective Properties of Random Heterogeneous Media (Sect. VI). In Sect. VI, we discuss theoretical and computational tools for studying transport and reaction in a random heterogeneous media (70). These media are composed of different materials or phases that are randomly mixed with each other. Conduction, diffusion and reaction processes in such media are often accompanied by a continuous alteration of the pore structure. One or two ingredients of the medium lose their macroscopic connectivity once the other component exhibits percolation-type behavior. In a porous catalyst layer of PEFC, as a random heterogeneous nanoporous medium, macroscopic properties such as diffusion and reaction rates are usually defined as averages of the corresponding microscopic quantities (71). These averages must be taken over a volume V, which is small compared to the systems size. At every point in a porous medium one uses the smallest such volume to fulfill requirement of transport equations such as Fick's law of diffusion and to treat the usual macroscopic variations of diffusivity, reaction rate, conductivity, connectivity, etc.

Structure and Water: Physical Models of Operation (Sect. VII). A poor balance of the coupled processes in MEAs at the macroscopic scale causes non-uniform distributions of reaction rates in the fuel cell electrodes, leading to losses in the fuel cell performance as well as inefficient utilization of expensive catalyst. Structural pictures of the pertinent materials serve as the basis for device-level models of operation. The general role of these models is to rationalize the coupled processes of transport and reaction in the different components and to relate local structures and processes to global performance rated in terms of voltage efficiency, effectiveness of catalyst utilization, water handling capabilities, and durability. Understanding for instance the structural factors that control the uniformity of reaction rate distributions in CLs is the basis for determining the overall effectiveness factor of catalyst utilization. As the main product of the fuel cell reaction, the presence of water is unavoidable in PEFCs (2). Water molecules determine the strength of molecular interactions, which control the formation and stability of self-organized structures in PEMs and CLs (63). In PEMs, liquid-like water acts as the pore former, pore filler and proton shuttle (72). Water distribution and the morphology of aqueous pathways determine proton conduction in PEMs and CLs. In the pore space of gas diffusion electrodes, the liquid water saturation determines the rate of convective and diffusive transport of reactants to/from the catalyst sites. A poorly balanced water distribution in the fuel cell can severely impair its performance and cause long-term degradation effects. If PEM or CLs are too dry, proton conductivity will be poor, potentially leading to excessive Joule heating, which could affect the structural integrity of the cell. Too much water in diffusion media (CLs and GDLs) could block the gaseous transport of reactants. Essentially, all processes in PEFCs are linked to water distribution and the balance of water fluxes. Establishing the linkages between microstructures and water balance in cell components requires fundamental understanding and knowledge of parameters for the following major aspects: How does the local water content in fuel cell media depend on material's microstructure and operating conditions? By which mechanisms does it local equilibrium? What are the relevant mechanisms and transport coefficients of water fluxes (diffusion, convection, hydraulic permeation, electro-osmotic drag)? What are the mechanisms and rates of phase changes (between liquid water, water vapour, surface water, strongly bound water)? What are the rates of transfer of water across interfaces between distinct fuel cell media, e.g., at interfaces between PEM and CL?

Effectiveness of Catalyst Utilization (Sect. VII). A plethora of publications in fuel cell research dwell on issues related to Pt utilization in PEFCs, *cf.* (73) for a review. Ambiguous definitions of catalyst utilization have been exploited and contradictory values have been reported, which could lead to a wrongful assessment as to how much the performance of fuel cells could be improved by advanced structural design of catalyst layers. Evidently, the accurate determination of catalyst utilization and effectiveness factors has a major impact on determining priorities in catalyst layer research. Most generally, the effectiveness factor of Pt utilization of a CL can be defined as the overall current conversion rate divided by the ideal rate if all Pt atoms were used equally in electrochemical reactions at given electrode potential and externally provided reactant concentrations. The effectiveness factor of Pt utilization is intimately linked to the uniformity of reaction rate distributions. The latter is affected by processes at all levels, including surface transport and kinetics at Pt nanoparticles as well as transport of reactants and product water at agglomerate level, catalyst layer level, MEA level, unit cell level, and stack level.

III. PROTON TRANSPORT IN PEMS

Presently the most widely used and tested polymer electrolyte membranes are perfluorinated sulfonated polymers like Nafion® (DuPont), Dow (Dow Chemicals), Flemion® (Asahi Glass) and Aciplex®. They vary in chemical structure, ion exchange capacity, and thickness. The membrane morphology and the basic mechanisms of proton transport are, however, similar. The base polymer, depicted schematically in Fig. 3, consists of Teflon-like backbones with randomly attached pendant sidechains, terminated by charged hydrophilic segments or endgroups (usually sulfonic acid groups, $-SO_3H$) (74–76).

Current membrane development focuses on perfluorinated ionomers, hydrocarbon and aromatic polymers and acid–base polymer complexes. Recently, polyether and polyketo polymers with statistically sulfonated phenylene groups such as sPEK, sPEEK and sPEEKK or polymers on the basis of benzimidazole (sulfonated PBI) have been applied as well. Recent reviews on membrane synthesis and experimental characterization can be found in (48, 77–82).

In this section, we will explore the role of the membrane environment on proton transport properties. In this context, it is insufficient to consider the membrane as an "inert container for water

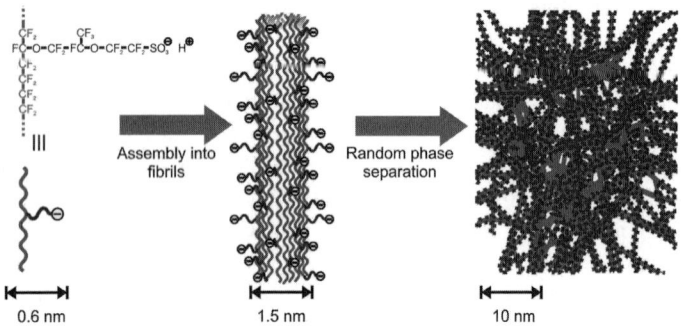

Figure 3. Schematic depiction of the structural evolution of Polymer Electrolyte Membranes. The primary chemical structure of the Nafion-type ionomer on the left with hydrophobic backbones, sidechains and acid head groups evolves into polymeric aggregates with complex interfacial structure (*middle*). Randomly interconnected phases of these aggregates and water-filled voids between them form the heterogeneous membrane morphology at the macroscopic scale (*right*).

Color Plates

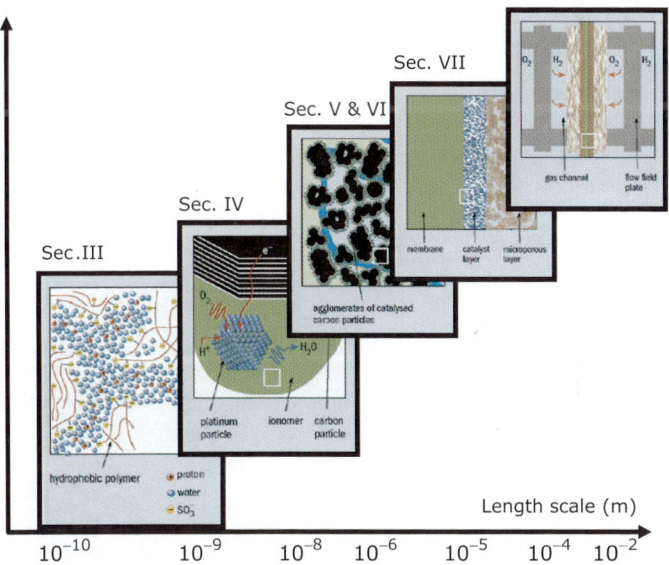

Figure 2. Multi-scale phenomena in PEFC, from fundamental proton transport in PEM (Sect. III), to kinetic mechanisms at nanoparticle electrocatalysts (Sect. IV), to structure formation (Sect. V) and effective properties (Sect. VI) of complex composite materials, to transport, reaction and performance at the macroscopic device level (Sect. VII).

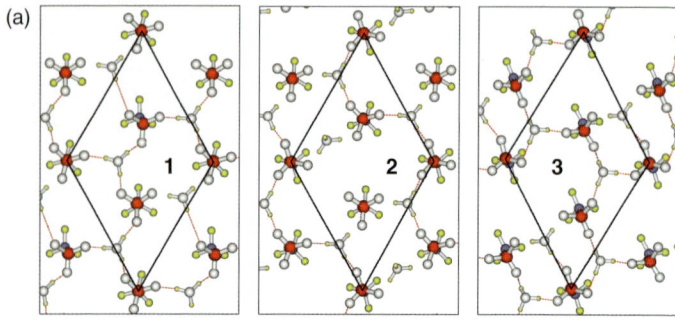

Figure 7. (**a**) Molecular mechanism of interfacial proton transfer at the minimally hydrated array.

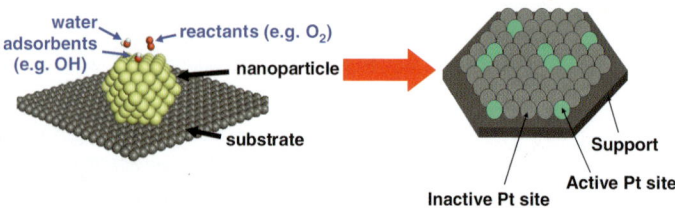

Figure 8. Mapping of surface structure of supported nanoparticle onto a 2D regular hexagonal array, distinguishing active and inactive sites.

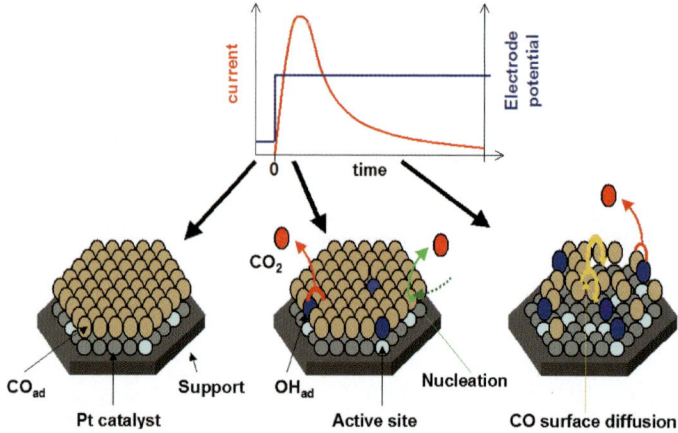

Figure 10. Chronoamperometric current transients and surface processes on catalyst model surface during CO_{ad} electrooxidation.

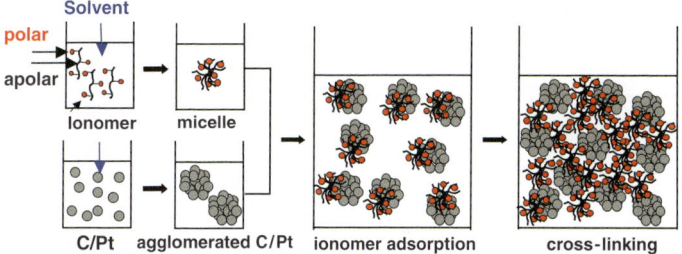

Figure 14. Schematic representation of structural formation processes during the fabrication of conventional catalyst layers in PEFC.

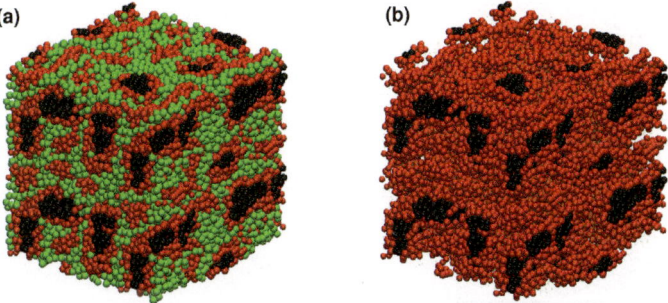

Figure 16. Equilibrium structure of a catalyst blend composed of Carbon (*black*), Nafion (*red*), Water (*green*) and implicit solvent. Hydrophilic domains are not shown in (**b**) for better visualization.

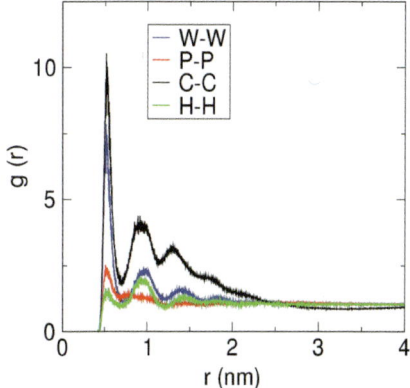

Figure 17. Site–site radial distribution functions for the CNWS system (*C* carbon; *P* polymer backbones; *W* water; *H* cluster containing hydronium).

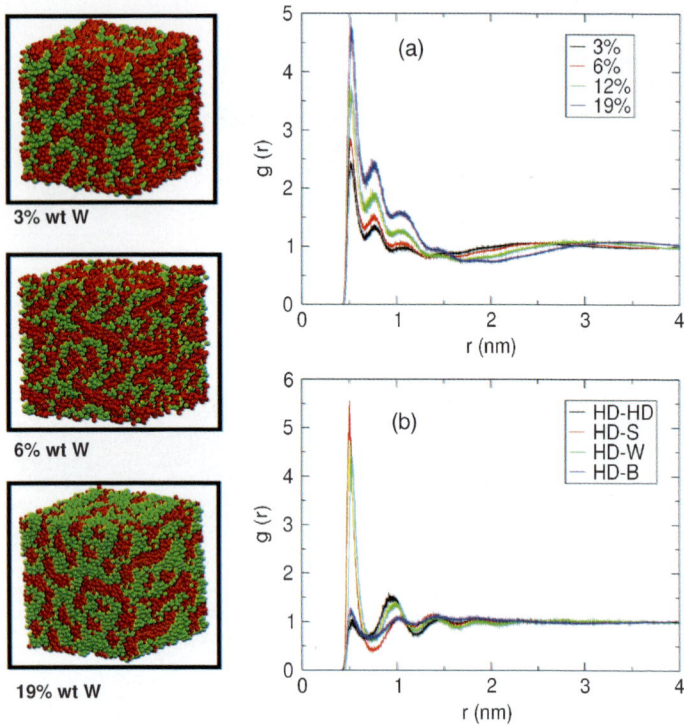

Figure 19. (a) Snapshots of the final microstructure in hydrated Nafion membrane at different water contents. Hydrophilic domain (water, hydronium and side chains) is shown in *green*, while hydrophobic domain is in *red*. (b) Site–site RDF showing the separation of hydrophilic and hydrophobic domains in Nafion membrane. *W* water; *S* side chain; *H* hydronium; *B* ionomer backbone.

pathways." Rationalizing the dependence of membrane conductivity on water content is a multiscale problem. Figure 3 depicts the three major levels in the structural evolution from primary chemical architecture to polymer aggregates at the nanometer scale to random heterogeneous morphology at the macroscale.

When exposed to water, self-organization of polymer backbones leads to the formation of a hydrophobic skeleton that consists of interconnected elongated fibrillar or lamellar aggregates (83, 84). The $-SO_3H$-groups dissociate and release protons as charge carriers into the aqueous sub-phase that fills the void spaces between polymer aggregates. Polymer sidechains remain fixed at the surfaces of those aggregates where they form a charged, flexible interfacial layer. The structure of this interface determines the stability of PEMs, the state of water, the strength of interactions in the polymer/water/ion system, vibration modes of sidechains, and mobilities of water molecules and protons. The charged polymer sidechains contribute elastic (entropic) and electrostatic terms to the free energy. This complicated interfacial region thereby largely contributes to differences in performance of membranes with different chemical architectures. Indeed, the picture of a 'polyelectrolyte brush' could be more insightful than the picture of a well-separated hydrophobic/-philic domain structure in order to rationalize such differences (85).

Proton conductivity depends on the distribution and structure of water and dynamics of protons and water molecules at multiple scales. In order to describe the conductivity of the membrane, one needs to take into account explicit polymer–water interactions at molecular level, phenomena at polymer/water interfaces and in pores at mesoscopic scale, and the statistical geometry and percolation properties of aqueous domains at macroscopic scale.

Well-humidified PEMs utilize the exceptional proton mobility in bulk-water, indicated by an activation energy of proton transport of ~ 0.12 eV (44, 86). The relevant mechanism of structural diffusion of a protogenic defect has been unraveled by *ab initio* molecular dynamics simulations (29–31, 87, 88).

A major incentive in membrane research for PEFCs is the development of advanced proton conductors that are suitable for operation at intermediate temperatures (120–200°C). The evaporation of weakly bound water at temperatures exceeding 90°C extinguishes the most favorable mechanism of proton transport through bulk water. Inevitably, aqueous-based PEMs for operation of PEFCs at such temperatures have to attain high rates of proton transport with a minimal amount of water that is tightly bound to a stable host polymer (47, 48, 78, 89). The development of new PEMs, thus, demands efforts in understanding of proton transport mechanisms under such conditions and their impact on effective proton conductivity at the macroscopic scale.

Minimally hydrated PEMs could only retain the strongly bound water molecules near polymeric aggregates. Sample-spanning pathways of proton transport along lowly hydrated acidic sidechains persist at small water contents, but proton conductivity decreases to a small residual value, with activation energies of proton transport rising to >0.35 eV (44, 86). The microscopic mechanism of proton transport changes, since narrow pores or necks in minimally hydrated PEMs could not perpetuate the high bulk-like proton mobility. In this regime, strong correlations at the polymer/water interface become vital. The compelling, yet unresolved questions are: What are the correlations and mechanisms of proton transport in the interfacial layers? Is high proton mobility possible under conditions of minimal hydration?

The complications for the theoretical description of proton transport in the interfacial region stem from fluctuations of the sidechains, their random distributions at polymeric aggregates, and their partial penetration into the bulk of water-filled pores. The importance of an appropriate flexibility of hydrated sidechains has been explored recently in extensive molecular modeling studies (50).

Several works in theory and molecular modeling have explored the role of acid-functionalized surface groups on proton conduction mechanisms in water-filled nanopores of PEMs (13, 37, 38). Empirical valence bond (EVB) approaches, introduced by Warshel and coworkers (90–92), are an effective way for incorporating environmental effects on bond-breaking and bond-making in solution. They are based on parameterizations of empirical interactions between reactant states, product states and where appropriate a number of intermediate states. The interaction parameters, corresponding to off-diagonal matrix elements of the classical Hamiltonian, are calibrated using *ab initio* potential energy surfaces in solution and relevant experimental data. This procedure significantly reduces the computational expenses of molecular level calculations in comparison to direct *ab initio* calculations. EVB approaches thus provide a powerful avenue for studying chemical reactions and proton transfer events in complex media, with a multitude of applications in catalysis, biochemistry, and proton-conducting PEMs.

Voth et al. (32, 52) and Kornyshev, Spohr and Walbran (13, 53, 93) adopted EVB-based models to study the effect of confinement in nanometer-sized pores and the role acid functionalized polymer walls on solvation and transport of protons in PEMs. The calculations by the Voth group displayed a strong inhibiting effect of sulfonate ions on proton motions. The EVB model by Kornyshev, Spohr and Walbran, was specifically designed to study effects of charge delocalization in SO_3^- groups, sidechain density, and of the fluctuations of sidechains and headgroups on proton mobility. It was found that proton mobility increases with increasing delocalization of the negative countercharge on SO_3^-. The motion of sulfonate groups increases the mobility of protons. Conformational motions of the sidechains facilitate proton motion as well. EVB-based studies were able to qualitatively explain the increase in proton conductivity with increasing water content in PEMs.

Continuum dielectric approaches have been used to study proton conductivity in model pores of PEMs with slab-like geometries. Polymer sidechains and anionic countercharges were represented as regular arrays of immobile point charges in (37) Poisson–Boltzmann theory was used to calculate the distribution of proton density $\rho^+(z)$ in pores and charge transfer theory was applied to determine electrostatic contributions to activation energies of proton transport in slab-like model pores. In the absence of correlations effects in proton transport, the conductance of a slab-like model pore is given by the product of proton mobility ($\mu^+(z)$) and proton density ($\rho^+(z)$), integrated over the pore thickness dimension,

$$\Sigma_p = \frac{L_x}{L} \int_{-z_0}^{z_0} dz \, \mu^+(z) \, \rho^+(z), \tag{1}$$

where L is the length of the pore and $\pm z_0$ denote the positions of the interfacial layers (39, 74). The main element of that work was the distinction between surface and bulk contributions to pore conductance. It was found that the region for surface conduction is confined to a layer corresponding to the thickness of about one monolayer of water near the charged interfaces. The bulk contribution is mainly affected by the concentration of protons $\rho^+(z)$, which increases from the pore centre towards the interfaces. On the other hand, surface mobility of protons is suppressed in the vicinity SO_3^- groups due to the existence of large Coulomb barriers. A higher density of SO_3^- groups diminishes Coulomb barriers and thus facilitates proton motion near the surface. With increasing water content in the pore, the trade-off between proton concentration and mobility was found to shift in favor of the bulk conduction. A refinement of this single pore model in (38) incorporated charge delocalization and thermal fluctuations of SO_3^- groups. These effects significantly reduce the interfacial Coulomb barriers, thereby facilitating proton motion near the interface.

EVB-based MD simulations as well as continuum dielectric approaches involve empirical correlations between the structure of acid-functionalized interfaces in PEMs and proton distributions and mobilities in aqueous domains. These approaches remain inconclusive with respect to the role of packing density, conformational fluctuations and charge delocalization of sidechains and SO_3^- groups on molecular mechanisms and rates of proton conduction. In particular, they fail in describing proton conduction in PEMs under conditions of low hydration with less than two water molecules added per sulfonate group, where interfacial effects prevail.

Overall, the effects of confinement and low hydration still represent great challenges for theory and molecular modelling. Due to their empirical foundation, the approaches described so far provide insufficient understanding of fundamental interactions between polymer groups, ionized side chains, water, and protons. It is not known how length, density, chemical structure and random distribution of sidechains and charged head groups determine water binding and molecular mechanisms of PT in hydrated channels or pores of nanoscale dimensions. As a result, theory still cannot explain the observed increase of the activation energy of PT from 0.12 eV for fully humidified PEMs to >0.35 eV for dehydrated PEMs.

On the other hand, the merits of such insights are obvious. It would become possible to evaluate the relative importance of surface and bulk mechanisms of PT. The transition from high to low proton mobility upon dehydration could be related to molecular parameters that are variable in synthesis. Conditions could be determined, for which high rates of interfacial PT could be attained with a minimal amount of tightly bound water. As an outcome of great practical value, this understanding could facilitate the deliberate design of proton conductors that operate well at minimal hydration and $T > 100°C$ (48).

Molecular modelling of PT at dense interfacial arrays of protogenic surface groups in PEMs warrants *ab initio* quantum mechanical calculations. In spite of the dramatic increase in computational resources it is yet a dream to perform full ab initio calculations of proton and water dynamics within realistic pores of PEMs. This venture faces two major challenges (i) structural complexity and (ii) the rarity of proton transfer events (39). The former defines a need for simplified model systems. The latter enforces the use of advanced computational techniques that permit an efficient sampling of rare events (94–98).

Ab initio molecular level simulations in PEM modelling based on density functional theory were employed by Paddison et al. (14, 50) in order to study sidechain correlations and examine direct proton exchange between water of hydration and surface groups. Detailed *ab initio* calculations by Paddison and Elliott of hydrated polymeric fragments, including several sidechains, are insightful in view of fundamental polymer–water interactions (49). These approaches ignore, however, correlation effects that arise in 2D interfacial conformations. As we will discuss below, such effects dramatically influence hydrogen-bond formation, acid dissociation, and flexibility of surface groups at hydrated interfaces.

A trifluoromethane sulfonic acid monohydrate (TAM) solid was explored in (51). The regular structure of the crystal (99) provides a proper basis for controlled ab initio molecular dynamics. The Vienna Ab initio Simulation Package (VASP) based on density functional theory (DFT) was used to study the dynamics in the system (100–103). Overall, an MD trajectory of >200 ps was simulated. This timeframe is still by far too short for direct observation of proton transfer events, which occur on time scales >1 ns. Intermittent introduction of a proton-hole defect triggered the transition from the native crystal structure, with localized positions of protonic charges, to an activated state with two delocalized protons, as indicated in Fig. 4. One of these protons resides within a Zundel-ion, $H_5O_2^+$ whereas the other one is accommodated between two SO_3^- groups, which approach each other at hydrogen-bond distance. The formation of the latter sulfonate $O \cdots H \cdots O$ complex requires a considerable re-arrangement of the crystal structure. The two almost simultaneously formed proton-complexes stabilize the intermediate

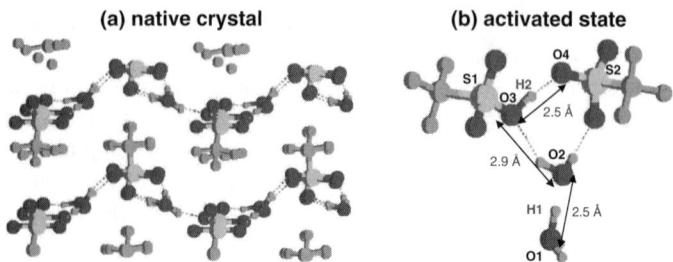

Figure 4. (a) The regular structure of the triflic acid monohydrate crystal as observed during microcanonical sampling. Each H_3O^+-ion is hydrogen-bonded to three neighboring sulfonate groups within a stable pyramidal arrangement. (b) Activated state with two delicalized protons (H1, H2).

state. The energy of formation for the defect state is approximately 0.3 eV. These calculations suggest that an appropriate flexibility of anionic sidechains could be vital for high proton mobility in PEMs under conditions of minimal hydration and high anion density. Furthermore a drift of the Zundel-ion was observed, which indicates its possible role as a relay group for proton shuttling between hydronium ions or sulfonate anions.

The model system considered in (51) is, however, rather stiff in the sense that chemical composition and water content of the crystal are fixed. Only minimal hydration of the system could be considered. A more recently started work aims explicitly at the understanding of structural correlations and dynamics at acid-functionalized interfaces between polymer and water in PEMs (33). It directly addresses the question how to increase proton conductivity in PEMs under conditions of minimal hydration.

The model in (33) emerges from the self-organized morphology of the membrane at the mesoscopic scale that is shown in Fig. 5a. The random array of hydrated and ionized sidechains is tethered to the surface of aggregated hydrophobic polymer backbones. Relevant structural properties include the shape, thickness, persistence length of aggregates, as well as the density and effective lengths of sidechains on their surface. In order to obtain a computationally feasible model for *ab initio* calculations, it is assumed that to a first approximation the highly correlated interfacial dynamics of sidechains, protons and water decouples from the dynamics of the polymeric aggregates. This implies that these supporting aggregates

Electrochemical Materials for PEM Fuel Cells

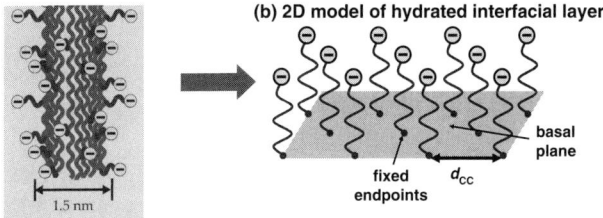

Figure 5. (a) Fibrillar aggregate of polymer backbones with charged sidechains tethered to the surface. (b) 2D array of sidechains with endpoints fixed at the positions of a regular hexagonal grid.

form a fixed frame of reference for the surface groups. The supporting aggregate layer is assumed to form an inert basal plane, anchoring the sidechains or surface groups (SGs). Terminal carbon atoms of the SGs are fixed at the positions of a regular hexagonal grid on this basal plane. The resulting primitive model consists, thus, of a regular hexagonal array of surface groups with fixed endpoints, as depicted in Fig. 5b. In spite of the highly simplified structure this model retains essential characteristics for studying structural conformations, stability and the concerted dynamics of polymer sidechains, water and protons at polymeric aggregates in PEMs. The approach implies that the effect of polymer dynamics on processes inside pores is primarily due to variations in chemical architecture, packing density, and vibrational flexibility of SGs.

Calculations in (33) focused on the shortest SGs, i.e., CF_3SO_3H, under conditions of minimal hydration, i.e., with one H_2O per group. The main parameter considered is the nearest neighbor distance of C atoms, d_{CC}, corresponding to the packing density of SGs. It was varied between $5\,\text{Å} \leq d_{CC} \leq 12\,\text{Å}$, which encompasses the range of variations for prototypical membranes (104–107). Upon deswelling of the membrane, sidechain separations are likely to decrease leading to more pronounced sidechain–sidechain correlations (108). On the other hand, experimental data also suggest that mean sidechain separations in the dry state should not drop significantly below $\sim 7\,\text{Å}$, since otherwise random ionomers could not form stable aggregates upon hydration.

The Vienna Ab initio Simulation Package (VASP) based on DFT was used (100–103). Figure 6 displays the formation energy,

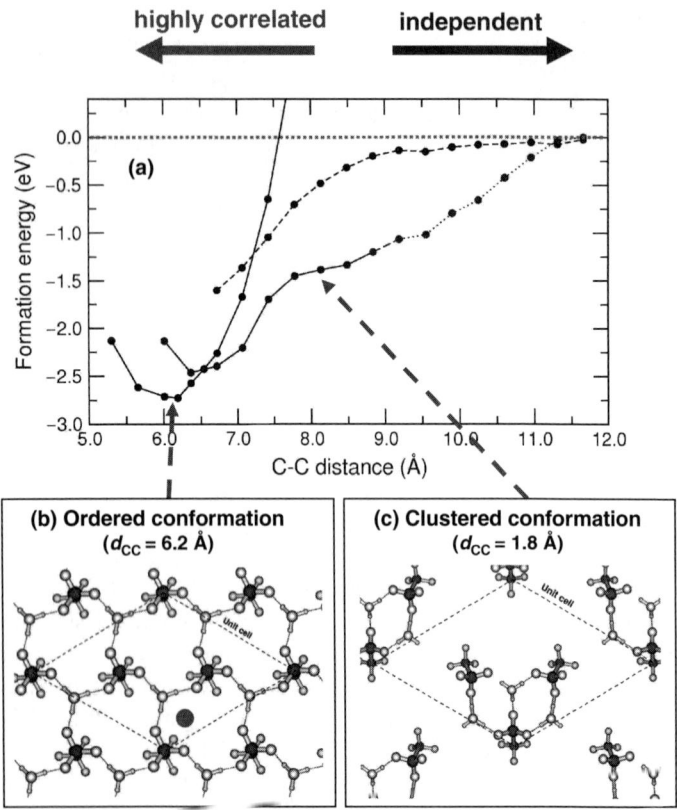

Figure 6. (a) Formation energy (per unit cell) of the minimally hydrated array of surface groups (CF$_3$SO$_3$H). Highly correlated and clustered conformation at two distinct packing densities of surface groups. Figure 6a is reprinted from A. Roudgar, S. P. Narasimachary, and M. Eikerling, *Hydrated Arrays of Acidic Surface Groups as Model Systems for Interfacial Structure and Mechanisms in PEMs*, J. Phys. Chem. B **110**, 20469, Copyright @ 2006 with permission from ACS.

E_f^uc, of stable interfacial conformations as a function of d_CC. E_f^uc is defined as the difference between the optimized total energy for a given value of d_CC, $E_\text{total}(d_\text{CC})$ and the total energy of a system of three independent surface groups, each with one water molecule, in the limit of infinite separation, E_total^∞,

$$E_\text{f}^\text{uc}(d_\text{CC}) = E_\text{total}(d_\text{CC}) - E_\text{total}^\infty. \tag{2}$$

As can be seen in Fig. 6, at high density, $d_{CC} < 6.7$ Å, ionized SGs and hydronium ions (H_3O^+) form an ordered "upright" conformation with full dissociation of all acid groups. At $d_{CC} > 6.7$ Å, cluster-like "tilted" conformations were found. The conformational transition at $d_{CC} \simeq 6.7$ Å is accompanied by a sharp transition from strong (>0.6 eV) to weak binding (<0.1 eV) of additional H_2O upon decreasing d_{CC}. In fact, these results suggest that the highly charged, minimally hydrated interface becomes hydrophobic at $d_{CC} < 6.7$ Å. This intriguing effect is due to strong interfacial correlations and the trigonal symmetry of H_3O^+ and ionized SO_3^- head groups. Notably, $d_{CC} \approx 7$ Å has been earmarked in experiment as a critical value for the occurrence of long-range proton conduction at lipid and stearic acid monolayers (36, 42, 43). According to the results of calculations in (33), this value represents a favorable trade-off between long-range correlations and flexibility of SGs. Arrays with longer SGs, more closely resembling the sidechains in real PEMs, exhibit similar interfacial conformations and structural transitions. For the longest SGs considered so far, which correspond to the sidechains in Dow membranes, 2D correlations and partial dissociation persist up to $d_{CC} \simeq 15$ Å.

Recently, *ab initio* Car-Parinello Molecular Dynamics (109, 110) calculations were used to study proton transfer at minimally hydrated interfaces with dense packing of proton-binding surface groups, corresponding the critical separation $d_{CC} \simeq 6.7$ Å. It was found that concerted tilting-rotation modes of the group that acts as acceptor/donor site for the transferring proton facilitate elementary interfacial proton exchange depicted in Fig. 7. The analysis of frequency spectra suggests that the local fluctuating modes couple only weakly to the relaxation of the remaining interfacial system. The analysis of potential energy surfaces provides the lowest energy path and the activation energy of the elementary proton exchange. These properties are currently being explored for varying surface group densities.

The goal of such calculations will be to discern the pathways of long-range PT along the minimally hydrated interface, for which the ionized SGs could act as relay groups. This structural diffusion mechanism will probably involve several steps of molecular rearrangements of the type in Fig. 7 and require the creation of proton-hole defects. At larger d_{CC} and higher degrees of hydration, surface groups will play a less active role in facilitating long

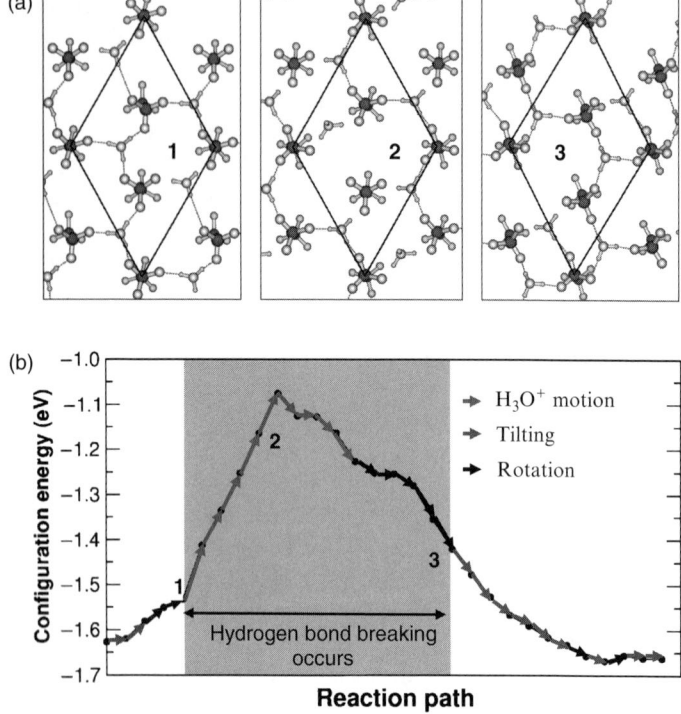

Figure 7. (a) Molecular mechanism of interfacial proton transfer at the minimally hydrated array and (b) configuration energy along reaction path. (*See Color Plates*).

range PT. Mechanisms of proton exchange between the interfacial network and adjacent water layers will be, nevertheless, of great importance for the interplay of bulk and surface transport as well as for the coupling between proton and water mobilities in well-hydrated membranes.

The overall objective is to unravel mechanisms of interfacial PT. This requires identification of collective coordinates (or reaction coordinates) and transition pathways of transferring protons. Differences in activation energies and rates of corresponding mechanism due to distinct polymer constituents, acid head groups, sidechain lengths, sidechain densities, and levels of hydration could be examined. Comparison with experimental data (e.g., NMR, IR, pulse experiments, and scanning electrochemical microscopy) on molecular mobilities in PEMs, biomembranes, and lipid bilayers and on conductivities (e.g., Arrhenius plots) could help to rationalize the effects of interfacial structure and segmental motions of sidechains on PT. In turn, calculations could stimulate new experiments on well-defined 2D structures. These studies could lead to constitutive relations between the dynamics at the interface and polymer architectures.

IV. NANOPARTICLE ELECTROCATALYSIS

The exchange current density j^0 is a measure of the capability of a CCL to convert chemical flux of reactants into an electronic current. It determines the value of the overpotential needed to attain the targeted fuel cell current density. This property links fundamental electrode theory with practical aspects of fuel cell performance. It is a major objective of efforts in materials science and fuel cell electrochemistry to develop novel catalyst systems and electrode designs that maximize the value of j^0. At the same time the catalyst loading, $[m_{Pt}]$, should be minimal in order to minimize costs of the fuel cell system, as explained in (68). The single most important hurdle to be cleared on the route to commercialization of PEFC technology is, therefore, to raise the ratio of current conversion rate to Pt loading, i.e., $j^0/[m_{Pt}]$, by a factor of 10–100. Detrimental factors for electrocatalytic activity and utilization of the catalyst could be identified at a hierarchy of scales, from nano- to macroscopic. Implications of the full interplay of structural effects at different scales will be discussed in Sect. VII.

In this section, we focus on the specific exchange current density, j^{0*}, normalized to the unit surface area of the electrocatalyst, which is the key target of fundamental studies in electrocatalysis (111, 112). It is not a mere materials constant, but it depends on size distributions of catalyst nanoparticles, their surface structure as well as the surface composition in the case of alloy catalysts like PtRu. In this section, we discuss approaches in theory and modeling that highlight particle size effects and the role of surface heterogeneity in fuel cell electrocatalysis.

Currently, the best-performing catalyst layers in PEFC contain highly dispersed Pt (or Pt-alloy) nanoparticles. Dispersion on a high surface area carbon support enhances the electrochemically active surface area. The synthesis of supported catalysts involves impregnation of high surface area carbon materials with molecular cluster precursors of Pt (e.g., H_2PtCl_6) and reduction of Pt on carbon (113, 114). Pt clusters attach to hydrophilic sites on the carbon surface (115). Adsorption energies of the metal clusters are characteristic of chemical adsorption processes, involving the formation of covalent bonds between particle and substrate.

It is known from numerous experimental studies that size, surface morphology, electronic structure of catalyst nanoparticles as well as the properties of the substrate determine the overall electrochemical activity (23). Particles with sizes in the range of 3 nm give high surface-to-volume ratios of the number of catalyst atoms, $\varepsilon^{S/V} > 50\%$. Since only surface atoms could be potentially active, this has a direct impact on catalyst utilization, a key target in fuel cell development. Predictive relations between particle size and activity are, however, difficult to establish, since the size of particles also affects electronic and geometric properties at their surface (23). A complex interplay of elementary surface processes, including molecular adsorption, surface diffusion, charge transfer, recombination, and desorption, determines observable rates of relevant reactions, i.e., reduction of oxygen and oxidation of hydrogen, methanol or carbon dioxide (56, 59, 116).

Not surprisingly then, all relevant electrochemical reactions in PEFC exhibit peculiar sensitivities to the surface structure of the catalyst (117). The abundances of the different surface sites, e.g., edge sites, corner sites, or sites on particular crystalline facets, are closely related to the size of the nanoparticles (118, 119). Substrate-particle interactions may alter the electronic structure of catalyst surface atoms at the rims with the substrate (120) and the substrate may serve as a source or sink of reactants via the so-called spillover effect (56, 121, 122). Finally, as shown by electronic structure calculations, the electronic structure on the surfaces of small metal clusters in the range of a few nanometer starts to deviate from the electronic structure of well-defined single-crystalline surfaces (123, 124), affecting the potential energy surface experienced by molecules on the catalyst surface.

All of these phenomena translate into observable dependences of the electrocatalytic activity on particle size. Diverging trends have been observed for different reactions. Oxygen reduction on Pt shows a maximum in mass activity for particle diameters of ∼3 nm

(125–129). In single nanoparticle experiments of hydrogen reduction kinetics on Pd particles, an increase by two orders of magnitude in turnover rates (normalized current per surface atom) was found upon reduction of the Pd cluster size from 10 to 1 nm (19, 56). On the other hand, for methanol electrooxidation a significant drop in activity was observed when the particle diameter was below 5 nm (130–134).

Due to the complex nature of structural effects in electrocatalysis only very few theoretical studies have been successful in establishing systematic trends in structure vs. reactivity relations. The outstanding example is the d-band model of Hammer and Nørskov (124, 135, 136). It manages to relate trends in chemisorption energies for various adsorbates on transition metal surface to the position of the d-band centre, the first moment of the density of states from the Fermi-level. Systematic DFT calculations and experiments on series of polycrystalline alloy films of the type $Pt_3 M$ (M = Ni, Co, Fe, and Ti) have confirmed predicted correlations between the position of the d-band center, oxygen chemisorption energies, and electrode activities for the oxygen reduction reaction (137). This success of the d-band model has fostered efforts in devising DFT-based high throughput combinatorial screening schemes for identifying highly active electrocatalyst materials (138). In view of their transferability to fuel cell electrocatalysis, it should be mentioned that d-band model and screening methods based thereon have been explored for idealized slab-geometries only, but not for nanoparticle systems.

Usually, insights gained from DFT calculations are not straightforwardly applicable to materials design. In electrocatalysis, direct DFT calculations have been applied predominantly for studying elementary surface processes on catalyst systems with well-defined periodic slab geometries that mimic single-crystalline surface structures. Due to computational costs, quantum mechanical calculations based on DFT are limited to model structures with a high degree of crystalline order (i.e., slab geometries) or small molecular clusters (atom numbers <100). Moreover, trajectories of the temporal evolution of such systems are in the range of 10–100 ps, complicating the calculation of thermodynamic properties.

Technical catalyst systems that use supported nanoparticles represent, special challenges, due to effects of quantum confinement, irregular surface structures with a large portion of low-coordination atoms, and the widely unexplored influence of the substrate (139, 140). Supported catalyst nanoparticles are relatively large systems, consisting of several 100s of atoms and exhibiting disordered surface structures. Moreover substrate effects have to be accounted for. On the other hand, the detailed atomistic surface structure and quantum effects of the confined particles, modified by the presence of the substrate, determine interactions between particles, the kinetics of processes on the catalyst surface, and they control the bonding strength and the kinetics of bond formation between particle and substrate.

Recent efforts employing DFT calculations started to explicitly explore morphologies and electrocatalytic properties of small metal nanoclusters (141, 142). In (143), calculations have explored the strongly modified chemical properties of Pd nanoclusters supported on Au(111). These calculations help to evaluate the electronic structure effects exerted by the substrate and the consequences of the low coordination of the cluster atoms on energetics and kinetics of binding small molecular or atomic species, e.g., OH_{ad}, CO_{ad}, H_{ad}.

Overall, the complexity of electronic and geometric effects on supported nanoparticle systems as well as the complexity of multistep reaction mechanisms still represent formidable challenges for predictive modeling of nanoparticle reactivity from first-principles. Several efforts in this direction have been reported recently. Neurock and coworkers have developed ab initio kinetic Monte Carlo algorithms to simulate complex surface reaction mechanisms under realistic electrochemiocal conditions. In consecutive steps, they simulated morphology, structure and composition of particles using first principles calculations and thereafter simulated coupled steps of adsorption, desorption, surface diffusion and charge transfer on the total reaction rate of nanoparticle systems (144, 145). Some progress has been made towards the development of multiscale modeling approaches (9). Realistically, it will still need vital refinements in calculation schemes and some growth in computational capabilities before multiscale modeling approaches could reach a mature level, at which they could be fully exploited as predictive tools in materials science.

One technique to overcome computational constraints of conventional DFT techniques is to employ Orbital Free DFT (OF-DFT) calculations. OF-DFT and recent modifications thereof including the coupling of OF-DFT with Kohn–Sham optimization steps provide highly accurate results for complex systems with up to 10,000 atoms due to linear scaling of the wavefunction minimizations with the number of atoms N in the system (146–149). Extension of these methods to *ab initio* molecular dynamics is under investigation. These capabilities of OF-DFT and its combinations will widen the scope of first principles calculations in electrocatalysis. Simulations of real particles with sizes ∼3 nm (∼400 atoms) on a substrate could become feasible. This would allow studying shapes of Pt nanoparticles, their surface morphology, the electronic structure of the Pt/substrate system, and the electronic density of states. Of particular interest will be the position of the d-band center and the distribution of excess charge density at the surface of the metal/substrate system. In addition, adsorption energies of small molecules (hydrogen, oxygen, water, CO) could be calculated as a probe of surface electronic structure effects. Results can be compared with predictions of the d-band model. Overall, this insight will be important for understanding the role of particle size and substrate for reactivity. Furthermore, characterizing the flux of electronic charge upon particle adsorption on the substrate and the corresponding binding energy will be insightful in view of evaluating the stability of particle-substrate bonds and the kinetics of bond-formation and breaking.

In the following part, we will discuss approaches in kinetic modeling that explore the links between particle size, surface heterogeneity and electrocatalytic activity of supported Pt nanoparticles (57, 59). At atomistic resolution, the rates of electrocatalytic processes may differ significantly among different surface sites. A strong differentiation in activity of catalyst surface atoms implies that only a fraction of sites constitute active sites, whereas the remaining surface atoms should be deemed "inactive" (150). This effect is most obvious for alloyed catalyst particles, when a second catalyst material is added to act as active sites through the bifunctional mechanism (151). For CO_{ad} electrooxidation on PtRu alloy catalysts, the Ru atoms on the surface constitute the active sites since they promote the formation of OH_{ad} via water splitting (152, 153), whereas the Pt surface atoms represent the inactive sites which act as a reservoir for adsorbed CO_{ad}. The distinct components of the alloy, thus, promote different reactions (154, 155).

CO_{ad} monolayer (ML) electrooxidation on Pt nanoparticles has a long history as a prototype electrochemical reaction. Moreover, irreversibly adsorbed CO_{ad} is a catalyst poison in PEFCs. Numerous studies have revealed strong effects of particle size and morphology on the electrocatalytic activity for CO_{ad} electrooxidation (57,59–62, 156–161).

Controversy has evolved in particular around the question, whether CO_{ad} surface mobility is a limiting process for the overall activity (162–164). Widely different values of CO_{ad} surface mobilities have been determined, with relatively high values found on flat surfaces (165) and considerably smaller values found on highly corrugated surfaces of small nanoparticles (166,167). Maillard et al. concluded that significant restrictions in CO_{ad} surface diffusivity arise on the smallest Pt nanoparticles (57).

Evidently, a heterogeneous surface model for CO_{ad} electrooxidation on Pt nanoparticles should incorporate the heterogeneous surface morphology and the limited mobility of CO_{ad}. The minimalist's modeling approach is to use a simple two-state model with a fraction $0 < \xi_{tot} \leq 1$ of electrocatalytically active sites as exclusive sites at which OH_{ad} can be formed by water splitting and a fraction $(1 - \xi_{tot})$ of inactive sites, which serve as a reservoir of CO_{ad}. Finite surface mobility of adsorbed CO_{ad} is included, which defines the time for CO_{ad} on inactive sites to reach the active sites. Therefore, a complex interplay of reactant surface mobility and on-site reactivity unfolds (59, 116). As a further simplification, the heterogeneous 3D nanoparticle surface is mapped onto a regular hexagonal 2D array with the two surface states as shown in Fig. 8. The total number of surface sites N is adjusted to match the total number of surface sites on the nanoparticle of a certain shape. A particle diameter of 3 nm corresponds to, e.g., $N = 397$ for a cubo-octahedral particle. Further structure-defining factors of the model are the fraction of active sites, ξ_{tot}, and the numbers of nearest neighbors (NN) of an active

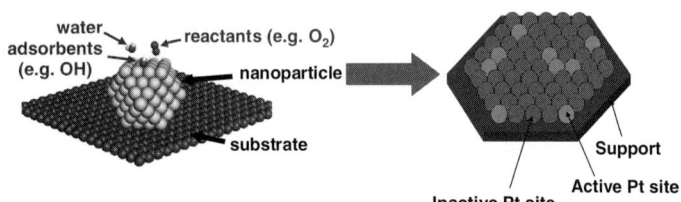

Figure 8. Mapping of surface structure of supported nanoparticle onto a 2D regular hexagonal array, distinguishing active and inactive sites. (*See Color Plates*).

site with other active sites, z_{aa}, or with inactive sites, z_{an}. These NN numbers represent the degree of active site clustering on the surface.

The considered reaction scheme of CO_{ad} electrooxidation follows the Langmuir–Hinshelwood mechanism. It distinguishes reaction steps of CO adsorption, OH_{ad} formation on active sites due to water splitting, surface mobility of CO_{ad}, and $COOH_{ad}$ formation and removal, as discussed in (59). The inclusion of finite surface mobility of adsorbed reactants to active sites leads to a complex interplay of reactant surface mobility and on-site reactivity. The state variables of the model are the surface coverage of CO_{ad}, θ_{CO}, on inactive sites, the CO_{ad}-free fraction of active sites, θ_ξ, and the fraction of active sites covered by OH_{ad}, θ_{OH}. In the mean field (MF) approach, these coverages represent averages, normalized to the disjoint surface fractions of active and inactive sites. The ranges of variation are $0 \leq \theta_{CO} \leq 1$, i.e., normalized to $(1 - \xi_{tot})$, as well as $0 \leq \theta_{OH} \leq 1$ and $0 \leq \theta_\xi \leq 1$, each normalized to ξ_{tot}. The model, moreover, involves a law of CO_{ad} removal from active sites (nucleation process) and it accounts for finite surface diffusion of CO_{ad}.

The balance equations for θ_ξ, θ_{OH}, and θ_{CO}, were formulated and solved with two approaches, an MF model with nucleation processes on active sites and kinetic Monte Carlo (kMC) simulations, as illustrated in Fig. 9. The calculated transient current is

$$j = e_0 \gamma_s \xi_{tot} \left(\nu_{ox}^{aa} + \nu_{ox}^{an} + m\nu_N + \nu_f - \nu_b \right), \tag{3}$$

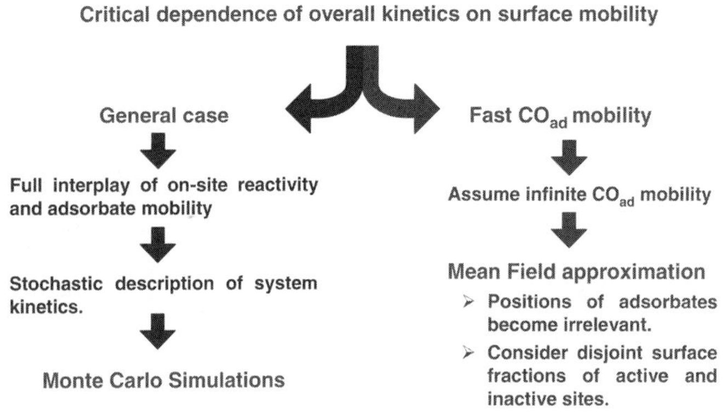

Figure 9. Scheme of approaches in kinetic modeling of nanoparticle electrocatalysis, depending on surface mobility of CO_{ad}.

where $\nu_{ox}^{aa}, \nu_{ox}^{an}$ are oxidation rates (between on OH_{ad} active and CO_{ad} on active sites or OH_{ad} on active and CO_{ad} on inactive sites), $\nu_N = k_N \left(1 - \theta_\xi\right)$ is the nucleation rate, and ν_f and ν_b are forward and backreaction rates of OH_{ad} formation.

It should be noted, that the active site model reduces to the well-known homogeneous MF models for the limit of fast CO_{ad} mobility and $\xi_{tot} = 1$, i.e., with all surface sites being equally active (116, 168–170). At the other end of the mobility scale, for vanishing CO_{ad} mobility on the homogeneous surface (i.e., for $\xi_{tot} = 1$), it can be shown that the active site model is equivalent to nucleation and growth models (171, 172). Accordingly, the active site model represents a generalization of homogeneous surface models towards structured surfaces with local variations in reactivity.

The active site model was solved under conditions resembling CO_{ad} monolayer oxidation. The experimental situation during measurements of chronoamperometric current transients is depicted in Fig. 10. The systematic comparison of model results with experimental transients for various particle sizes and electrode potentials allows to reliably determine structural and kinetic parameters, providing vital insights into particle size effects.

The general solution of the model, that incorporates the effects of heterogeneous surface geometries and finite surface mobilities

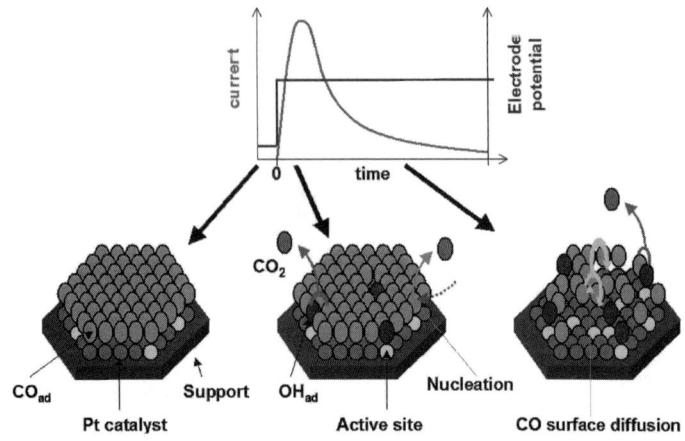

Figure 10. Chronoamperometric current transients and surface processes on catalyst model surface during CO_{ad} electrooxidation. (*See Color Plates*).

of reactants, requires kMC simulations, cf. Fig. 9. This stochastic method has been successfully applied in the field of heterogeneous catalysis on nanosized catalyst particles (173, 174). The temporal evolution of the system takes the form of Markovian random walk through configuration space. This method reflects the probabilistic nature of the electrochemical system as a many-particle system. Since these simulations permit atomistic resolution, any level of structural detail may easily be incorporated. Moreover, kMC simulations proceed in real time. The simulation of current transients or cyclic voltammograms is, thus, straightforward (170). In order to simulate chronoamperometric current transients a variable time step algorithm, as described in (175) was implemented.

As a limiting case of the general active site model, a MF approximation with active sites in the limit of fast CO_{ad} surface diffusion, was considered in (59). The model in this limit accounts for heterogeneous surface structure of the catalyst but assumes uniform coverages of adsorbates on the disjunct surface fractions of active and inactives sites. This simplification permits a straightforward deterministic formulation of the kinetic equations. It provides full analytical solutions in the experimentally relevant case when water splitting is fast in comparison with CO_{ad} removal. Systematic fitting procedures can be used with this model, which can provide starting values for more elaborate fitting of experimental data with the kMC approach. As discussed in (116), correlation effects cause a mismatch between MF approximation and kMC solution in the case of extensive active site clustering.

The model was used to analyze dependences of CO_{ad} electrooxidation on structural and kinetic parameters for various particle sizes and over a wide range of electrode potentials. The good agreement with experimental data is demonstrated in Fig. 11. In comparison with alternative surface models (homogeneous MF approach and nucleation and growth approach), the active site model clearly shows superior agreement with experiment, as illustrated in Fig. 12.

For small particles (1.8–3.3 nm) the fraction of active sites found is $\xi_{tot} \sim 10\%$. It increased only for larger particles with nanograined structure. These findings suggest that active sites are likely to be related to defect sites rather than low-coordination sites of an idealized crystallite structure. All electrochemical steps, i.e., water splitting, nucleation and recombination of OH_{ad} and CO_{ad} could be described well by Tafel-laws, as shown in Fig. 13. Thereby,

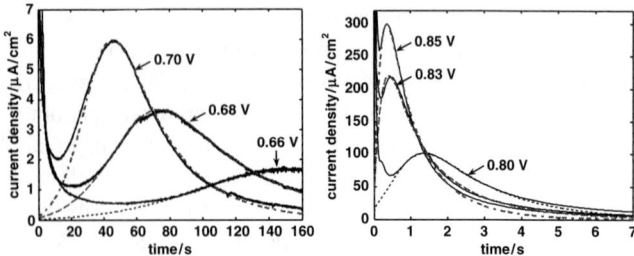

Figure 11. Comparison of experimental chronoamperometric current transients in CO_{ad} monolayer oxidation for particle diameter 3.3 nm with results calculated in the active site model. Fits at small potentials (*left*) were obtained with the mean field approach. At large potentials (*right*), kinetic MC simulations were used to reproduce the data. Reprinted from B. Andreaus, F. Maillard, J. Kocylo, E. Savinova, and M. Eikerling, *Kinetic Modeling of CO_{ad} Monolayer Oxidation on Carbon-Supported Platinum Nanoparticles*. J. Phys. Chem. B **110**, 21028, Copyright @ 2006 with permission from ACS.

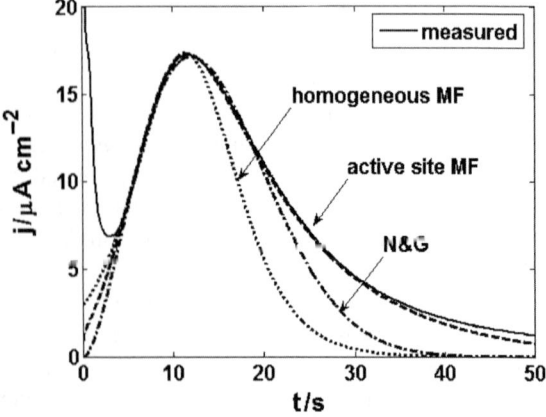

Figure 12. Comparison of active site model with homogeneous mean field model and nucleation and growth model. The active site model exhibits the best agreement with experimental data. Reprinted from B. Andreaus, F. Maillard, J. Kocylo, E. Savinova, and M. Eikerling, *Kinetic Modeling of CO_{ad} Monolayer Oxidation on Carbon-Supported Platinum Nanoparticles*. J. Phys. Chem. B **110**, 21028, Copyright @ 2006 with permission from ACS.

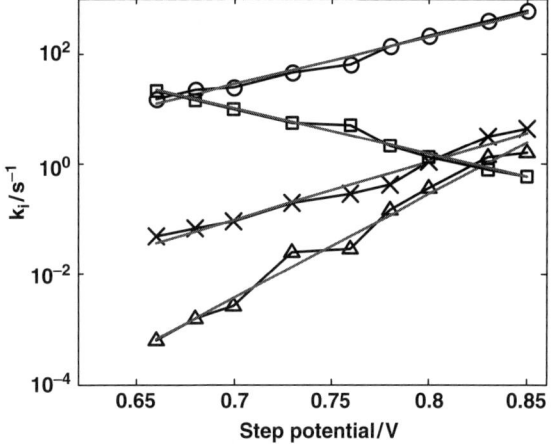

Figure 13. Tafel-plots of kinetic parameters for CO_{ad} monolayer oxidation at Pt nanoparticles of diameter 3.3 nm. Reprinted from B. Andreaus, F. Maillard, J. Kocylo, E. Savinova, and M. Eikerling, *Kinetic Modeling of CO_{ad} Monolayer Oxidation on Carbon-Supported Platinum Nanoparticles.* J. Phys. Chem. B **110**, 21028, Copyright @ 2006 with permission from ACS.

the equilibrium potential of water splitting, E_{OH}^{eq}, electron-transfer coefficients for the various steps, and generic rate constants could be determined. In general it was found that the nucleation process is the limiting step, whereas water splitting (i.e., formation of OH_{ad}) is the fastest process, with the recombination step in-between. Most importantly, CO_{ad} mobility was found to decrease by at least two decades for decreasing particle size, from $> 10^{-14}$ cm^2 s^{-1} for decreasing particle size with diameter >5 nm to $\sim 10^{-16}$ cm^2 s^{-1} for particles with diameter 1.8 nm.

In general the simple approach to modeling electrocatalytic activity on corrugated nanoparticle surfaces reveals a remarkable consistency with experimental data. It establishes vital links between particle size, surface morphology and kinetic processes. Applied to systematic experimental data, the model provides useful diagnostic capabilities for analyzing surface structure and kinetic parameters of surface processes. Extension of the model to alloy PtRu nanoparticles is straightforward. In this case, active sites can be identified with Ru surface atoms. Structural analysis of results provides important information on the fraction and clustering of Ru surface atoms.

Future studies have to clarify the role of active sites and active site clustering as well as the mechanism of nucleation. In this respect, the model could clearly benefit from more detailed electronic structure calculations for systems that could mimic heterogeneous nanoparticle surfaces. These calculations could, moreover, form the basis for refinements in terms of surface structure representation, reaction pathways, surface mobility, and kinetics of charge transfer steps. The role of adsorbate interactions and anion adsorption effects should be evaluated.

V. SELF-ORGANIZATION IN PEMS AND CATALYST LAYERS

Despite recent progress in developing multiscale modeling approaches (9), enormous challenges remain in bridging between atomistic simulations of realistic structures and continuum models that describe the operation of functional materials for PEFC applications. While full multiscale methods will not be available in the near future, meso-scale simulation techniques can close the gap between atomistic simulations and macroscopic properties of the system. Such simulations provide vital insight into aspects that have to be considered in fabrication and operation of advanced materials for PEFCs.

To improve structure–performance relationships of CLs, preparation methods exploit variations in terms of applicable solvent, particle sizes of primary carbon powders, wetting properties of carbon materials, and composition of the catalyst layer ink (64–66). These factors determine the complex interactions between Pt/carbon particles, ionomer molecules and solvent molecules and, therefore, control the catalyst layer formation process. Mixing the ionomer with dispersed Pt/C catalysts in the ink suspension prior to deposition, as depicted in Fig. 14, will increase interfacial surface area between the ionomers and Pt/C nanoparticles. The choice of a dispersion medium determines whether ionomer is to be found in the solubilized, colloidal or precipitated forms. This influences the microstructure and pore size distribution of the CL (66, 176). In general, high catalytic yields hinge on compositions and porous structures with well-attuned pore size distributions and wetting properties of the pore network.

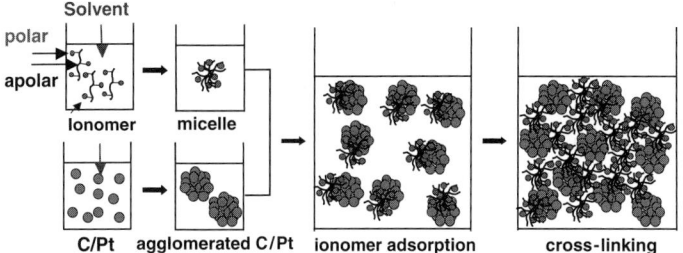

Figure 14. Schematic representation of structural formation processes during the fabrication of conventional catalyst layers in PEFC. (*See Color Plates*).

Agglomerates are mesoscopic structural elements of CLs, which consist of Pt/C particles. They are vital for catalyst utilization and the overall voltage efficiency of the layer (66–68, 177). Apart from characterizing the size, internal porosity and wetting properties of internal/external surfaces of such agglomerates, it is critical to understand composition inside agglomerates, i.e., whether ionomer penetrates into their internal pores. The latter question is of significant fundamental interest as well as of practical importance for performance modeling of CLs and PEFCs at the macroscopic device level.

Mesoscale calculations provide insights into segregation, structural correlations and dynamical behavior of different phases in complex CL. They contribute to furnishing relations between structure, transport properties, and reactivity (4,54,176). Despite an enormous number of phenomenological models, less effort has focused on exploring effects of the microstructure of the CL on transport and reactivity.

Advanced experimental and theoretical tools can address the correlation of transport-reaction processes with structural details at meso-to-micro- and down to atomistic scale. Coarse-grained molecular dynamics (CG-MD) techniques can describe the system at the micro-to-meso level, while still being able to capture the morphology at long time and length scales (178–180). The application of CG models, however, requires special care. Due to the reduced number of degrees of freedom, CG simulations may not be able to accurately predict physical properties that directly rely upon time correlation functions (e.g., diffusion) (180).

A significant number of meso-scale computational approaches have been employed to understand the phase-segregated morphology and transport properties of water-swollen Nafion membranes (8, 13, 45, 46, 53, 178–183). Because of computational limitations, full atomistic models are not able to probe the random morphology of these systems. However, as demonstrated by these simulations and applications to other random composite media, meso-scale models are computationally feasible to capture their morphology. Several approaches have been used such as termed cellular automata (181) and coarse-grained meso-dynamics based on self-consistent mean field theory (178, 179). For Nafion, most of these simulations support the idea that irregularly shaped, nanometer-size clusters of ionic head groups and water form the proton-conducting network that is embedded into the hydrophobic polymer matrix.

Structural complexity is even more pronounced in CLs, since they consist of a mixture of Nafion ionomer, Pt clusters supported on carbon particles, solvent, and water. Independent computational strategies are generally needed in order to simulate the microstructure formation in CLs. For instance, the structure of the ionomer-phase in CLs cannot be trivially inferred from that in membrane simulations, as there are distinct correlations between Nafion, water and carbon particles in CLs. The computational approach based on CG-MD simulations is developed in two major steps. In the first step, Nafion chains, water and hydronium molecules, other solvent molecules and carbon/Pt particles are replaced by corresponding spherical beads with pre-defined sub-nanoscopic length scale. In the second step, parameters of renormalized interaction energies between the distinct beads are specified.

We consider four main types of spherical beads: polar, nonpolar, apolar, and charged beads (184). Clusters including a total of four water molecules or three water molecules plus a hydronium ion are represented by polar beads of radius 0.43 nm. The configuration of an unfolded Nafion chain is shown in Fig. 15. A sidechain unit in Nafion ionomer has a molecular volume of 0.306 nm^3, which is comparable to the molecular volume of a four-monomeric unit of polytetrafluoroethylene PTFE (0.325 nm^3) (178). Therefore, each four monomeric unit (–[–CF_2–CF_2–CF_2–CF_2–CF_2–CF_2–CF_2–CF_2–]–) and each sidechain (represented by a charged bead) are coarse-grained as spherical beads of volume

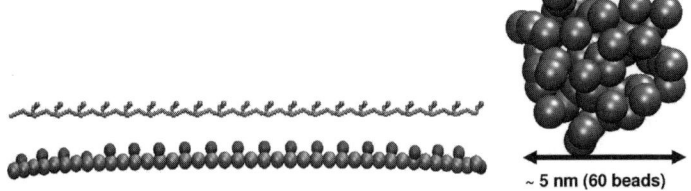

Figure 15. Coarse-grained model of Nafion 20-unit oligomer (*left*), length ~ 30 nm, and carbon particles (*right*), diameter ~ 5 nm.

$0.315 \, \text{nm}^3 (r = 0.43 \, \text{nm})$. The hydrophobic Nafion backbone is replaced by a coarse-grained chain of 20 apolar beads, as illustrated in Fig. 15.

Carbonaceous particles can be coarse-grained in various ways, building upon a new technique called multi-scale coarse graining (MS-CG) (185–187). In this method, the CG potential parameters are systematically obtained from atomistic-level interactions (188). Using this technique, a model was built for semispherical carbon particles based on CG sites in the C60 system, see Fig. 15.

The interactions between non-bonded beads are modeled by the Lennard–Jones (LJ) potential

$$U_{\text{LJ}}(r) = 4 D_{ij}^0 \left[\left(\frac{r_{ij}}{r}\right)^{12} - \left(\frac{r_{ij}}{r}\right)^6 \right], \quad (4)$$

where the effective bead diameter (r_{ij}) is 0.43 nm for side-chain, backbone and water beads. The strength of interaction (D^0) is limited to five possible values ranging from weak ($1.8 \, \text{kJ mol}^{-1}$) to strong ($5 \, \text{kJ mol}^{-1}$) (184). The electrostatic interactions between charged beads are described by the Coulombic interaction

$$U_{\text{el}}(r) = \frac{q_i q_j}{4\pi \epsilon_0 \epsilon_r r} \quad (5)$$

with relative dielectric constant ϵ_r. The effect of solvent is incorporated by changing ϵ_r as well as by varying the degree of dissociation of Nafion sidechains. Interactions between chemically bonded beads (in Nafion chains, for example) are modelled by harmonic potentials for the bond length and bond angle.

$$V_{\text{bond}}(R) = \frac{1}{2} K_{\text{bond}} (R - R_{\text{bond}})^2,$$

$$V_{\text{angle}}(\theta) = \frac{1}{2} K_{\text{angle}} \{\cos(\theta) - \cos(\theta_0)\}^2, \qquad (6)$$

where the force constants are $K_{\text{bond}} = 1{,}250\,\text{kJ}\,\text{mol}^{-1}\,\text{nm}^{-2}$ and $K_{\text{angle}} = 25\,\text{kJ}\,\text{mol}^{-1}$ angle, respectively. R_0 and θ_0 are the equilibrium bond length and angle. The size of the simulation box can vary from $50 \times 50 \times 50\,\text{nm}^3$ to $500 \times 500 \times 500\,\text{nm}^3$, depending on system and composition.

CG-MD simulations in the presence of solvent provide insights into the effect of dispersion medium on microstructural properties of the catalyst layer (63). Due to the slow dynamics of the system, special care should be taken for the equilibration procedure. During simulations, potential energies are monitored to ensure that the system reached to the equilibrium state. Simulations started from a random configuration, where a 500 ps thermalization was first performed.

To explore the interaction of Nafion and solvent in the catalyst ink mixture, simulations were conducted in the presence of carbon particles, water, implicit polar solvent (with different dielectric constant ϵ_r) and ionomer. The snapshot of the final microstructure was analyzed in terms of density map profiles, radial distribution functions (RDFs), pore size distributions, and pore shapes. Figure 16 depicts a snapshot of the Carbon–Nafion–Water–Solvent (CNWS) blend. The interaction parameters of the carbon particles are selected to mimic the properties of VULCAN-type C/Pt particles.

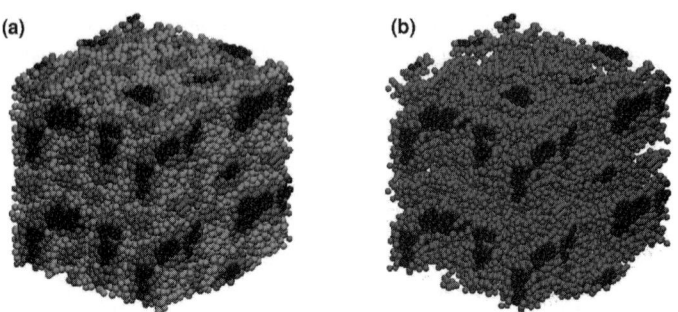

Figure 16. Equilibrium structure of a catalyst blend composed of Carbon (*black*), Nafion (*red*), water (*green*) and implicit solvent. Hydrophilic domains are not shown in (**b**) for better visualization. (*See Color Plates*).

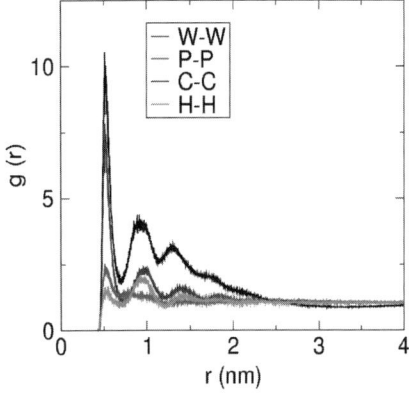

Figure 17. Site–site radial distribution functions for the CNWS system (C carbon; P polymer backbones; W water; H cluster containing hydronium). (*See Color Plates*).

They are hydrophobic, with a repulsive interaction with water and Nafion sidechains, and semi-attractive interactions with other carbon particles as well as with Nafion backbones.

Structural analysis based on site–site RDFs is shown in Fig. 17. The RDFs reveal general features of phase segregation, corresponding to the microstructures in Fig. 16. There is a strong correlation between carbon particles. Ionomer forms a random structure. As expected, hydrated protons (H) and water (W) behave similarly. The correlation between hydrophilic species (H and W) and ionomer (N) is significantly stronger than that between those species and Carbon (C). The autocorrelation functions g_{SS} and g_{HH} exhibit a similar structure as g_{WW}. This indicates a strong clustering of sidechains and hydronium ions due to the aggregation and folding of polymer backbones. The primary S–S and H–H peaks are, however, suppressed compared to the primary peak in g_{WW} due to electrostatic repulsion between these charged beads. g_{BB} and g_{CC} exhibit upturns towards r, superimposed on the primary bead–bead correlations, which correspond to the characteristic dimensions of carbon particles (∼5 nm diameter) and backbone clusters (∼2–3 nm).

More detailed analysis of RDFs reveals a strong correlation of carbon particles and polymer backbones (g_{CB}) (63). This correlation suggests that polymer backbones are attached to the surfaces of carbon particles, while sidechains strive to maximize their separation from the surface of carbon agglomerates. Overall, the correlation functions provide valuable structural information at the nanometer scale that allows refining the picture of the phase-segregated catalyst layer morphology. Ionomer backbones form clusters or fibers that are attached to the surface of carbon particles. Sidechains are expelled from the vicinity of carbon. Water and hydronium ions tend to maximize their separation from the carbon while trying to stay in the vicinity of the sidechains. Structural conformations demonstrate that sidechains, hydronium ions and water molecules form interconnected clusters inside of ionomer domains. A key finding is that iononmer molecules do not penetrate into the carbon agglomerates. Sidechains are buried inside hydrophilic domains with a weak contact to carbon domains.

The effect of the solvent dielectric constant on structural correlations was examined in (63). Overall, the magnitude of the peaks decreases from low to high polar solvent. The polar solvents ($\epsilon_r = 20, 80$) behave similarly, while the effect of the apolar solvent ($\epsilon_r = 2$) is markedly different. Low ϵ implies stronger correlations between C and hydrophobic polymer backbones, exhibiting a stronger tendency to phase-segregate into hydrophobic and hydrophilic domains. The magnitude of short-range interactions and the carbon agglomerate size steadily decrease by increasing ϵ_r. Therefore, the carbon agglomerates become more separated, as the pore size between agglomerates increases. The peak positions for long-distance correlations are shifted to large distances for the apolar solvent. A possible explanation is that a low dielectric solvent causes the formation of a separate clustered phase of carbon particles confined between thick fibrous aggregates of ionomer. In the presence of apolar solvent, the structural organization into separate hydrophilic and hydrophobic domains spans to higher distances compared to polar solvents.

The interaction parameters are validated by calculating the energy of mixture for each pair of solvent–water, solvent–Nafion and water–Nafion components and comparing to that in the experiments. Moreover, site–site W–W radial distribution function obtained from CG-MD simulation are compared with those of the atomistic MD simulation using the force-matching procedure, as shown in Fig. 18.

In order to obtain cluster size distributions of different species in our CG-MD simulations, the clusters of water, polymer and

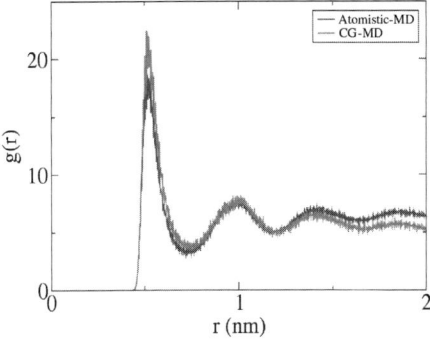

Figure 18. Site–site W–W radial distribution function obtained from CG-MD simulation and compared with that of the atomistic MD simulation using the force-matching procedure.

carbon-particles are first determined by a Monte Carlo procedure. For all simulated morphologies in the presence of three different solvents ($\epsilon_r = 80$, 20 and 2), phase-segregated clusters are produced. Morphologies are typically equilibrated in $<0.5\,\mu s$. The size distribution of bead-clusters, the largest cluster and the total number of clusters is computed inside the simulation box, using periodic boundary conditions. The output is given in the form of a density map (XPM) file. It was shown that the more polar solvent leads to smaller cluster sizes of carbon particles (63). Notice that the clusters formed of carbon particles do not necessarily represent single agglomerates. More analysis together with the simulations at larger length scales is generally required in order to characterize the agglomerates and to explore structural properties of the catalyst ink in terms of dispersion medium.

CG-MD simulations can also predict the mesoscopic structure of the hydrated Nafion membrane with different water contents. Membrane simulations were performed with water contents of 3, 6, 12 and 19 wt%. The mesoscopic structure of the hydrated membrane is visualized in Fig. 19, revealing a sponge-like structure. Water beads together with hydrophilic beads of Nafion sidechains form aggregated clusters, which are embedded in the hydrophobic phase of the Nafion backbones. Our analysis shows that the hydrophilic sub-phase is composed of a three dimensional network of irregular

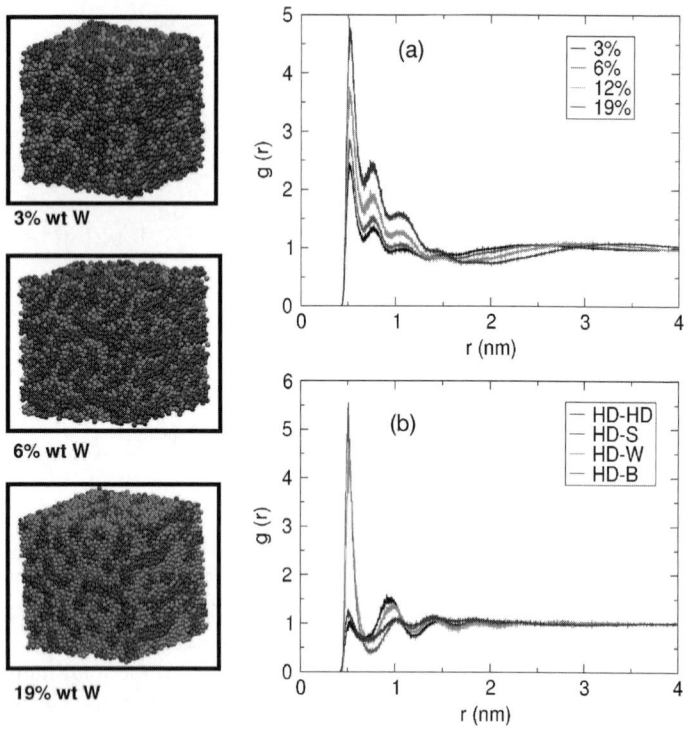

Figure 19. (a) Snapshots of the final microstructure in hydrated Nafion membrane at different water contents. Hydrophilic domain (water, hydronium and side chains) is shown in *green*, while hydrophobic domain is in *red*. (b) Site–site RDF showing the separation of hydrophilic and hydrophobic domains in Nafion membrane. *W* water; *S* side chain; *H* hydronium; *B* ionomer backbone. (*See Color Plates*).

channels. The channel size increases with water content from 1, 2 to 4 nm at 3, 6 and 19 wt% water, respectively. The site-site RDFs obtained from CG-MD simulations match perfectly to those from atomistic MD simulations. The RDF between the sidechain beads and the components of the mixture show that sidechains are surrounded with water and hydrated protons. Water and hydrated protons show similar interactions with sidechains and with ionomer backbones, Fig. 19b.

Proper parameterizations of interactions between Pt/C particles in colloidal dispersion are essential for these calculations. In the calculations reported above, Lennard–Jones-type interaction potentials were employed to describe these interactions. Different interaction potentials, based on Derjaguin–Landau–Verwey–Overbeek (DLVO) theory, are more appropriate for understanding agglomeration phenomena (stability, kinetics) in colloidal dispersion (189). Parameterizations of DLVO-type interactions between Pt/C in the form

$$V(r) = \varepsilon \left[a_1 \left(\frac{\sigma}{r}\right)^{36} - a_2 \left(\frac{\sigma}{r}\right)^6 + a_3 \exp\left(-\lambda \left(\frac{r}{\sigma} - 1\right)\right) \right] \quad (7)$$

include a screened, exponentially decreasing electrostatic repulsion term between charged surfaces in solution (last term). Using these interactions, properly parameterized, in CG-MD simulations, is expected to provide a refined picture of self-organization during structure formation of CLs and PEMs.

VI. EFFECTIVE PROPERTIES OF RANDOM HETEROGENEOUS MEDIA

The operation of Membrane Electrode Assemblies (MEAs) that form the central operating unit of PEFCs involves transport of gases in diffusion media, transport of protons and electrons in conduction media, production of water, two-phase flow and phase transformation of water in porous media, electro-osmotic effects, and electrochemical processes at highly dispersed Platinum-based catalyst nanoparticles. The components of the MEA, i.e., the proton-conducting Polymer Electrolyte Membrane (PEM) and gas diffusion electrodes (GDEs), including catalytic layers (CLs), have to provide some (in PEM: proton transport, electro-osmotic drag, water transport) or all of these functions (in GDE) simultaneously. The materials that could provide this intricate coupling of distinct processes belong to the class of random heterogeneous media. In general, it is a challenge for theory and diagnostics to unravel the structure–property relations of such materials.

Composite porous catalyst layers are the key-components of PEFCs. A major contribution to the potential drop and losses of efficiency in these fuel cells is due to transport limitations of the gaseous reactants in the catalyst layers. A number of models, with different approaches and degrees of complexity, have been presented in literature. In most of those, the frequently used modeling technique is to consider the composite catalyst layer as an effective, pseudo-homogeneous medium, where the flow of gases and water products is described by effective approximations such as Navier–Stokes equation or Darcy's law.

A conventional cathode catalyst layer (CCL) is often modeled as an agglomerated structure with bimodal pore size distribution (Sect. VII). Although, such continuum models avoid the use of local boundary conditions and, to some extent, provide the effects of porosity, more realistic network models of random heterogeneous media are generally preferred. Together with the effect of network topology, and in particular pore interconnectivity, the effect of pore heterogeneity (e.g., roughness) should be accounted for as well.

In a heterogeneous catalyst, reactant molecules diffuse through the pore network, collide with pore walls, and react on active sites on these walls. This implies that the topology of the pore network and the morphology of pores affect the molecular movement and the accessibility of the active sites. Hence the diffusivities of the components and therefore the conversion and product distributions of the reaction depend on the catalyst geometry. On the other hand, porous amorphous catalysts and supports often have a random (fractal) internal surface down to molecular scales (190–193). This has been confirmed by small-angle X-ray scattering and adsorption studies, and is a consequence of the specific preparation conditions of these materials, which are often based on a sol-gel synthesis method.

Numerous works on the effects of this fractal surface morphology on diffusion and reaction phenomena have been performed, from the catalyst fractal pore scale, Fig. 20, all the way up to the scale of industrial reactors (194, 195). The attention was initially focused on Knudsen diffusion, since it is expected that it is the dominant diffusion mechanism in many gas reactions in mesoporous catalysts. It has been shown how the gradientless (self-)diffusion is strongly influenced by surface roughness (193, 196). Model reactions have also been added to study the effective activity and selectivity as a function of surface roughness (197). Especially interesting is to find out whether the diffusivities depend on the rates of reactant conversion. This is important for applications, because it shows when it is allowed to use the easier computations based on traditional continuity equations, using (non-fractal) reaction calculations and fractal diffusivities, which were calculated separately in the absence of reactions. The results have been qualitatively compared to what is known for diffusion in microporous materials. Microporous

Figure 20. Part of a typical 3D pore with a rough fractal internal surface.

materials like zeolites or silicalite, are frequently used for separation, adsorption and catalysts supports. Since these processes are often diffusion limited, it is important to know the molecular diffusivities. The correlation between the chemical heterogeneity and diffusion still remains poorly understood.

It is not feasible, if not impossible, to make exact predictions for the effective properties of composite materials with the simple models of morphology (70, 71, 198). Both continuum and discrete models have been applied to estimate the effective properties of random heterogeneous materials. As described above, continuum models represent the classical approach to describing and analyzing transport processes in materials of complex and irregular morphology. Thus, the effective properties of materials are defined as averages of the corresponding microscopic quantities. The shortcoming of discrete models such as random network models, Bethe lattice models etc. compared to continuum modes, is the demand for large computational effort to represent a realistic model capable of describing the material and simulating its effective properties. It is of course a challenging problem to generate a realization of a heterogeneous material, for which limited microstructure information is available.

VII. RECONCILING THE SCALES

EFFECTS OF STRUCTURE AND WATER IN PEFCs

Since currently used PEMs and CLs in PEFCs are random heterogeneous media, tailoring their physical properties in view of optimized performance of PEFCs is a multiscale problem. It ranges from the molecular dynamics of proton transport in PEMs and elementary chemical processes at the catalyst surface to the macroscopic effective properties of composite materials that govern gas diffusion, water permeation, charge transport, and reaction. In the previous sections, we reviewed several of the major issues arising in these materials and highlighted approaches in theoretical modeling and computational materials science that relate molecular and morphological structures of PEMs and CLs to their performance. At the macroscopic device level, the performance of the PEFC is a complex function of all of these structural effects and processes. The task at hand in view of design and fabrication of advanced functional materials is to reconcile the scales and derive predictive capabilities from resulting structure vs. properties relationships.

There is no principal problem with the level of performance of current PEMs. Proton conductivities (~ 0.1 S cm^{-1}) and stability of Nafion-type membranes meet the requirements of fuel cell developers. Aside from the current cost, the major issue with state-of-the-art PEMs is their reliance on water as the working liquid. The dependence of the operational characteristics of current PEMs on water content incurs the problem of water management at all scales and in all components of the cell (2,4). It is well known that aqueous-based PEMs operate poorly at high operating temperatures due to evaporative loss of weakly bound water, which is needed to facilitate proton conduction. Moreover, in an operating PEFC parts of the PEM close to the anode side could attain a poorly hydrated state at high current densities due to the electro-osmotic drag effect (4–6, 199). This could cause critical current density effects in the fuel cell voltage.

There are two principle solutions to the critical issue of membrane dehydration. At the materials science front, efforts focus on design and fabrication of anhydrous proton conductors that are completely independent of water for their performance (200–205) or membranes that retain sufficient proton conductivity with a minimal amount of water that is stably bound to the host polymer, e.g., employing block-copolymer systems (81, 206, 207). Such materials would help expanding the operational range flexibility of fuel cells with proton-conducting electrolytes, providing options for sustained operation at $T > 100°C$.

The alternative to advanced materials solutions would be an engineering solution to mitigating critical performance affects with aqueous-based PEMs, based on a consistent picture of the physics of membrane and fuel cell operation. The key to accomplishing this task lies in an in-depth understanding of mass transport phenomena, in particular of the processes involving equilibration, transport, and transformation of water. At the moment, it is not clear whether successful future generations of fuel cells will be based on anhydrous or aqueous-based proton conductors or, depending on the requirements of the application, both. But in this context it may be insightful to consider the following analogy: similar to fuel cells, nature needs protons as mediators of biological energy transformations; faced with the same problem of shuttling protons between spatially separated sites for oxidation and reduction processes, nature's preferred solution is based on aqueous proton conduction; organisms have evolved in such a way that they can provide sufficient rates of proton transfer under all relevant conditions.

Even if new materials could be identified that could abolish the dependence of PEM operation on hydration, the presence of water is still unavoidable as the product of the fuel cell reaction. A poorly balanced water distribution in the fuel cell can severely impair its performance and cause long-term degradation effects. In state-of the-art PEFCs, proton conductivity will be poor if PEM or CLs are too dry, potentially leading to excessive Joule heating, which could affect the structural integrity of the cell. Too much water in diffusion media (CLs and GDLs) could block the gaseous transport of reactants. Essentially, all processes in PEFCs are linked to water distribution and the balance of water fluxes in all parts.

Early models of PEFCs ignored the formation of liquid water and assumed that water is transported entirely in vapor form (208–211). Many simplistic modeling approaches dealt with the water management problem within the scope of a single-phase approximation, assuming for instance ideal gas diffusivities in porous gas diffusion electrodes or saturation conditions in PEM. At high current densities, however, extensive liquid water accumulation in pores impairs the accessibility of gaseous reactants to reaction sites in the cathode side. Moreover, dehydration in the anode and membrane worsens the transport of hydrated protons through membrane and anode catalyst layer resulting in a dramatic loss in the fuel cell performance (199,212–215). More recently numerous two-phase models have been proposed to take into account formation and transport of liquid water in gas diffusion media (216–223). Liquid water was introduced into the models assuming a capillary equilibrium between liquid and gas phases in the porous media and employing empirical relations between liquid water saturation and relevant parameters, such as capillary pressure, gas diffusivity, and liquid permeability (68). Recently, design optimization of PEM fuel cells, especially Cathode Catalyst Layers (CCLs), was both experimentally and numerically approached based on the available single- or two-phase models.

The structure-based model proposed in (4, 67) and exploited in optimization studies in (224, 225) correlates the performance of the CCL with the volumetric amounts of Pt, carbon, ionomer, and pores. The model predicts the best performance with ∼33 vol.% of ionomer, as validated in experiment. The theory was also used to explore novel design ideas (226, 227). As predicted in theory and later on confirmed in experiment, functionally graded layers (enhanced ionomer content near membrane and reduced ionomer content near GDL) raise the performance by ∼5–10% compared to standard CCLs with uniform composition.

Higher order structural information, involving surface areas of interfaces, orientations, sizes, shapes and spatial distributions of the phase domains, connectivity, etc. has not been involved in optimization studies. Interactions among distinct phases have been neglected as well up to date. More detailed structural representation in modeling, beyond percolation theory and effective media concepts, is only warranted if microstructures of relevant media could be characterized experimentally and if nanostructures with desired characteristics could be fabricated, e.g., considering regular arrays of well-defined structural elements or hierarchical laminates (70,228).

The issue of water management in PEFCs relates to the external operating conditions as well as to the microstructure of each component. Experiments and recent coarse-grained MD simulations (Sect. V) have shown that different ink compositions result in microstructures CLs with different pore size distributions (PSDs)

(63–66). Recently, a PSD-related model was developed to study the two-phase behavior in PEFCs but the effect of the component microstructure on water management has not been discussed (10, 11, 229, 230). Structure-based models of water management focusing separately on PEMs or CLs have been considered in (4, 68, 199). The role and function of CCL composition, pore space morphology and wetting properties in controlling water balance and performance were thoroughly studied in (68). Subsequently, we will review a few of the main results from structural models.

WATER MANAGEMENT IN PEMs

Under ideal operation of PEFCs, the membrane would retain a uniform level of hydration at the value of the water content that gives the highest level of proton conductivity, σ_p^s. It would therefore perform like a linear Ohmic resistance, with irreversible voltage losses η_{PEM} related to the fuel cell current density j_0 by the simple relation $\eta_{PEM} = \frac{L_{PEM}}{\sigma_p^s} j_0$. This ideal behavior is only observed in the limit at sufficiently small j_0. At high currents, the electro-osmotic coupling between proton and water fluxes causes non-linear deviations from the ideal performance. In the extreme case, these deviations could result in a critical current density, j_{pc}, at which the increase in η_{PEM} causes the failure of the complete cell. It is thus crucial to develop membrane models that could predict critical current densities on the basis of experimental data on structure and effective properties.

In order to be able to rationalize the role of the PEM in regulating the water fluxes through the complete cell, it is vital to understand spatial distributions of water and water fluxes. Modeling approaches that focus on membrane water management have been recently reviewed in (4, 5). We will focus here on the general structure of the models and the complicating traits.

The physical mechanism of membrane water balance and the formal structure of modeling approaches are straightforward. Proton flow from anode toward cathode induces water transport through electro-osmotic coupling. The phenomenological electro-osmotic coupling coefficient, n_d, incorporates contributions of molecular effects due to protons being bound in diffusing protonated clusters like H_3O^+ or $H_9O_4^+$, and of the hydrodynamic coupling in nanometer-sized water channels (231,232). Typical values of n_d are in the range $n_d \sim 1$–3 (12).

Under stationary operation, electro-osmotic flux has to be compensated by a backflux of water from cathode to anode, driven by concentration and pressure gradients. These gradients generate a profile of water distribution in PEMs, with decreasing water contents from cathode to anode. With increasing j_0, the water content gradients will increase as well. At the critical current density, the water content would fall below the percolation threshold of proton conduction near the anode. As a consequence, η_{PEM} would dramatically increase, extinguishing the positive voltage output of the fuel cell.

The molar flux of liquid water in the membrane is

$$\vec{N}_{1\lambda} = n_d(\lambda)\frac{\vec{j}}{F} - D^m(\lambda)\nabla c_1 - \frac{k_p^m(\lambda)}{\mu_1}\nabla p_1, \qquad (8)$$

where c_1, p_1 and μ_1 are liquid water concentration, pressure and viscosity, respectively, and $D^m(\lambda)$ is the membrane diffusivity, $k_p^m(\lambda)$ the hydraulic permeability, and $n_d(\lambda)$ the electroosmotic drag coefficient. It is indicated that the transport parameters are generally functions of the water content λ, which is defined as the number of water molecules per sulfonic acid site in the membrane. In the membrane water is neither produced nor consumed. Due to mass conservation we thus have

$$\nabla \cdot \vec{N}_l = 0. \qquad (9)$$

Proton current density is determined by Ohm's law

$$\vec{j} = -\sigma_p(\lambda)\nabla\varphi \text{ and } \nabla \cdot \vec{j} = 0 \qquad (10)$$

with the membrane conductivity $\sigma_p(\lambda)$. Proton current in the membrane is transported mainly in the through-plane direction. It is expedient to consider a one-dimensional problem with scalar variables j and N_l. Specifying boundary conditions on vapor pressures, gas pressures, water and proton fluxes, solutions of this problem can be straightforwardly obtained. In spite of the apparent simplicity of the problem, many different solutions and ambiguous results have been reported. Calculated water content profiles, Ohmic losses η_{PEM} and critical current densities j_{pc} are strongly affected by the volume-averaged physical properties $D^m(\lambda)$, $k_p^m(\lambda)$, $n_d(\lambda)$, and $\sigma_p(\lambda)$ and their dependences on membrane morphology and λ. It is impossible to calculate these properties from first principles. They have to be obtained from experiments.

Ambiguity in membrane models is mainly due to varying assumptions on membrane micromorphology, water distribution and equilibration with the polymer, and swelling properties. The well-established genuine structural picture of the PEM as a self-organized, phase-segregated polymer is inappropriate for the modeling of PEM operation in PEFCs. Performance modeling on the basis of this structure would warrant predictive multiscale modeling capabilities. Two simpler structural pictures have been employed, which highlight the different mechanisms of water transport. The so-called diffusion models consider the membrane as a continuous non-porous phase, in which water of hydration is dissolved (208). The prevailing mode of water transport in this model is molecular diffusion, $D^m(\lambda)$, whereas $k_p^m(\lambda) = 0$. Structural models of membrane operation, on the other hand, treat the membrane as a heterogeneous porous medium with percolating water networks for proton and water transport. The relevant mechanism by which water equilibrates with the polymer material is capillary condensation. Analogous to water flux in porous rocks, the capillary pressure controls water filling in the membrane, and gradients in capillary pressure are responsible for water flux from cathode to anode. The hydraulic permeation model, developed in (199) and reviewed in (4) is based on these assumptions. Of course, swelling and reorganization of polymer matrix and water-containing pathways should be accounted for. A phenomenological model of swelling upon water uptake, based on a random network model of pores, was suggested in (72). An external gas pressure gradient applied between anode and cathode sides of the fuel cell may be superimposed on the internal gradient, providing a means to control the water distribution under conditions in operational PEFCs. Hydraulic permeation models rely on measured capillary isotherms or pores size distributions of the membrane. Main dependencies of the critical current density on membrane parameters are given by

$$j_{pc} \sim \frac{(w_s - w_c)r_1}{n_d L_{PEM}}, \quad (11)$$

where w_s is the saturation water content, w_c is the percolation threshold of proton conduction, r_1 is the first moment of the pore size distribution (i.e., average pore size), L_{PEM} is the thickness of the PEM, and n_d is the electro-osmotic coefficient (assumed constant for simplicity). This expression, suggests that membranes with higher water uptake (i.e., higher ion-exchange capacity), larger pores, lower percolation threshold (higher connectivity), reduced electro-osmotic drag, or reduced thickness are less prone to dehydration.

Diffusion model and hydraulic permeation model differ markedly in their predictions of water content profiles and critical current densities (4). In the hydraulic permeation model, water content profiles are highly non-linear with strong dehydration arising only in the interfacial regions close to the anode. Overall, severe dehydration occurs only at current densities closely approaching j_{pc}. The hydraulic permeation model is consistent with experimental data on water content profiles and differential membrane resistance (233, 234). Recently, it was demonstrated that this model could explain the response of membrane performance in operating fuel cells on changing external pressure conditions (235). Figure 21 shows

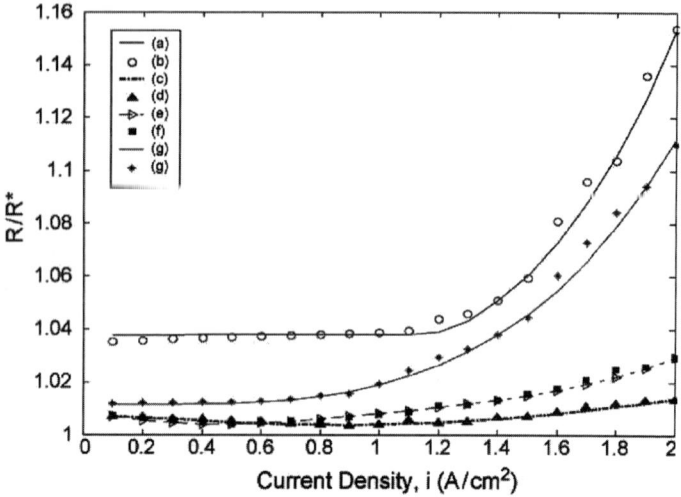

Figure 21. PEM resistance in operational PEFC as a function of the fuel cell current density, comparing experimental data (*dots*) and calculated results from a performance model based on the hydraulic permeation model. Reprinted from S. Renganathan, Q. Guo, V.A. Sethuraman, J.W. Weidner and R.E. White, *Polymer Electrolyte Membrane Resistance Model*, J. Power Sources 160, 386, Copyright @ 2006 with permission from Elsevier Science.

experimental data for the PEM resistance in an operational PEFC for various applied gas pressures in anode and cathode compartments (from (235)). The experimental data (dots) are reproduced very well within the framework of the hydraulic permeation model (solid lines).

Water sorption data provide information on the strength and mechanism of water binding to the PEM material. More generally, the Gibbs free energy of water sorption can be calculated from vapor sorption isotherms, i.e., $G_s^a = -RT \ln(P/P^s)$, where R is the molar gas constant, P is the vapor pressure and P^s is the saturated vapor pressure. Assuming that the mechanism of water uptake is capillary condensation, a Gibbs free energy of water binding can be calculated independently from the measured capillary radius r^c (236), i.e., $G_c^a = 2\sigma \cos(\theta) V_m/r^c$, where σ is the surface tension of water, θ is the wetting angle of pores in PEMs, and V_m is the molar volume of water. On the basis of data for Nafion (236), such a comparison demonstrates that capillary condensation is the mechanism of water uptake at $\lambda > 3 (G_s^a = G_c^a)$, whereas at smaller λ a large discrepancy was found ($G_s^a > G_c^a$), indicative of strong attractive interactions between water and charged interfaces in lowly hydrated pores of PEMs. As described in detail in Sect. III, *ab initio* calculations for lowly hydrated polymer fragments or densely packed arrays of acid functionalized surface groups, could provide direct information on energies of water binding at low λ (33, 49, 50).

The previous discussion suggests that hydraulic permeation should be the dominant mode of water transport at sufficiently large λ, whereas a diffusive contribution to water transport will dominate at low λ. This change in the predominant mechanism of water transport with λ could explain the peculiar transition in transverse water concentration profiles through operating PEMs observed in recent neutron scattering experiments (26). Water management models that could account for this interplay of diffusion and hydraulic permeation have been considered in (4, 199, 230).

POROUS STRUCTURE AND WATER ACCUMULATION IN CLS

The toughest competitions between random composite morphology and complex coupled processes unfold in the cathode catalyst layer (CCL) (68). Reactions proceed at Pt nanoparticles (cf. Sect. IV), which are randomly dispersed on a high-surface-area carbon matrix. During fabrication, the colloidal solution of carbon/Pt and ionomer self-organizes into a phase-segregated composite with interpenetrating percolating phases for the transport of electrons, protons, and gases (cf. Sect. V).

The purpose of catalyst layer modeling is to evaluate how the emerging structure, specified by composition, pore size distribution, and wetting properties of pores, determines the CCL performance, rated in terms of voltage efficiency, water handling capability, and catalyst utilization.

Liquid water arrives in the CCL either via transport through the PEM or it is generated in the electrochemical reaction. Invariably, PEFCs require a medium that is highly effective in converting liquid water into water vapor, since otherwise liquid water will clog pores and channels in gas diffusion layers and flow fields that are needed for the gaseous supply of reactants.

The net rate of vaporization of a porous electrode with thickness L is given by

$$Q^{lv}(z) = \frac{e_0 \kappa^e}{L} \xi^{lv}(S_r) \{q_r^s(T) - q(z)\} \quad (12)$$

with the saturated vapor pressure in pores of capillary radius r_c,

$$q_r^s(T) = q^{s,\infty}(T) \exp\left(-\frac{2\sigma \cos(\theta) V_m}{RT r^c}\right),$$

$$q^{s,\infty}(T) = q^0 \exp\left(\frac{G^a}{k_B T}\right), \quad (13)$$

where $q^{s,\infty}(T)$ is the saturation pressure for a planar vapor/liquid interface, σ the surface tension, θ the wetting angle of catalyst layer pores, and e_0 the elementary charge. κ^e is an intrinsic rate constant of evaporation (237) and ξ^{lv} is the ratio of the real liquid/vapor interfacial area to the apparent electrode surface area. For a porous medium with total porosity X_{p,r_c}, a simple estimate of this factor gives $\xi^{lv} \sim X_{p,r_c} L / 2 r_c$, assuming cylindrical pores with radius r_c and lengths $l_c \sim 4 r_c$. An activation energy of evaporation $G^a \approx 0.44$ eV and a pre-exponential factor $q^0 \simeq 1.18 \times 10^6$ atm reproduce the saturation pressure of water in the range of 0–100°C. Assuming a typical catalyst layer thickness of $L \approx 10\,\mu m$ it was demonstrated in (68) that evaporation rates in CCLs are high enough to keep up with rates of water production. The comparison depends, however, strongly on T and on the porous structure. Higher temperatures and smaller pore sizes help increasing evaporation rates.

Obviously, the CCL not only determines the rate of current conversion and the major fraction of irreversible voltage losses in the PEFC, but it also plays a key role for the water balance of the whole cell. Indeed, due to a benign porous structure with a large portion of pores in the nanometer range the CCL emerges as the PEFC's favorite water exchanger. Once liquid water arrives in gas diffusion layers or flow fields, they are unable to handle it.

For the purpose of physical modeling of catalyst layer operation, it is expedient to consider the layer essentially as a two-scale system. At the scale <100 nm, operation is determined by the properties of agglomerates of carbon and Pt, which are surrounded by a film of ionomer (cf. Sect. V). Primary pores in agglomerates should be maximally wetted in order to provide a large active Pt surface area. Due to the requirement of proton supply through water-filled pathways in carbon pores and in aqueous domains in ionomer, only wetted Pt particles can be active. At this scale, mass transport limitations due to diffusion of oxygen and hydrogen ions through water filled pores inside agglomerates are insignificant (177). At the macroscopic scale (100 nm−10 µm), gaseous diffusion of oxygen is vital in order to ensure uniform utilization of catalyst throughout the complete layer thickness. Agglomeration and bimodal pore size distribution are thus vital in view of the disparate functional requirements. Ideally, pores inside agglomerates should be made hydrophilic and larger pores between agglomerates should be hydrophobic. Structural control during fabrication should thus focus not only on composition, but also on agglomeration, porous morphology, and wetting properties of pores.

The challenge for modeling the water balance in CCL is to properly link the composite, porous morphology with liquid water accumulation, transport phenomena, electrochemical kinetics, and performance. An approach for establishing these links was explored in (68). This task involves two vital steps, as indicated in Fig. 22. At the materials level, it requires relations between composition, porous structure, liquid water accumulation and effective properties, including proton conductivity, gas diffusivities, liquid permeabilities, electrochemical source term, and vaporization source term. Catalyst layer composition is specified in terms of volume fractions of the solid carbon/Pt phase, X_{PtC}, ionomer phase X_{el}, and the remaining

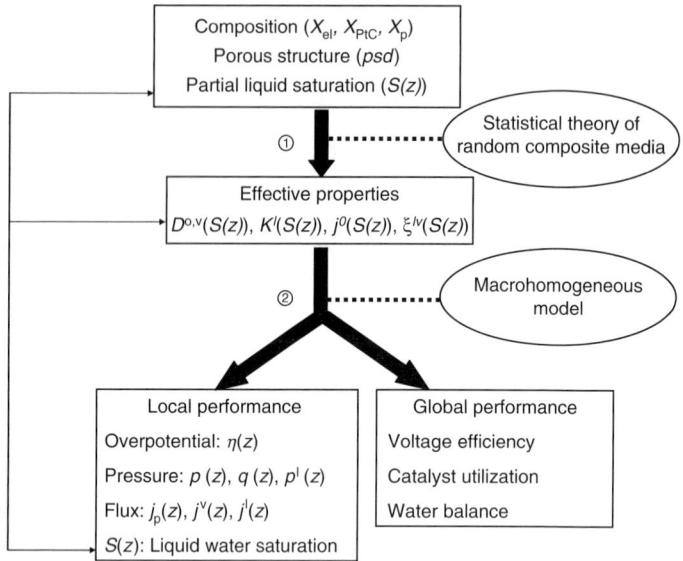

Figure 22. Scheme of the structure-based physical model of catalyst layer operation, linking structure/composition, effective properties and performance.

pore space, $X_p = 1 - X_{PtC} - X_{el}$. In view of the random phase-segregated, agglomerated microstructure of current CCLs, the pore space is subdivided into primary pores with sizes in the range of 1–10 nm inside agglomerates of carbon/Pt particles and secondary pores between those agglomerates with sizes in the range of 10–100 nm. The volume fractions of primary and secondary pores are X_μ and X_M, respectively.

Functional relationships between effective properties and structure have been discussed in (67, 68). Since the liquid water saturation $S_r(z)$ is a spatially varying function at $j_0 > 0$, these effective properties also vary spatially in an operating cell, warranting a self-consistent solution for effective properties and performance. These correlations have been exploited with some success in systematic optimization studies of catalyst layer composition (225, 226).

Theoretical predictions on the composition-dependent performance of CCLs were found to be in good agreement with experimental data, although these works did not take into account the pore space morphology and liquid water formation in pores. These assumptions may work well at small current densities, for which liquid water accumulation is not critical.

It is assumed that capillary forces at the liquid–gas interfaces in pores equilibrate the local water content in the catalyst layer. Pore-filling under stationary conditions is, therefore, expressed through the Young–Laplace equation, which relates capillary pressure, p^c, and capillary radius, r^c, to local gas pressure, p^g, and liquid pressure, p^l,

$$p^c = \frac{2\sigma \cos(\theta)}{r^c} = p^g - p^l. \quad (14)$$

On the cathode side of PEFC under usual operating conditions the problem at hand is to deal with excessive amounts of liquid water in pores and of gaseous supply channels due to a net electro-osmotic flux through PEM and the production of water on this side. Under these conditions, it can be assumed that the ionomer phase in the CCL is fully hydrated. Moreover, an assumption was made that small primary pores inside agglomerates are hydrophilic, ensuring that these pores are filled completely with water under normal operating conditions (p^g, p^l). The liquid water front advances in secondary pores between agglomerates. Wettability of these pores is vital for controlling the water formation. Assuming that these pores are still partially hydrophilic (typical wetting angles of carbonaceous materials are found in the range $\theta \sim 80$–$90°$), the saturation is given by

$$S = \frac{1}{X_p} \int_0^{r^c} dr' \frac{dX_p(r')}{dr'}. \quad (15)$$

Equations (14) and (15) establish relations between local water content, operating conditions, wetting properties, and pore space morphology in the CCL.

Finally, continuity equations that account for mass conservation laws, i.e.,

$$\frac{\partial \rho(\vec{r}, t)}{\partial t} + \nabla \cdot \vec{j}(\vec{r}, t) = Q(\vec{r}, t) \quad (16)$$

including fluxes of protons, gases, and water as well as transformations between these species, have been established to relate the CCL structure and effective properties to performance.

Capabilities of this approach have been demonstrated in (68). A simple representation of the pore space by a bimodal δ-distribution revealed the role of the CCL as a "watershed" in the PEFC. For this case, a full analytical solution could be found. This solution still captures essential physical processes, critical phenomena, operation conditions and major structural features such as typical pore sizes (r_μ, r_M) and distinct contributions to porosity from primary and secondary pores (X_μ, X_M).

In terms of liquid water saturation and water management in the CCL the bimodal δ-distribution leads to a three-state model. The three states that any representative elementary volume element could attain are depicted in Fig. 23. Effective properties are constant in each of these states. In the dry state, the porous structure is water-free ($S_r \approx 0$). Gaseous transport is optimal. Electrochemical reaction and evaporation rates are, however, poor. In the optimal wetting state ($S_r = X_\mu/X_p$), primary pores are completely water-filled while secondary pores are water-free; catalyst utilization and exchange current density are high; the surface area for evaporation, ζ^{lv}, is large as well; moreover, diffusion coefficients will still be high, since secondary pores remain water-free. In the fully flooded

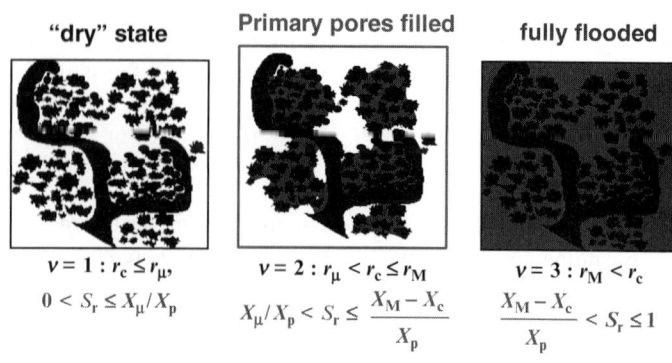

Figure 23. The three principal states, in which the CCL could operate, if a bimodal δ-function like pore size distribution is assumed. The layer will poorly perform in the fully flooded and in the "dry" states. The optimum is attained in the intermediate state in which primary pores are fully flooded and secondary pores provide the space for gaseous diffusion of reactants and water vapor. Reprinted from M. Eikerling, *Water Management in Cathode Catalyst Layers of PEM Fuel Cells: A Structure-Based Model*, J. Electrochem. Soc. **153**, E58, Copyright @ 2006 with permission from ACS.

state ($S_r = 1$), all pores are filled with water and the affected parts of the CCL are deactivated due to impeded gas transport at macroscopic scale.

An expression for the characteristic current density, below which complete liquid-to-vapor conversion is possible, was obtained. This quantitative parameter is related to saturated vapor pressure and vapor diffusivity. Moreover, the CCL fulfils an important function in regulating opposite hydraulic fluxes towards PEM and GDL sides. The results also strongly suggest the CCL as a critical fuel cell component in view of excessive flooding that could give rise to critical effects in η_0 (j_0)-relations. Under certain conditions ($j_{crit}^{fl} > j_{crit}^{lv}$), critical liquid water formation arises first in the interior of the layer, close to the CCL/GDL interface and not at the PEM/CCL boundary.

The model reveals sensitive dependencies of CCL operation on porous structure, thickness, wetting angle, total cathodic gas pressure, and net liquid water flux from the membrane. With rather favorable parameters (10 μm thickness, 5 atm cathodic gas pressure, 89° wetting angle), the critical current density of CCL flooding is found in the range 2–3 A cm^{-2}. For increased thickness, smaller gas pressure or slightly reduced wetting angles of secondary pores, CCLs could be flooded at current densities well below 1 A cm^{-2}. The contact angle is an important parameter in this context, highly influential but difficult to control in fabrication (238, 239).

Solutions of the physical model of CCL operation for general continuous pore size distributions were presented in (240). With this decisive extension the full coupling between composite porous morphology, liquid water accumulation, transport of reactants and products, and electrochemical conversion in the oxygen reduction reaction could be explored, cf. Fig. 22. Continuous PSD's allow relating global performance effects (limiting currents, bistability) to local distributions of water, fluxes, concentrations of reactants, and reaction rates in the layer. It was found that a CCL alone cannot give rise to limiting current behavior in voltage-current curves. The explanation is simple and intuitive: in the fully saturated state, the CCL retains a residual oxygen diffusivity through liquid-water filled pores $D_{fl}^{o} = 2.0 \times 10^{-6}$ cm^2 s^{-1}; the main effect will be a corresponding drastic decrease in the reaction penetration depth to values in the range of ∼100 nm; the overvoltage of the CCL will increase accordingly so that the smaller fraction of active Pt atoms

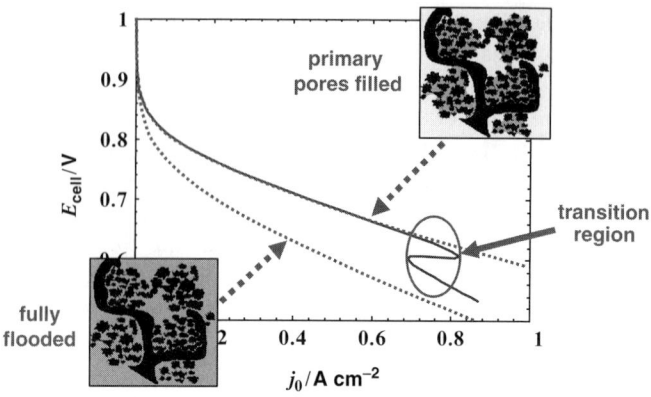

Figure 24. Current-voltage curve calculated with the CCL model that includes the full coupling between porous structure, liquid water accumulation and performance affects. The two reference curves (*red dotted lines*) correspond to the ideally wetted state (only primary pores filled with liquid water) and the fully saturated state. Bistability occurs in the transition region.

in the active part can maintain the fixed total current; the voltage increase depends roughly logarithmically on the reaction penetration depth, following the dependence imposed by the Butler–Volmer equation.

Upon increasing the current density generated by the fuel cell, a transition between two principal states of operation occurs, as illustrated in Fig. 24. The ideally wetted state at low current densities exhibits levels of liquid water saturation well below the critical value for pore blocking, corresponding to uniform distributions of reactants and reaction rates. In the fully saturated state, liquid water saturation exceeds the critical value in parts of the layer. These parts could sustain only low residual gas diffusivity. Corresponding reactant and reaction rate distributions are highly non-uniform. The main part of the CCL will be inactive. The transition between the two states of operation can occur monotonously or it could involve bistability as a signature of nonlinear coupling between liquid water accumulation, gaseous transport, and electrochemical conversion rate. Bistability means that two steady state solutions of the continuity equations coexist in the transition region.

The critical current density of the transition from ideally wetted state to transition region or fully saturated state is the major

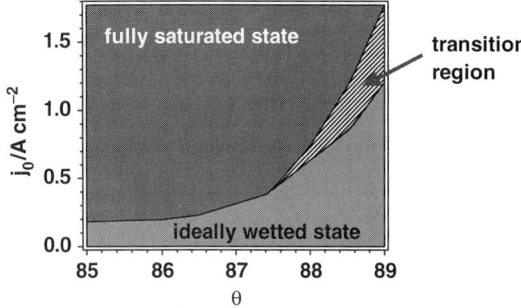

Figure 25. Stability diagram of CCL, indicating operation in ideally wetted state, fully saturated state, or transition region as a function of wetting angle and fuel cell current density.

optimization target of CCLs in view of their water handling capabilities. A larger critical current density allows extracting higher voltage efficiency and power density from PEFCs. Critical current densities depend on structural parameters and operating conditions. Stability diagrams have been introduced for assessing effects of these parameters on CCL performance. The stability diagram in Fig. 25 displays the effect of the wetting angle on fuel cell operation, distinguishing between ideally wetted state, bistability region, and fully saturated state.

The task of water management in CCL is to push back capillary equilibrium to small enough pores so that liquid water formation cannot block gaseous transport in secondary pores. Beneficial conditions for extending the range of the ideally wetted state are a high total porosity, large volume fraction of secondary pores, a wetting angle that closely approaches $90°$, a high total gas pressure and high temperature of operation.

EFFECTIVENESS OF CATALYST UTILIZATION

Evaluation of the effectiveness of catalyst utilization is the most impressive example of the multiscale nature of challenges in fuel cell research. The effectiveness factor of Pt utilization is intimately linked to the uniformity of reaction rate distributions. The latter is affected by processes at all levels, including surface transport and kinetics at Pt nanoparticles as well as transport of reactants and product water at agglomerate level, catalyst layer level, MEA level, unit cell level, and stack level.

The challenges in catalyst layer research can be best explained by considering the exchange current density j^0. It is the major physical property that links structure and activity of the catalyst layer with the working voltage of the fuel cell. A parameterization that relates j^0 to catalyst loading $[m_{Pt}]$ was proposed in (68)

$$j^0 = 2 \times 10^3 \, [m_{Pt}] \, j^{0*} \varepsilon^{S/V} \Gamma_a g\,(S_r) \frac{f\,(X_{PtC}, X_{el})}{X_{PtC}}. \tag{17}$$

The factor 2×10^3 approximates the total Pt surface area per real electrode surface area that would be obtained by spreading a catalyst loading of 1 mg cm^{-2} as an ideal monoatomic layer. In order to minimize irreversible voltage losses of the oxygen reduction reaction (ORR), j^0 should be as large as possible. On the other hand, cost reduction demands lowest possible $[m_{Pt}]$ (in units mg cm^{-2}). The best compromise of these competing objectives can be achieved if the remaining factors on the right hand side of (17) are maximized.

At the nanoparticle scale, this involves improving the specific activity of the catalyst (j^{0*}) and the surface-to-volume atom ratio of catalyst nanoparticles ($\varepsilon^{S/V}$). Additional statistical effects include the possible existence of active sites on the catalyst surface.

At the agglomerate level, the effectiveness factor Γ_a is mainly determined by electrostatic effects due to the non-linear distribution of proton concentration and electrostatic potential inside of agglomerates. In (177), a model of ionomer-free, water-filled agglomerates was considered. Based on simple size considerations and recently confirmed by coarse-grained MD simulations (63) it was assumed that ionomer is not able to penetrate nanometer-sized primary pores inside of agglomerates. Protons are delivered through a film of ionomer phase to the surface of agglomerates, from which they diffuse into water-filled primary pores. Poisson–Nernst–Planck equation and the oxygen diffusion equation were solved, including a sink term to account for the oxygen reduction reaction.

Proton concentration, $c_{H^+}(r)$, decreases from the surface towards the centre of agglomerates while the electrostatic potential, $\phi(r)$, increases. In the Faradaic current density,

$$j_F^z(r) \propto c_{H^+}(r) \exp\left[\frac{\alpha_c F}{RT}\phi(r)\right] \tag{18}$$

the two effects counterbalance each other. As can be seen from Equation (18) the cathodic transfer coefficient α_c plays a vital role in controlling this interplay. For negligible oxygen diffusion limitations, an analytical expression for the effectiveness factor of ideal spherical agglomerates was found,

$$\Gamma_a = 3 \int_0^1 \rho^2 d\rho \exp\left[-(1-\alpha_c)\tilde{\Phi}(\rho)\right], \quad (19)$$

with the normalized radial coordinate $\rho = r/R_a$, where R_a is the agglomerate radius. The normalized potential, $\tilde{\Phi}(\rho)$, is obtained from the solution of the Poisson–Boltzmann-like expression

$$\frac{1}{\rho^2}\frac{d}{d\rho}\left(\rho^2\frac{d\tilde{\Phi}}{d\rho}\right) = -K_2 \exp\left(-\tilde{\Phi}\right), K_2 = \frac{R_a^2}{\lambda_D^2}, \quad (20)$$

where $\lambda_D = \left(\frac{\varepsilon\varepsilon_0 RT}{F^2 c_{H^+}^{ref}}\right)^{1/2}$ is the Debye-length, characterizing the depth of proton penetration into the agglomerate. It can be seen from Equation (19) that the effectiveness factor at the agglomerate level would be $\Gamma_a = 1$ for $\alpha_c = 1$ and negligible oxygen diffusion limitations. Figure 26 shows the effects of α_c and of the agglomerate radius R_a on Γ_a.

At the macroscopic scale, the wetted fraction of catalyst particles in pores of the CCL ($g(S_r)$), (68) and the statistical fraction of

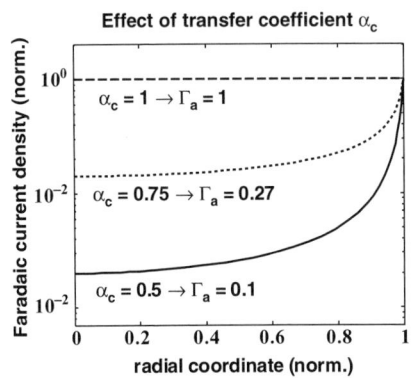

Figure 26. Effectiveness factor agglomerates. The figure on the *left-hand side* shows the distribution of Faradaic current density in agglomerates and corresponding effectiveness factors for varying cathodic transfer coefficients α_c, at fixed $R_a = 200$ nm. The *table on the right* shows the effect of the agglomerate radius at fixed $\alpha_c = 0.75$.

catalyst particles at the triple phase boundary in random composite media ($f\left(X_{\text{PtC}}, X_{\text{el}}\right)/X_{\text{PtC}}$) further reduce j^0 in Equation (17) (4, 67).

The total effectiveness factor of Pt utilization of a CCL can be defined as the overall current conversion rate divided by the ideal rate if all Pt atoms were used equally in electrochemical reactions at the given electrode overpotential and externally provided reactant concentrations. This definition accounts for effects due to statistical inhomogeneity at multiple scales, as accounted for in Equation (17), as well as for non-uniform distributions of reactants, electrostatic potentials, and rates of electrochemical reactions at agglomerate and CCL levels due to limited transport of reactants, electrons, and protons.

Including detrimental factors at all scales, an estimate obtained in (73) reveals that this total effectiveness of catalyst utilization in typical conventional CCLs at normal operating conditions lies in the range of ~3–5%. This suggests that catalyst layers could be made much better. Higher performance at significantly reduced Pt loading could be achieved, if major design constraints related to the random three-phase composite morphologies and relatively high thickness (~10 μm) of currently used layers could be overcome. Overall, maximizing the ratio of the $j^0/[m_{\text{Pt}}]$ is a joint challenge for electrocatalysis and for understanding statistical inhomogeneity in random composite media with hierarchical structure.

New architectures based on nano-template or random nanoporous electrodes are unanimously seen as the future in electrode technology. An example is the design developed by 3M. It utilizes a nanostructured substrate of organic whiskers, onto which a continuous layer of Pt is sputter-deposited (241). Resulting CLs are very thin (<0.5 μm) two-phase composites of Pt (on an inert substrate) and they utilize water-filled voids between the whiskers as the diffusion medium for reactants and protons. In this design, the utilized Pt surface can reach up to 100%. However, due to large Pt grain sizes (~10 nm) only a small fraction of Pt atoms is at the catalyst surface, compromising the high surface-to-volume atom ratios of nanoparticles. Indeed, a recent study revealed very similar effectiveness of catalyst utilization in conventional and 3M type CLs (242).

As discussed in this section, the porous matrix of the CCL is not just an inert container of catalyst nanoparticles that are embedded into it, but it plays an important role in controlling the uniformity of reaction rate distributions and thus effectiveness of catalyst utilization as well as the water management of the entire cell. In the future, it will be critical to link the modeling work with systematic experimental studies on correlations between structure, effective properties, operating conditions, and performance. Further exploration hinges on the detailed characterization of pore size distributions, wetting properties of pores, distribution and state of water in pores, transport properties, vaporization exchange, and distributed electrochemical activity. This work is underway. Other efforts focus on integration of CL models into complete structural models of water management in MEA, including porous composite electrode compartments on anode and cathode sides as well as the polymer electrolyte membrane.

VIII. CONCLUDING REMARKS

Design and development of advanced materials plays a key role in vigorous efforts to enhance performance and cost-effectiveness of PEFCs (243). Experience over the past two decades has shown that the dramatic improvements in power density, durability, and cost would be impossible without a breakthrough in the concept of proton conducting membranes and catalyst layers. Moreover, all components of operational PEFCs have to cooperate well in order to optimize the interplay of transport and reaction. Current membranes and catalyst layers are fabricated as random heterogeneous media. Their properties evolve on a hierarchy of scales, from Ångstrom to meters, exhibiting complex morphologies and an intricate interplay of fundamental processes. Evidently, the rational design of functionally optimized materials demands strategies that are based on systematic understanding provided by physical theory, molecular modeling, and performance modeling at the device level.

Here we focused on the main challenges in theoretical fuel cell research, including molecular mechanisms of proton transport in polymeric membranes, complex reaction mechanisms at nanoparticle electrocatalysts, proton and electron transport in random composites, reactant supply through porous gas diffusion media, and water balances in all parts. Recent efforts in theory, modeling and diagnostics have explicitly identified constraints and reserves for improvements of the current design of catalyst layers and membranes in PEFCs.

If theory is to help in the design of advanced proton conductors, mechanisms of proton conduction in aqueous-based polymer membranes have to be understood at the same level of molecular detail as the well-studied mechanisms of proton transport in bulk water. The key question in this realm is, whether sufficient rates of proton transport could be attained under conditions of minimal hydration, with immobilized solvent, and at elevated temperatures. This problem hinges on understanding the role of polymer chemical structure and interfacial structure in phase-segregated polymers. The principal viability of interfacial proton transport to sustain high proton mobility is corroborated by the multitude of experimental studies of lateral PT at surfaces of biomembranes and at lipid monolayers.

Current calculations, exploiting DFT and CPMD methods are on the way to establishing fundamental steps of interfacial proton transfer that could be of similar significance as the transformation between $H_9O_4^+$ and $H_5O_2^+$ for proton transport in bulk water. The packing density of protogenic surface groups transpires as the critical parameter for achieving high mobility of protons at lowly hydrated interfaces. If the packing density is too low, surface groups are only weakly correlated, prohibiting direct proton transfer between them. If the density is too high, strong hydrogen bonds will suppress the dynamics of protons and water at the interface. A critical separation distance of surface groups that could represent a good compromise of long-range correlations and flexibility is ~ 7 Å. Further important criteria, which depend on the chemical architecture of the polymer (backbones, sidechains and acid head groups), include the ease of formation and breaking of H-bonds and the conformational flexibility of sidechains and acid groups.

Systematic studies in theory and molecular modeling on proton mobilities at 2D hydrated structures with controlled molecular architectures could identify collective coordinates (or reaction coordinates) and transition pathways at the interface. Differences in activation energies and rates of proton transfer, calculated with appropriate sampling techniques (e.g., transition path sampling) could be evaluated in relation to distinct polymer constituents, acid head groups, lengths and densities of sidechains, and levels of hydration. Comparison with experimental data (e.g., NMR, IR, pulse experiments, and scanning electrochemical microscopy) on molecular mobilities and conductivities (e.g., Arrhenius plots) in PEMs, biomembranes, and lipid bilayers will help to furnish effects of interfacial structure and segmental motions of sidechains on PT. New mechanisms could be exploited together with opportunities to control membrane structure via block-copolymer self-assembly (244, 245). Moreover, theory and modeling efforts should also address proton transport in anhydrous systems, which offer promising alternatives for membrane materials.

With potential improvements in intrinsic electrocatalytic activity of Pt surface atoms and current statistical limitations of catalyst effectiveness arising at all scales, catalyst layers offer the most compelling opportunities for innovation. These prospects drive efforts in the design of nanostructured catalysts and nanostructured substrates. In electrocatalysis, nanoparticles with Pt surface atoms provide highest accessibilities and activities for the reactions in PEFCs. Particle size has an effect on stable particle conformations, surface morphology, as well as on the nature and abundance of active sites. The electronic structure of the particles, affected by the presence of the substrate, controls the interplay of elementary surface processes and the stability of particle-substrate bonding. Recent systematic studies in experiment and theory have demonstrated strong enhancements due to the Pt-skin effect (111, 112). All of these structural factors together determine apparent rates of complex multistep reactions in catalyst layers. Understanding the role of the substrate represents an additional challenge in this context. While the substrate is primarily seen as an electronic conductor with no catalytic activity, it may influence the overall activity by affecting the Pt electronic structure, or by acting as a reservoir of reactants.

Developing predictive structure vs. reactivity relationships of electrocatalyst systems is a true multidisciplinary and multiscale effort. Sizes of phase domains of carbon/Pt and proton-conducting ionomer ($\sim$10–100 nm) and their connected pathways as well as the porous architecture (pore sizes 5–100 nm) determine transport of electrons, protons, reactant gases, and water, which meet and react at electrochemically active sites of Pt. The integration of Pt nanoparticles is crucial for obtaining large Pt surface areas. A sufficient nanoporosity of the substrate is needed to ensure a large wetted surface fraction of Pt and, thus, a large electrochemically active surface area. Evidently, understanding the interplay of these structural effects warrants a well-devised set of methods in physical theory, molecular modeling and macro-scale modeling of performance.

The future work includes a systematic approach towards understanding structure formation, electrokinetics, and degradation phenomena in fuel cell catalysts and catalyst layers. At the nanopartile scale, combined computational and theoretical methods (DFT, OF-DFT, kinetic Monte Carlo simulations, MD simulations) are needed to overcome the size and time-scale limitations of conventional DFT techniques. Simulations of physical properties of realistic Pt/C nanoparticle systems could provide interaction parameters needed in models of self-organization phenomena (agglomeration and phase segregation). Theoretical models of electrokinetic mechanisms, on the other hand, are needed to establish trends between catalyst structure and oxygen reduction kinetics.

We have discussed statistical computational approaches to modeling of microstructure formation and self-organization phenomena in porous catalyst layers and membranes of PEFCs. Dynamic simulations of structural changes during formation processes lend themselves equally well to the study of mechanisms underlying structural degradation. Coarse-Grained MD simulations of self-organization in catalyst layers suggest that the resulting structures are inherently unstable. Applicable solvents with different dielectric constants correspond to different stable conformations in terms of agglomerate sizes, sizes of ionomer domains, and pores space morphology. The replacement of low-dielectric solvents used during fabrication of catalyst layers by water as the working liquid in the operating fuel cell will thus destabilize the initially formed structures, causing a drift towards a new stable conformation.

Structural models presented here, provide guidelines for adjusting thickness, composition, porosity, and operating conditions in view of high performance and effective water management. Despite the simplicity and limitations of meso-scale simulations, these studies rationalize the main features of microstructure formation in catalyst layers of PEFCs. They help to furnish the basic picture of agglomeration and phase-segregation phenomena in complex media. Important findings dwell on the role of the solvent in determining agglomerate sizes and porous structures in catalyst layers. Moreover, the finding of ionomer exclusion from agglomerates is vital for rationalizing the effectiveness of catalyst utilization. The quantitative comparison of mesoscale simulations with experimental data on size distributions of pores and phase domains can be performed on the basis of structural correlation functions. Thereby interactions parameters needed in simulations can be validated and refined.

Consistent structural pictures form the basis for understanding water management in catalyst layers and membranes. Experimental information on water distribution and strength of water binding is needed to validate existing macrohomogeneous models. Water management in PEMs can be explained on the basis of the hydraulic permeation model. This model is consistent with experimental data on membrane resistance, critical current densities, and spatial distribution of water under operation. Moreover, it explains the dependence of membrane performance on the applied external gas pressure.

In cathode catalyst layers, porous structure and water accumulation control the spatial distribution of reaction rates. Excessive accumulation of liquid water leads to rather nonuniform reaction rate distributions, which induce high electrode overvoltages. Moreover, the local reaction rates and, thus, the overall performance are affected by statistical factors at all scales. Improvements of these factors can be addressed by introducing nanostructured substrates, which are ultrathin, hydrophilic and provide a high internal surface area for the deposition of catalyst. However, advanced structures cannot focus solely on optimizing catalyst utilization and minimizing catalyst loading. The function of the cathode catalyst layer as a water exchanger has to be accounted for.

Although, several classes of realistic morphological models are available, it should be noted that there is no "universal" morphological model that can be used for every type of random heterogeneous or porous material (198). It depends on the complexity of the microstructure, the available microstructure-experimental information, and the computational resources that one may choose the appropriate model to represent the microstructure of catalyst layers and membranes in PEFCs.

REFERENCES

1. *Handbook of Fuel Cells: Fundamentals, Technology, Applications*, Ed. by Vielstich, A. Lamm, and H. Gasteiger, VCH-Wiley, Weinheim, 2003.
2. M. Eikerling, A.A. Kornyshev, and A.R. Kucernak, *Phys. Today* **59** (2006) 38.
3. E.J. Carlson, *DOE Hydrogen Program Annual Merit Review*, May 16–19, 2006.

4. M. Eikerling, A.A. Kornyshev, and A.A. Kulikovsky, "*Encyclopedia of Electrochemistry*", Ed. by A.J. Bard and M. Stratmann, Vol. 5, *Electrochemical Engineering*, vol. Ed. by Digby D.M. and P. Schmuki, chapter 8.2, pp. 447–543, VCH-Wiley, Weinheim, 2007.
5. A.Z. Weber and J. Newman, *Chem. Rev.* **104** (2004) 4679.
6. T.E. Springer, T.A. Zawodzinski, and S. Gottesfeld, *J. Electrochem. Soc.* **138** (1991) 2334.
7. S. Gottesfeld and T.A. Zawodzinski, in *Advances in Electrochemical Science and Engineering*, Vol. 5, Ed. by R.C. Alkire, H. Gerischer, D.M. Kolb, and C.W. Tobias, Wiley-VCH, Weinheim, 1997, pp. 195–301.
8. S.S. Jang, V. Molinero, T. Çağin, and W.A. Goddard III, *J. Phys. Chem. B* **108** (2004) 3149.
9. W. Goddard, B. Merinov, A. van Duin, T. Jacob, M. Blanco, V. Molinero, S.S. Jang, and Y.H. Jang, *Mol. Sim.* **32** (2006) 251.
10. A.Z. Weber and J. Newman, *J. Electrochem. Soc.* **151** (2004) A311.
11. A.Z. Weber and J. Newman, *J. Electrochem. Soc.* **151** (2004) A326.
12. K.-D. Kreuer, S.J. Paddison, E. Spohr, and M. Schuster, *Chem. Rev.* **104** (2004) 4637.
13. E. Spohr, *Mol. Sim.* **30** (2004) 107.
14. S.J. Paddison, *Annu. Rev. Mater. Res.* **33** (2003) 289.
15. A. Hickner and B.S. Pivovar, *Fuel Cells* **5** (2005) 213.
16. K.A. Mauritz and R.B. Moore, *Chem. Rev.* **104** (2004) 4535.
17. L. Rubatat, G. Gebel, and O. Diat, *Macromolecules* **37** (2004) 7772.
18. G. Gebel and O. Diat, *Fuel Cells* **5** (2005) 261.
19. J. Meier, J. Schiotz, P. Liu, J.K. Norskov, and U. Stimming, *Chem. Phys. Lett.* **390** (2004) 440.
20. J. Meier, K.A. Friedrich, and U. Stimming, *Faraday Discuss.* **121** (2002) 365.
21. P.K. Babu, E. Oldfield, and A. Wieckowski, in *Modern Aspects of Electrochemistry*, Vol. 36, Ed. by C.G. Vayenas, B.E. Conway, and R.E. White, 2003, pp. 1–50.
22. Y.Y. Tong, E. Oldfield, and A. Wieckowski, "NMR Spectroscopy in Electrochemistry" in *Encyclopedia of Electrochemistry*, Vol. 2, Interfacial Kinetics and Mass Transport, Wiley, Chichester, UK, 2002.
23. A. Wieckowski, E.R. Savinova, and C.G. Vayenas, *Catalysis and Electrocatalysis at Nanoparticle Surfaces*, Marcel Dekker, New York, 2003.
24. I.A. Schneider, D. Kramer, A. Wokaun, and G.G. Scherer, *Electrochem. Commun.* **7** (2005) 1393.
25. I.A. Schneider, S.A. Freunberger, D.K. Wokaun, and G.G. Scherer, *J. Electrochem. Soc.* **154** (2007) B383.
26. F. Xu, G. Gebel, O. Diat, and A. Morin, *J. Electrochem. Soc.*, **154** (2007) 1389.
27. M. Eikerling, A.A. Kornyshev, and A.A. Kulikovsky, *Fuel Cell Rev.* **1** (2005) 15.
28. D. Wang, J.S. Wainright, U. Landau, and R.F. Savinell, *J. Electrochem. Soc.* **143** (1996) 1260.
29. D. Marx, M.E. Tuckerman, J. Hutter, and M. Parinello, *Nature* **397** (1999) 601.
30. M.E. Tuckerman, K. Laasonen, M. Sprik, and M. Parrinello, *J. Phys. Chem.* **99** (1995) 5749.

31. M.E. Tuckerman, K. Laasonen, M. Sprik, and M. Parrinello, *J. Chem. Phys.* **103** (1995) 150.
32. M.K. Petersen, F. Wang, N.P. Blake, H. Metiu, and G.A. Voth, *J. Phys. Chem. B* **109** (2005) 3727.
33. A. Roudgar, S.P. Narasimachary, and M. Eikerling, *J. Phys. Chem. B* **110** (2006) 20469.
34. G. Hummer and C. Dellago, *Phys. Rev. Lett.* **97** (2006) 245901.
35. S. Braun-Sand, M. Strajbl, and A. Warshel, *Biophys. J.* **87** (2004) 2221.
36. A.Y. Mulkidjanian, J. Heberle, and D.A. Cherepanov, *Biochim. Biophys. Acta* **1757** (2006) 913.
37. M. Eikerling and A.A. Kornyshev, *J. Electroanal. Chem.* **502** (2001) 1.
38. P. Commer, A.G. Cherstvy, E. Spohr, and A.A. Kornyshev, *Fuel Cells* **2** (2002) 127.
39. M. Eikerling, A.A. Kornyshev, A.M. Kuznetsov, J. Ulstrup, and S. Walbran, *J. Phys. Chem. B* **105** (2001) 3646.
40. M. Kato, A.V. Pisliakov, and A. Warshel, *Proteins Struct. Funct. Genet.* **64** (2006) 829.
41. J. Heberle, J. Riesle, and G. Thiedemann et al., *Nature* **370** (1994) 379.
42. J. Zhang and P.R. Unwin, *Phys. Chem. Chem. Phys.* **4** (2002) 3814.
43. V.B.P. Leite, A. Cavalli, and O.N. Oliveira Jr., *Phys. Rev. E* **57** (1998) 6835.
44. M. Cappadonia, J.W. Erning, S.M.S. Niaki, and U. Stimming, *Solid State Ionics* **77** (1995) 65.
45. A. Vishnyakov and A.V. Neimark, *J. Phys. Chem. B* **105** (2001) 9586.
46. A.S. Ioselevich, A.A. Kornyshev, and J.H.G. Steinke, *J. Phys. Chem. B* **108** (2003) 11953.
47. S.M. Haile, *Mater. Today* **6** (2003) 24.
48. M. Schuster, T. Rager, A. Noda, K.D. Kreuer, and J. Maier, *Fuel Cells* **5** (2005) 355.
49. J.A. Elliott and S.J. Paddison, *Phys. Chem. Chem. Phys.* **9** (2007) 2602.
50. S.J. Paddison and J.A. Elliott, *Phys. Chem. Chem. Phys.* **8** (2006) 2193.
51. M. Eikerling, S.J. Paddison, L.R. Pratt, and T.A. Zawodzinski Jr., *Chem. Phys. Lett.* **368** (2003) 108.
52. M.K. Petersen and G.A. Voth, *J. Phys. Chem. B* **110** (2006) 18594.
53. E. Spohr, P. Commer, and A.A. Kornyshev, *J. Phys. Chem. B* **106** (2002) 10560.
54. M. Eikerling, A.S. Ioselevich, and A.A. Kornyshev, *Fuel Cells* **4** (2004) 131.
55. M. Boudart, *Adv. Catal.* **20** (1969) 153.
56. M. Eikerling, J. Meier, and U. Stimming, *Z. Phys. Chem.* **217** (2003) 395.
57. F. Maillard, M. Eikerling, O.V. Cherstiouk, S. Schreier, E. Savinova, and U. Stimming, *Faraday Discuss.* **125** (2004) 357.
58. L.B. Hansen, P. Stoltze, J.K. Nørskov, B.S. Clausen, and W. Niemann, *Phys. Rev. Lett.* **64** (1990) 3155.
59. B. Andreaus, F. Maillard, J. Kocylo, E. Savinova, and M. Eikerling, *J. Phys. Chem. B* **110** (2006) 21028.
60. O.V. Cherstiouk, P.A. Simonov, and E.R. Savinova, *Electrochim. Acta* **48** (2003) 3851.
61. F. Maillard, S. Schreier, M. Heinzlik, E.R. Savinova, S. Weinkauf, and U. Stimming, *Phys. Chem. Chem. Phys.* **7** (2005) 385.

62. F. Maillard, E.R. Savinova, and U. Stimming, *J. Electroanal. Chem.* **599** (2006) 221.
63. K. Malek, M. Eikerling, Q. Wang, T. Navessin, and Z. Liu, *J. Phys. Chem. C.* **111** (2007) 13627.
64. M. Uchida, Y. Aoyama, E. Eda, and A. Ohta, *J. Electrochem. Soc.* **142** (1995) 463.
65. M. Uchida, Y. Aoyama, E. Eda, and A. Ohta, *J. Electrochem. Soc.* **142** (1995) 4143.
66. M. Uchida, Y. Fuuoka, Y. Sugawara, N. Eda, and A. Ohta, *J. Electrochem. Soc.* **143** (1996) 2245.
67. M. Eikerling and A.A. Kornyshev, *J. Electroanal. Chem.* **453** (1998) 89.
68. M. Eikerling, *J. Electrochem. Soc.* **153** (2006) E58.
69. R. Fernandez, P. Ferriera-Aparicio, and L. Daza, *J. Power Sources* **151** (2005) 18.
70. S. Torquato, *Random Heterogeneous Mater.*, Springer, New York, Berlin, Heidelberg, 2002.
71. M. Sahimi, G.R. Gavalas, and T.T. Tsotsis, *Chem. Eng. Sci.* **45** (1990) 1443.
72. M. Eikerling, A.A. Kornyshev, and U. Stimming, *J. Phys. Chem. B* **101** (1997) 10807.
73. Z. Xia, Q. Wang, M. Eikerling, and Z. Liu, *Can J Chem.* **86** (2008) 657.
74. M. Eikerling, A.A. Kornyshev, and E. Spohr, "Proton-Conducting Polymer Electrolyte Membranes: Water and Structure in Charge" in *Advances in Polymer Science, Polymers for Fuel Cells*, Ed. By G.G. Scherer, DOI 10.1007/12_2008_132.
75. D. Seeliger, C. Hartnig, and E. Spohr, *Electrochim. Acta* **50** (2005) 4234.
76. S. Tanimura and T. Matsuoka, *J. Poly. Sci. B* **42** (2005) 1905.
77. B. Smitha, S. Sridhar, and A.A. Khan, *J. Memb. Sci.* **259** (2005) 10.
78. S.M. Haile, *Acta Mater.* **51** (2003) 5981.
79. O. Savadogo, *J. Power Sources* **127** (2004) 135.
80. M.A. Hickner, H. Ghassemi, Y.S. Kim, B.R. Einsla, and J.E. McGrath, *Chem. Rev.* **104** (2004) 4587.
81. Y. Yang and S. Holdcroft, *Fuel Cells* **5** (2005) 171.
82. M.A. Hickner and B.S. Pivovar, *Fuel Cells* **5** (2005) 213.
83. G. Gebel, *Polymer* **41** (2000) 5829; L. Rubatat, G. Gebel, and O. Diat, *Macromolecules* **37** (2004) 7772.
84. G. Gebel and O. Diat, *Fuel Cells* **5** (2005) 261.
85. R.R. Netz and D. Andelmann, *Phys. Rep.* **380** (2003) 1.
86. M. Cappadonia, J.W. Erning, and U. Stimming, *J. Electroanal. Chem.* **376** (1994) 189.
87. A.A. Kornyshev, A.M. Kuznetsov, E. Spohr, and J. Ulstrup, *J. Phys. Chem. B* **107** (2003) 3351.
88. N. Agmon, *Chem. Phys. Lett.* **244** (1995) 456.
89. H.G. Herz, K.D. Kreuer, J. Maier, G. Scharfenberger, M.F.H. Schuster, and W.H. Meyer, *Electrochim. Acta* **48** (2003) 2165.
90. A. Warshel and R.M. Weiss, *J. Am. Chem. Soc.* **102** (1980) 6218.
91. A. Warshel, in *Computer Modeling of Chemical Reactions in Enzymes and in Solutions*; Wiley, New York, 1991.
92. J. Åqvist and A. Warshel, *Chem. Rev.* **93** (1993) 2523.

93. S. Walbran and A.A. Kornyshev, *J. Chem. Phys.* **114** (2001) 10039.
94. P.G. Bolhuis, D. Chandler, C. Dellago, and P.L. Geissler, *Annu. Rev. Phys. Chem.* **53** (2002) 291.
95. C. Dellago, P.G. Bolhuis, and D. Chandler, *J. Chem. Phys.* **108** (1998) 9236.
96. W.E and E. Vanden-Eijnden, *J. Stat. Phys.* **123** (2006) 503.
97. G. Torrie and J. Valleau, *Chem. Phys. Lett.* **28** (1974) 578.
98. G. Mills, H. Jonsson, and G.K. Schenter, *Surf. Sci.* **324** (1995) 305.
99. J.B. Spencer and J.O. Lundgren, *Acta Cryst. B* **29** (1973) 1923.
100. G. Kresse and J. Hafner, *Phys. Rev. B* **47** (1993) 558.
101. G. Kresse and J. Hafner, *Phys. Rev. B* **49** (1994) 14251.
102. G. Kresse and J. Furthmüller, *Phys. Rev. B* **54** (1996) 11169.
103. G. Kresse and J. Hafner, *J. Phys. Condens. Mater.* **6** (1994) 8245.
104. G. Gebel, Ph.D. thesis, Université Joseph Fourier, 1989.
105. G. Gebel and R.B. Moore, *Macromolecules* **33** (2000) 4850.
106. A.L. Rollet, O. Diat, and G. Gebel, *J. Phys. Chem. B* **106** (2002) 3033.
107. B. Loppinet and G. Gebel, *Langmuir* **14** (1998) 1977.
108. T.D. Gierke, G.E. Munn, and F.C. Wilson, *J. Polym. Sci. Polym. Phys.* **19** (1981) 1687.
109. R. Car and M. Parrinello, *Phy. Rev. Lett.* **55** (1985) 2471.
110. CPMD code by J. Hutter et al., MPI für Festkörperforschung and IBM Zurich Research Laboratory, 1995–2002.
111. V.R. Stamenkovic, B.S. Mun, M. Arenz, K.J. Mayrhofer, C.A. Lucas, G.F. Wang, P.N. Ross, and N.M. Markovic, *Nat. Mater.* **6** (2007) 241.
112. V.R. Stamenkovic, B. Fowler, B.S. Mun, G. Wang, P.N. Ross, C.A. Lucas, N.M. Markovic, and V.R. Stamenkovic, *Science* **315** (2007) 493.
113. H.G. Petrow and R.J. Allen, US Patent, 3992512 (1976).
114. S.H. Joo, S.J. Choi, I. Oh, J. Kwak, Z. Liu, O. Terasaki, and R. Ryoo, *Nature* **412** (2001) 169.
115. X. Hao, W.A. Spieker, and J.R. Regalbuto, *J. Coll. Int. Sci.* **267** (2003) 259.
116. B. Andreaus and M. Eikerling, *J. Electroanal. Chem.* **607** (2007) 121.
117. M. Boudart, *Adv. Catal.* **20** (1969) 153.
118. R. Van Hardefeld and F. Hartog, *Surf. Sci.* **15** (1969) 189.
119. K. Kinoshita, *J. Electrochem. Soc.* **137** (1990) 845.
120. S. Mukerjee, in *Catalysis and Electrocatalysis at Nanoparticle Surfaces*, Ed. by A. Wieckowski, E.R. Savinova, and C.G. Vayenas, M. Dekker, New York, 2003 pp. 501–530.
121. V.P. Zhdanov and B. Kasemo, *Surf. Sci. Rep.* **39** (2001) 25.
122. W.-J. Liu, B.-L. Wu, and C.-S. Cha, *J. Electroanal. Chem.* **476** (1999) 101.
123. B. Hammer, O.H. Nielsen, and J.K. Nørskov, *Catal. Lett.* **46** (1997) 31.
124. B. Hammer and J.K. Nørskov, *Adv. Catal.* **45** (2000) 71.
125. M.L. Sattler and P.N. Ross, *Ultramicroscopy* **20** (1986) 21.
126. A. Gamez, D. Richard, P. Gallezot, F. Gloaguen, R. Faure, and R. Durand, *Electrochim. Acta* **41** (1996) 307.
127. Y. Takasu, N. Ohashi, X.-G. Zhang, Y. Murakami, H. Minawaga, S. Sato, and K. Yahikozawa, *Electrochim. Acta* **41** (1996) 2595.
128. H.A. Gasteiger, S.S. Kocha, B. Sompalli, and F.T. Wagner, *Appl. Catal. B* **56**, (2005) 9.

129. P.N. Ross, in *Handbook of Fuel Cells* Ed. by W. Vielstich, A. Lamm, and H. Gasteiger, Wiley, Chichester, GB, 2003, Vol. 2, pp. 465–480.
130. A. Kabbabi, F. Gloaguen, F. Andolfatto, and R. Durand, *J. Electroanal. Chem.* **373** (1994) 251.
131. T. Frelink, W. Visscher, and J.A.R. Van Veen, *J. Electroanal. Chem.* **382** (1995) 65.
132. K. Yahikozawa, Y. Fujii, M. Yoshiharu, N. Katsunori, and Y. Takasu, *Electrochim. Acta* **36** (1991) 973.
133. Y. Takasu, T. Iwazaki, W. Sugimoto, and Y. Murakami, *Electrochem. Commun.* **2** (2000) 671.
134. O.V. Cherstiouk, P.A. Simonov, and E.R. Savinova, *Electrochim. Acta* **48** (2003) 3851.
135. J.K. Norskov and C.H. Christensen, *Science* **312** (2006) 1322.
136. B. Hammer and J.K. Nørskov, *Nature* **376** (1995) 238.
137. V.R. Stamenkovic, B.S. Mun, K.J.J. Mayrhofer, P.N. Ross, N.M. Markovic, J. Rossmeisl, J. Greeley, and J.K. Nørskov, *Angew. Chem. Int. Ed.* **18** (2006) 2815.
138. J. Greeley, T.F. Jaramillo, J. Bonde, I. Chorkendorff, and J.K. Norskov, *Nat. Mater.* **5** (2006) 909.
139. A. Gross, *Topics in Catalysis* **37** (2006) 29.
140. D.M. Kolb, G.E. Engelmann, and J.C. Ziegler, *Angew. Chem. Int. Ed.* **39** (2000) 1123.
141. L. Xiao and L. Wang, *J. Phys. Chem. A* **108** (2004) 8605.
142. C. Song, Q. Ge, and L. Wang, *J. Phys. Chem. B* **109** (2005) 22341.
143. A. Roudgar and A. Gross, *Surf. Sci.* **559** (2004) L180.
144. D. Mei, E.W. Hansen, and M. Neurock, *J. Phys. Chem.* **107** (2003) 798.
145. M. Neurock and D. Mei, *Top. Catal.* **20** (2002) 5.
146. S.C. Watson and E. Carter, *Comp. Phys. Commun.* **128** (2000) 67.
147. Y.A. Wang and E.A. Carter, "Orbital-Free Kinetic-Energy Density Functional Theory," in *Theoretical Methods in Condensed Phase Chemistry*, a volume in *Progress in Theoretical Chemistry and Physics*, Ed. by S.D. Schwartz, pp. 117–184, Kluwer, Dordrecht, 2000.
148. B. Zhou and Y.A. Wang, *J. Chem. Phys.* **124** (2006) 081107 (1–5).
149. W.Q. Tian, L.V. Liu, and Y.A. Wang, *Phys. Chem. Chem. Phys.* **8** (2006) 3528.
150. J. Solla-Gullón, F.J. Vidal-Iglesias, E. Herrero, J.M. Feliu, and A. Aldaz, *Electrochem. Commun.* **8** (2006) 189.
151. M. Watanabe and S. Motoo, *J. Electroanal. Chem.* **60** (1975) 275.
152. N.M. Marković and P.N. Ross, *Surf. Sci. Rep.* **45** (2002) 117.
153. H.A. Gasteiger, N. Marković, P.N. Ross, and E.J. Cairns, *J. Phys. Chem.* **97** (1993) 12020.
154. N.M. Marković and P.N. Ross, *Surf. Sci. Rep.* **45** (2002) 117.
155. P. Liu and J.K. Nørskov, *Fuel Cells* **1** (2001) 192.
156. K.A. Friedrich, F. Henglein, U. Stimming, and W. Unkauf, *Electrochim. Acta* **45** (2000) 3283.
157. O.V. Cherstiouk, P.A. Simonov, V.I. Zaikovskii, and E.R. Savinova, *J. Electroanal. Chem.* **554** (2003) 241.
158. F. Maillard, E.R. Savinova, P.A. Simonov, V.I. Zaikovskii, and U. Stimming, *J. Phys. Chem. B* **108** (2004) 17893.

159. K.J.J. Mayrhofer, M. Arenz, B.B. Blizanac, V.R. Stamenkovic, P.N. Ross, and N.M. Marković, *Electrochim. Acta* **50** (2005) 5144.
160. M. Arenz, K.J.J. Mayrhofer, V.R. Stamenkovic, B.B. Blizanac, T. Tomoyuki, P.N. Ross, and N.M. Marković, *J. Am. Chem. Soc.* **127** (2005) 6819.
161. J. Solla-Gullón, F.J. Vidal-Iglesias, E. Herrero, J.M. Feliu, and A. Aldaz, *Electrochem. Commun.* **8** (2006) 189.
162. N.P. Lebedeva, M.T.M. Koper, J.M. Feliu, and R.A. van Santen, *J. Phys. Chem. B* **106** (2002) 12938.
163. M.T.M. Koper, N.P. Lebedeva, and C.G.M. Hermse, *Faraday Discuss.* **121** (2002) 301.
164. T. Kobayashi, P.K. Babu, L. Gancs, J.H. Chung, E. Oldfield, and A. Wieckowski, *J. Am. Chem. Soc.* **127** (2005) 14164.
165. P.J. Feibelman, B. Hammer, J.K. Nørskov, F. Wagner, M. Scheffler, R. Stumpf, R. Watwe, and J. Dumesic, *J. Phys. Chem. B* **105** (2001) 4018.
166. J.P. Ansermet, Ph.D. thesis, University of Illinois, Urbana-Champaign, II, 1985.
167. L.R. Becerra, C.A. Klug, C.P. Slichter, and J.H. Sinfelt, *J. Phys. Chem.* **97** (1993) 12014.
168. M. Bergelin, E. Herrero, J.M. Feliu, and M. Wasberg, *J. Electroanal. Chem.* **467** (1999) 74.
169. A.V. Petukhov, *Chem. Phys. Lett.* **277** (1997) 539.
170. M.T.M. Koper, A.P.J. Jansen, R.A. van Santen, J.J. Lukkien, and P.A.J. Hilbers, *J. Chem. Phys.* **109** (1998) 6051.
171. C. McCallum and D. Pletcher, *J. Electroanal. Chem.* **70** (1976) 277.
172. A. Bewick, M. Fleischmann, and H.R. Thirsk, *Trans. Faraday Soc.* **58** (1962) 2200.
173. V.P. Zhdanov and B. Kasemo, *Surf. Sci. Rep.* **39** (2000) 25.
174. V.P. Zhdanov and B. Kasemo, *Surf. Sci.* **545** (2003) 109.
175. D.T. Gillespie, *J. Comp. Phys.* **22** (1976) 403.
176. P. Gode, F. Jaouen, G. Lindbergh, A. Lundblad, and G. Sundholm, *Electrochem. Acta* **48** (2003) 4175.
177. Q. Wang, M. Eikerling, D. Song, and Z. Liu, *J. Electroanal. Chem.* **573** (2004) 61.
178. J.T. Wescott, Y. Qi, L. Subramanian, and T.W. Capehart, *J. Chem. Phys.* **124** (2006) 134702.
179. D.Y. Galperin and A.R. Khokhlov, *Macromol. Theory Simul.* **15** (2006) 137.
180. R.D. Groot and P.B. Warren, *J. Chem. Phys.* **107** (1997) 4423.
181. A. Vishnyakov and A.V. Neimark, *J. Phys. Chem.* **104** (2000) 4471.
182. D.A. Mologin, P.G. Khalatur, and A.R. Kholhlov, *Macromol. Theory Simul.* **11** (2002) 587.
183. P.G. Khalatur, S.K. Talitskikh, and A.R. Khokhlov, *Macromol. Theory Simul.* **11** (2002) 566.
184. S.J. Marrink, H.J. Risselada, S. Yefimov, D.P. Tieleman, and A.H. de Vries, *J. Phys. Chem. B.* **111** (2007) 7812.
185. S. Izvekov and A. Violi, *J. Chem. Theory Comput.* **2** (2006) 504.
186. S. Izvekov and A. Violi, *J. Phys. Chem. B* **109** (2005) 2469.
187. S. Izvekov, A. Violi, and G.A. Voth, *J. Phys. Chem. B* **109** (2005) 17019.
188. G.A. Voth, *J. Phys. Chem. B* **109** (2005) 17019.

189. A. Coniglio, L. De Arcangelis, E. Del Gado, A. Fierro, and N. Sator, *J. Phys. Condens. Mater.* **16** (2004) S4831.
190. M. Sahimi, *Rev. Mod. Phys.* **65** (1993) 1393.
191. D. Avnir, D. Farin, and P. Pfeifer, *Nature* **308** (1984) 261.
192. M.-O. Coppens, in *Fractals in Engineering,* Ed. by J. Lévy-Vehel, E. Lutton, and C. Tricot, Springer, New York, Heidelberg, Berlin, 1997, 2nd ed., p. 336.
193. K. Malek and M.-O. Coppens, *Phys. Rev. Lett.* **87** (2001) 125505.
194. M. Sahimi, G.R. Gavalas, and T.T. Tsotsis, *Chem. Eng. Sci.* **45** (1990) 1443.
195. M.-O. Coppens, *Colloids Surf. A* **187–188** (2001) 257.
196. M.-O. Coppens and G.F. Froment, *Fractals* **3** (1995) 807.
197. K. Malek and M.-O. Coppens, *Chem. Eng. Sci.* **58** (2003) 4787.
198. M. Sahimi, *Heterogeneous Mater.* 2003, Part I and Part II, Berlin, New York, Heidelberg, Springer.
199. M. Eikerling, Yu.I. Kharkats, A.A. Kornyshev, and Yu.M. Volfkovich, *J. Electrochem. Soc.* **145** (1998) 2684.
200. M.F.H. Schuster, W.H. Meyer, M. Schuster, and K.D. Kreuer, *Chem. Mater.* **16** (2004) 329.
201. G. Scharfenberger, W.H. Meyer, G. Wegner, M. Schuster, K.D. Kreuer, and J. Maier, *Fuel Cells* **6** (2006) 237.
202. A.B.H. Susan, M. Yoo, H. Nakamoto, and M. Watanabe, *Chem. Lett.* **32** (2003) 836.
203. D.A. Boysen, T. Uda, C.R.I. Chisholm, and S.M. Haile, *Science* **303** (2004) 68.
204. T. Uda and S.M. Haile, *Electrochem. Solid State Lett.* **8** (2005) A245.
205. L.X. Xiao, H.F. Zhang, E. Scanlon, L.S. Ramanathan, E.W. Choe, D. Rogers, T. Apple, and B.C. Benicewicz, *Chem. Mater.* **17** (2005) 5328.
206. H. Ghassemi, J.E. McGrath, and T.A. Zawodzinski Jr., *Polymer* **47** (2006) 4132.
207. J. Ding, C. Chuy, and S. Holdcroft, *Macromolecules* **35** (2002) 1348.
208. T.E. Springer, T.A. Zawodzinski, and S. Gottesfeld, *J. Electrochem. Soc.* **138** (1991) 2334.
209. D.M. Bernardi and M.W. Verbrugge, *AIChE J.* **37** (1991) 1151.
210. D.M. Bernardi, *J. Electrochem. Soc.* **137** (1990) 3344.
211. T.F. Fuller and J. Newman, *J. Electrochem. Soc.* **140** (1993) 1218.
212. R. Mosdale and S. Srinivasan, *Electrochim. Acta* **40** (1995) 413.
213. A.A. Kulikovsky, *Electrochem. Commun.* **4** (2002) 845.
214. M.L. Perry, J. Newman, and E.J. Cairns, *J. Electrochem. Soc.* **145** (1998) 5.
215. C.C. Boyer, R.G. Anthony, and A.J. Appleby, *J. Appl. Electrochem.* **30** (2000) 777.
216. Y. Wang, *Chem. Rev.* **104** (2004) 4727.
217. U. Pasaogullari and C.Y. Wang, *Electrochim. Acta* **49** (2004) 4359.
218. P. Mukherjee, C.Y. Wang, and K.S. Chen, *J. Electrochem. Soc.* **154** (2007) B823.
219. D. Natarajan and T.V. Nguyen, *J. Power Sources* **115** (2003) 66.
220. M. Noponen, E. Birgersson, J. Ihonen, A. Vynnycky, A. Lundblad, and G. Lindbergh, *Fuel Cells* **4** (2004) 365.
221. P. Berg, K. Promislow, J. St-Pierre, J. Stumper, and B. Wetton, *J. Electrochem. Soc.* **151** (2004) A341.
222. T. Berning and N. Djilali, *J. Electrochem. Soc.* **150** (2003) A1589.
223. A.A. Kulikovsky, *Electrochim. Acta* **49** (2004) 5187.

224. D. Song, Q. Wang, Z. Liu, T. Navessin, and S. Holdcroft, *Electrochim. Acta* **49** (2004) 731.
225. D. Song, Q. Wang, Z. Liu, M. Eikerling, Z. Xie, T. Navessin, and S. Holdcroft, *Electrochim. Acta* **50** (2005) 3347.
226. Q. Wang, M. Eikerling, D. Song, Z. Liu, T. Navessin, Z. Xie, and S. Holdcroft, *J. Electrochem. Soc.* **151** (2004) A950.
227. Z. Xie, T. Navessin, K. Shi, R. Chow, Q. Wang, D. Song, B. Andreaus, M. Eikerling, Z. Liu, and S. Holdcroft, *J. Electrochem. Soc.* **152** (2005) A1171.
228. G. ten Brinke and O. Ikkala, *Chem. Rec.* **4** (2004) 219.
229. A.Z. Weber and J. Newman, *J. Electrochem. Soc.* **150** (2003) A1008.
230. A.Z. Weber, R.M. Darling, and J. Newman, *J. Electrochem. Soc.* **151** (2004) A1715.
231. A. Lehmani, O. Berard, and P. Turq, *J. Stat. Phys.* **90** (1997) 379.
232. C.L. Rice and R. Whitehead, *J. Phys. Chem.* **69** (1965) 4017.
233. F.N. Büchi and G.G. Scherer, *J. Electrochem. Soc.* **148** (2001) A183.
234. R. Mosdale, G. Gebel, and M. Pineri, *J. Memb. Sci.* **118** (1996) 269.
235. S. Renganathan, Q. Guo, V.A. Sethuraman, J.W. Weidner, and R.E. White, *J. Power Sources* **160** (2006) 386.
236. J. Divisek, M. Eikerling, V. Mazin, H. Schmitz, U. Stimming, and Yu.M. Volfkovich, *J. Electrochem. Soc.* **145** (1998) 2677.
237. K.J. Beverley, J.H. Clint, and P.D.I. Fletcher, *Phys. Chem. Chem. Phys.* **1** (1999) 149.
238. J. Alcaniz-Monge, A. Linares-Solano, and B. Rand, *J. Phys. Chem. B* **105** (2001) 7998.
239. M.L. Studebaker and C.W. Snow, *J. Phys. Chem.* **59** (1955) 973.
240. J. Liu and M. Eikerling, *Electrochim. Acta* **53** (2008) 4435.
241. M.K. Debe, in *Handbook of Fuel Cells: Fundamentals, Technology, and Applications*, 3, 576–589, Ed. by W. Vielstich, A. Lamm, and H. Gasteiger, Wiley, Chichester, UK, 2003.
242. Q. Wang, M. Eikerling, D. Song, and S. Liu, *J. Electrochem. Soc.* **154**, (2007) F95.
243. A.S. Arico, P. Bruce, B. Scrosatti, J.-M. Tarascon, and W. van Schalwijk, *Nat. Mater.* **4** (2005) 366.
244. S. Förster and M. Anotnietti, *Adv. Mater.* **10** (1998) 195.
245. O. Ikkala and G. ten Brinke, *Science* **295** (2002) 2407.

6

Modeling of Catalyst Structure Degradation in PEM Fuel Cells

Jeremy P. Meyers

Department of Mechanical Engineering, The University of Texas at Austin, Austin, TX, USA

Abstract In this chapter, the requirements of a high-performance catalyst layer are examined in order to understand the ways in which the structure might degrade with operation. The formation of oxide species on the surface of the platinum is examined to understand its role in both oxygen reduction and in setting the equilibrium state of the surface as a function of both pH and potential. Models of dissolution of platinum and subsequent migration through the cell are reviewed, along with an analysis of experimental studies aimed at determining the evolution of particle-size distributions with operation of a fuel cell. The mechanism of corrosion of the carbon supports is also examined, specifically under off-design points such as cell reveral and localized fuel starvation. Implications for material development and future modeling needs are reviewed.

I. INTRODUCTION

The worldwide effort to commercialize proton-exchange membrane fuel cells (PEMFCs) has been inspired in no small part by the optimization of the microstructure of the catalyst layer, which has allowed for drastic reductions in the amount of platinum that is required to construct a practical fuel cell and has convinced both investors and developers that a fuel cell might be designed and built at sufficiently low cost to compete with incumbent technologies. The catalyst layer is designed to ensure facile transport of all reactants and products through the thickness of the catalyst layer to the surface of the dispersed electrocatalyst. As the only electrocatalysts with sufficient activity to construct a practical fuel cell are composed of precious metals with high cost per unit mass, catalyst-layer design hinges on the enhancement of the surface-to-volume ratio of these catalysts, and upon dispersion of these catalysts in such a way that they are highly utilized over the entire range of operation, and so that access to the electrochemically active surfaces is provided for all reacting species. This effective dispersion has been achieved by ensuring contiguous and continuous pathways for ionic, electronic, and gas transport, specifically by dispersing platinum nanoparticles on carbon supports and mixing with ionomer in an ink that can be bonded to the proton-exchange membrane that acts as a separator. The nanometer length scale platinum and platinum-alloy particles in the catalyst layers provide theoretical surface areas of over 100 m^2 g^{-1} and fuel cell active areas of 25–40 m^2 g^{-1}.

In order for the fuel cell to retain the advantages of this optimized structure, however, that structure must be able to resist to changes in both morphology and surface properties. In truth, however, the catalyst structure and properties tend to degrade rather drastically with operation, and electrocatalyst stability may be a determining factor in the useful lifetime of PEM-based fuel cells. Performance loss, particularly in the low-to-medium current density range of operation, under both steady-state and cycling conditions, has been attributed to a loss of electrochemically active surface area (ECSA) of the high-surface-area carbon-supported platinum electrocatalyst. Mass-transport losses, which can be attributed to degradation of surface properties and structure of both the catalyst layer and of the gas-diffusion media, tend to dominate at higher current densities. As we consider degradation of these structures, it is helpful to begin with a review of the structure and function of a well-designed catalyst layer, and then to consider how changes from that baseline might affect performance.

II. CARBON-SUPPORTED PLATINUM CATALYSTS

Carbon-supported Pt nanoparticle catalysts in aqueous acid electrolyte form oxides as byproducts of the electrochemical oxidation of Pt by water, which can be formed on different sites depending on both the coverage and electrode potential (1). At low coverages, OH tends to adsorb on Pt sites at edges of the particles. At higher coverages, adsorbed O shifts to populate bridged Pt sites, and at higher coverages still, for potentials above 0.95 V in non-adsorbing electrolytes (as one expects an ionomer such as Nafion to behave), O moves into sites beneath the surface of the particle formed by the place-exchange of O and Pt.

As a result of these multiple oxide states, cyclic voltammograms (CV's) of platinum reveal features attributed to the formation and reduction of a multi-layer platinum oxide film. These same features are reproduced in CV's of commercially available membrane-electrode assemblies (2, 3). The observation of increasing charge under the oxide reduction voltammetric peak in a study of a cathode catalyst layer with time at potential suggests that platinum is slowly oxidized at voltages relevant to fuel cell operation (4). The oxygen on the surface can be formed with H_2O as the only source of oxygen species, but the presence of gas-phase oxygen accelerates oxidation (4). Indeed, PEMFC's maintained at high voltages near idle conditions reveal a rapid loss of activity, attributed to the blocking of platinum surface sites from participation in the oxygen reduction reaction by the formation of platinum oxide (3). This loss, however, is completely and quickly recoverable by a "cathode reduction" by allowing the cathode potential to drop to allow the cathode potential to drop to ~ 0.5 V, consistent with reduction of the oxide layer formed at the higher cathode voltages (3).

III. CHEMICAL STATE OF PLATINUM IN ACIDIC CONDITIONS

One can map out the envelope of stability of chemical species in a system composed of platinum and acid electrolyte with changes in potential and pH. Pourbaix diagrams indicate the thermodynamically-stable phase and oxidation state of a metal as a function of electrolyte pH and electrode potential (5). Even a simplified model of platinum equilibrium states in acidic conditions notes the possibility of formation of a dissolved platinum ionic species Pt^{2+}, as well as the formation of an oxide layer in the presence of water (5, 6). There are differences between sources in the literature, however, and illustrate the need for additional thermodynamic data, especially for the equilibrium between solid and liquid phases, to clarify the effect of potential on the equilibrium phases of Pt in acid electrolytes.

While fuel cell technologists are most concerned with the solubility of platinum in the ionomer of interest, as the ionomer will be in direct contact with the electrocatalyst and specify the chemical environment at the catalyst surface, it is an experimentally difficult proposition to determine platinum concentration in a solid polymer electrolyte: one cannot enhance rates of mass transport through forced convection, so it is difficult to determine whether measured concentrations in the polymer will be equilibrated with the catalyst or whether the measurement will be corrupted by mass-transport limitations. As such, imperfect but guiding insights into the dissolution behavior of platinum can be found by measuring is its solubility in aqueous electrolyte. This solubility is governed by the chemical state of the platinum surface and of the platinum species in solution; these chemical states are, in turn, governed by solvent, pH, temperature, and potential (7–12).

The solubility of both bulk and highly dispersed Pt has been determined under oxidizing conditions in various aqueous (7, 8). One clear trend is that nanoparticles of platinum are found to have a higher solubility than bulk platinum (7); furthermore, in the acidic region, solubility increased with decreasing pH (12). This increase suggests that the excess surface energy of formation of the crystallite has a de-stabilizing effect and while the precise state of water in ionomer is unclear, drier operating conditions should result in a lower pH and thereby increase solubility of platinum in the ionomer. There have been, to date, only limited studies of the effect of temperature on Pt solubility, despite the importance of temperature in the operation of the fuel cell, particularly in the search for high-temperature membranes to ease thermal management in automotive systems. In the studies that have been completed, however, the solubility of Pt appears to increase with temperature following the Arrhenius relationship (12).

The solubility of platinum has been found to increase with the potential up to 1.1 V vs. a reversible hydrogen electrode. The variation of solubility with potential suggests a 2-electron dissolution reaction, and so, if equilibrated with bulk (neutral) platinum, it is reasonable to suggest a +2 ion (Pt^{2+}).

More recent results for bulk and high-surface-area carbon-supported platinum in sulfuric and perchloric acid electrolytes showed slopes lower than what one would expect for a 2-electron process (10, 11). These results suggest a dissolution mechanism involving dissolution of the oxide rather than Pt metal, or a different dissolved species.

IV. PT DISSOLUTION UNDER NON-EQUILIBRIUM CONDITIONS

Platinum dissolution plays a major role in the loss of electrochemically active surface area, particularly at the cathode, where higher potentials are encountered (10, 13, 14). After the platinum dissolves, it can then either deposit on existing platinum particles to form larger particles (10, 15). or diffuse into electrochemically inaccessible portions of the membrane-electrode assembly; while still present in the fuel cell, the platinum that deposits in the membrane is electrically isolated from the external circuit and therefore cannot contribute to the production of electricity (10). Platinum dissolution can occur both as a result of potential cycling and under stack "idling" conditions where the cathode potential is highest (13).

A fuel cell subject to dynamic loads will undergo many rapid changes over the course of its lifetime. As the fuel cell cycles between high and low current, its cell potential will also vary, generally between 0.6 and 1.0 V. For cells operating with relatively pure hydrogen as a fuel, the anode will remain close to the reversible hydrogen potential implying that the positive electrode will experience potential swings as the cell potential changes to match variable power demands. The variation of the cathode potential will change several properties of the electrode materials, notably the degree of oxide coverage of both platinum and carbon, and the hydrophobicity of the surfaces.

It must be noted the oxide appears to protect the platinum surface from dissolution at higher potentials. When the cathode potential rises rapidly to higher values, the platinum can dissolve at a rapid rate until an oxide layer is formed to curtail further dissolution. Patterson initially presented data on the rapid loss of electrochemically active area with potential cycling (16), and the phenomenon was subsequently modeled by Darling and Meyers, using simple

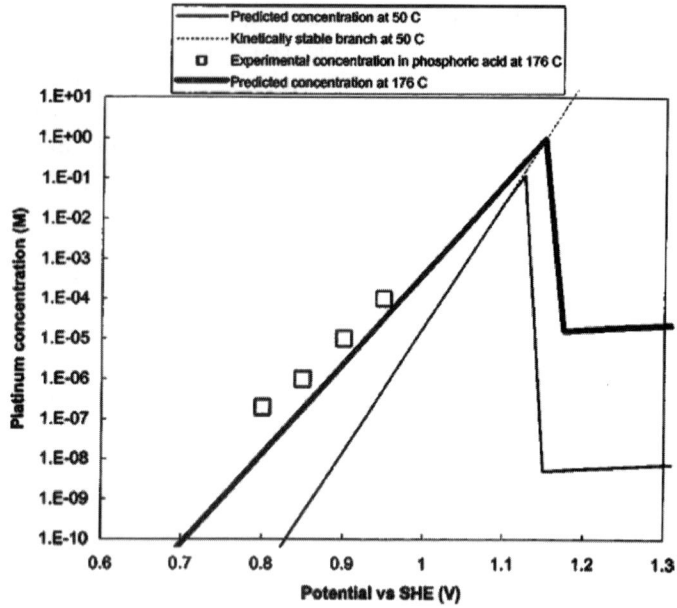

Figure 1. Equilibrium concentration of dissolved platinum vs. electrode potential from mathematical model. Image is from (14), Robert M. Darling and Jeremy P. Meyers, "Kinetic Model of Platinum Dissolution in PEMFC's," *Journal of the Electrochemical Society*, **2003**, *150*, A1523–A1527. Reproduced by permission of ECS – The Electrochemical Society.

models to describe the rates of platinum dissolution and oxide formation and the subsequent movement of soluble platinum species through the cell (14, 17). This model predicts that platinum is fairly stable at either low or high potentials, but there is a kinetically stable branch where platinum will dissolve rapidly when transitioning from low to high potentials. This is shown schematically in Fig. 1. The general trend of platinum solubility was subsequently measured in liquid electrolytes, and the same trend was discovered, although the equilibrium concentrations differed greatly from the model behavior proposed (18). These results are shown in Fig. 2. The model suggests a trend, and while the qualitative agreement with data is encouraging, considerably more effort must be made to fully describe the process fully and to provide predictive guidance in the search for new materials.

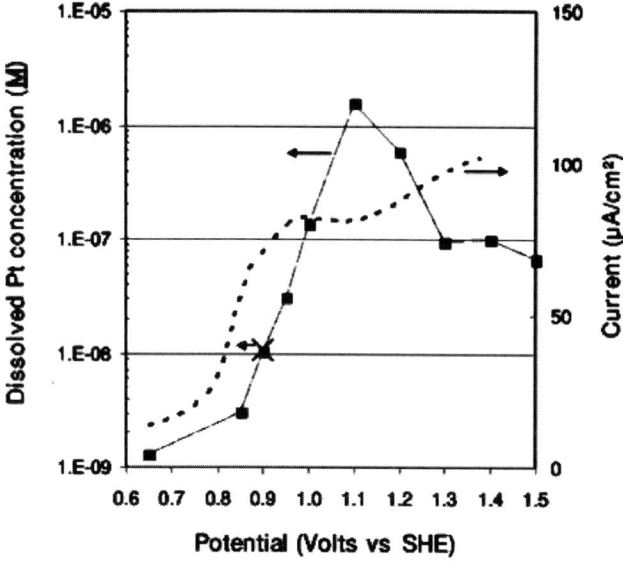

Figure 2. Measured dissolved platinum concentration in solution and initial dissolution rates in $0.57\,M$ perchloric acid. Image is from (18). V. Komanicky et al., "Stability and Dissolution of Platinum Surfaces in Perchloric Acid," *Journal of the Electrochemical Society*, **2006**, *153*, B446–B451. Reproduced by permission of ECS – The Electrochemical Society.

It is clear that any attempt to develop stable catalysts for fuel cell applications must consider not only the stability of the catalysts under constant potential conditions, but also under potential cycling. To design catalysts that are robust to this degradation mode, considerably more information is needed about the nature of the oxide, the kinetics of its formation, and its ability to protect the catalyst from dissolution over the entire range of potentials.

There are data on transient experiments, though they have yet to be reduced to a comprehensive model. Attempts to understand how platinum changes under transient conditions have included measurements of platinum dissolution during potential cycling (19–23), square-wave potentials steps (21, 24–26), and both galvanostatic and potentiostatic holds (11, 19, 27). A rotating ring-disk electrode (RRDE) study (used to detect by-products of an electrochemical reaction at a center disk as they are carried by a convection to a ring located just outside the disk) detected generation of soluble platinum species during potential cycling in acid electrolytes (20). During the cathodic scan, Pt^{2+} was detected at approximately 0.5 V and was associated with reduction of the oxide film.

Rand and Woods (23) detected both Pt^{2+} and Pt^{4+} species after potential cycling, and confirmed that the charge difference between anodic and cathodic sweep corresponds to the amount of dissolved as long as potentials are sufficiently low to avoid oxygen evolution. Ota et al. (27) investigated the corrosion rate of platinum in a potential range of oxygen evolution and found that in this region the corrosion rate is proportional to the oxygen evolution current.

Kinoshita et al. measured the amount of platinum dissolved during potential cycling of three different platinum samples: platinum sheet, high-surface-area platinum black, and carbon-supported platinum (21). They reported that the amount of platinum dissolved per cycle was independent of the upper potential limit at potentials >1.2 V. Their studies revealed very similar dissolution rates per potential cycle with widely varying scan rates, indicating that the dissolution reaction is more dependent on number of oxidation-reduction cycles than on the length of the cycle or time at oxidizing potentials, suggesting that transitions between low and high potential are more deleterious than holds at any single potential.

In recent years, several studies examined platinum dissolution in actual PEMFC electrodes. Ex situ transmission electron microscopy (TEM) analysis of cathode catalyst layers after long-term steady-state and potential cycling operation have shown dramatic changes in platinum particle size and distribution (10, 15, 28, 29). The fact that these techniques are ex-situ do complicate the measurements and effectively prevent one from watching particle-size distributions evolve in real time in an operating cell, but suggest that there are several mechanisms taking place, which result in different shifts in number, mean size, and distribution of platinum particles. The possible mechanisms for nanoparticle growth include local coalescence of agglomerated particles, agglomeration of particles through dissolved platinum migration, and dissolution of the catalyst and subsequent reprecipitation of platinum elsewhere in the cell. The particle growth rates and mechanisms may change as a function of many key conditions, including electrode potential, rate and amplitude of cycling, state of hydration of the membrane and operating temperature. Ferreira et al. (10) classified coarsened platinum particles into two groups: spherical particles still in contact with the carbon support and non-spherical particles removed from the carbon support; perhaps most interesting is the large difference in morphologies depending upon whether or not the particle is electrically contacting the carbon support (or perhaps, the difference in nucleation site).

Guilminot et al. (30) reported detection of Pt^{z+} ions ($z = 2, 4$) in the membrane of a cycled MEA using ultraviolet spectroscopy, which served to confirm the dissolution mechanism in a proton-exchange membrane fuel cell. Yasuda et al. (29, 31, 32) studied electrochemically-cycled MEAs using both TEM and cyclic voltammetry. They presented evidence for platinum dissolution, diffusion of the dissolved platinum into the membrane, and reduction of the dissolved platinum as particles in the membrane near the membrane-cathode interface by dissolved hydrogen.

Darling and Meyers developed mathematical models for dissolution and re-deposition of platinum in PEMFCs (14, 17). It was assumed that platinum dissolution is determined by potential, particle size, and coverage ratio by oxide. In this model, the oxide layer can protect the platinum from dissolution, but the kinetics of oxide formation are slow relative to the rate of dissolution, so rapid changes in potential can expose the bare platinum to corrosive potentials in the interim between the potential step and coverage of the surface with oxide. Changes of the ECSA, particle size distribution, and concentration of ionic species during potential holding or cycling were calculated. Their models did not account for nucleation sites other than existing platinum particles; as such, their model provided no means for redeposition in the membrane. The mechanism of redeposition has been explored in a model put forth by Bi et al.; they note that the location of the platinum redeposition band in the membrane depends explicitly upon the rate of hydrogen crossover and the point at which the potential at a hypothetical surface drops sufficiently to promote the platinum reduction reaction.

V. PARTICLE GROWTH

As mentioned in the previous section, imaging can be used for quantifying the changes in catalyst particle size, distribution, and morphology (particularly in the cathode) following electrochemical aging; in this way, particle data measured directly from the TEM/STEM images can be used to elucidate the mechanisms of particle coarsening that contribute to reduction of the catalyst electrochemically active surface area and the resulting fuel cell performance degradation. Ultra-microtomy sample preparation has enabled direct imaging of intact recast ionomer, carbon/Pt, and pore network surfaces within MEA porous catalyst layers via TEM. This technique has proven valuable for the imaging of catalyst particles post-mortem, including the ability to determine the locations where particles are located pre- and post-test (33). In a fresh sample, Pt particles are evident in the ionomer region of the MEA, indicating that there is only a weak bond between platinum and carbon support, and that particles might already be electrically isolated before the cell is constructed. After cycling, larger particle agglomerates can be found in the ionomer region (33).

Potential cycling, such as that encountered during transient operation, increases the rate of Pt particle growth in the cathode compared with steady state operation (34, 35). At high cathode potentials (potentials between 0.9 and 1.2 V), Pt particles become unstable and the equilibrium concentration of mobile Pt species in solution will increase significantly, accelerating the coarsening of Pt particles. The Pt particle size distributions become wider as a function of increasing potential, as does the nominal Pt particle size.

There are three primary Pt particle coarsening mechanisms that are believed to be important for PEM fuel cells (1) Ostwald ripening occurs when small particles dissolve, diffuse, and redeposit onto larger particles, resulting in reduced Pt particle surface area via a minimization in surface energy; (2) reprecipitation occurs when Pt dissolves into the ionomer phase within the cathode and then precipitates out again as newly formed Pt particles; (3) particle coalescence occurs when Pt particles are in close proximity and sinter together to form a larger particle. Each particle coarsening mechanism is characterized by distinctive elements in the evolution of the particle size distribution. For example, Ostwald ripening of the Pt particles will result in the growth of larger Pt particles at the expense of the smaller Pt particles, and the typical gaussian particle size distribution/profile will shift to higher and wider size distributions during electrochemical aging (36).

In the case of Pt dissolution and reprecipitation, the mobile platinum species that dissolve into the ionomer phase will nucleate both within the ionomer phase and onto the carbon support surface; this process will result in a distribution of particle sizes depending on the localized platinum concentrations within the ionomer. The particle size distribution for the case PEMFC particle growth does not shift to an entirely different range, as would occur for complete Ostwald ripening. Instead, the size distribution gets broader due to the nucleation of small, highly separated Pt particles within the polymer, as well as larger particles, which form due to the coarsening via coalescence of Pt particles nucleating close together.

Furthermore, dissolved platinum in solution in the ionomer phase may redeposit onto existing nuclei, resulting in growth of existing Pt particles. A change in the form/profile of the resulting Pt particle size distribution occurs when a Pt dissolution/reprecipitation mechanism is coupled with particle coalescence. This distribution is characterized by a change from a single, narrow, gaussian Pt particle size distribution to bimodal particle size distribution, where some fraction of small Pt particles are retained in the cathode while other Pt particles grow larger.

Ferreira et al. found that at the interface between the cathode catalyst layer and the gas-diffusion layer, Ostwald ripening dominated, most likely due to the limited hydrogen crossing through the membrane for Pt ion deposition. At the region of the catalyst layer closer to the interface with the membrane separator, however, Pt deposition in the ionomer was affected by the hydrogen crossing through the membrane, resulting in a significantly less positive mixed potential at a nucleation site, resulting in reduction of the mobilePt ions, which resulted in a bimodal particle-size distribution (10).

It is still difficult to conclude which Pt growth process is dominant. Separately counting Pt particles on or off carbon supports in TEM images could help to gain more details of Pt particle growth. Mathematical simulation including Pt ions transport could also assist to understand cathode Pt degradations. It is likely that both modes of change to the particle-size distribution, but we do not yet understand which experimental variables will accelerate (or mitigate) which modes.

The knowledge of particle-size distribution statistics provides important information in studying particle coarsening mechanisms, as Ostwald ripening and coalescence mechanisms produce very different distributions. Understanding of the mechanisms and development of comprehensive models will certainly provide direction for the development of new materials and new catalyst layer constructions.

VI. CORROSION OF CATALYST SUPPORT

In addition to loss of the platinum, the carbon supports that anchor the platinum crystallites and provide electrical connectivity to the gas-diffusion media and bipolar plates are also subject to degradation. In phosphoric acid systems, graphitized carbons are the standard because of the need for corrosion resistance in high-temperature acid environments (37), but PEM fuel cells have not employed fully graphitized carbons in the catalyst layers, due in large part to the belief and hope that the extra cost of graphitization could be avoided. As durability targets are scrutinized, however, it is important to consider the importance of carbon stability, and the use of graphitized carbons are being considered for use in PEMFC's to provide the required corrosion stability (38).

The electrochemical reaction of carbon to form carbon dioxide is thermodynamically permitted at the potentials at which the fuel cell cathode operates, but is believed to be almost negligibly slow in that potential range because of the lower temperatures of PEM cells compared with phosphoric acid. However, even if it proceeds very slowly, it can affect the long-term durability of PEMFCs. Electrochemical corrosion of carbon materials as catalyst supports of PEMFCs will cause electrical isolation of the catalyst particles as they are separated from the support, or lead to aggregation of catalyst particles, both of which will result in a decrease in the electrochemical active surface area of the catalyst. Furthermore, electrochemical oxidation of the carbon, even to intermediate degrees of oxidation resulting in the formation of a surface oxide instead of completely to carbon dioxide, can increase in hydrophilicity of the carbon surface, which, in turn, results in a decrease in effective gas permeability as the layer becomes more prone to filling with liquid water.

VII. CARBON CORROSION IN PEMFCS

In the last few years, several articles have been devoted the study of carbon corrosion in actual gas diffusion electrodes of PEMFCs (39, 40). Roen et al. (40) detected CO_2 in the cathode exhaust gas during potential cycling in an inert atmosphere of humidified helium. Their studies revealed enhancement rates of CO_2 evolution at potentials more positive than 0.9 V vs. a hydrogen reference electrode, and they also found a peak at about 0.6 V during anodic sweeps when only platinum catalyst existed. Intensity of this peak increased with the amount of platinum catalyst. It was proposed that CO adsorbed on platinum during this portion of the sweep, below 0.6 V, and it was subsequently oxidized. Mathias et al. (39) reported the dependence of carbon corrosion current on the potential, material, temperature, and time. The potential dependence displayed Tafel behavior, and the temperature dependence generally followed Arrhenius behavior

Fundamental studies have been carried out on carbon corrosion in aqueous acid and neutral solutions to study PEMFC degradation mechanism (41–46). Peaks during anodic sweeps of carbons appear to be enhanced in the presence of platinum catalyst. The enhancement of carbon corrosion implies that carbon corrosion is enhanced by the platinum catalyst under open-circuit conditions. The behavior is consistent with a model of the surface in which some species, such as absorbed CO, are generated at relatively low potentials. These species can move to the platinum catalyst by surface diffusion, and are subsequently oxidized at the platinum to form CO_2. While potential holds below 1.0 V at room temperature did not change surface oxide compositions significantly, the authors found an increase of oxide at holds above 0.8 V at 65°C, suggesting that temperature plays an important role in mediating the oxide coverage. As mentioned above, surface oxide determines hydrophobicity, which in turn affects the effective gas permeability of the electrode. If, however, the changes to hydrophobicity of the carbon support occur only at the carbon ionomer interface, these changes might not influence the rates of gas transport through the gas diffusion electrode. Therefore, further investigations to connect the electrochemical surface oxidation of the carbon support with the hindrance of gas diffusion will be required.

Wettability and gas diffusivity of the electrode changes by surface oxidation of carbon supports, and it is expected that these surface properties will, in turn, affect liquid-phase saturation levels and gas-phase transport in the composite catalyst layers. Supported platinum-on-carbon catalysts remain the state of the art for fuel cells under development today, but considerably more information is needed about the controlling factors of reactivity and stability as new catalyst/support structures are considered. Of course, alternatives to supported platinum are under evaluation, but present theoretical and practical challenges of their own, and it is likely that very different mechanisms of degrdation will be at work in these alternative catalyst systems are employed.

VIII. FUEL STARVATION

Potential cycling can also take another form in situations where the positive electrode rises far above the value of the hydrogen reference electrode. Full-sized cells, on the order of several hundred square centimeters in area, will experience different conditions between inlet and outlet, and this can lead to current distributions that cannot easily be simulated in subscale testing. Furthermore, cells arranged in a stack configuration can experience different flows of fuel, air, and coolant resulting from imperfect manifolding. Therefore, adjacent cells in a stack can experience different conditions in terms of hydrogen and oxygen content, but will be forced to carry the same current as their neighboring cells, as they are connected in series.

Several authors have noted that in the case of gross fuel starvation, cell voltages can become negative, as the anode is elevated to positive potentials and the carbon is consumed instead of the absent fuel. In the case of gross fuel starvation, for multiple cells in a stack, fuel maldistributions can result in some cells having insufficient fuel to carry the current that is being pushed through it by adjacent cells. In the absence of a sufficient anodic current source from hydrogen, the cell potential climbs higher until oxidation occurs- in this case, the oxidation of the carbon support of the catalyst layer. A diagram illustrating the change of electrode potentials under starvation conditions is shown in Fig. 3.

For reversals of this type, the anodic current is generally provided by carbon corrosion to form carbon dioxide, and results in permanent damage to the anode catalyst layer. Proper reactant distribution is critical to avoid this problem, and stack developers have accordingly sought to monitor the voltage of each cell to avoid such

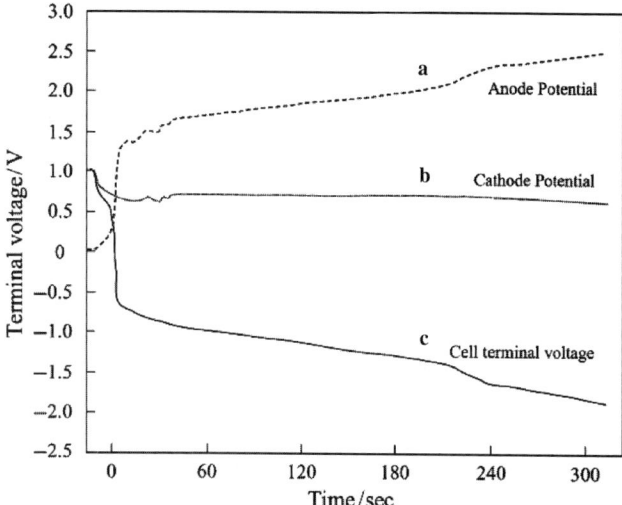

Figure 3. Electrode potentials for a cell driven to pass current after hydrogen flow is interrupted. Reprinted from *Journal of Power Sources*, 130, Taniguchi Akira et al., "Analysis of electrocatalyst degradation in PEMFC caused by cell reversal during fuel starvation," 746–752, Copyright 2004, with permission from Elsevier.

a problem (47). Obviously, such an extensive monitoring system will add considerable cost and complexity to the fuel cell stack and control scheme.

IX. LOCALIZED FUEL STARVATION

A more subtle form of fuel starvation was proposed in a paper by Reiser et al. (48). They suggest that transient conditions, or localized fuel starvation can induce local potentials on the air electrode significantly higher than 1 V, and thereby induce corrosion of the carbon supports that results in permanent loss of electrochemically active area. The cell potential can remain in the range of expected conditions even as this condition persists, and that this "reverse" current mechanism can induce damage to the cathode without being directly observable.

The mechanism suggests that the highly conductive bipolar plates of the fuel cell allow for sufficient redistribution of current in the plane of the current collectors that all regions of the cell experience the same potential difference. In the regions of the cell where fuel is present on the anode, the fuel cell behaves normally; the fact that the hydrogen reaction is so facile implies that the potential in the fuel-rich regions will stay close to its equilibrium voltage and is capable of delivering high currents until the hydrogen is consumed. In the regions of the cell where there is no fuel present, there is no proton or electron source at lower potentials, so the electrodes must shift to significantly higher potentials to maintain the potential difference imposed by the active part of the cell while still conserving current. Thus, a reverse current is established, and current is driven from the positive electrode to the negative electrode in the fuel-starved region, opposite the direction of normal current flow in the active portion of the cell. The only reactions that can sustain this current in the fuel-starved region are oxygen evolution and carbon corrosion on the positive electrode, and oxygen reduction from crossover on the negative electrode. The fact that such severe corrosive conditions can be induced without a direct means of measuring the onset of the problem has led some developers to reach the conclusion that conventional carbon support for Pt are unlikely to meet automotive durability targets and that implementation of corrosion-resistant supports combined with system-level strategies to carefully control the distribution of fuel on the anode are required (49).

This mechanism is shown schematically in Fig. 4 This problem can be induced not only by poor cell-to-cell flow distributions, but also by local blockages, differences in channel depth tolerances, and by water blockage, if water vapor condenses in the anode channels and fills the channels. A good example of this is given by Patterson and Darling (50). This phenomenon has been visualized by neutron imaging, and suggests that improper water management can cause major problems with fuel distribution and, consequently, damage to the cathode catalyst layers (51).

X. START/STOP CYCLING

While the problem of localized starvation under normal operation can perhaps be mitigated through careful control of reactants

Modeling of Catalyst Structure Degradation in PEM Fuel Cells

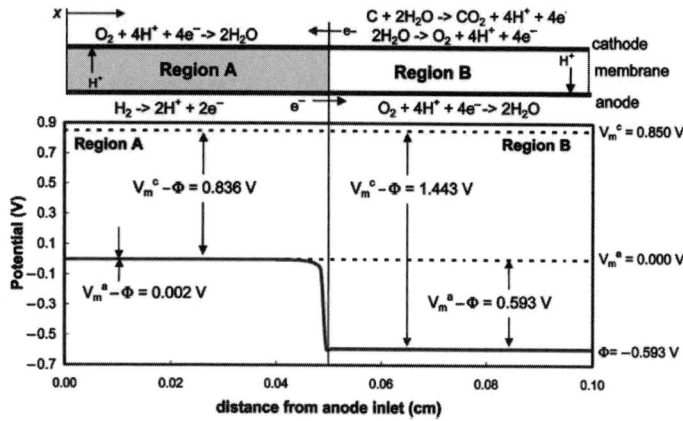

Figure 4. Schematic diagram illustrating the "reverse current" mechanism. Image taken from (48). Carl A. Reiser et al., "A Reverse-Current Decay Mechanism for Fuel Cells," *Electrochemical and Solid-State Letters*, **2005**, 8, A273–A276. Reproduced by permission of ECS – The Electrochemical Society.

and water management (design, operating conditions and materials choices), there is an aspect of fuel cell operation where such maldistributions must almost certainly exist, at least for a short time: namely, start-up and shut down. Under conditions of a prolonged shutdown, unless the stack is continually provided with fuel, hydrogen crossover from anode to cathode will eventually empty out the anode chamber and result in an air-filled flow channel. In this case, starting flow of fuel will induce a transient condition in which fuel exists at the inlet but is still fuel-starved at the exit. As a result, starting and stopping the fuel cell can induce considerable damage to the cell. This phenomenon has been modeled and reveals that an unprotected start can induce local potentials on the cathode in excess of 1.8 V relative to a hydrogen reference electrode (52). This modeling study suggests that potential control (voltage clipping) is by far the most effective means of minimizing this effect. A review of system-level strategies to minimize this mode of degradation has been reported by Perry et al. (53).

XI. TEMPERATURE AND RELATIVE HUMIDITY

Another aspect of fuel cell operation that is likely to affect the integrity of the cell are the changes in temperature and relative humidity that are associated with transitions between low and high power. In general, for cells that operate at fixed stoichiometric ratios, operation at low current implies a relatively cool, and wet cell; higher currents imply a hotter, drier cell (54). The fact that the ionomer swells with water uptake suggests that increases in water uptake as the membrane is exposed to high RH conditions can lead to compressive stresses in the membrane that then yield to tensile residual stresses during drying (55). These stresses are suggested as a significant contributor to mechanical failures of the membrane. Another recent study suggests that drying can considerably strain the membrane-electrode assembly, and that mechanical failure of membranes can result from gradual reduction in ductility combined with excessive strains induced by constrained drying of the MEA (56). Both temperature and relative humidity have been shown to effect rate of catalyst surface area loss due to platinum particle growth (57). These studies suggest that more needs to be learned about material properties and how they change over the course of fuel cell operation.

Accelerated fuel cell life tests have become common, because life tests under standard conditions can often last thousands of hours. In accelerated tests, membrane degradation can occur much faster than under normal operating conditions (58). However, the appropriateness of a given accelerated condition as a gauge of general durability is not necessarily clear. To date, four different accelerated parameters or a combination of these parameters have been employed in accelerated life testing: (1) elevated temperature (2) reduced humidity (3) open circuit voltage (OCV) and (4) cycling (relative humidity (RH), temperature, potential, freeze/thaw or start/stop) conditions. Membrane degradation is often monitored by changes in gas crossover rate or fluoride-ion emission rate (FER) during the *in-situ* test. While membrane degradation generally results in failures in the cell associated with crossover, rather than degradation of the catalyst layer, the catalyst layer can play an important role in controlling the rates of chemical attack in the membrane.

XII. ALLOY EFFECTS

Platinum alloys have been used in both phosphoric acid (37) and PEM fuel cell systems (59) to improve the activity of the oxygen reduction reaction. UTC developed alloy catalysts for use in PAFC powerplants in the 1970s, and suggested in their search that in phosphoric acid systems, the dissociation of the O–O bond was the rate-determining step, and consequently sought opportunities to change the spacing of Pt–Pt and thereby affect the cleavage of the O–O bond. As data from various alloys were analyzed, it was shown that the specific activity of the catalysts could be correlated with the platinum-platinum interatomic distance (60).

Several platinum alloys on high-surface-area carbon in acid electrolytes show high degrees of crystallinity, and enhancement of exchange current densities ($\sim$2–3 x) relative to Pt/C (37, 60, 61) It has been shown that the kinetics of the reaction correlate not only to Pt–Pt bond distance, but also to Pt d-band vacancies in the alloy; in studies on PtCo alloys, activities were correlated to inhibition of adsorbed OH formation (62). Which of these effects is the primary driver of catalyst performance is still unknown. While fundamental studies of the oxygen reduction reaction continue (63–65), the competing reaction pathways and considerable complexities of the oxide-covered alloy surfaces (not to mention the implications of ternary, and quaternary alloys and beyond) require a considerable amount of experimental and theoretical effort to understand.

While alloys have been shown to enhance performance in some configurations, attention must be paid to the stability of the alloys, particularly in light of the fact that Pt–Ru catalysts, which have been used in direct methanol and reformate cells to enhance CO tolerance on the anode, have been revealed to be quite unstable under normal operation (66). Generally, the metals which are co-alloyed with platinum would tend to be less stable as pure metals than platinum, (5) so one might naturally be concerned about introducing a less stable metal to a structure that has already been shown to degrade in the fuel cell environment.

Antolini et al. have written a review of the stability of Pt-alloy catalysts (67) and note discrepancies among various studies about the stability of alloys relative to platinum catalysts. They note that PtCr and PtCo tend to exhibit greater stability than PtV, PtNi, and PtFe, though there are still questions about whether the means of preparation of the catalyst matters. As previously discussed, changes in platinum crystallite morphology and carbon corrosion contribute strongly to the rates of degradation of PEM fuel cells in normal operating conditions (15, 68).

It has been shown that some platinum alloy catalysts, particularly those containing cobalt, show markedly improved stability to potential cycling relative to unalloyed platinum (69). Yu et al. show significant improvements in catalyst stability with PtCo catalysts after a protocol of square-wave potential cycling between 0.87 and 1.2 V vs. RHE. While other studies have shown considerable loss of Co into the membrane after cycling (70), data suggests that this is perhaps due to imperfect formation of the alloy in the synthesis step, and that leaching of the alloy after preparation can mitigate the problem (59). TEM analysis of MEA cross-sections after cycling show clear evidence of Pt migration towards and into the PEMFC membrane for Pt/C cathodes, while PtCo/C materials show minimal deterioration (71). However, the PtCo crystallite sizes were larger than their Pt/C counterparts; thus, an alloy produces higher mass activity and confers a higher stability to cycling and resistance to dissolution, despite a lower metal surface area compared to Pt/C. While it does appear that alloying can have a positive effect upon catalyst stability, is not clear from these studies, whether the improvement in stability is due to improved thermodynamic stability, or due to differences in kinetics of platinum dissolution or passivating oxide formation. If we are to design catalysts specifically to increase their stability, more information is needed about the mechanism of degradation and the role that the alloying element plays on fundamental properties of the catalyst.

XIII. CONCLUSIONS

The mechanisms of catalyst layer degradation are many in number, and most modeling studies to date have considered simplified versions of degrdadation in isolation. In reality, the mechanisms probably include more subtlety, and one mode of degradation likely influences others. Researchers are still trying to isolate the mechanisms and determine which physical or chemical properties can be modified to minimize degradation.

REFERENCES

1. Teliska, A.; O'Grady, W. E.; Ramaker, D. E. *Journal of Physical Chemistry B* **2005**, *109*, 8076–8084.
2. Paik, C. H.; Jarvi, T. D.; O'Grady, W. E. *Electrochemical and Solid State Letters* **2004**, *7*, A82–A84.
3. Uribe, F. A.; Zawodzinski, T. A. *Electrochimica Acta* **2002**, *47*, 3799–3806.
4. Paik, C. H.; Saloka, G. S.; Graham, G. W. *Electrochemical and Solid State Letters* **2007**, *10*, B39–B42.
5. Pourbaix, M. *Atlas of electrochemical equilibria* **1966**, Oxford, New York, Pergamon Press 1st edition.
6. Lee, J. B. *Corrosion* **1981**, *37*, 467–481.
7. Azaroul, M.; Romand, B.; Freyssinet, P.; Disnar, J. R. *Geochimica et Cosmochimica Acta* **2001**, *65*, 4453–4466.
8. Bowles, F. W.; Gize, A. P.; Vaughan, D. J.; Norris, S. J. 1995. no 520: 65–73. *Chronique de la Recherche Miniere* **1995**, *529*, 65–73.
9. Bindra, P.; Clouser, S. J.; Yeager, E. *Journal of the Electrochemical Society* **1979**, *126*, 1631–1632.
10. Ferreira, P. J.; la O', G. J.; Shao-Horn, Y.; Morgan, D.; Makharia, R.; Kocha, S.; Gasteiger, H. A. *Journal of the Electrochemical Society* **2005**, *152*, A2256–A2271.
11. Wang, X. P.; Kumar, R.; Myers, D. J. *Electrochemical and Solid State Letters* **2006**, *9*, A225–A227.
12. Mitsushima, S.; Koizumi, Y.; Ota, K.-I.; Kamiya, N. *Electrochemistry Communications* **2007**, *75*.
13. Mathias, M.; Gasteiger, H.; Makharia, R.; Kocha, S.; Fuller, T.; Pisco, J. *Abstracts of Papers of the American Chemical Society* **2004**, *228*, 002–FUEL.
14. Darling, R. M.; Meyers, J. P. *Journal of the Electrochemical Society* **2003**, *150*, A1523–A1527.
15. Xie, J.; Wood, D. L.; More, K. L.; Atanassov, P.; Borup, R. L. *Journal of the Electrochemical Society* **2005**, *152*, A1011–A1020.
16. Patterson, T. 2002 AIChE Spring National Meeting, 2002; pp. 313–318.
17. Darling, R. M.; Meyers, J. P. *Journal of the Electrochemical Society* **2005**, *152*, A242–A247.
18. Komanicky, V.; Chang, K. C.; Menzel, A.; Markovic, N. M.; You, H.; Wang, X.; Myers, D. *Journal of the Electrochemical Society* **2006**, *153*, B446–B451.
19. Benke, G.; Gnot, W. *Hydrometallurgy* **2002**, *64*, 205–218.
20. Johnson, D. C.; Napp, D. T.; Bruckenstein, S. *Electrochimica Acta* **1970**, *15*, 1493–1509.
21. Kinoshita, K.; Lundquist, J. T.; Stonehart, P. *Journal of Electroanalytical Chemistry* **1973**, *48*, 157–166.
22. Mitsushima, S.; Kawahara, S.; Ota, K. I.; Kamiya, N. *Journal of the Electrochemical Society* **2007**, *154*, B153–B158.
23. Rand, D. A. J. a. R. W. *Journal of Electroanalytical Chemistry* **1972**, *35*, 209–218.
24. Arvia, A. J.; Canullo, J. C.; Custidiano, E.; Perdriel, C. L.; Triaca, W. E. *Electrochimica Acta* **1986**, *31*, 1359–1368.

25. Canullo, J. C.; Triaca, W. E.; Arvia, A. J. *Journal of Electroanalytical Chemistry* **1986**, *200*, 397–400.
26. Egli, W. A.; Visintin, A.; Triaca, W. E.; Arvia, A. J. *Applied Surface Science* **1993**, *68*, 583–593.
27. Ota, K. I.; Nishigori, S.; Kamiya, N. *Journal of Electroanalytical Chemistry* **1988**, *257*, 205–215.
28. Schulze, M.; Schneider, A.; Gulzow, E. *Journal of Power Sources* **2004**, *127*, 213–221.
29. Yasuda, K.; Taniguchi, A.; Akita, T.; Ioroi, T.; Siroma, Z. *Physical Chemistry Chemical Physics* **2006**, *8*, 746–752.
30. Guilminot, E.; Corcella, A.; Charlot, F.; Maillard, F.; Chatenet, M. *Journal of the Electrochemical Society* **2007**, *154*, B96–B105.
31. Akita, T.; Taniguchi, A.; Maekawa, J.; Sirorna, Z.; Tanaka, K.; Kohyama, M.; Yasuda, K. *Journal of Power Sources* **2006**, *159*, 461–467.
32. Yasuda, K.; Taniguchi, A.; Akita, T.; Ioroi, T.; Siroma, Z. *Journal of the Electrochemical Society* **2006**, *153*, A1599–A1603.
33. More, K. L.; Reeves, K. S. *DOE Hydrogen Program Review* **2005**, www.hydrogen.energy.gov/pdfs/review05/fc39_more.pdf.
34. Borup, R.; Davey, J.; Garzon, F. H. *ECS Transactions* **2006**, *3*, 879.
35. More, K. L.; Borup, R; Reeves, K. S. *ECS Transations* **2006**, *3*.
36. Ascarelli, P., V. Contini, R. Giorgi *Journal of Applied Physics* **2002**, *91*, 4556.
37. Landsman, D. A.; Luczak, F. J. In *Handbook of Fuel Cells-Fundamentals, Technology, and Applications*; Vielstich, W., Gasteiger, H. A., Lamm, A., Eds.; Wiley: West Sussex, England, 2003; Vol. 4.
38. Yu, P. T.; Gu, W.; Makharia, R.; Wagner, F. T.; Gasteiger, H. A. *ECS Transactions* **2007**, *3*, 797.
39. Mathias, M. F.; Makharia, R.; Gasteiger, H. A.; Conley, J. J.; Fuller, T. J.; Gittleman, C. J.; Kocha, S. S.; Miller, D. P.; Mittelsteadt, C. K.; Xie, T.; Yan, S. G.; Yu, P. T. *Interface* **2005**, *14*, 24.
40. Roen, L. M.; Paik, C. H.; Jarvi, T. D. *Electrochemical and Solid-State Letters* **2004**, *7*, A19–A22.
41. Willsau, J.; Heitbaum, J. *Journal of Electroanalytical Chemistry* **1984**, *161*, 93–101.
42. Kangasniemi, K. H.; Condit, D. A.; Jarvi, T. D. *Journal of the Electrochemical Society* **2004**, *151*, E125–E132.
43. Siroma, Z.; Ishii, K.; Yasuda, K.; Miyazaki, Y.; Inaba, M.; Tasaka, A. *Electrochemistry Communications* **2005**, *7*, 1153–1156.
44. Kinumoto, T.; Takai, K.; Iriyama, Y.; Abe, T.; Inaba, M.; Ogumi, Z. *Journal of the Electrochemical Society* **2006**, *153*, A58–A63.
45. Chaparro, A. M.; Mueller, N.; Atienza, C.; Daza, L. *Journal of Electroanalytical Chemistry* **2006**, *591*, 69–73.
46. Wang, X.; Li, W.; Chen, Z.; Waje, M.; Yan, Y. *Journal of Power Sources* **2006**, *158*, 154–159.
47. Barton, H. R.; Ballard Power Systems Inc.: United States of America, 2004, US patent #6,724,194. "Cell Voltage Monitor for a Fuel Cell Stack" issued April 20, 2004.
48. Reiser, C. A.; Bregoli, L.; Patterson, T. W.; Yi, J. S.; Yang, J. D.; Perry, M. L.; Jarvi, T. D. *Electrochemical and Solid-State Letters* **2005**, *8*, A273–A276.

49. Makharia, R.; Kocha, S. S.; Yu, P. T.; Sweikart, M. A.; Gu, W.; Wagner, F. T.; Gasteiger, H. A. *ECS Transactions* **2006**, *1*, 3.
50. Patterson, T. W.; Darling, R. M. *Electrochemical and Solid-State Letters* **2006**, *9*, A183–A185.
51. Pekula, N.; Heller, K.; Chuang, P. A.; Turhan, A.; Mench, M. M.; Brenizer, J. S.; Unlu, K. *Nuclear Instruments & Methods in Physics Research Section a-Accelerators Spectrometers Detectors and Associated Equipment* **2005**, *542*, 134–141.
52. Meyers, J. P.; Darling, R. M. *Journal of the Electrochemical Society* **2006**, *153*, A1432–A1442.
53. Perry, M. L.; Patterson, T. W.; Reiser, C. *ECS Transactions* **2006**, *3*, 783–795.
54. Yan, X. Q.; Hou, M.; Sun, L. Y.; Cheng, H. B.; Hong, Y. L.; Liang, D.; Shen, Q.; Ming, P. W.; Yi, B. L. *Journal of Power Sources* **2007**, *163*, 966–970.
55. Kusoglu, A.; Karlsson, A. M.; Santare, M. H.; Cleghorn, S.; Johnson, W. B. *Journal of Power Sources* **2006**, *161*, 987–996.
56. Huang, X. Y.; Solasi, R.; Zou, Y.; Feshler, M.; Reifsnider, K.; Condit, D.; Burlatsky, S.; Madden, T. *Journal of Polymer Science Part B-Polymer Physics* **2006**, *44*, 2346–2357.
57. Borup, R. L.; Davey, J. R.; Garzon, F. H.; Wood, D. L.; Inbody, M. A. *Journal of Power Sources* **2006**, *163*, 76–81.
58. Knights, S. D.; Colbow, K. M.; St-Pierre, J.; Wilkinson, D. P. *Journal of Power Sources* **2004**, *127*, 127–134.
59. Gasteiger, H. A.; Kocha, S. S.; Sompalli, B.; Wagner, F. T. *Applied Catalysis B: Environmental* **2005**, *56*, 9–35.
60. Jalan, V.; Taylor, E. J. *Journal of the Electrochemical Society* **1983**, *130*, 2299–2302.
61. Mukerjee, S.; Srinivasan, S.; Soriaga, M. P.; McBreen, J. *Journal of the Electrochemical Society* **1995**, *142*, 1409–1422.
62. Stamenkovic, V.; Schmidt, T. J.; Ross, P. N.; Markovic, N. M. *Journal of Physical Chemistry B* **2002**, *106*, 11970–11979.
63. Norskov, J. K.; Rossmeisl, J.; Logadottir, A.; Lindqvist, L.; Kitchin, J. R.; Bligaard, T.; Jonsson, H. *Journal of Physical Chemistry B* **2004**, *108*, 17886–17892.
64. Anderson, A. B. *Electrochimica Acta* **2002**, *47*, 3759–3763.
65. Roques, J.; Anderson, A. B.; Murthi, V. S.; Mukerjee, S. *Journal of the Electrochemical Society* **2005**, *152*, E193–E199.
66. Piela, P.; Eickes, C.; Brosha, E.; Garzon, F.; Zelenay, P. *Journal of the Electrochemical Society* **2004**, *151*, A2053–A2059.
67. Antolini, E.; Salgado, J. R. C.; Gonzalez, E. R. *Journal of Power Sources* **2006**, *160*, 957–968.
68. Mathias, M. F.; Makharia, R.; Gasteiger, H. A.; Conway, J. J.; Fuller, T. J.; Gittleman, C. J.; Kocha, S. S.; Miller, D. P.; Mitelsteadt, C. K.; Xie, T.; Yan, S. G.; Yu, P. T. *Interface* **2005**, *14*, 24–35.
69. Mukundan, R.; Kim, Y. S.; Garzon, F.; Pivovar, B. *ECS Transactions* **2005**, *1*, 403–413.
70. Colon-Mercado, H. R.; Popov, B. N. *Journal of Power Sources* **2006**, *155*, 253–263.
71. Ball, S.; Hudson, S.; Theobald, B.; Thompsett, D. *ECS Transactions* **2006**, *1*.

7

Modeling Water Management in Polymer-Electrolyte Fuel Cells

Adam Z. Weber, Ryan Balliet, Haluna P. Gunterman and John Newman

Environmental Energy Technologies Division, Lawrence Berkeley National Laboratory, Berkeley, CA, USA
Department of Chemical Engineering, University of California, Berkeley, CA, USA

I. INTRODUCTION

Fuel cells may become the energy-delivery devices of the twenty-first century with realization of a carbon-neutral energy economy. Although there are many types of fuel cells, polymer-electrolyte fuel cells (PEFCs) are receiving the most attention for automotive and small stationary applications. In a PEFC, hydrogen and oxygen are combined electrochemically to produce water, electricity, and waste heat.

During the operation of a PEFC, many interrelated and complex phenomena occur. These processes include mass and heat transfer, electrochemical reactions, and ionic and electronic transport. Most of these processes occur in the through-plane direction in what we term the PEFC sandwich as shown in Fig. 1. This sandwich comprises multiple layers including diffusion media that can be composite structures containing a macroporous gas-diffusion layer (GDL) and microporous layer (MPL), catalyst layers (CLs), flow fields or bipolar plates, and a membrane. During operation fuel is fed into

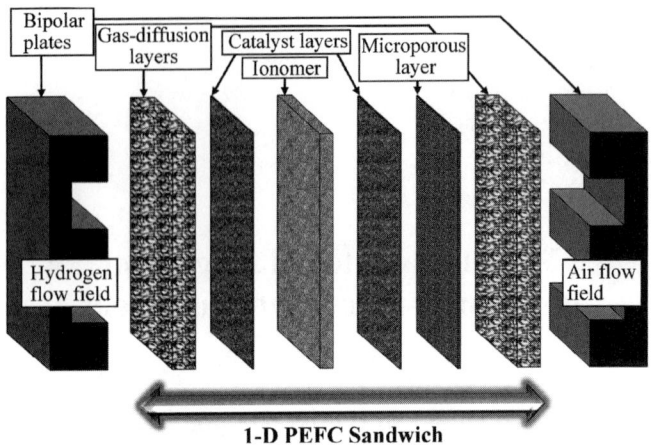

Figure 1. 3-D fuel-cell schematic showing the different layers in the PEFC sandwich or through-plane direction.

the anode flow field, moves through the diffusion medium, and reacts electrochemically at the anode CL to form hydrogen ions and electrons. The oxidant, usually oxygen in air, is fed into the cathode flow field, moves through the diffusion medium, and is electrochemically reduced at the cathode CL by combination with the generated protons and electrons. The water, either liquid or vapor, produced by the reduction of oxygen at the cathode exits the PEFC through either the cathode or anode flow field. The electrons generated at the anode pass through an external circuit and may be used to perform work before they are consumed at the cathode.

The performance of a PEFC is most often reported in the form of a polarization curve, as shown in Fig. 2. Roughly speaking, the polarization curve can be broken down into various regions. First, it should be noted that the equilibrium potential differs from the open-circuit voltage due mainly to hydrogen crossover through the membrane (i.e., a mixed potential on the cathode) and the resulting effects of the kinetic reactions. Next, at low currents, the behavior of a PEFC is dominated by kinetic losses. These losses mainly stem from the high overpotential of the oxygen-reduction reaction (ORR). As the current is increased, ohmic losses become a factor in lowering the overall cell potential. These ohmic losses are mainly from ionic losses in the electrodes and separator. At high currents, mass-transport limitations become increasingly important.

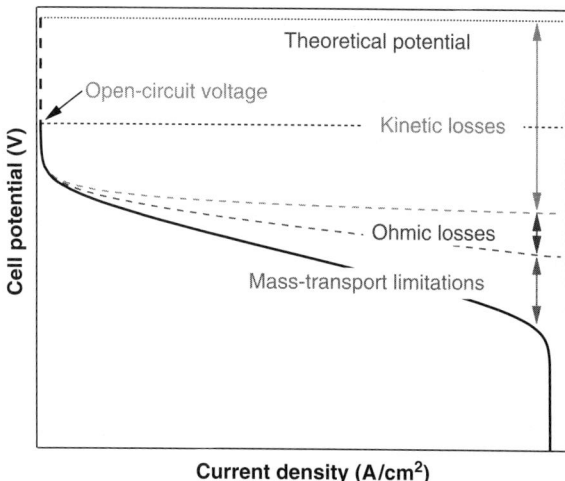

Figure 2. Schematic of a polarization curve showing the typical losses in a PEFC. The curve demonstrates a severe onset of flooding at high current densities.

These losses are due to reactants not being able to reach the electrocatalytic sites.

Key among the issues facing PEFCs today is water management. Due to their low operating temperature ($<100°C$), water exists in both liquid and vapor phases. Furthermore, state-of-the-art membranes require the use of water to provide high conductivity and fast proton transport. Thus, there is a tradeoff between having enough water for proton conduction (ohmic losses), but not too much or else the buildup of liquid water will cause a situation in which the reactant-gas-transport pathways are flooded (mass-transfer limitations). Figure 3 displays experimental evidence of the effects of water management on performance. In Fig. 3a, a neutron image of water content displays flooding near the outlet of the cell due to accumulation of liquid water and a decrease in the gas flowrates. The serpentine flow field is clearly visible with the water mainly underneath the ribs. Figure 3b shows polarization performance at 0.4 and 0.8 V and high-frequency resistance at 0.8 V as a function of cathode humidification temperature (1). At low current densities, as the inlet air becomes more humid, the membrane resistance decreases, and the performance increases. At higher current densities, the same effect occurs; however, the higher temperatures

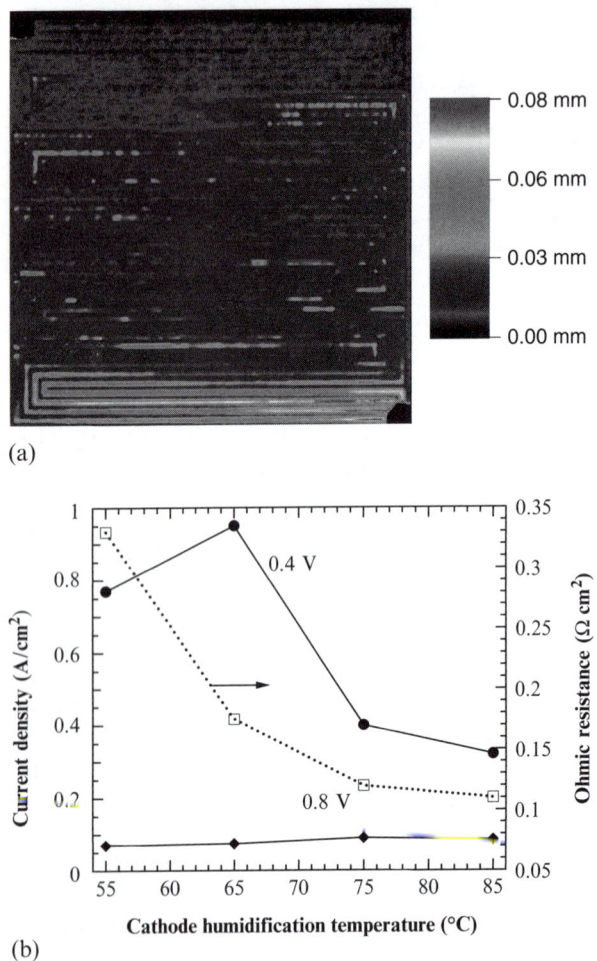

(a)

(b)

Figure 3. Experimental data showing water-management impacts on performance. (a) Neutron image of water thickness at 60°C, 1 A cm^{-2}, saturated inlet gases. (b) Cell performance and area-specific resistance as a function of humidifier temperature; the cell temperature is 80°C and the humidifier temperatures correspond to inlet relative humidities of 33, 53, 81, and 122%, respectively. (Figure (a) is courtesy of Dr. Michael Hickner, and figure (b) is adapted from (1) with permission of Elsevier.)

and more humid air also results in a lower inlet oxygen partial pressure. This later effect is also one of water management and is why the performance shows a maximum as a function of humidifier temperature.

Due to the complex and coupled nature of the underlying physical phenomena and the lack of definitive experimental evidence, fundamental modeling provides one of the only avenues to understand PEFCs fully and thoroughly. Modeling allows one to parse and explain the different regions in the polarization curve, elucidate optimal designs and operating conditions, and explore the governing physics and water-management aspects. A good PEFC model should have a physical basis, be predictive and agree with experimental data and trends, have a minimum of fitting parameters, and adequately model the dominant transport phenomena. Macroscopic modeling of PEFCs has been recently reviewed (2–9). The most noteworthy of those reviews are those by Weber and Newman (2) and Wang (3), who examined models of transport phenomena up to the end of 2003. This chapter serves to update (through June 2007) and append those reviews by focusing on more recent modeling trends and developments with a theme of water management. This article is also a stand-alone entity that is perhaps more pedagogic than previous reviews, and seeks to explain the current state of understanding of water management and its modeling.

The focus of this chapter is on the macroscopic modeling of PEFC water management. The structure is based on water-management phenomena, with emphasis on what the models have taught and shown, and not an encyclopedic list of the recently published models. Section II contains the major modeling approaches and governing equations and is the basis for the majority of models. Section III deals with GDL-related simulation studies including more microscopic investigations. Section IV is on design strategies for water management and focuses primarily on flow-field designs and their interaction with the PEFC sandwich. Section V examines transient analysis, especially that during load changes. Sections VI and VII detail some special applications and models regarding subzero operation and freeze phenomena and higher-temperature phenomena, respectively. Before proceeding to the governing equations, it is worthwhile to discuss what experiments tell about water management.

1. *In Situ* Visualization of Water

Water management has been a ripe opportunity for study through mathematical modeling because of its complex nature in PEFCs, as well as the fact that there is only limited direct experimental validation of it. While one can easily obtain a polarization curve or even a segmented-cell current density, relating these more global results to specific phenomena requires mathematical models. The corollary of course is that validation of the models can be done only through these averaged or tangential results; thus, various models can fit the data with the same accuracy, but come up with different limiting factors depending on how the model is biased. Luckily, the realm of experimental imaging is starting to allow for direct comparisons of predicted water contents and water management in an operating PEFC.

The field of view captured by imaging techniques spans areas ranging from that of a full- or subscale-size cell – via magnetic resonance imaging (MRI), nuclear magnetic resonance (NMR), or neutron radiography – to the midrange scale that can be thought of as focusing more on a single channel – via study of transparent cells or fluorescent microscopy – down to the micrometer scale – via X-ray tomography. Although the visualization methods with regards to PEFC imaging are still in the development phase, their strengths and weaknesses may be leveraged to create a more complete picture of relevant and limiting processes.

MRI, NMR, and neutron imaging techniques are often employed to image test cells and have resolutions between 10 and $100\,\mu m$. MRI and NMR both exploit the signal generated by disturbing a magnetic field with an electromagnetic force; the main difference between the two methods being that MRI also tracks the geometric source of resonance. Because the two methods utilize similar phenomena, they both exhibit a strong resolution-to-run-time tradeoff. As a point of reference, a minute-long scan can give roughly $50\,\mu m$ in resolution (10). These resonance techniques are nonoptimal for imaging conductive material, such as the GDL and CL, because the signal from most paramagnetic materials decays too quickly for analysis. Alternatively, this quick decay means that the water content of the membrane may be studied in near isolation. Several groups have used resonance techniques to study the water distribution in the membrane of operating fuel cells (11–13).

Findings include confirmation of the link between proper membrane hydration and performance, and validation of the model predictions of using counterflow, rather than coflow, to promote a more uniform liquid-water distribution in the cell (11). The main strength of resonance imaging is its accessibility and state of development as a field in which the physics is well understood. It is a convenient imaging technique for studying water content in nonconductive media provided that the information sought does not require speed or layer-by-layer resolution. Furthermore, unlike some of the other imaging techniques, it can simultaneously provide both chemical and geometric information.

Neutron-imaging experiments (see Fig. 3a) also treat large areas of a PEFC at a time, more so than MRI and NMR (14). They also are gaining popularity because of neutron imaging's short temporal resolution, on the order of seconds (15). Neutron imaging is similar to X-ray imaging, but instead of bombarding a sample with X-rays, a neutron source is utilized. Image masking may be used in conjunction with neutron radiography results to differentiate the PEFC components into CL and membrane, GDLs, and flow fields. Image masking involves keeping only image data from a specific depth and for a PEFC system has a resolution of 100 μm, although that should be decreasing with more advanced detectors (16). Transient results indicate the importance of incorporating temperature effects in order to understand how the water distribution reaches steady state. For example, Hickner et al. (15) found that the cell achieved steady state about 100–200 s after the current load underwent a step increase from 0 to 1,000 mA for their particular set-up. They also confirmed the competition between water production and heat generation, where increasing current density generates water, but eventually the water content declines as local heating effects become important. Other neutron-radiography studies treat the effect of changing pressure on systems operating under varying humidification conditions, and provide further checks on what constitute physically accurate modeling results (17, 18). Neutron imaging is a highly powerful tool, especially for transient analysis of PEFC systems, but lack of neutron sources severely limits accessibility. A commonality between all three methods discussed thus far is their use in studying in-plane movement and distribution of water.

Midrange imaging techniques treat mm^2 sized areas and provide insight into and validation of through-plane flow patterns.

Imaging results elucidate possible material-structure changes and also present ideas on the correct boundary condition at the GDL/gas-channel interface. Transparent-PEFC imaging entails the replacement of the bipolar plate with a transparent material and then direct observation of the system. This method is useful for seeing water-droplet formation at the GDL/gas-channel interface and movement down the channel (see Fig. 19) (19–22). Water droplets have been seen to grow until they become large enough to be wicked to the side walls, along which water then moves (of course, the transparent material undoubtedly has thermal and wetting characteristics different from actual bipolar plates). Litster et al. (23) used another midrange imaging technique, fluorescent microscopy, to propose a fingering-and-channeling transport method from ex situ imaging of GDLs. From their observations, they derived a movement mechanism where the water within the GDL is pulled by capillary action along paths that begin to merge into each other. Then, once a dominant pathway forms, water from nearby channels is siphoned into the dominant conduit, and a droplet forms at the GDL surface.

Determining with absolute certainty the dominant water-movement mechanism, requires finer resolution of the internal PEFC environment. Synchrotron X-ray radiography employs a particle accelerator to generate a high-energy electron beam, which is impacted with a target to form X-rays that are subsequently focused onto an object of interest. Resolution is on the order of 10 μm when applied to a PEFC system, although with more sophisticated treatments a better resolution is obtainable (24). By taking a series of images from different angles and reconstructing them, synchrotron radiography can be used to generate a 3-D tomographic image of a GDL. The rendered image distinguishes the GDL structure – carbon and PTFE appear the same because of their similar electron density – from the water found within (24). Initial results show that many pockets of water can remain in the GDL even after 2 min of purging (24). Manke et al. (25) also show that there is a periodicity in which water droplets form, grow, and move away from the initial break-through point at the GDL/gas-channel interface. The temporal and spatial regularity of the cycle led to the suggestion that liquid water within the GDL is pulled by capillary action along paths that begin to merge into each other and form one larger water conduit to the surface of the GDL. A drawback of synchrotron tomography is that a full 3-D image can take up to 30 min or an hour depending on the number of pictures taken; therefore, this technique has limited usefulness for the study of transient phenomena like the initial liquid-water percolation with the GDL. Improvements in resolution and sensitivity may someday uncover the GDL microstructure (e.g., pores of Teflon and carbon) in enough detail to provide the underlying structure for Lattice-Boltzmann or pore-network models.

II. BASIC PHENOMENA, METHODOLOGY, AND GOVERNING EQUATIONS

To model water management inside a PEFC, one must be cognizant of the underlying physical phenomena which are occurring. These phenomena require knowledge not only of water transport but also of transport of the other species, governing thermodynamic and kinetic relations, etc. Furthermore, there are different global modeling methodologies for modeling PEFCs, and in particular, the PEFC sandwich or through-plane direction. The easiest division to make is between macroscopic and microscopic models. The microscopic models seek to model transport on an individual pore level, whereas the macroscopic ones are a continuum and average over this level. Although the microscopic models may provide more realistic conditions and factors, they require a lot more knowledge of the microstructure and are much more expensive in terms of computation time. Macroscopic models are more common for PEFCs, although it is the current trend to try to incorporate more microscopic details into them.

Most of the current macroscopic models utilize a macrohomogeneous approach, wherein the exact geometric details of the modeling domain are neglected. Instead, the domain is treated as a randomly arranged porous structure that can be described by a small number of variables such as porosity and surface area per unit volume. Furthermore, transport properties within the domain are averaged over the volume of it. Thus, all variables are defined at all positions within the domain. Averaging is performed over a region that is small compared to the size of the domain, but large compared to its microstructure.

A model can be classified based on its geometric dimensionality as shown in Fig. 4. Zero-dimensional (0-D) models are mainly empirical and model a PEFC with a simple equation; these are typically

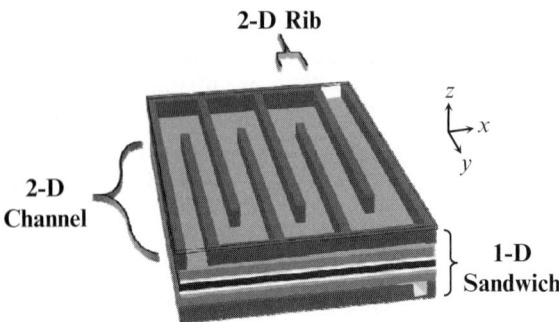

Figure 4. Schematic showing the different model dimensionalities. 0-D models are simple equations and are not shown, the 1-D models comprise the sandwich (z direction), the 2-D models comprise the 1-D sandwich and either of the two other coordinate directions (x or y), and the 3-D comprise all three coordinate directions.

used to fit data and get a general idea of the relative magnitude of the various phenomena. 1-D models treat the PEFC sandwich in varying degrees of complexity, ranging from simple equations to complex expressions derived from physical models. Furthermore, they can incorporate other (nongeometric) dimensional effects in terms of size, i.e., microscopic and macroscopic effects (e.g., consumption of reactant in a pore of a particle which is within a porous electrode). 2-D models deal with effects in the PEFC which occur in the sandwich as well as in another direction, either across or along the gas channel. Finally, 3-D models include the 1-D sandwich and consider effects in both directions in the flow field.

Pseudo-dimensional models can also be used where one or more directions are treated rigorously and another direction is treated simplistically. A classic example is a pseudo 2-D or 1 + 1-D model where multiple 1-D sandwich models are run and tied together through their external boundary conditions to account for flow along the channel. Based on scale-separation arguments and the additional computational cost and complexity of running higher-dimensional models, we believe that a pseudo 3-D model (2-D sandwich with rib (land) and channel effects, and a separate along-the-channel model) provides the best compromise in terms of reality and complexity; however, 1-D models are very good starting points for investigating specific phenomena (e.g., carbon corrosion during startup).

Although the number of PEFC models is large, the number of modeling groups and approaches is significantly smaller. The obvious reason is that as a group becomes more familiar with a model, they continually upgrade it in terms of complexity to make it more physically realistic. For an approach, if it is general, then the community adopts and alters it. Furthermore, the models can generally be categorized based on what they attempt to model. For example, there are those that account for two-phase flow and flooding versus those that focus instead on membrane dehydration and low-relative-humidity operation. With the advancement of computational efficiency and speed and the physical understanding of PEFC operation, models currently in use are multidimensional, account for most water-management aspects such as flooding and dehydration, and are nonisothermal. The use of transient models is also coming on-line as discussed later. While it is interesting to examine the historical route for the modeling of some phenomena and to examine each modeling group's contribution to the field, such a review is outside the purview of this article, and can be found in our review article of macroscopic PEFC transport modeling (2).

In this section, the general governing equations are presented and discussed. The discussion is loosely arranged by the various PEFC layers as shown in Fig. 1. This section is to serve as a primer for the following sections wherein more detailed analyses are made concerning the movement of water in GDLs, flow fields, and specific applications. Therefore, it is more of a how-to section than demonstrating modeling results. The treatment of the CLs and membrane are also contained within this section, as is a general discussion of two-phase flow in the GDLs, but first, the fundamental governing equations are presented.

1. Fundamental Governing Equations

A PEFC is governed by thermodynamics, kinetics, and transport phenomena as described by conservation equations. In this section, the relevant equations are presented. These basic equations form the basis of all macroscopic PEFC models. The differences, as discussed in this chapter, are due to how one defines fluxes (i.e., transport equations) and the relevant source and sink terms.

i. Thermodynamics

As shown in Fig. 2, the theoretical potential represents the highest voltage obtainable for a single cell as derived from thermodynamics. The overall fuel-cell reaction can be broken down into the two overall electrode reactions. If hydrogen is the primary fuel, it oxidizes at the anode according to the reaction

$$H_2 \rightarrow 2H^+ + 2e^-. \quad (1)$$

At the cathode, oxygen is reduced

$$4H^+ + 4e^- + O_2 \rightarrow 2H_2O. \quad (2)$$

Adding equation (1) and equation (2) yields the overall reaction

$$2H_2 + O_2 \rightarrow 2H_2O. \quad (3)$$

The potential of the overall cell is given by a Nernst equation (26,27)

$$U = U^\theta + \frac{RT}{2F} \ln\left(\frac{p_{H_2}\sqrt{p_{O_2}}}{p_w}\right). \quad (4)$$

where subscript w stands for water, R is the ideal-gas constant, T is the absolute temperature, F is Faraday's constant, and U^θ is the standard cell potential, a combination of appropriately chosen reference states that is a function of temperature and can be unit dependent. U^θ can be related to the Gibbs free energy of the reaction

$$\Delta G = -2FU^\theta. \quad (5)$$

Similarly, an enthalpy potential can be defined as

$$U_H = \frac{\Delta H}{2F} = U^\theta - T\frac{\partial U^\theta}{\partial T}. \quad (6)$$

This potential is also known as the thermoneutral potential, or the potential at which there is no net heat generation.

Using the first law of thermodynamics yields an expression for the heat generation of the PEFC (28, 29)

$$Q = i(U_H - V), \quad (7)$$

where Q is the total heat generated per superficial area, i is the superficial current density, and V is the (observed) cell potential. The above heat generation can also be broken down into reversible and irreversible parts, which are given by

$$Q_{\text{rev}} = i\left(U_H - U^\theta\right) \tag{8}$$

and

$$Q_{\text{irrev}} = i\left(U^\theta - V\right), \tag{9}$$

respectively. It is worth mentioning that the PEFC community typically defines efficiency using the deviation of the operating potential from the Gibbs free energy or reversible potential. This definition does not account for the intrinsic reversible losses. The correct definition should be from the enthalpy and not the reversible potential; such a definition allows for a fair comparison of PEFCs with other energy-conversion devices using the higher heating value of the fuel.

ii. Kinetics

The initial drop in the polarization curve (Fig. 2) is due to the sluggish kinetics of the ORR at the temperatures normally used for current PEFC operation (<100°C). A typical electrochemical reaction can be expressed as

$$\sum_k \sum_i S_{i,k,h} M_i^{z_i} \to n_h e^-, \tag{10}$$

where $s_{i,k,h}$ is the stoichiometric coefficient of species i residing in phase k and participating in electron-transfer reaction h, n_h is the number of electrons transferred in reaction h, and $M_i^{z_i}$ represents the chemical formula of i having valence z_i.

The rate of an electrochemical reaction depends upon the concentrations of the various species and the potential drop across the reaction interface between phases k and p, which are normally the electrode and electrolyte, respectively. In general, a Butler–Volmer expression can be used to describe the kinetics

$$i_h = i_{0_h} \left[\exp\left(\frac{\alpha_a F}{RT}\left(\Phi_k - \Phi_p - U_h^{\text{ref}}\right)\right) \prod_i^a \left(\frac{p_i}{p_i^{\text{ref}}}\right)^{s_{i,k,h}} \right.$$
$$\left. - \exp\left(\frac{-\alpha_c F}{RT}\left(\Phi_k - \Phi_p - U_h^{\text{ref}}\right)\right) \prod_i^c \left(\frac{p_i}{p_i^{\text{ref}}}\right)^{-s_{i,k,h}} \right], \quad (11)$$

where i_h is the transfer current between phases k and p due to electron-transfer reaction h, the products are over the anodic and cathodic reaction species, respectively, α_a and α_c are the anodic and cathodic transfer coefficients, respectively, p_i and p_i^{ref} are the partial pressure and reference partial pressure for species i, respectively, and i_{0_h} and U_h^{ref} are the exchange current density per unit catalyst area and the potential of reaction h evaluated at the reference conditions and the operating temperature, respectively. In the above expression, the composition-dependent part of the exchange current density is explicitly written, with the multiplication over those species in participating in the anodic or cathodic direction. The reference potential can be determined using a Nernst equation (e.g., see equation (4)); if the reference conditions are the same as the standard conditions (i.e., 100 kPa pressure for the different gas species), then U^{ref} has the same numerical value as U^θ.

The term in parentheses in equation (11) can be written in terms as an electrode overpotential

$$\eta_h = \Phi_k - \Phi_p - U_h^{\text{ref}}. \quad (12)$$

In this chapter, the reference electrode used is defined as a platinum metal electrode exposed to hydrogen at the same temperature and electrolyte (e.g., Nafion®) as the solution of interest. With this reference electrode, the electrode overpotential defined in equation (12) is the same as having the reference electrode located next to the reaction site but exposed to the reference conditions (i.e., it carries its own extraneous phases with it). Typical values for the reference conditions are those in the gas channels. If the reference electrode is exposed to the conditions at the reaction site, then a surface or kinetic overpotential can be defined

$$\eta_{s_h} = \Phi_k - \Phi_p - U_h, \quad (13)$$

where U_h is the reversible potential of reaction h. The surface overpotential is the overpotential that directly influences the reaction rate across the interface. Comparing equations (13) and (12), one can see that the electrode overpotential contains both a concentration and a surface overpotential for the reaction; the reader is referred to Neyerlin et al. (30) for a very good discussion of the different overpotentials and related kinetic expressions for the ORR.

For the hydrogen-oxidation reaction (HOR) at the anode equation (11), becomes, in the absence of poisons,

$$i_{HOR} = i_{0_{HOR}} \left[\frac{p_{H_2}}{p_{H_2}^{ref}} \exp\left(\frac{\alpha_a F}{RT}(\eta_{HOR})\right) - \exp\left(\frac{-\alpha_c F}{RT}(\eta_{HOR})\right) \right], \quad (14)$$

where 1 and 2 denote the electron- and proton-conducting phases, respectively. Because the electrolyte is a polymer of defined acid concentration, the proton concentration does not enter directly into equation (14). However, if one deals with contaminant ions, then the activity of protons should explicitly enter into equation (14) either through the equilibrium potential or the kinetic equation, depending on the reference state used. Also, it has recently been shown that the HOR may proceed with a different mechanism at low hydrogen concentrations; in this case, the kinetic equation is altered through the use of a surface adsorption term (31). Due to the choice of reference electrode, the reference potential and reversible potential are both equal to zero.

Unlike the facile HOR, the oxygen-reduction reaction (ORR) is slow and represents the principal inefficiency in many fuel cells. Due to its sluggishness, the ORR is modeled reasonably well with Tafel kinetics with a dependence on oxygen partial pressure, m_0, of between 0.8 and 1 (30, 32–34)

$$i_{ORR} = -i_{0_{ORR}} \left(\frac{p_{O_2}}{p_{O_2}^{ref}}\right)^{m_0} \exp\left(\frac{-\alpha_c F}{RT}(\eta_{ORR})\right). \quad (15)$$

For the kinetic region, the values of the theoretical and experimental Tafel slopes have been shown to agree with α_c equal to 1 (30, 32, 34–41). As with the case of the HOR, the dependence of the reaction rate on the hydrogen ion activity is not shown explicitly. While this is typically reasonable, as discussed in Sect. IV.1, under low humidity conditions, the change in the proton concentration and especially its activity coefficient necessitate accounted explicitly for the proton activity (42).

While the ORR and HOR are the principal reactions occurring in PEFCs, it is worth noting the possibility of side reactions that can occur in the CLs. These stem from durability and degradation analyses and, although mentioned below, are not covered in this chapter on water management. One of these other reactions include the two-electron reduction of oxygen crossing over to the anode to hydrogen peroxide (43). In addition, hydrogen peroxide also forms at the cathode as part of the ORR (44). Also, hydrogen in the membrane that is crossing over can reduce platinum ions to metal, forming a platinum band in the membrane (45). Platinum itself undergoes oxide formation and stripping, which includes possible dissolution and movement as ions (46, 47). Finally, oxygen evolution (the anodic term to the ORR equation) and carbon oxidation at the cathode can also occur due to fuel starvation at the anode (48, 49).

iii. Conservation equations

The conservation equations stem from the underlying fundamental physics. There are three principal equation types that are of interest: mass, energy, charge. These are presented in turn below.

For conservation of mass, it is necessary to write a material balance for each independent component in each phase. For PEFCs, the differential form of the material balance for species i in phase k is (2)

$$\frac{\partial \varepsilon_k c_{i,k}}{\partial t} = -\nabla \cdot \mathbf{N}_{i,k} - \sum_h a_{l,k} s_{i,k,h} \frac{i_{h,1-k}}{n_h F} + \sum_l s_{i,k,l} \sum_{p \neq k} a_{k,p} r_{l,k-p}. \quad (16)$$

The term on the left side of the equation is the accumulation term, which accounts for the change in the total amount of species i held in phase k within a differential control volume. The first term on the right side of the equation keeps track of the material that enters or leaves the control volume by mass transport. The remaining two terms account for material that is gained or lost due to chemical reactions. The first summation includes all interfacial electron-transfer reactions, the second summation accounts for non-electrochemical interfacial reactions (e.g., evaporation/condensation).

In the above expression, $c_{i,k}$ is the concentration of species i in phase k, and $s_{i,k,l}$ is the stoichiometric coefficient of species i in phase k participating in heterogeneous reaction l (see equation

(10)). a_h is the specific surface area (surface area per unit total volume) of the interface for the electrochemical reactions. In the above expression, Faraday's law

$$N_{i,k} = \sum_h s_{i,k,h} \frac{i_h}{n_h F} \tag{17}$$

was used to change the interfacial current density into an interfacial flux quantity. $r_{l,k-p}$ is the rate of the heterogeneous reaction l per unit of interfacial area between phases k and p.

For the conservation of charge, the equation is similar to the mass balance above. Because a large electrical force is required to separate charge over an appreciable distance, a volume element in the electrode will, to a good approximation, be electrically neutral; thus one can assume electroneutrality for each phase

$$\sum_i z_i c_{i,k} = 0. \tag{18}$$

The assumption of electroneutrality implies that the diffuse double layer, where there is significant charge separation, is small compared to the volume of the domain, which is normally the case. The general charge balance, assuming electroneutrality becomes

$$\frac{\partial \rho_e}{\partial t} = \sum_k \nabla \cdot \mathbf{i}_k, \tag{19}$$

where ρ_e is the charge density that can be substituted with the double-layer capacity and the potential as is done for transient or impedance analyses. For steady-state cases, there is no accumulation of charge, and the conservation of charge becomes the divergence of the total current density is zero.

For conservation of energy, if one desires to account only for the total heat generation, Eq. (7) can be used. However, if the specific heat-generation locations and the thermal gradients are desired, a conservation equation can be used. For PEFCs, the governing thermal-energy conservation equation becomes (2, 26, 50).

$$\sum_k \rho_k \hat{C}_{pk} \frac{\partial T}{\partial t} = -\sum_k \rho_k \hat{C}_{pk} \mathbf{v}_k \cdot \nabla T + \nabla \cdot \left(k_T^{\text{eff}} \nabla T \right) + \sum_k \frac{\mathbf{i}_k \cdot \mathbf{i}_k}{\kappa_k^{\text{eff}}}$$
$$+ \sum_h i_h \left(\eta_h + \Pi_h \right) - \Delta H_{\text{evap}} r_{\text{evap}}, \tag{20}$$

where it has been assumed that the temperatures in the various phases (i.e, membrane, gas, liquid, and solid) are in equilibrium with each other. If such an assumption is undesirable, which could be the case, then similar energy equations can be used for each phase (for example, see Hwang et al. (51)). The first term on the left side of equation (20) is the accumulation of energy, where $\hat{C}_{p_k}$ and ρ_k are the (average) heat capacity and density of phase k, respectively. The first term of the right side represents convection of energy, where $\mathbf{v}_k$ is the mass-averaged velocity of phase k, respectively. The second term on the left side represents heat transfer due to conduction, where k_T^{eff} is the effective thermal conductivity of the system. The third term is due to ohmic heating where κ_k^{eff} is the effective electronic or ionic conductivity of phase k. The fourth term is the heat generation due to the electrochemical reactions, where the irreversible generation is given by the overpotential, η, and the reversible part is given by the Peltier coefficient, Π (52). The last term is due to evaporation/condensation of water, where ΔH_{evap} is the heat of vaporization and r_{evap} is the rate of evaporation. Finally, unlike the other conservation equations, that of energy expands the energy flux explicitly into its convective and conductive parts.

2. Membrane Modeling

One of the most important parts of the PEFC is the electrolyte or membrane, especially in terms on water management since drier feeds cause the membrane to lose water and thus become more resistive and ohmically limit the cell performance. The PEFC membrane is a proton conductor where the anions (typically sulfonic acid moieties) are tethered to the polymer backbone. There are numerous studies of the various membranes' properties, structure, etc., many of which are contradictory. A main problem is that the current state-of-the-art membranes are random copolymers and are thin, thereby making characterization difficult. Furthermore, pretreatment of the membrane can have a profound effect on its morphology and hence its properties. In fact, depending on how one pretreats the membrane, there can be large differences in the water uptake or water content (known as λ, moles of water per mole of sulfonic acid site) depending on the reservoir phase in contact with the membrane (53). This discrepancy, known as Schröder's paradox, can be as large as a difference between $\lambda = 14$ for a vapor-equilibrated membrane and

$\lambda = 22$ for a liquid-equilibrated one. With the corresponding differences in water content, the membrane microstructure and hence its transport parameters and maybe even transport phenomena may change (54). For more detailed discussions please see the relevant literature including very good reviews on Nafion® (the current polymer of choice) (55) and alternative hydrocarbon membranes (56).

Due to its importance and complexity, the membrane's behavior has been simulated with a whole range of models, from the atomistic and molecular through to the macroscopic. The microscopic models try to predict the membrane microstructure and phase separation due to water uptake, as well as examine transport through it at a fundamental level. The macroscopic models are often more empirical and focus on describing the transport and relevant parameters of the membrane in a macrohomogeneous fashion. As per the overall approach of this chapter, discussion is made on the macroscopic models; for microscopic analyses, see the review in this volume as well as that of Kreuer et al. (57) The discussion below is focused mainly on the governing transport equations using a concentrated-solution-theory approach, and developments in the last few years. For more detailed historical and other modeling approaches, the reader is referred to recent reviews on this subject (2, 58, 59).

i. Concentrated solution theory

Concentrated solution theory takes into account all binary interactions between all of the species, and it uses a more general driving force, namely, that of chemical potential. In this fashion, it is similar to the Stefan–Maxwell multicomponent diffusion equations (see equation (30)). In fact, there is a direct analog of those equations and the dusty-gas model that is used for PEFC membrane modeling, which is termed the binary friction model (59, 60)

$$\nabla \mu_i = \nabla (RT \ln x_i - z_i F \Phi) = \sum_{j \neq i} \frac{RT x_j}{D_{i,j}^{\text{eff}}} \left(\frac{\mathbf{N}_j}{c_j} - \frac{\mathbf{N}_i}{c_i} \right) - \frac{RT}{D_{i,\text{m}}^{\text{eff}}} \left(\frac{\mathbf{N}_i}{c_i} \right), \tag{21}$$

where the m denotes the interaction with the membrane and eff denotes an effective property of the membrane. As discussed by Fimrite et al. (59, 61) and Carnes and Djilali (60), this treatment is similar to that of the dusty fluid model applied to the membrane (62, 63), but accounts for the bulk movement of water in a more consistent manner using a different reference frame. The binary friction model assumes that hydronium ions and water act as separate species within the membrane microstructure. Furthermore, the electrochemical potential is used instead of the chemical potential as a driving force. The mole fractions and diffusion coefficients in the above equation can be related to the water content of the membrane (60, 64).

A very similar treatment to that above can be reached by starting with the original equation of multicomponent transport (65)

$$\mathbf{d}_i = c_i \nabla \mu_i = \sum_{j \neq i} K_{i,j} \left(\mathbf{v}_j - \mathbf{v}_i\right), \tag{22}$$

where $\mathbf{d}_i$ is the driving force per unit volume acting on species i and can be replaced by a chemical potential gradient of species i, and $K_{i,j}$ are the frictional interaction parameters between species i and j. Instead of introducing the concentration scale, one can invert the above set of equations and relate the inverted $K_{i,j}$s to experimentally measured transport properties using a set of three orthogonal experiments (65, 66). Doing this results in the proton and water governing transport equations,

$$\mathbf{i}_2 = -\frac{\kappa \xi}{F} \nabla \mu_w - \kappa \nabla \Phi_2 \tag{23}$$

and

$$\mathbf{N}_w = \xi \frac{\mathbf{i}_2}{F} - \alpha_w \nabla \mu_w \tag{24}$$

respectively, where α_w is the transport coefficient of water (64, 66) and ξ is the electroosmotic (drag) coefficient (67). The chemical-potential driving force can either be used as is or substituted by a mole-fraction or water-content expression, depending on how one wants to express the transport properties. The concentrated-solution-approach governing equations remain valid for all water contents assuming that the correct interaction parameters are known as a function of water content, and there is a methodology to calculate the water content as mentioned below.

It is worth mentioning some special simplifications that have been and continue to be used for membrane modeling. All of these other approaches use Ohm's law for proton movement,

$$\mathbf{i}_2 = -\kappa \nabla \Phi_2 \tag{25}$$

where κ is the ionic conductivity of the membrane. Thus, they do not account for the streaming current term in equation (23). For water movement, these other approaches differ as follows. The first is for membranes at lower water contents where one can use a dilute-solution analog to the above equations (i.e., the Nernst–Planck equation) (26). This approach results in equation (24) for water where a concentration driving force is used for the chemical-potential one. For liquid-equilibrated membranes, a more empirical approach is to use Schlögl's equation for water movement (68, 69)

$$\mathbf{v}_w = -\left(\frac{k}{\mu}\right)\nabla p_L - \left(\frac{k_\Phi}{\mu}\right) z_f c_f F \nabla \Phi, \qquad (26)$$

where k and k_Φ are the effective hydraulic and electrokinetic permeability, respectively, p_L is the hydraulic or liquid pressure, μ is the water viscosity, and z_f and c_f refer to the charge and concentration of fixed ionic sites, respectively. Finally, a straightforward, albeit not rigorous, approach is to combine linearly the expected driving forces for water movement

$$\mathbf{N}_w = \xi \frac{\mathbf{i}}{F} - D_w \nabla c_w - c_w \frac{k}{\mu} \nabla p_k, \qquad (27)$$

where p_k can be the gas- or liquid-phase pressure. While this equation can describe water movement, it is on a tenuous basis in terms of the underlying physics and the separability of the driving forces.

ii. Water content and properties

Essentially, all of the models center around the same or very similar governing equations as those described above. The difference is in how one relates the various gradients and model parameters to the water content of the membrane. The chemical-potential driving force has been used directly, changed into λ or the concentration or mole fraction of water, or separated into pressure and concentration terms; different approaches that are all not equivalent, as discussed in the previous subsection.

The simplest analysis for water content is to fix the anode and cathode boundary values of the water content using a water-uptake isotherm (i.e., λ versus water activity in contact with the membrane), and assume a linear gradient between the values. While this is insufficient in many circumstances, it does allow for analytic solutions to be generated (see, for examples, Okada and coworkers (70, 71), Carnes and Djilali (60), and St-Pierre (14)), which may be utilized in system and stack models.

More complicated analyses try and predict the water content using a submodel that describes the believed physics with a minimum number of fitting parameters. While there are various models for predicting the water-uptake isotherm (58), the comments below focus on those models which encompass the entire experimentally observed water content range from dry to liquid-equilibrated. The most prominent types of these models are those of Eikerling and coworkers (72, 73), Weber and Newman (64), and Choi and Datta (74, 75), which have been modified by various authors. All of these models try to account for the water uptake and water content using macroscopic approaches based on flow-through-porous-media theory, where there are defined water pathways through the membrane. In terms of driving forces, Eikerling et al. uses both a concentration and a pressure, although the focus is more on the convective, pressure-related movement, Weber and Newman use the chemical potential directly, and Choi and Datta use essentially the binary friction model equation (21).

Both Eikerling et al. and Weber and Newman assume that there are pores within the membrane that are either liquid-equilibrated or vapor-equilibrated. Eikerling et al. assumes a random network of pores that are either filled with bulk-like water or bound water, and impregnation by liquid water is easier than condensation. They use effective-medium theory to predict conductivity results from impedance data. Their model is more of a microscopic one in which λ is calculated by changing the number of pores that are filled and examining the types of liquid-film bonds between pores. Weber and Newman also assume a pore-size distribution, but use an "interaction" coefficient to relate whether the pore is liquid-equilibrated or vapor-equilibrated. The "interaction" coefficient is said to be physically related to the microstructure and surface and elastic energies within the polymer, although it is a fitting parameter in practice. Furthermore, Weber and Newman assume equilibrium between protons and water within the membrane to predict λ for the vapor-equilibrated part of the membrane.

Unlike the two models above, the one by Choi and Datta is more of an interface model. While it is also more rigorous physically than the above ones, it is not clear how one can predict the water content changes within the membrane and how well the model can be used in a full-cell simulation. For example, it is unclear how important the interfaces are since in a full cell the CLs contain membrane tendrils and those will provide the protons into the membrane. The Choi and Datta model calculates the extra energy stored in the vapor–liquid interface at the membrane surface, resulting in a lower water content for a vapor-equilibrated membrane than a liquid-equilibrated one. They also utilize a chemical-equilibrium model to predict proton concentrations and water uptake in the vapor-equilibrated state.

Basically, all of the above models are using a construct of capillary condensation and phenomena to predict water contents. While this is not truly the physical representation of the membrane, it does serve as a way of organizing and visualizing the experimental data. It may be that such approaches are limited in their ability to predict water content since they average over the microstructure, which is key in determining the water content. However, more sophisticated molecular-dynamic-type models, which predict the water content and microphase separation of these membranes better, cannot be used in full-cell simulations. The current belief seems to be that the mechanical properties and microstructure of the membrane are the important relations that must be considered to come up with an accurate membrane model; the challenge is to find a way to do this in a macroscopic fashion.

As a related aside, it is worth examining the impact of membrane constraint on water content. Inside of an operating PEFC, the membrane is constrained due to the clamping pressure applied on the stack. The impact of this constraint is mainly unknown, especially on the transport properties since none of them have been really measured under constraint conditions. Furthermore, constraint can lead to membrane thinning and perhaps physiochemical degradation. There have been two macroscopic models that examine this issue in terms of water content and PEFC performance. The first, by Weber and Newman (76), shows that the water content will decrease due to the constraint, although a stress balance shows that the membrane does not feel much constraint since its swelling pressure will compress the GDLs. A more detailed and rigorous treatment of constraint, especially in terms of mechanical-property analysis was done recently by Nazarov and Promislow (77). They also show that the membrane will be only slightly constrained, but this is enough to affect the water transport through it, as can be seen

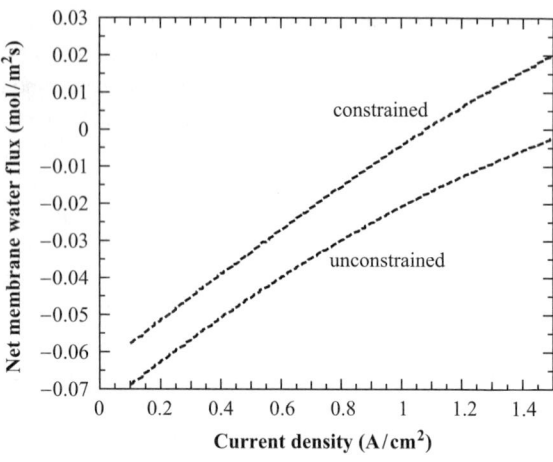

Figure 5. Simulated net membrane water as a function of current density for an unconstrained and a constrained membrane with a liquid-equilibrated cathode and a vapor-equilibrated anode. (The figure is adapted from (77) with permission of The Electrochemical Society).

in Fig. 5. In the figure, the net water flux through the membrane increases (more flow from anode to cathode) around 20% due to the more uniform and lower water content in the membrane. The curves naturally increase with current density due to the larger impact of electroosmotic flow compared to the back diffusion. Both of the constraint studies show that water management can be affected by membrane constraint and there is a need to study this issue in more detail both theoretically and experimentally, especially in how it pertains to chemical-mechanical degradation and PEFC durability.

The overall guiding issue for the membrane models is to predict transport of the various species. Regardless of what set of governing equations is used, one must utilize the experimentally measured parameters. While the conductivity and electroosmotic coefficient have been well characterized with regard to their temperature and water-content dependences, the transport coefficient is slightly more complicated. Due to the intricacies of Schröder's paradox, some models will utilize a permeability, some a diffusion coefficient, some a binary interaction parameter, and some a transport coefficient. The value of those coefficients should be more-or-less interchangeable under the same conditions (i.e., temperature and water content), and thus many models can get by with using nonphysical values for the diffusion coefficient (e.g., step changes and values at supersaturated conditions) for example. While this might yield satisfactory trends and data predictions, it is probably best to use diffusion coefficients for vapor-equilibrated membranes and permeabilities for liquid-equilibrated ones as done by Weber and Newman (64) for their transport coefficient for example. It is worth noting that for all of the transport parameters, their values increase with both temperature and humidity (i.e., a liquid-equilibrated value is higher than the corresponding vapor-equilibrated one).

iii. Other transport through the membrane

In terms of membrane modeling and understanding full-cell behavior including water management, one must recognize that other species may be transported through the membrane. Of largest interest is the transport of hydrogen and oxygen. The crossover of these gases results in a mixed potential at the electrode – thus explaining the difference between the observed open-circuit potential and the equilibrium potential (see Fig. 2) – and a chemical short of the cell. Although the crossover is normally only a small efficiency loss, it does limit the thickness of the membrane (78), and can become important if pinholes or membrane thinning occur. Furthermore, crossover is attributed to carbon corrosion during fuel starvation (49), platinum band formation (45), and peroxide generation (43). In addition, recent studies have also shown that the dilution effect by crossover of nitrogen can be important (79).

For the above reasons, membrane modeling should account for gas crossover. The easiest method to do this is to use experimentally measured permeation coefficients (which increase with water content and temperature)

$$\mathbf{N}_i = -\psi_i \nabla p_i, \qquad (28)$$

where ψ_i and p_i are the permeation coefficient and partial pressure of species i, respectively. A dilute solution approach can be used since the gases are minor components inside the membrane. Also, permeation coefficients are used instead of separate diffusion and solubility coefficients since it simplifies the analysis and the need for experimental data.

Besides gases, the other species' transport not addressed above is that of ions besides protons. Positive valence contaminant ions can occur in the membrane due to such issues as platinum and cobalt dissolution from the cathode (46), ruthenium dissolution from the anode (80), air impurities (81,82), and contamination from the other PEFC components (e.g., bipolar plates) (81,83). All of these cations will ion-exchange with the protons to a certain degree and thus decrease the conductivity of the membrane. While modeling these effects is out of the purview of this chapter, a brief modeling approach is as follows. While one can use dilute-solution approaches due to their low concentrations, it is suggested that more concentrated-solution-theory equations be used since the ions interact strongly with the proton and possibly water movement. Thus, equations in the form of equation (21) or equation (22) should be added for each ion, and rate and/or equilibrium affinities between the ions and membrane included. In addition, since the membrane no longer holds only a single type of positive charge, electroneutrality equation (18) must also be included. Finally, the resulting binary interaction parameters will result in the need to measure such transport properties as transference numbers for each ion. (26) It should be noted that the above approach is also required in any case where there are multiple ions (either anion or cation) that are mobile, such as ionic-liquid electrolytes, impregnated membranes (e.g., PBI), etc.

3. Two-Phase Flow

It is well known that water and specifically liquid-water management is crucial in performance optimization and perhaps durability mitigation. Fuel cells that operate below 100°C have the problem that water exists both in vapor and liquid forms. This two-phase-flow problem is a critical aspect for PEFC modeling. In fact, recent trends in PEFC modeling show a focus on understanding two-phase flow more than any other phenomena. One problem is that the necessary parameters related to two-phase flow in PEFCs are still mainly unknown due to inadequate experimental methods that can probe the complex materials used. Although progress on this front is being made, such as advanced imaging techniques, there is still a long way to go.

Simultaneous flow of both liquid and gas occurs within the GDLs and CLs, although most modeling studies focus on the former due to the fact that the CLs are much thinner than the GDLs and also contain a membrane phase and electrochemical reaction that complicate the transport picture. Furthermore, most GDLs are composite structures with a relative thick macroporous layer combined with one or more microporous layers of tailored properties such as wettability. In this section, the general, macroscopic treatments and governing equations of two-phase flow are presented. Specifically, the transport equations for the two fluids and their interaction with each other are discussed since the mass balances of the gas-phase species and liquid water can be deduced from equation (16). In later sections, more detailed analyses of water movement in GDLs in terms of parameter expressions, specific phenomena, simulation results, and microstructure are given.

Before proceeding to the introduction of the governing equations, some general comments should be made. Although GDLs and two-phase flow have been getting more interest in terms of their ability to tune water management, the macroscopic modeling methodology is essentially at the same state-of-the-art as when it was last reviewed (2,3). A lone exception is the model of Promislow et al. (84) that provides a mathematically less intensive methodology to account for vapor–liquid interfaces, and the so-called dry to liquid transition either along the channel or within the GDL itself. Since those reviews, several aspects dealing with water management have been explored, but the methodologies have remained essentially the same. The most noteworthy aspects are the examination of composite and even graded structures (i.e., GDL and MPLs), the coupling between thermal and water management (e.g., heat-pipe effect), examination of anisotropic and in-plane properties, inclusion of more microstructural details through microscopic models, the examination of interactions between the GDL and the flow channel, and the inclusion of a wettability distribution within the porous matrix. All of these aspects are discussed in other sections of this chapter with the exception of the last. The idea of having separate hydrophilic and hydrophobic pores was popularized for PEFCs by Weber et al. (85) and Nam and Kaviany (86). Since then, it has become much more common to measure both types of distributions and use them in modeling analyses (see Gostick et al. (87) for example). However, although one can measure hydrophilic distributions, it is noted that typical GDLs are more hydrophobic than hydrophilic on average (e.g., one must initially apply a pressure to wet the material (88)).

i. Liquid-phase transport

There are various methodologies to treat the liquid water. The first and simplest is to treat it as a mist or fog flow in that it has a defined volume fraction but moves with the same superficial velocity of the gas. While this could be satisfactory for flow fields, it does not make physical sense within a porous medium. The more common method is to use a separate transport equation for the liquid phase. Typically, this is done using the empirically based Darcy's law

$$\mathbf{N}_{w,L} = -\frac{k}{\bar{V}_w \mu} \nabla p_L, \tag{29}$$

where $\bar{V}_w$ is the molar volume of water, k is the effective permeability, μ is the viscosity, and all of the properties are valid for pure water. Some models also account for water movement using the Navier–Stokes equations, although Darcy's law is typically added as a source term that dominates the transport. Finally, some of the extensions of Darcy's law, such as the Brinkman equation, which allow for the no-slip condition to be met at the particle surfaces (i.e., a second derivative of pressure is used) have been used in simulations (89), although for the most part Darcy's law is used. While Darcy's law is a simple equation to implement, the challenge comes in how one calculates the effective permeability. This issue, along with saturation, is at the core of two-phase-flow models, and is discussed briefly below after introducing the gas phase transport equations.

ii. Gas-phase transport

To treat the gas-phase transport, the generalized multicomponent Stefan–Maxwell equations are used,

$$\nabla x_i = -\frac{x_i}{RT}\left(\bar{V}_i - \frac{M_i}{\rho_G}\right)\nabla p_G + \sum_{j \neq i} \frac{x_i \mathbf{N}_j - x_j \mathbf{N}_i}{\varepsilon_G c_T D_{i,j}^{\text{eff}}}, \tag{30}$$

where one of the equations is dependent on the others since the sum of the mole fractions is unity. In the above equation, ρ_G is the density of the gas phase, x_i and M_i are the mole fraction and molar mass of species i, respectively, and the first term accounts for pressure diffusion. This term is often neglected, although it could be important on the anode side of the cell due to the vast differences in molar mass between hydrogen and water (90). In the second term, c_T is the total concentration or molar density of all of the gas species, ε_G is the volume fraction of the gas phase, and $D_{i,j}^{\text{eff}}$ is the effective binary interaction parameter between i and j; by the Onsager reciprocal relationships, $D_{i,j}^{\text{eff}} = D_{j,i}^{\text{eff}}$ for ideal gases. The effective diffusion coefficient is defined as

$$D_{i,j}^{\text{eff}} = \frac{1}{\tau_G} D_{i,j}, \tag{31}$$

where τ_G is the tortuosity of the gas phase. Both the gas-phase volume fraction and tortuosity depend on the saturation, S, or pore volume fraction of liquid. While this is straightforward for the gas-phase volume fraction

$$\varepsilon_G = \varepsilon_o (1 - S), \tag{32}$$

where ε_o is the porosity of the medium, the tortuosity is another story. Typically, a Bruggeman expression is used for the tortuosity (91–94)

$$\tau_G = \varepsilon_G^{-0.5}. \tag{33}$$

However, it is believed that the above expression underpredicts the tortuosity and more complicated expressions or analyses are required or as is often the case, the tortuosity is used as a fitting parameter.

The Stefan–Maxwell equations stem from looking at the velocity of the individual species relative to a reference state. This reference state is typically assumed to be the laboratory reference frame (i.e., stationary), which allows for the Stefan–Maxwell equations to account for not only diffusive fluxes but also convection. For example, for a two-component system, the Stefan–Maxwell equations will result in the equation of convective diffusion,

$$D_i \nabla^2 c_i = \mathbf{v}_G \nabla c_i \tag{34}$$

which is sometimes used in the simulation of PEFCs. In the above expression, $\mathbf{v}$ is the mass-averaged velocity of the gas phase

$$\mathbf{v}_G = \frac{\sum_{i \neq s} M_i \mathbf{N}_i}{\rho_G}. \tag{35}$$

As the pore size decreases, molecules collide more often with the pore walls than with each other. This movement, intermediated by these molecule-pore-wall interactions, is known as Knudsen diffusion (95). In this type of diffusion, the diffusion coefficient is a direct function of the pore radius (50). In the models, Knudsen diffusion and Stefan–Maxwell diffusion are treated as mass-transport resistances in series (50, 96), and combined to yield

$$\nabla x_i = -\frac{\mathbf{N}_i}{c_\mathrm{T} D_{K_i}^{\mathrm{eff}}} + \sum_{j \neq i} \frac{x_i \mathbf{N}_j - x_j \mathbf{N}_i}{c_\mathrm{T} D_{i,j}^{\mathrm{eff}}}, \qquad (36)$$

where the $D_{K_i}^{\mathrm{eff}}$ is the effective Knudsen diffusion coefficient. In effect, the pore wall, with zero velocity, constitutes another species with which the diffusing species interact, and it determines the reference velocity used for diffusion (97). The above equation also can be derived from a dusty-gas analysis (98).

From an order-of-magnitude analysis, when the mean-free path of a molecule is <0.01 times the pore radius, bulk diffusion dominates, and when it is >10 times the pore radius, Knudsen diffusion dominates. This means that Knudsen diffusion is significant when the pore radius is less than about 0.5 μm, which occurs in MPLs, CLs, and macroporous GDLs where there is a high saturation thereby resulting in only the small hydrophobic pores being open for gas flow.

Although the Stefan–Maxwell equations account for convection, another relation is necessary to determine the pressure drop within the porous media. This is typically accomplished in the same fashion as liquid-water flow above, i.e., Darcy's law for the gas phase

$$\mathbf{v}_\mathrm{G} = -\frac{k_\mathrm{G}}{\mu_\mathrm{G}} \nabla p_\mathrm{G}, \qquad (37)$$

where k_G is the effective gas permeability.

Equation (37) can be either used as a separate momentum equation to determine the pressure, or it can be thought of as an additive term to the Stefan–Maxwell equations a la the dusty-gas model (98)

$$\nabla x_i = -\frac{x_i k_\mathrm{G}}{D_{K_i}^{\mathrm{eff}} \mu_\mathrm{G}} \nabla p_\mathrm{G} + \sum_{j \neq i} \frac{(x_i \mathbf{N}_j - x_j \mathbf{N}_i)}{c_\mathrm{T} D_{i,j}^{\mathrm{eff}}} - \frac{\mathbf{N}_i}{c_\mathrm{T} D_{K_i}^{\mathrm{eff}}}. \qquad (38)$$

However, this treatment is not rigorously correct since there is no strong justification for being able to combine the bulk-fluid velocity with the transport equations linearly in general.

iii. Coupling between liquid and gas phases

It is well known that gas and liquid interact to a certain extent in a porous medium (99–101). This interaction is embedded in terms of the transport parameters and how they depend on the saturation of the medium (e.g., see equation (3)). For variables such as permeability, everything from empirically determined dependences from soil studies to cut-and-random-rejoin bundle-of-capillary models have been used; typically, a more-or-less cubic dependence is utilized. It is worth noting that to date no one has been able to measure successfully the permeability functionality for PEFC materials, due in part to their thinness and very complicated, chemically heterogeneous microstructure. Saturation and the saturation dependences of the various transport parameters are the main way in which flooding is accounted.

While saturation is a key concept, it is actually a dependent and not an independent variable. To determine the saturation, one uses the independent variables of gas and liquid pressures, which are characterized by a capillary pressure (100–103)

$$p_C = p_L - p_G = -\frac{2\gamma \cos\theta}{r}, \tag{39}$$

where γ is the surface tension of water, r is the pore radius, and θ is the internal contact angle that a drop of water forms with a solid. Equation (39) is based on how liquid water wets the material; hence, for a hydrophilic pore, the contact angle is $0° \leq \theta < 90°$, and for a hydrophobic one, it is $90° < \theta \leq 180°$. To calculate the saturation from the capillary pressure, there are various methodologies: one can use it as a fitting parameter; one can use empirically determined functions (e.g., Leverett J-function (86, 103)), although these usually stem from hydrophilic soil analyses; one can develop detailed microscopic and/or pore-network models; or one can use macroscopic idealizations such as a bundle of capillaries.

Before proceeding, it is of interest to examine the issue of whether a high capillary pressure and hence flooding is more a

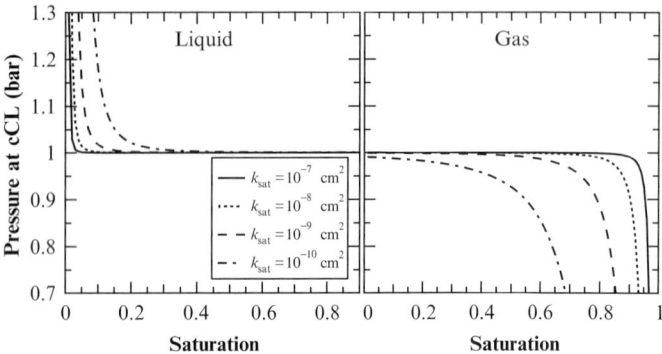

Figure 6. Liquid- (*left*) and gas-phase (*right*) pressure at the cathode catalyst layer as a function of average liquid saturation and saturated (absolute) permeability for a 0.25 cm cathode GDL, a gas-channel liquid- and gas-phase pressure of 1 bar, and conditions of 1 A cm^{-2} and 65°C.

result of an increasing liquid pressure relative to the gas one or a decreasing gas pressure relative to the liquid one. Doing a back-of-the-envelope calculation yields the results shown in Fig. 6. For the calculation, a current density of 1 A cm^{-2} is used to determine the fluxes by Faraday's law and typical water crossover values, saturated air at 65°C is fed, and a cubic dependence of the permeability on saturation is assumed. The figure displays the expected pressure at the cathode GDL/CL interface as a function of the average saturation (assumed uniform) of the GDL and the absolute permeability, which is a function of the GDL microstructure alone. First, it should be pointed out that this simple calculation only examines the respective pressure drops and does not consider how they impact the saturation (i.e., the saturation is taken as an independent variable that is not connected to the capillary pressure). From the calculations, the gas velocity is three-orders of magnitude higher than the liquid one due to the low gas density. However, as discussed in Sect. III.2.iii, when nonisothermal effects are accounted for, the water-vapor flux will switch direction for fully humidified conditions and the gas-phase velocity can decrease substantially due to this heat-pipe effect. Due to the velocity differences, flooding due to a relative decrease in the gas pressure results in a wider saturation window than flooding due to a relative increase in the liquid pressure. Therefore, gas-phase pressure drops should be accounted for. The figure also gives rough design guidelines for the GDL. For example, for a given permeability, it is apparent that the GDL should operate at a low but not too low (e.g., 15%) saturation to enable good gas and liquid transport; of course, as noted, the simple analysis does not account for the feedback between the capillary pressure and the saturation. Finally, the increase in liquid pressure with lower absolute permeability displays the fact that small-pore layers (e.g., MPLs) can be used to pressurize the liquid, as long as they remain at relatively low saturations (i.e., very hydrophobic) to avoid the decrease in gas pressure; this is discussed in more detail in Sect. III.2.ii.

While the above two-phase flow equations are sufficient for modeling purposes, their implementation can result in convergence and stability issues. For this reason, various simplifications and alternative methodologies have been used. The first such methodology is to use the saturation as the driving force, resulting in a governing equation of

$$\mathbf{N}_{w,L} = -D_S \nabla S, \qquad (40)$$

where D_S is a so-called capillary diffusivity

$$D_S = \frac{k}{\mu \bar{V}_w} \frac{dp_C}{dS}. \qquad (41)$$

Although the above equation is valid, it gives the false impression that the saturation is the driving force for fluid flow, and that a saturation condition should be used as a boundary condition. Furthermore, care must be taken in the interpretation of the capillary diffusivity.

Another simplification is to assume that the liquid and water vapor are in equilibrium, which is not a bad assumption since they have a large interfacial contact area within the porous medium. This assumption allows one to combine the two material balances so that there is only one for water, and the evaporation/condensation rate does not have to be explicitly calculated. One of the material-balance equations is then replaced by the equilibrium expression given by the Kelvin equation (100)

$$p_0^{\text{vap}} = p_{0,\text{o}}^{\text{vap}} \exp\left(\frac{p_C \bar{V}_w}{RT}\right), \qquad (42)$$

where $p_{0,\text{o}}^{\text{vap}}$ is the uncorrected (planar) vapor pressure of water and is a function of temperature. The treatment of water in this manner greatly enhances the convergence and stability of the numerical simulation.

Related to the above equilibrium methodology is the multiphase mixture model (104, 105) typically used in computational-fluid-dynamics models. This model uses algebraic manipulations to convert the two-phase flow equations to a pseudo single phase. Thus, although the two-phase mixture moves at a calculated mass-average velocity, interfacial drag between the phases and other conditions allow each separate phase velocity to be determined. The liquid-phase velocity is found by (104, 106)

$$\mathbf{v}_L = \lambda_L \frac{\rho_m}{\rho_L} \mathbf{v}_m + \frac{k\lambda_L (1 - \lambda_L)}{\varepsilon_o \rho_L v_m} [\nabla p_C + (\rho_L - \rho_G) g], \qquad (43)$$

where the subscripts m stands for the mixture, ρ_k and ν_k are the density and kinematic viscosity of phase k, respectively, and λ_L is the relative mobility of the liquid phase

$$\lambda_L = \frac{k_{r,L}/\nu_L}{k_{r,L}/\nu_L + k_{r,G}/\nu_G}. \qquad (44)$$

In equation (43), the first term represents a convection term, and the second comes from a mass flux of water that can be broken down as flow due to capillary phenomena and flow due to interfacial drag between the phases. The velocity of the mixture is basically determined from Darcy's law using the properties of the mixture. The appearance of the mixture velocity is a big difference between this approach and other pseudo-one-phase models. While the use of the multiphase mixture model does speed computational time and decreases computational cost, problems can arise if the equations are not averaged correctly. Also, this approach does not necessarily agree with literature data and the physical picture. For example, it is unclear whether the pseudo one-phase treatment can allow for variable pore-size distribution and mixed wettability effects to be considered.

4. Electron Transport

Although not directly tied to water management, for completeness of the governing equations, electron transport needs to be modeled. For all of the electronically conducting materials, electron transport is modeled with Ohm's law and an effective conductivity that accounts for the volume fraction of the electronically conducting phase and its tortuosity.

$$\mathbf{i}_1 = -\sigma_o \varepsilon_1^{1.5} \nabla \Phi_1, \tag{45}$$

where ε_1 and σ_o are the volume fraction and electrical conductivity of the electronically conducting phase, respectively. The above equation has been adjusted for porosity and tortuosity using a Bruggeman correction. For most PEFC components, carbon is the conducting phase, with water, air, membrane, and Teflon® being insulating. Although most of the relevant PEFC layers are conductive enough not to warrant too much concern with ohmic drops, full-size cells or thin GDLs with low in-plane conductivity may cause situations wherein there are nonuniformities that are caused by the electron transport (107–109).

5. Catalyst-Layer Modeling

The CLs are the thinnest layers, but the most complex in a PEFC. Inside the CLs, the electrochemical reactions take place in an environment where all of the various phases exist. Thus, the membrane and two-phase-flow models must be used in the CL along with additional expressions related to the electrochemical kinetics on the supported electrocatalyst particles. A schematic of a typical PEFC CL is shown in Fig. 7, where the electrochemical reactions occur at the

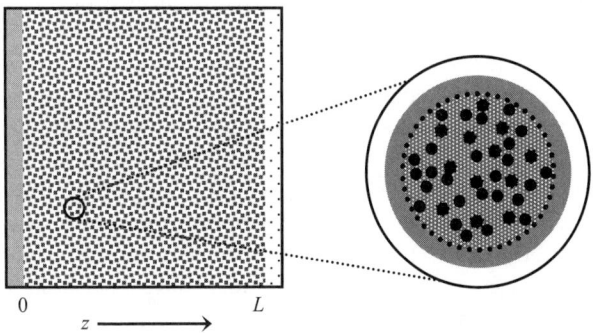

Figure 7. Idealized schematic of the cathode catalyst layer (going from $z = 0$ to $z = L$) between the membrane and cathode diffusion medium showing the two main length scales: the agglomerate and the entire porous-electrode. *Grey, white*, and *black* indicate membrane, gas, and electrocatalyst, respectively, and the *grey region* outside of the *dotted line* in the agglomerate represents an external film of membrane or water on top of the agglomerate.

two-phase interface between the electrocatalyst (in the electronically conducting phase) and the electrolyte (i.e., membrane). Although a three-phase interface between gas, electrolyte, and electrocatalyst has been proposed as the reaction site, it is now not believed to be as plausible as the two-phase interface, with the gas species dissolved in the electrolyte. This idea is backed up by various methods or experimentally derived evidence, such as microscopy, and a detailed description is beyond the scope of this chapter. Experimental evidence also supports the picture in Fig. 7 of an agglomerate-type structure where the electrocatalyst is supported on a carbon clump and is covered by a thin layer of membrane, which may then be covered by a thin film of liquid water (110–114). Figure 7 is an idealized picture, and the actual structure is probably more of a "spaghetti and meatball" structure, where the carbon agglomerates are connected to each other and covered by thin tendrils of membrane.

As discussed in our recent review (2), various modeling approaches have been used for the CLs. In this chapter, we focus only on the most relevant ones. In accordance with the experimental picture, the modeling consensus is that an embedded agglomerate model is required for the CLs (see, for example (2), and (58)). In fact, recent studies have clearly shown that treating the cathode CL as an interface with uniform properties leads to several erroneous conclusions, especially due to the impact of channel-rib effects that distribute the electrons, water, heat, and oxygen unevenly at the CL boundary (115–118). While an embedded agglomerate model is now utilized in most models, there is still an effort towards including more microstructural details. Without such inclusions, optimization studies and analysis become too far removed from reality. While most models do this inclusion using a more macrohomogeneous approach as detailed below, there are two notable exceptions. The first, by Wang et al. (119, 120), assumes a random microstructure and solves the macroscopic equations through such a network. These results provide a nice link between the macroscopic and the microscopic analyses; however, the models are still too computationally costly to be used in complete full-cell simulations without requiring simplifications of the other layers. That being said, the model allows one to get a handle on such effects as tortuosity and inactive regions (whether that is catalyst, ionomer, or gas pores) in the layer. For tortuosity, they predict a Bruggeman coefficient (see equation (33)) for each phase equal to 3.5, which does change with phase volume fraction. However, the deviations can be minimized if the effective or active phase volume fraction is used instead of the overall value.

The idea of active phase volume fraction was also examined by Farhat (121), who did a statistical analysis to determine what percentage of the three phases (ionomer, platinum, and gas) are in contact with each other and thus where reaction can proceed. Such an analysis is interesting, but the low platinum site utilization number it provides (22%) assumes that an exact three-phase contact needs to exist for reaction. This is not necessarily the case since if the ionomer film over the catalyst is thin, it may be that it is a two-phase contact, and the gas-phase just needs to be near the covered reaction site. Also, it is possible that the ionomer also only needs to be near the platinum for protons to react. The truth is that the exact microstructure is dynamic and currently unknown. What is established is that 100% utilization of the platinum does not occur even with no gas-transport limitations since the platinum may be isolated and no longer in electrical contact with the carbon, or it could be far inside the primary pores of the carbon and thus inaccessible to protons and gas (122).

Similar to the approach of Wang et al., Durand and coworkers (123–127) use spherical agglomerate structures in a regular (not random) 3-D hexagonal arrays. In between the agglomerates, there are either gas pores or the region is flooded with electrolyte. The equations solved are mainly Ohm's law and Fick's law with kinetic expressions, which is a simpler analysis than that of Wang and coworkers. The results of the models show the concentration contours around a particle and agree with experimental current densities and trends. Such a model also allows for the detailed placement of the electrocatalyst particles to be studied and the various performance gains realized, even though it may not yet be possible to make such an arrangement experimentally.

i. Modeling equations

The kinetic equations for the main HOR and ORR reactions were introduced in Sect. II.2.ii, including some mention about possible side reactions. As noted above, in addition to the electrode and kinetic interactions, two-phase flow and membrane models must be used. This is typically done by utilizing the models discussed in

Sects. II.2 and II.3 above, but accounting for the fact that their volume fractions are not unity. For the two-phase flow equations, this does not really change their expressions, except that the material balances must be altered to account for the reaction rates as discussed below. One change may be in the concept of electrode flooding, which is also discussed below. For the membrane equations, one must now account for the fact that its volume fraction is not unity. Thus, the transport properties must be altered to account for the dispersed phase by something like a Bruggeman relation. In addition, the correct superficial fluxes must be used through the use of membrane volume fractions. Finally, it is still unknown whether the ionomer in the CLs behaves in the exact same fashion as that in the separator. For example, do the ionomer tendrils in the CL swell in the same manner as in the membrane, or is the CL ionomer dominated by interfacial and surface effects? Research both through first-principle modeling and detailed experimentation is still ongoing to answer these and similar questions. From the macroscopic modeling perspective, with few exceptions, the same membrane equations and properties are used in both the separator and the CLs, and swelling is ignored in the CLs.

The kinetic expressions result in transfer currents that relate the potentials and currents in the electrode (platinum on carbon) and membrane phases as well as govern the consumption and production of reactants and products,

$$\nabla \cdot \mathbf{i}_2 = -\nabla \cdot \mathbf{i}_1 = a_{1,2}\mathbf{i}_\text{h}, \qquad (46)$$

where $-\nabla \cdot \mathbf{i}_1$ represents the total anodic rate of electrochemical reactions per unit volume of electrode and i_h is the transfer current for reaction h between the membrane and electronically conducting solid (i.e., equations (14) and (15) for the HOR and ORR, respectively). The above charge balance assumes that faradaic reactions are the only electrode processes (i.e., it neglects crossover and other side reactions); double-layer charging is neglected (as is appropriate under steady-state conditions). This equation can be used in the conservation-of-mass equation (16) to simplify it. For example, if the ORR is the only reaction that occurs at the cathode, the following mass balance results

$$\nabla \cdot \mathbf{N}_{O_2,G} = -\frac{1}{4F} a_{1,2} i_{0_{ORR}} \left(\frac{p_{O_2}}{p_{O_2}^{ref}}\right) \exp\left(-\frac{\alpha_c F}{RT}(\eta_{ORR})\right) = \frac{1}{4F} \nabla \cdot \mathbf{i}_1.$$
(47)

Before discussing the models in more depth, a note should be made concerning catalyst loading. Many models use platinum loading in their equations, especially for optimizing designs and in normalizing the current produced (equivalent to a turnover frequency in catalysis). In this respect, the catalyst loading, m_{Pt}, is the amount of catalyst in grams per PEFC geometric area. If a turnover frequency is desired, the reactive surface area of platinum, A_{Pt}, can be used (usually given in $cm^2 \, g^{-1}$). This area can be related to the radius of a platinum particle assuming perhaps a certain roughness factor, but more often is experimentally inferred using cyclic voltammetry measuring the hydrogen adsorption. These variables can usually be determined and then used to calculate the specific interfacial area between the electrocatalyst and electrolyte,

$$a_{1,2} = \frac{m_{Pt} A_{Pt}}{L}, \tag{48}$$

where L is the thickness of the catalyst layer. This assumes a homogeneous distribution of electrocatalyst in the CL.

A factor closely related to the catalyst loading is the efficiency or utilization of the electrode. This tells how much of the electrode is actually being used for electrochemical reaction and can also be seen as a kind of penetration depth. In order to examine ohmic and mass-transfer effects, sometimes an effectiveness factor, E, is used. This is defined as the actual rate of reaction divided by the rate of reaction without any transport (ionic or reactant) losses. As noted above, a value of 100% efficiency ($E = 1$) does not necessarily correspond to the loading of catalyst but instead to the electrochemically active catalyst area.

In the CLs, there are two main length scales and both are important. The two scales are the whole layer and the agglomerate (see Fig. 7). To account for both the local agglomerate level as well as effects across the porous electrode, an embedded agglomerate model is used. In this type of model, the traditional porous electrode equations are used to calculate the gas composition and the overpotential change across the CL due to ohmic, mass-transfer, and reaction effects, and the agglomerate model is used for the reaction site to determine the correct transfer current density. In this fashion, the embedded agglomerate model is essentially a pseudo 2-D model where one dimension is the electrode and the other is into the agglomerate (obviously if one is doing a multiple dimensional model, then the agglomerate is an additional pseudo dimension which is a microscopic-scale dimension).

In terms of the porous electrode equations, no new relations are required. As noted above, the membrane equations and two-phase flow equations are used with appropriate scaling factors, and the reaction rates are determined from the agglomerate model presented below. From a historical and reference perspective, Euler and Nonnenmacher (128) and Newman and Tobias (129) were some of the first to describe porous-electrode theory. Newman and Tiedemann (130) review porous-electrode theory for battery applications, wherein they had only solid and solution phases. The equations for when a gas phase also exists have been reviewed by Bockris and Srinivasan (131) and DeVidts and White (132), and porous-electrode theory is also discussed by Newman and Thomas-Alyea (26) in more detail.

The main function of the agglomerate model is to obtain the correct transfer or reaction current density. One of the most detailed applications of this model is that of Shah et al. (133) In their model, they account for such impacts as membrane swelling, inactive catalyst in the agglomerate pores, surface films of both ionomer and water if the vapor phase is saturated, and the number and dispersion of agglomerates. Furthermore, they do everything in a geometrically and material-balance consistent manner. Such an in-depth model allows for detailed analysis to be done in terms of impacts of flooding and other CL resistances and structural parameters on performance.

For the agglomerate model, the characteristic length scale is the radius of the agglomerate, R_{agg}, and all of the agglomerates are assumed to be the same shape and size. This assumption does not necessarily agree with reality, and it would be better to have a distribution or even a discrete few agglomerates with different radii. In the agglomerate model, the reactant or product diffuses through the electrolyte film surrounding the particle and into the agglomerate, where it diffuses and reacts. Hence, there is a concentration and possibly a potential and temperature distribution within the agglomerate. The equations for modeling the agglomerate are similar to those presented above (i.e., mass balances, kinetics, energy balance, etc.) in spherical coordinates. As mentioned above, the role of the agglomerate model is to determine how the transfer current density should be altered, and this is typically done using an effectiveness factor, resulting in

$$\nabla \cdot \mathbf{i}_2 = a_{1,2} i_h E. \tag{49}$$

As an example, if one takes the ORR to be a first-order reaction following Tafel kinetics, the solution of the mass-conservation equation in a spherical agglomerate yields an analytic expression for the effectiveness factor of (50, 134)

$$E = \frac{1}{3\phi^2}(3\phi \coth(3\phi) - 1), \tag{50}$$

where ϕ is the Thiele modulus for the system (135)

$$\phi = \frac{R_{\text{agg}}}{3}\sqrt{\frac{k'}{D_{O_2,\text{agg}}}}, \tag{51}$$

where k' is a rate constant given by

$$k' = \frac{a_{1,2} i_{0_{\text{ORR}}}}{4F c_{O_2}^{\text{ref}}} \exp\left(-\frac{\alpha_c F}{RT}(\eta_{\text{ORR}})\right), \tag{52}$$

where the reference concentration is that concentration in the agglomerate that is in equilibrium with the reference pressure

$$c_{O_2}^{\text{ref}} = p_{O_2}^{\text{ref}} H_{O_2,\text{agg}}, \tag{53}$$

where $H_{O_2,\text{agg}}$ is Henry's constant for oxygen in the agglomerate. While the above analytic solution is nice, if the reaction is not first order or if one wants to account for varying potential and/or temperature within the agglomerate, the relevant governing equations must be solved numerically with the correct surface boundary conditions to determine E.

If external mass-transfer limitations can be neglected, then the surface concentration in equation (61) can be set equal to the bulk concentration, which is taken from solving the porous electrode equations. Otherwise, the surface concentration is unknown and must be calculated. To do this, an expression for the diffusion of oxygen to the surface of the agglomerate is written

$$W_{O_2}^{\text{diff}} = A_{\text{agg}} D_{O_2,\text{film}} \frac{c_{O_2}^{\text{bulk}} - c_{O_2}^{\text{surf}}}{\delta_{\text{film}}}, \tag{54}$$

where $W_{O_2}^{\text{diff}}$ is the molar flow rate of oxygen to the agglomerate, A_{agg} is the specific external surface area of the agglomerate, and the film can be either membrane or water (if two or more films are desired, similar expressions can be written for each film). The above expression uses Fick's law and a linear gradient, which should be valid due to the low solubility of oxygen and thinness of the film. At steady state, the above flux is equal to the flux due to reaction and diffusion in the agglomerate (as well as the flux through any other films), and thus the unknown surface concentration(s) can be replaced. Doing this and using the resultant expression in the conservation equation (47) yields

$$\nabla \cdot \mathbf{i}_1 = 4F c_{O_2}^{\text{bulk}} \left(\frac{1}{\frac{\delta_{\text{film}}}{A_{\text{agg}} D_{O_2,\text{film}}} + \frac{1}{k'E}} \right). \tag{55}$$

This equation is the governing equation for the agglomerate models for the cathode under the assumptions of first-order reaction, isothermal and isopotential agglomerate. One also can write the above factor as an overall effectiveness factor. If desired, this factor could be used as a fitting parameter, thereby avoiding the necessity of detailed calculations and perhaps multiple fitting parameters on the agglomerate scale. A final analysis would be to assume that $E = 1$ and just calculate the effect of the covering films on the reaction rate. Physically, such an analysis assumes that only the platinum on the agglomerate surface is active, or in other words, the buried platinum is inactive perhaps due to inadequate contact with the ionomer.

Before examining some of the modeling results in terms of impacts on water management and optimization, it is worthwhile to mention CL flooding. The way in which CL flooding is accounted for is by two different approaches. The first, as noted above, is to assume a liquid film that forms and provides an extra mass-transfer resistance to the reactant gas. The second is more of a macrohomogeneous approach wherein the two-phase-flow equations are used to alter the value of the transfer current using the saturation

$$a_{1,2} = a_{1,2}^o (1 - S), \tag{56}$$

where $a_{1,2}^o$ is the maximum or dry specific interfacial area. In comparing the two approaches, it seems that the saturation approach allows for greater reaction rates (higher current densities). The reason is that the CLs have small pores that are at least partially hydrophobic, and thus it takes a high liquid pressure to flood them (depending on the assumed contact angle), whereas even a thin film can effectively shut down the reaction. Of course, the film is spread over a much larger surface area and depends on the agglomerate radius. It is tough to say which approach is better as they both have their advantages and disadvantages, with the agglomerate-film perhaps more physically realistic assuming that the agglomerate parameters are well known. Finally, one should be cognizant that it is hard to say whether flooding in the GDLs or the CLs is more dominant, and one can tailor the relative influence of each by changing the underlying model used; more experimental evidence is required on this front before a definitive conclusion can be reached.

ii. Optimization analyses

A good embedded agglomerate model can help to predict optimal microstructural parameters for improved performance. However, since even some of the more complicated models still make several assumptions such as uniform agglomerate shape and size, the resulting optimizations provide only future experimental research directions. That being said, the results of such models do help to guide intuition, design experiments and structures, and examine how CLs operate (136). Recent models mainly examine distributions of platinum, Nafion®, operational changes, and material properties such as agglomerate wettability and CL thickness (133, 137–140). While most of the models deal with experimentally-based values, some look at possible structures that are more ordered and perhaps experimentally unobtainable currently (141).

A mentionable model is that of Eikerling (140), who does a comprehensive macrohomogeneous approach using structural properties of the CL. His model is similar to that of Weber et al. (85), but goes beyond it in terms of analyzing the effect of water content on both the primary and secondary pores within the agglomerate. Specifically, Eikerling calculates the critical saturation and conditions for optimal performance (i.e., primary pores flooded and secondary ones empty). Although the model does have some drawbacks based on its simplifications (such as a single contact angle and no membrane or GDL models), it does do a good job in demonstrating the intricate balance needed in water and thermal management.

The model of Shah et al. (133) does a detailed analysis of CL flooding showing how relative humidity, temperature, and water exit towards the channel cause nonuniform and suboptimal performance. They also investigate the effect of CL and GDL capillary properties, showing that changing the GDL properties has a larger impact than those of the CL on overall saturation and performance. The model of Wang et al. (138) demonstrates that functionally gradient materials can have a significant impact on performance. While they state that 35% Nafion is the optimal loading due to a competition between ionic and gas transports, they show that having more ionomer nearer to the membrane improves performance by about 10% while having the opposite gradient results in substantially lower performance due to a much lower overall oxygen concentration within the layer. Thus, oxygen transport has a significantly larger impact on performance than proton conduction.

While the above examples and many others optimize the CL properties individually, there are two noteworthy examples that do a multivariable optimization. The results of such studies indicate that one should not optimize a single variable without considering the others since the optimum can change. For example, Song et al. (139) demonstrate that while both ionomer and platinum loadings exhibit optimum values that increase from the membrane to the GDL interface (in agreement with Wang et al.), when one considers both loadings, the optimum ionomer loading still remains linear, but the platinum loading adopts a convex shape. A more detailed optimization routine was conducted by Djilali and coworkers (137), who examined multiple variables such as ionomer volume fraction, platinum-to-carbon ratio, platinum loading, volume fraction of ionomer in an agglomerate, and GDL porosity. Their analyses also examined the impact of operating variables on performance. They show that at high current densities, the optimum structure actually has lower platinum loading than at low current densities so that the CL has a higher porosity and hence enhanced gas flow. Therefore, one must be aware of the expected operating conditions when one does an optimization, and multivariable optimization should be done to realize the true ideal structure. While doing a multivariable optimization can be laborious, there is perhaps opportunity to

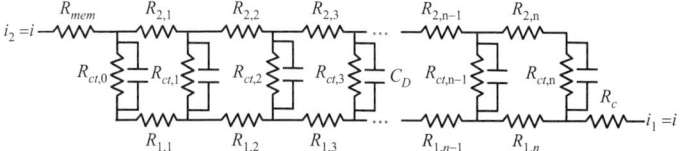

Figure 8. Simple equivalent-circuit representation of a porous electrode. The total current density, i, flows through the membrane and then the electrolyte phase (2) and the solid phase (1) and a contact resistance at each respective end. In between, the current is apportioned based on the resistances in each phase and the charge-transfer resistances and double-layer charging. The charge-transfer resistances can be nonlinear because they are based on kinetic expressions.

use such methods as Monte-Carlo algorithms to reach design space previously ignored.

iii. Impedance models

To get a handle on the controlling phenomena and to characterize the CL and the entire PEFC experimentally, AC impedance or electrochemical impedance spectroscopy (EIS) is often used. The idea is that by applying only a small perturbation to the current during operation, the system response can be studied in situ and in a noninvasive way. Typically, a frequency range is scanned in order to acquire signatures for the different phenomena that occur with different time constants; however, the very long time constant for water rearrangement inhibits the efficacy of EIS for mapping these phenomena.

To analyze the resulting output, a model of the system is required. These models typically assume an equivalent circuit (which can be relatively complicated) for the various physical processes occurring in the PEFC (142–147). Figure 8 shows an example of such a circuit for a porous electrode where the membrane resistance is also considered. The use of equivalent-circuit analysis is really inadequate for studying operation in detail; however, it is very useful for characterizing the CL and membrane resistances and similar properties. These EIS studies allow one to determine the overall resistances in the PEFC, and notably, those of both the ionic and the electronic pathways in the CL (93, 146). Most of these studies show increased high-frequency resistance as the membrane dehydrates and an increased low-frequency loop as flooding occurs. EIS can also be used to map the changes occurring in the PEFC as a function of time. Such analysis allows for signatures to be determined for degradation concerns, such as those dealing with membrane hydration (148) or increased flooding due to loss of hydrophobicity (149).

While a good equivalent-circuit representation of the transport processes in a PEFC can lead to an increased understanding, it is not as good as taking a physics-based model and taking it into the frequency domain. These models typically analyze the cathode side of the PEFC (150–152). An exception is the model of Wiezell et al. (153) that analyzed the anode side and the membrane. In their analysis, they show that the HOR mechanistic steps give arise to various loops in the complex domain. In addition, water electroosmotic flow and impact of water on conductivity can also give rise to low-frequency loops that are semi-inductive and can indicate microstructural relaxation of the polymer. Of the cathode models, those of Springer et al. (151) and Guo and White (152) are perhaps the most complete. Guo and White utilize an embedded agglomerate model and develop extensive expressions for the various loops and time constants. They focus mainly on gas transport and show how it impacts the EIS spectra. The model of Springer et al. also includes a relatively simple membrane model and is based on their previous modeling work (154), thereby allowing a nice comparison to the predicted governing phenomena and changes within the EIS spectra.

The use of impedance models allows for the calculation of parameters, like gas-phase tortuosity, which cannot be determined easily by other means, and can also allow for the separation of diffusion and migration effects. Overall, impedance is a very powerful experimental tool, especially for characterization and trends, but its results are only as meaningful as the model used for its analysis.

6. Model Implementation and Boundary Conditions

To finish this section, it is worthwhile to mention modeling implementation and boundary conditions. Almost all of the models utilize a control-volume approach to solving the equations. This approach is based on dividing the modeling domain into a mesh that determines the control volumes. Using Taylor series expansions, the governing equations are cast in finite-difference form, and typically the governing transport equations have been combined with the conservation equations to yield a set of second-order equations. In this fashion, one is performing conservation equations within each control volume. The exact details of the numerical methods can be found elsewhere (for example, see (155)).

The various PEFC layers or domains are linked to each other through boundary conditions. There are two main types of boundary conditions, those that are internal and those that are external. The internal boundary conditions occur between layers inside the modeling domain, and the external are the conditions at the boundary of the entire modeling domain. Typically, coupled conditions are used for internal boundaries wherein the superficial flux and interstitial concentration of a species are made continuous. However, as mentioned above, boundary conditions between the membrane and electrode can involve the fact that there is only ionic current in the membrane and electronic in the GDL. Another common boundary condition is to have a change in concentration because a species dissolves. This is similar to the internal boundary condition in the membrane and is used sometimes where phases are not continuous across the boundary.

The external boundary conditions specify the concentrations and values for all of the species and variables or their fluxes at the boundary. Examples include specifying the inlet conditions such as gas feed rates, composition, temperature, and humidity, or specifying the current density or potential or specifying the thermal flux to the coolant stream. The external boundary conditions are often the same as operating conditions, and therefore are very similar for most simulations, although there can be differences such as what condition is used for two-phase flow (i.e., zero saturation or zero capillary pressure). One of the most important and perhaps most complex boundary conditions is that between the GDL and the flow channel, which can have a substantial impact on water management and performance (for example, see (133)); this condition is studied in more detail in Sect. IV.2.3.

III. WATER MOVEMENT IN GAS-DIFFUSION LAYERS

Section II introduced the governing equations for water movement. While the recent membrane and CL modeling results were discussed in Sects. II.2 and II.5, respectively, for the GDLs, only the two-phase-flow equations were mentioned (see Sect. II.3). Furthermore, the impact of GDL design and optimization is now becoming more important than ever. The reasons are that the GDL has traditionally been a relatively ignored layer, many of the other layers are somewhat set in their designs, and the impact of GDL properties on water management is very significant. In this section, the functioning of the GDL is discussed. The discussion is separated into two parts. The first part focuses on more microscopic and pore-level treatments of liquid and gas transport in GDLs, and the second part discusses some specific modeling analyses with respect to GDL operation and optimization.

1. Microscopic Treatments

It is known that a GDL is comprised of carbon fibers that have been treated to be made hydrophobic. The actual microstructure is currently unknown, although imaging techniques such as X-ray tomography as described in Sect. I.1, are getting closer. Liquid movement through the layer is similarly hard to quantify experimentally. Figure 9 shows two ideas as to how water moves throughout the GDL microstructure. In both mechanisms, liquid water within the GDL forms preferential pathways that begin to merge into each other and form one larger water conduit to the interface of the GDL with the gas channel. These pathways form through the carbon-fiber interces, and the formation gives rise to a tree-like water distribution (Fig. 9a). Based on their experimental data, Litster et al. (23) propose a fingering and channeling transport method, as seen in Fig. 9b, where instead of small water branches coalescing to form one large break-through path, several water pathways develop in parallel. Once a dominant pathway forms, water from nearby channels is siphoned into the dominant conduit, and a droplet forms at the GDL/gas-channel interface, whereupon it is carried away, and the process begins anew. Although both mechanisms ensure that water moves toward the GDL surface, the initialization points and method vary greatly between the two explanations and can therefore change the creation and results of models. The capillary-tree mechanism will depend strongly on condensation effects within the medium because the initial braches are small. Alternatively, consideration of the channeling mechanism is less dependent on distributed sources throughout the medium and rather on water build-up at the CL/GDL interface. The constant advancing and receding of

Modeling Water Management in Polymer-Electrolyte Fuel Cells 321

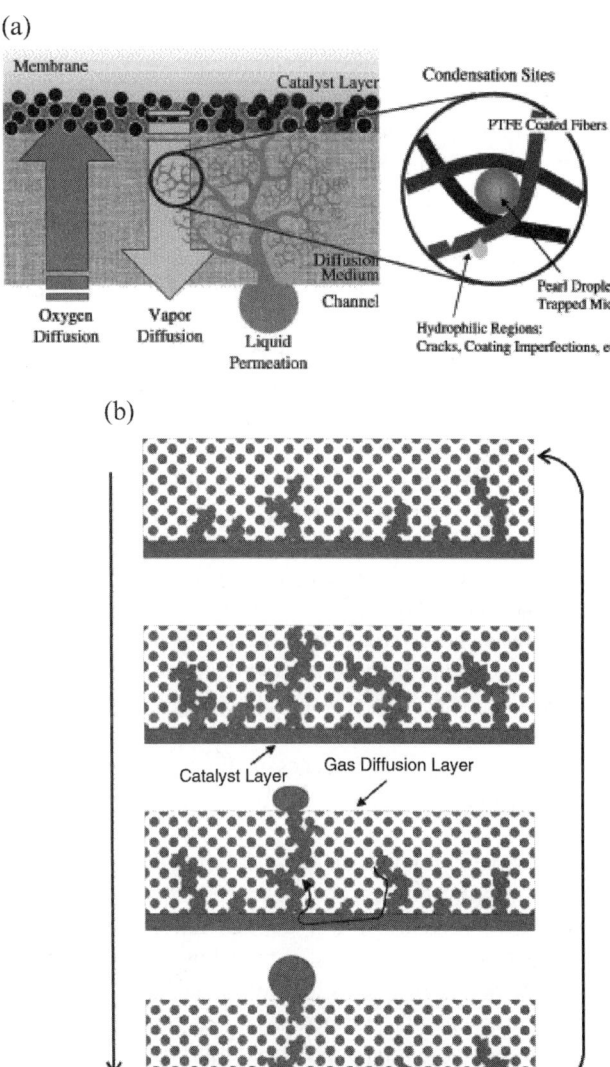

Figure 9. Depiction of (**a**) capillary-tree and (**b**) channeling mechanisms of water movement through a GDL. (Figure (**a**) is from (86) and (**b**) is from (23) with permission of Elsevier.)

water also suggests that wetting hysteresis in the system could play a major role in what channels become dominant water pathways.

The end result of the above analysis is that it is hard to predict fluid movement in the physiochemical heterogeneous structure that is a GDL; a macrohomogeneous approach, as discussed in Sect. II, is often utilized. Bulk-flow parameters and constitutive relations offer a simple means to capture average fluid movement and simplify the underlying complex geometry of the medium. The Carman–Kozeny equation (100) for determining absolute permeability, Wyllie equation for determining relative permeability (100), and Leverett J-function for determining the capillary-pressure-saturation relationship (103, 156) are the most commonly employed relations for modeling water movement through the GDL. The constitutive relations are typically empirically derived but their ability to capture bulk system characteristics is dependent on the assumption that the tested sample size is large enough for one to obtain a representative average and neglect end effects. However, the difficulty of procuring accurate measurements for GDL properties may be gleaned from the spread in the parameter values (64, 116, 133, 157–161). Furthermore, hysteresis and heterogeneities in the medium complicate quantifications and compromise predictive capabilities of macroscopic models with respect to transient phenomena. Microscopic models are thus becoming necessary to elucidate governing flow mechanisms and to predict differences in flow pathways for yet uncharacterized materials or changes due to GDL manufacturing or PEFC design.

Lattice-Boltzmann simulations have been extended to multiphase flow (162–164), and several full-morphology and network models have started to analyze flow through fibrous materials in particular (165–169). Such microscopic modeling in general is still in the infancy stage and mechanistic understanding of how water moves through different porous media, i.e., as thin films, slugs, droplets, or some combination thereof, is still being studied for both steady-state and transient conditions (170, 171). Nonetheless, some preliminary attempts have been made to derive constitutive relations – namely permeability and capillary-pressure-saturation curves – via direct treatment of microphysiochemical structure.

With respect to derivation of relative permeability for GDLs, Markicevic and Djilali (172) developed a two-scale model for flow around obstacles using saturation and phase length scales as variable parameters. Relative permeability was found to be dependent upon the relative sizes of the saturation and phase length scales. Relative permeability was seen to vary from a linear to nonlinear dependence on saturation depending on whether flow was in the Darcy, Brinkman, or Stokes regime.

The determination of capillary-pressure-saturation curves can be accomplished through the use of pore-network, full-morphology, and Lattice-Boltzmann modeling with pore-network simulations being most common (167–169). Vogel et al. (167) compared the three techniques in terms of computational intensity and predictive capabilities. Pore-network modeling is the most simple of the three, and involves the idealization of a medium and assumption of the pore-size distribution and connectivity of the pores. Provided that these inputs were available, pore-network modeling was found to capture the same trend as the Lattice-Boltzmann model for saturation levels above 0.1. The full-morphology approach incorporates the next level of complexity in that the explicit microstructure of the medium is treated, but imbibition is idealized to proceed by advancement of spheres (173). A consequence of the approach was consistent overprediction of saturation due to the artificially high intrusion of water into the pore because of the assumption of intrusion by spherical fronts. The Lattice-Boltzmann model was the most rigorous and accounted for interfacial phenomena best, but the detail comes at the price of being computationally expensive and also having the limitation of being grid-spacing dependent (167).

Because of the relative ease with which pore-network modeling can be executed, attempts have been made to generate capillary-pressure-saturation curves using pore-network models. Sinha and Wang (169) developed an alternative expression for use with the Leverett J-function from that proposed by Udell (156) (see equation (61)) by generating an idealized GDL structure and solving for the flow pattern as dictated by minimizing capillary pressure for each advancement. Curve fitting resulted in a similar function as that of Udell, except multiplied by a factor of 2.3 and with an additive constant. Schulz et al (168). noted even closer agreement between their pore-network results and the Leverett J-function, which is surprising considering that the Leverett J-function was derived for soil systems (103). Schulz et al. also went on to predict through-plane permeability at varying compression ratios, and observed reasonable agreement with experimental data despite the fact that neither breaking of fibers nor PTFE coating is taken into account in the simulation (168).

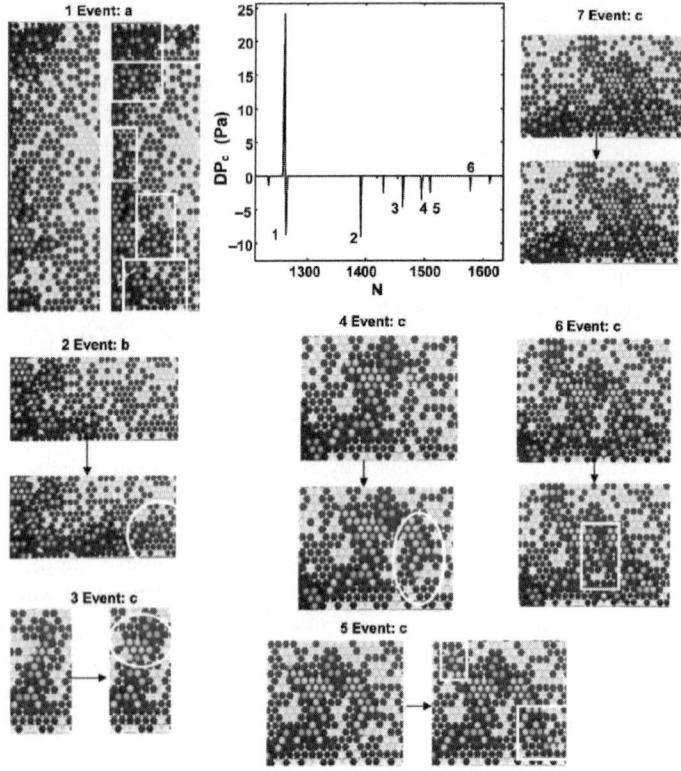

a. Penetration in large clusters of pores

b. Capillary Fingering

c. Penetration in small clusters of pores

Figure 10. Imaging and correlation of penetration events in a mixed-wettability system and corresponding changes in capillary pressure. (The figure is reproduced from (174) with permission of the American Institute of Physics.)

The next step toward creating realistic GDL domains requires treatment of the wettability heterogeneities in the system. Simulations have yet to conquer the task of solving for constitutive relations for flow in porous media of mixed wettability in any system, let alone in fibrous systems and GDLs, but initial modeling and experimental advances are underway, as seen in Fig. 10 (169, 174).

2. Macroscopic Analyses

The above microscopic treatments allow for a much greater understanding of two-phase flow in the chemical heterogeneous and complex structure of a GDL. However, they are currently too detailed and computationally costly to be linked to the other PEFC layers and used for full-cell analyses. Hence, these analyses typically use the more macroscopic equations introduced in Sect. II.3. While most recent models include GDLs using those governing equations in some form or another, many cell-level models now focus on various effects and properties within GDLs. Some of the more important effects and the corresponding simulation studies are detailed below, but before that begins, mention should be made concerning the different macroscopic approaches towards the determination of the transport parameters.

i. Determining two-phase-flow parameters

As noted in Sect. II.3, the main point of two-phase-flow models is the determination of the liquid saturation and hence the gas-phase tortuosity or effective diffusion coefficient. Key among these parameters is the effective permeability of both the liquid and gas phases. As discussed in our previous review (2), the effective permeability in a PEFC has not been measured accurately experimentally, and there is a multitude of expressions and models to determine it and associated two-phase-flow parameters. Below, a short discussion is given on the two-phase-flow parameters for completeness; the reader is referred to the respective references for more in-depth discussion and analysis.

The effective permeability of a system may be broken into two parts, the saturated or absolute and the relative permeabilities. The saturated permeability, or the permeability at complete saturation, of the medium, k_{sat}, is a function of geometry and microstructure alone, whereas the relative permeability, k_r, accounts for interactions between two or more fluids in a medium. The effective permeability is then taken to be the product of the two

$$k = k_r k_{\text{sat}}. \tag{57}$$

Experimentally, the absolute permeability may be found from Darcy's law equation (29) by measuring the flowrate across a medium with a known pressure drop. The experiment is simple in principle, but prone to error and edge effects and thus should be performed at multiple flowrates and pressure drops to find the best fit of Darcy's law and ensure complete filling of the pore space by the fluid. The permeability of anisotropic materials is more difficult to ascertain since the three permeabilities are harder to decouple. Another common method for determining absolute permeability is use of the Carman–Kozeny equation (100)

$$k_{\text{sat}} = \frac{\varepsilon^3}{k_K \varepsilon (1-\varepsilon)^2 S_0^2},\tag{58}$$

where k_K is the Kozeny constant, which depends on the medium and represents a shape and tortuousity factor, and S_0 is the specific surface area based on the solid's volume. The Carman–Kozeny equation is based on Poiseulle's equation and thus on laminar flow through capillary tubes; any derived permeability is still for idealized, isotropic conditions. Some attempts to incorporate the fibrous nature of GDLs have been made via use of the Ergun equation and fractal theory (178, 179), but without a definitive determination method, absolute permeability values used in models will probably continue to vary and be based on experimental values. The typical range of experimentally and theoretically determined values for k_{sat} is from 10^{-11} to 10^{-5} cm^2, with most values lying between 10^{-9} and 10^{-8} cm^2 (64, 113, 116, 157, 158, 160, 161, 177).

More uniformity between models is seen with respect to the constitutive relation chosen to define the relative permeability. The most common expression is the Wyllie expression and is based on a cut-and-rejoin model of tubes and is used due to its simple form,

$$k_r = S^3.\tag{59}$$

Other options for ascertaining relative permeability include using the Corey (178) (Brooks–Corey) (179) relation, Van Genuchten relation (180), and a statistical derivation (181). One point of note is that the above relations are based on permeability experiments and calculations which have been predominantly studied in the context of soil science and oil reclamation. Their applicability to PEFC systems is nebulous because unlike a bed of sand, the GDL comprises a highly porous, mixed wettability, irregularly shaped, and interconnected fibrous microstructure. Nonetheless, most models utilize expressions and relations as developed in hydrological studies, since there are not any for the PEFC components due to experimental difficulties.

The most notable example of a hydrological expression that is applied frequently to PEFC systems is the Leverett expression that relates capillary pressure, P_C, to saturation, S (103)

$$P_C = \gamma \cos\theta \sqrt{\frac{\varepsilon}{k}} J(S), \qquad (60)$$

where $J(S)$ is found empirically and known as the Leverett J-function. The most commonly used expression for this function is that of Udell (156)

$$J(S) = 1.417(1-S) - 2.120(1-S)^2 + 1.263(1-S)^3 \text{ for } \theta < 90°$$
$$J(S) = 1.417S - 2.120S^2 + 1.263S^3 \qquad \text{for } \theta > 90°.$$
$$(61)$$

The applicability of the Leverett J-function has been debated (87, 158). One of the main limitations of the Leverett approach in describing a wide range of systems is that the square-root term does not scale across systems of varying topology (100). Another source of concern has been the differences between the conditions under which the Leverett J-function was derived and the characteristics of the GDLs. The Leverett J-function was developed for an isotropic soil of uniform wettability and a small particle aspect ratio, whereas a standard GDL is anisotropic, of mixed wettability due to nonuniform PTFE coating – which also complicates the definition of a contact angle – and a large particle aspect ratio as is common for fibrous materials. Besides using a Leverett J-function, the only other alternatives are those based on tangential experimental results (182) or those that use an idealized construct (85). Overall, despite concerns regarding the applicability of various constitutive relations on permeability and pressure-saturation curves, viable alternatives are yet lacking in the field. However, with further development of physically accurate pore-network or Lattice-Boltmann models, probing permeabilities and pressure-saturation curves with simulations is a possibility.

ii. Microporous layers

With the increased understanding of the importance of GDL design, PEFC manufacturers have begun to examine composite GDLs of various layers tuned for specific tasks. The most common design is a bilayer one in which there is a macroporous layer (which is similar to a traditional GDL) and a microporous layer (MPL) (see Fig. 1). MPLs are made from carbon black pressed with hydrophobic binder (they are usually hydrophobic) and are anywhere from 0.1 to 20 μm thick, with permeabilities an order of magnitude or two less than GDLs, small pore sizes, and various porosities (157, 183). MPL properties can be tailored by use of different carbons, different carbon-to-binder ratios, use of pore formers, etc (183).

The idea behind the addition of the microporous layer is that if a less permeable material is placed between the cathode CL and GDL, water movement into the GDL would decrease and reactant transport increase. Another possible benefit would be to use a partially hydrophilic MPL to wick water away from the cathode CL, thereby decreasing flooding in it. Beyond the duty of water management, MPL incorporation may have the extra benefit of reducing contact resistance between the GDL and CL, increasing the usable active area of the CL, promoting efficient gas redistribution, and protecting the CL from pinhole formation caused by piercings from the GDL fibers. MPLs have improved PEFC performance consistently in experiments (184–192), with initial empirical optimizations demonstrating increased peak power for composite GDL/MPLs that balance the strengths of high gas permeability and high hydrophobicity (183). MPL traits that lead to better performance include higher hydrophobicity (184–186), thinner layers (191), and micron-sized pores (192), just to name a few. Such experiments have provided valuable insight into how currently used MPLs compare. The next step underway is to use models to fine tune these general rules of thumb and elucidate what dictates MPL efficacy and how to optimize the MPL material properties for various operating conditions and optimization criteria.

General models that compare strictly the difference in PEFC performance between having a MPL and not show that addition of a MPL leads to more uniform membrane hydration and gas and current distribution over the entirety of the PEFC (193–195). One of the first MPL-specific works is that of Pasaogullari and Wang (196), who used a 1-D, half-cell, isothermal, and two-phase model to study the effects of MPL porosity, wettability, and thickness on performance. An interesting result stemmed from a comparison between their model and one that assumed a uniform gas pressure. In the case of nonuniform gas pressure, liquid flow from cathode to channel leads to a counterflow of gas from channel to cathode and higher capillary pressure and gas transport. Their model predicts as much as a 50% increase in oxygen transport to the reactive surface, thus intimating the importance of accounting for varying gas pressure. Pasaogullari and Wang proposed that the discontinuity in saturation levels over the CL/MPL interface became so small as to take away the driving force to pull water into the MPL. One drawback of the Pasaogullari and Wang model is that by considering only the half-cell domain, the effect of back diffusion cannot be studied. Water distribution was ascertained based only on the idea that the MPL would wick water away from the cathode CL and neglected the possibility that the MPL would push water back toward the anode.

Weber and Newman examine this possibility by treating the entire PEFC sandwich; however they assume isobaric and isothermal conditions (197). Through inclusion of the other half of the PEFC, Weber and Newman found that one of the major effects of the MPL is to promote back diffusion and membrane hydration. This is seen in Fig. 11, which shows the predicted water pressure and saturation profiles throughout the PEFC sandwich. It is readily apparent that the MPL is acting to pressurize the liquid stream without flooding itself, thereby increasing the back flux of water through the membrane and decreasing the cathode GDL saturation level. Of interest would be what performance difference would have been observed if the gas pressure had not been kept constant. Applying Pasaogullari and Wang's theory of water movement triggering counterflow of the gas phase, the decrease in outward cathode-side water flux induced by including a MPL would also increase oxygen mass-transfer limitations inside the GDL, similar to that seen when nonisothermal phenomena are accounted for (see Sect. III.2.v). In the model of Weber and Newman, oxygen-transport limitations dominate and dictate the extent that power output can be improved under different conditions and material properties. The main lesson to be gleaned is that for system-level predictions, the perceived limiting mechanism has a large influence on which conditions benefit most from including a MPL and its optimal characteristics.

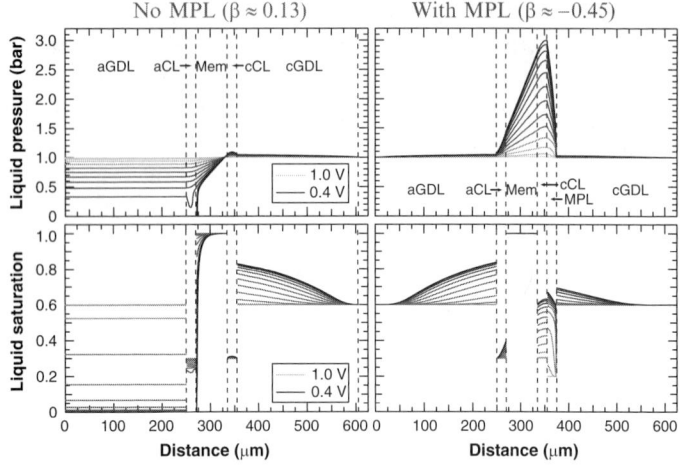

Figure 11. Simulations of a PEFC with and without a MPL. Also given are the membrane net-water-flux-per-proton-flux results. The various PEFC sandwich layers are noted, and the curves correspond to changes in potential going from 1 to 0.4 V in 50 mV increments. The simulation was at 60°C, saturated feed gases. (The figure is reproduced from (197) with permission of The Electrochemical Society.)

Increasing MPL thickness has been observed both in experiments and simulations to increase performance up to a critical thickness. After surpassing this critical thickness, the precise value depending on the operating conditions, performance steadily decreases (191, 197). The critical thickness arises mainly due to the trade-offs between increasing the liquid pressure and back flux versus increasing oxygen mass-transport limitations and ohmic drop (191, 197). Essentially, one is trying to minimize the overall saturation in Fig. 11 of the composite MPL and GDL structure without significantly increasing the composite's resistance or thickness. Changing the fraction of hydrophilic pores in the MPL so as to balance the advantage of repelling water and keeping PTFE loadings low enough not to compromise electrical contact between the CL and GDL, also demonstrates a maximum at low values (197). However, performance sensitivity is shown to be low until the MPL became more hydrophilic than the GDL, in which case the MPL begins to flood rapidly with changes in wettability. In terms of providing manufacturing guidance, this means that specifying the exact fraction of hydrophilic pores is not of utmost concern provided that the MPL is more hydrophobic than the GDL.

Karan et al. (187) sought to resolve two theories proposed at the time on how MPLs impact water management. The first theory centers on the idea that the MPL wicks away water from the cathode CL while the second claims that the MPL pushes water toward the anode. Note that the former hypothesis originates from half-cell models that could not allow for back-diffusion, and the second from full-cell models. However, upon accounting for variance in PEFC assemblies and measurement errors, they found no statistically significant change in the membrane net water with and without a MPL. Whether this result is general or specific to the operating conditions and MPL properties being examined is unclear. However, while their results seem to favor the first hypothesis of changes in the gas flow, a later full-cell model by Pasaogullari et al. (177) based on their half-cell model showed similar results of significant back flux as that of Weber and Newman, especially with a highly hydrophobic and dense MPL. Pasaogullari et al. also demonstrate that, the thicker the membrane, the better the performance gain with subsaturated feeds because there is a larger increase in the average membrane water content with a MPL than without one. Overall, the optimal MPL properties are seen to be highly dependent on the coupled physical phenomena in the cell, and so their simulation requires careful consideration of all mechanisms and trade-offs involved. Finally, there is still a need to validate fully the role of a MPL, especially the impact it has on nonisothermal phenomena and vice versa.

iii. Temperature-gradient (heat-pipe) effect

Almost all of the recent models are nonisothermal. Furthermore, most of them also account not only for heat generation, but also for the existence of temperature gradients. Nonisothermal modeling in this fashion is a change from what had been done earlier, where, due to the thinness of the PEFC sandwich, one assumed that at least the sandwich was at a uniform temperature. However, both experiments and modeling have shown that the low thermal conductivities and overall efficiency of a PEFC can result in temperature gradients through plane. Furthermore, these effects are expected to become larger as the PEFC is more humidified since, if liquid water is produced rather than water vapor, there is much more heat generated at the cathode CL (see equation (7)). Finally, water management and thermal management are shown to be strongly coupled phenomena. (52, 198–204)

The equations for treating nonisothermal phenomena are discussed in Sect. II, and some of the nonisothermal effects are discussed throughout Sect. IV. In this section, we want to emphasize the coupling between water and thermal management in GDLs. Unlike the other PEFC layers, GDLs can sustain relatively large temperature gradients due to their relative thickness, and somewhat low thermal conductivity. Besides impacting transport properties, gas concentrations, etc., the GDL temperature gradient can create a heat-pipe effect, as shown in Fig. 12. In the figure, a temperature gradient induces phase-change and net mass-transfer of water and thermal transfer of heat. Hence, water is evaporated in the cathode CL due to the heat of reaction and moves in the vapor phase down the temperature gradient. The water condenses as it moves along the gradient due to the change in vapor pressure with temperature. Obviously, the heat-pipe effect is more significant at higher temperatures and with larger gradients due to the change in water vapor pressure with temperature; this latter issue also causes much more dilution of reactant gases by water vapor at higher temperatures (see Fig. 3b, for example).

Two of the most extensive modeling studies of this effect are those of Wang and Wang (204) and Weber and Newman (52). In both, simulations are completed to investigate the heat-pipe effect on both water and thermal movement. Both show that the heat-pipe effect can result in at least 15% of the total heat transfer in the GDL,

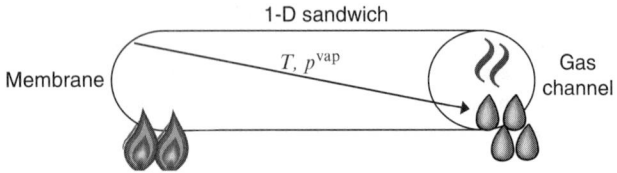

Figure 12. Schematic representation of a heat pipe on the cathode side of a PEFC. Water is evaporated in the CL, moves in the vapor phase, and condenses down the temperature (vapor-pressure) gradient.

while simultaneously providing a means for water movement from the cell in the vapor phase. In fact, it may be that this water-vapor movement is the dominant method to remove water from the cell, and could be a reason why MPLs (which are much more thermally insulating) help in water management.

Wang and Wang show comparisons between nonisothermal cases where the heat-pipe effect and water phase change were and were not considered. The liquid-saturation contours from this comparison are shown in Fig. 13. It is apparent that the heat-pipe effect causes a higher liquid condensation amount and hence saturation under the rib (land), which is the coolest part of the domain. This also results in larger liquid-pressure gradients in-plane, with more water moving from the rib to the channel. Not shown is the result that, when water phase change is considered, the temperature profile becomes more uniform and lower throughout the whole 2-D domain.

Weber and Newman use a 1-D model and not a 2-D one, and, similarly to that of Wang and Wang, they demonstrate that the heat-pipe effect can cause a substantial amount of water movement in the vapor instead of liquid phase, thereby causing overall lower liquid saturations in the GDL. Furthermore, the water vapor will move down the temperature gradient, which is opposite the incoming reactant gas flow, thereby resulting in an additional mass-transport limitation in the system. In other words, the temperature and hence water vapor-pressure gradient results in a retardation of the gas flow,

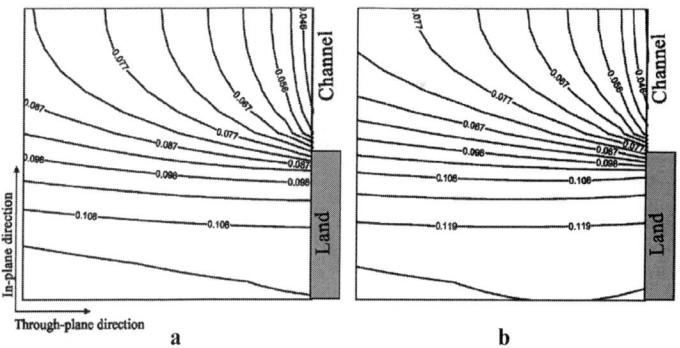

Figure 13. 2-D liquid saturation contours near the gas-inlet region for the case where the heat-pipe effect is neglected (**a**) and considered (**b**). The inlet gases are fed saturated at $80°C$, and the current density is around 1.3 A cm^{-2}. (The figure is reproduced from (204) with permission of The Electrochemical Society.)

which can even cause bulk gas-flow reversal (i.e., convective flow is out of the system) in the anode GDL. Overall, the impact of temperature gradients inside the GDL, especially with saturated feeds, results in significant water-management aspects that should be considered.

iv. Anisotropic properties

Proportionately few PEFC models treat anisotropies due to the paucity of definitive experimental data, additional computational complexity added by solving for extra dimensions, and the difficulty of then incorporating anisotropic values at the risk of losing convergence. The majority of PEFC models have arbitrarily applied in- or through-plane values. However, as more models incorporate higher dimensionalities, inclusion of anisotropies becomes increasingly important, particularly in light of design and limiting-behavior considerations. Strong anisotropies in any direction may completely reroute flow patterns and thereby overhaul the PEFC landscape being modeled. Anisotropies in GDL parameters (i.e., diffusion coefficient, electronic and thermal conductivities, and permeability) are to be expected even if considering only manufacturing effects, i.e., how carbon fibers are pressed to form paper GDLs or woven to form cloth GDLs. Compression effects can further exacerbate permeability anisotropies (205). For these reasons and because of more efficient computational algorithms and processors, groups are starting to study the effects of property anisotropies on PEFC operation (206–208). In all cases, in-plane values are typically believed to be larger than through-plane values.

Tomadakis and Sotirchos (TS) developed an expression to determine relative Knudsen diffusivities of fibrous networks that accounts for both in-plane, defined in relation to the main flat surface of the material, and through-plane differences in the material (208). The effective diffusivity, D_{eff}, is given by

$$D_{\text{eff}} = \varepsilon \left(\frac{\varepsilon - \varepsilon_p}{1 - \varepsilon_p}\right)^\alpha D_{\text{abs}}, \tag{62}$$

where D_{abs} is the absolute molecular diffusivity and the constants ε_p and α are 0.11 and 0.521 and 0.11 and 0.785 for the in- and through-plane directions, respectively (209). Note that under TS theory, in-plane values will always be greater than through-plane values. For GDLs with porosities >0.2, this is typically the case, with through-plane values about twice those of in-plane ones and approaching them as the porosity increased (210). Diffusion-coefficient anisotropies are the least extreme of those that are treated.

The in-plane electronic conductivity has also been found to be higher than in the through-plane by up to a factor of 10 (206, 211). Electronic conductivity values directly influence current distribution, and incorporating 2-D effects can alter optimum design, e.g., rib-to-channel ratio, because the current-collector width should be minimized to promote oxygen flow while not impeding electron transport.

GDL thermal conductivities are also presumed to be higher in the in-plane direction than the through-plane direction with estimated values ranging from order 1 to $10\,\text{W}\,\text{m-K}^{-1}$ (210). Khandelwal and Mench (212) observed a near 50% asymptotic decrease in through-plane conductivity by increasing PTFE loading from 0 to 20% for their system, but whether the same magnitude of change or direction of trend would be seen in the in-plane direction is unknown. As discussed above, temperature and temperature-gradient effects are intricately coupled to water management and PEFC performance.

The final transport parameter to be discussed, permeability, enjoys the distinction of being the most treated in the literature. As discussed in Sect. III.2.i, the range of absolute permeabilities is relatively wide in the literature, both from experiment and model fits. In terms of anisotropy, Gostick et al. (213) found the in-plane permeability to be up to two times greater than through-plane permeability, in general agreement with the literature. However, the spread in measurements is great enough that in-plane and through-plane permeability values could be chosen such that through-plane is greater than in-plane permeability (207). A counterintuitive consequence of incorporating anisotropies is that inhomogeneities in a medium increases the permeability (214). While the term "anisotropic system" has typically been used to describe an orthotropic system for which the in-plane values in the along-the-channel and perpendicular-to-channel directions have been the same (x and y in Fig. 4), Pharoah (215) and Williams et al. (216) suggest that anisotropy between these values impacts the extent of convective transport in flow under the rib. Pharoah found that convective flows that traverse serpentine flow channels become significant for through-plane permeabilities $>1 \times 10^{-8}$ cm^2. The work portends the need for a 3-D model to capture fully water movement and distribution. As currently published anisotropic models deal primarily with the channel-rib-catalyst-layer plane, subsequent discussion will be duly simplified to this plane. The link between permeabilities and water movement is clear. But again, the net result of anisotropies on water management and performance optimization is quite nebulous, especially when considered in conjunction with previously discussed transport phenomena.

Several methods of tackling the question of what occurs in an anisotropic system have been proposed. One approach is to introduce a resistance at the boundaries of the GDL to compensate for lower through-plane values. This approach has been carried out for solving the cathode-side electron profile and suggested to be of use for elucidating temperature profiles as well because both flow patterns originate from the CL interface and feed into the current collector (211). However, the approach is more or less empirical in terms of the resistance values to use. The success of this method may also be compromised due to the more complex and coupled movement of liquid and water vapor in the system.

As an initial step toward creating an anisotropic 2-D model, Pharoah et al. (206) studied separately the influence of thermal and electronic anisotropies on the current-density distribution in a cathode half-cell model. The results are shown in Fig. 14 for the current

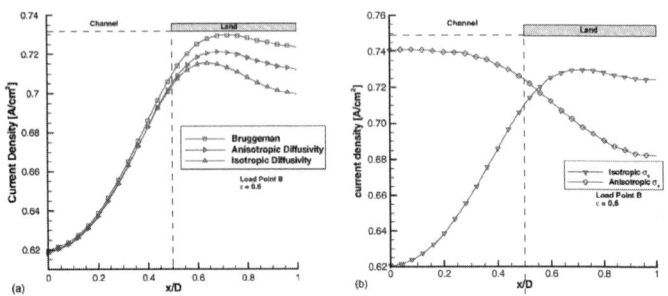

Figure 14. Current-density distribution at 0.65 V with saturated air at the cathode CL/GDL interface as a function of GDL in-plane dimensions using anisotropic (**a**) gas diffusivities or (**b**) electronic conductivity. (The figure is adapted from (206) with permission of Elsevier.)

density along the CL/GDL boundary. For Fig. 14a, the model tested three values for D_{eff}, the first found by applying the Bruggeman relation to in-plane permeability, the second by applying the in-plane TS expression, and the third by incorporating anisotropy by using both in- and through-plane values from TS theory. For all scenarios studied, the current density using the isotropic Bruggeman expression was greatest, followed by the anisotropic expression, and lastly the in-plane isotropic expression. The trend is explained by noting two points. First, the Bruggeman expression results in values greater than both values derived from TS theory (208). Second, the anisotropic case has higher current densities even though the through-plane permeability is less than its isotopic counterpart because nonuniformities in void networks lead to less resistance to movement (214). A general observation is that, while the magnitude of current density differs for each case, the general shape of the current-distribution curves remain the same as seen in Fig. 14a. The anisotropic and isotropic TS results remain similar under the channel but differ increasingly as one moves under the rib, where diffusion limitations become more severe. The maximum current density shifts toward the rib upon accounting for anisotropy, but the peaks align as the system handles larger loads since oxygen consumption becomes dominant over lateral diffusion for oxygen transport.

Figure 14b shows the results of the case when the diffusivity is kept isotropic but the electronic conductivity is either isotropic or anisotropic with a higher value in plane. From the figure, it is clear that not only does the anisotropic case have a higher current density, but it also follows the opposite trend from the isotropic case of having higher current density under the collector than under the channel. The current distribution becomes more strongly dependent on the oxygen distribution as electronic conductivity increases and facilitates movement of electrons toward the collector. Higher in-plane conductivity enables current to spread out more evenly over the rib, thus reducing primary-current effects and increasing efficient use of the collector. It is worth noting that in both the diffusivity and conductivity anisotropy studies, the overall polarization behavior did not change significantly until very high current densities. This shows that the overall polarization often masks local fluctuations and heterogeneities.

Pasaogullari et al. (207) input anisotropic values for diffusivity, electronic and thermal conductivity, and permeability simultaneously. Their results also demonstrate a much smaller effect on the global polarization scale than on the local distributions. It is worth noting that for their isotropic case, through-plane values were used everywhere, as opposed to using in-plane values, which is more typical in the literature. Diffusion values were taken from TS theory, electronic and thermal conductivity from experiments – with in-plane values being greater than through-plane values – and permeabilities taken from experiments but with through-plane permeability being an order of magnitude higher than in-plane permeability. Effects of anisotropies in electronic conductivity are similar to those obtained by Pharaoh et al. The temperature profiles mirror the potential profiles, which is not too surprising since they have similar boundary conditions and anisotropies.

Of more interest is examining the effect of collective anisotropies on water management (Fig. 15). Due to the additional coupling of transport phenomena and treatment of multiphase transport, one sees changes in the liquid pressure and saturation. The impact of the anisotropies leads to a lower overall temperature and smaller temperature gradients, which in turn with the anisotropic permeability causes higher observed levels of saturation and spreading of liquid water over the entirety of the GDL. As noted, the values used by Pasaogullari et al. for the isotropic case are smaller than the anisotropic one, and it would be of interest to see how different values (e.g., average between the two) changes the impact of the anisotropic properties. In any event, there is a need for further theoretical and experimental studies to determine accurately the effects and values of anisotropic transport properties in the GDL.

v. *Compression*

A key difference between the state of a GDL in and ex situ is the existence of a compressive force applied by the flow fields and PEFC assembly. PEFC components are held together with a compressive load to ensure contact between the layers, but the additional force also affects porosity, pore-size distribution, conductivities, and contact resistances (198, 217–224). Regions under the rib or land are expected to have decreased diffusivities and permeability in the in-plane direction and increased though-plane conductivity due to the vertical compaction of fibers (221, 223). Experimentally, most

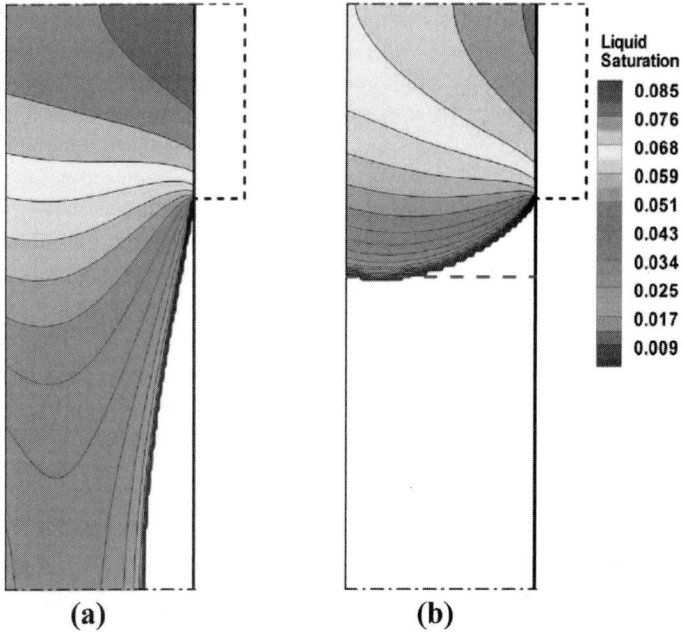

Figure 15. Liquid saturation profiles at 0.6 V with 80°C, fully humidified feeds for (**a**) anisotropic and (**b**) isotropic GDL properties. The *dashed lines* on the right correspond to the flow-field rib, and the *solid lines* correspond to the liquid front. (The figure is reproduced from reference (207) with permission of The Electrochemical Society.)

GDLs exhibit an ideal compression value that balances the benefit from decreasing contact resistance with the penalty of impeding gas and liquid flow (217, 218, 222, 223). The level of trade-off observed has been seen to be dependent on current density; the greater the current density, the more severe the performance decrease with increasing pressure (222).

Another interesting consequence of GDL compression is the changing of ratios between through-plane and in-plane transport parameters and subsequent augmentation or diminishment of parameter anisotropies (205). Several groups found that electrical conductivity increased primarily in the though-plane direction and similar results are postulated for thermal conductivity (218, 223). Because through-plane conductivities are assumed to be smaller than their in-plane counterparts, increases in conductivities due to compression may reduce anisotropies under ribs (206, 210, 211). The extent to which compression alters GDL properties is highly dependent upon the GDL material (218, 222). For example, carbon cloth is mechanically less rigid and therefore suffers greater decreases in porosity than does carbon paper (225). Decreased rigidity of carbon cloth GDLs also implies that they cannot spread pressure from the rib toward the region under the channel as well as carbon paper can, thus leading to distinct flow properties in the medium.

With respect to modeling the compression, cues can be taken from experimental results. Based on their experiments, Ihonen et al. (218) and Ge et al. (222) conclude that thermal and electrical contact resistances are the main source of impedance and therefore should be incorporated into PEFC models. Although accounting for clamping effects by tuning contact resistances allows for their quick incorporation into fuel-cell models (198, 221), adjusted contact resistances do not account for deformations nor changes to the internal properties of the GDL.

Wu and colleagues have been one of the few groups to treat both changes in contact resistance and deformation effects (219, 220). Based on the observation that the GDL has the lowest compressive modulus of the PEFC components, all deformation was assumed to occur in the GDL (219, 226). GDL porosity was adjusted according to deformation theory (227)

$$\varepsilon = \frac{\varepsilon_0 - 1 + e^{\varepsilon_V}}{e^{\varepsilon_V}}, \tag{63}$$

where ε_0 is the uncompressed porosity and ε_V is the bulk strain. In their models, Wu and coworkers examined flooding and performance on the cathode side of the cell in both crossflow and flow-under-the-rib situations. Figure 16 demonstrates the compression of the GDL and its expansion into the gas channel. In terms of water management, compression effects result in higher liquid saturations and a penetration of the high-saturation front from under the channel towards the rib. The flow-under-the-rib simulations of Wu and coworkers demonstrate a lower saturation when compression is considered. This is opposite to that seen in Fig. 16, and begs the question as to the dominant water-movement pathway and the correct modeling geometry. However, even though the saturation is lower upon compression, the current density is lower everywhere as well,

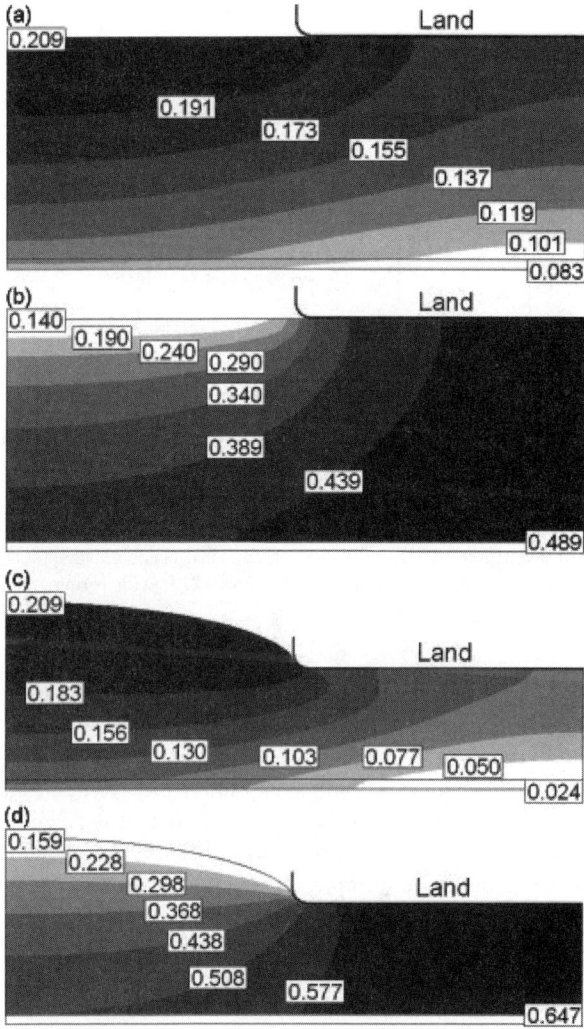

Figure 16. Distributions at 0.3 V of (**a**) oxygen mole fraction and (**b**) liquid saturation with 0% compression ratio and (**c**) oxygen mole fraction and (**d**) liquid saturation with 50% compression ratio. (The figure is reproduced from (220) with permission of Elsevier.)

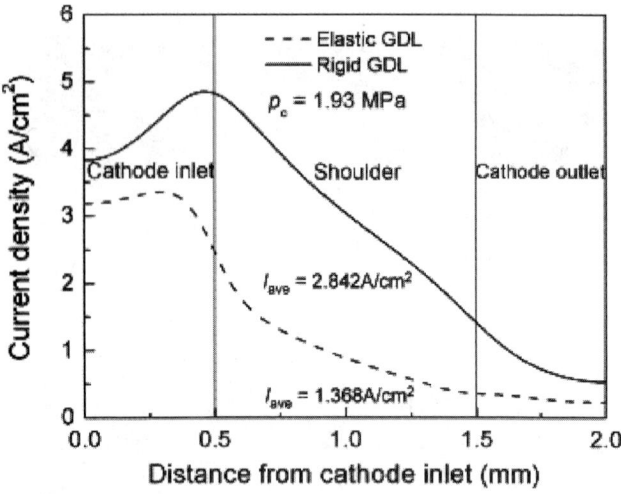

Figure 17. Current-density distribution at the cathode CL along the channel as a function of GDL elasticity. The nominal clamping compression is 1.93 MPa. (The figure is reproduced from (219) with permission of Elsevier.)

as shown in Fig. 17. The reason is because the changes in the effective oxygen diffusion coefficient due to compression effects on porosity, etc. outweigh the decreased flooding. These results affirm that PEFC performance is dominated primarily by oxygen-transport limitations rather than just by the amount of liquid water. This observation is in agreement with experimental results from Ihonen et al. wherein the same value for optimum clamping pressure is obtained for differing humidity conditions (218).

An additional complexity when modeling water content stems from temperature variations that arise due to compression. In their simulation, Hottinen and Himanen (224) observe higher temperatures under the rib due to a decreased thermal contact resistance with the rib upon compression. Higher temperatures may partially offset the increase saturation levels under the channel and thereby impede the onset of flooding. Alternatively, the preferential formation of water pathways under regions of compression – cause by formation of hydrophilic pathways due to breaking of fibers and PTFE coating – may dominate saturation profiles (228). To understand fully the underlying mechanisms and physics of how compression changes PEFC performance, there is a need for improved imaging techniques (see Sect I.2) and/or microscopic, mechanical models to account for fluid flow through packed, and possibly broken, fibers.

IV. DESIGN CONSIDERATIONS

In most operating cells, even at steady state the same materials must operate over a range of conditions due to nonuniformities in a variety of critical parameters including temperature, current density, reactant concentration, and relative humidity. These nonuniformities arise primarily due to system limitations. For example, isothermal conditions are not achievable in-plane in most applications because a finite coolant flow rate must be used. Similarly, oxygen concentrations change substantially in-plane because low air flow rates are desirable from the perspective of parasitic power and cell humidification. Of course, these two variables are also coupled to one another along with the other parameters listed above.

Such nonuniformities play a critical role in cell power output and durability, largely due to their impact on cell hydration. If a region of the cell is operated with a low relative humidity, ionic conductivity in that region is reduced, and the local current density goes down. Such operation has also been shown to lead to membrane failure (229). On the other hand, if a region of the cell operates with too much liquid water present, mass transport will be impeded and the local current density will drop. Because it is likely that at least one of these undesirable hydration states will exist somewhere on the cell planform, it is important to be able to predict their effects accurately. In this section, therefore, modeling approaches to low-relative-humidity operation are covered as are modeling techniques for the presence of liquid water in the flow field. In addition, models for common strategies that have been developed for dealing with the potential for nonoptimal cell hydration are reviewed.

1. Low-Relative-Humidity Operation

Humidification of reactant streams onboard a power plant can be onerous due to the additional mass, complexity, and cost required. This is especially true on the cathode side where a large amount of inert gas must be humidified along with the oxygen. For these reasons, developers often try minimize or eliminate humidification. As a result, the cell is susceptible to dryout, particularly at the reactant inlets. Predicting the location and magnitude of cell dehydration, as well as the impact on performance, is the subject of this section. As noted, many multidimensional models now allow for the study of unsaturated feeds; example examinations since the last reviews (2,3) are discussed below.

Due to the nonuniformities described above, it is difficult to use either a 1-D through-plane model or a channel/rib 2-D model by itself to predict the effects of low relative humidity operation on overall cell performance. For this reason $1 + 1$-D or $1 + 2$-D models are used to determine the current-density distribution down the channel, assuming low-relative-humidity gas streams at the inlet.

An example of a $1 + 2$-D model of this type is presented by Guvelioglu and Stenger (230, 231). In this case a linear channel/rib combination is divided into control volumes along the length of the channel. For the through-plane direction the model is a 2-D, isothermal, single-phase CFD simulation which takes inlet molar flow rates, cell temperature, and cell voltage, and computes the local cell current density and average reactant concentrations. These concentrations are then used as inlet conditions for the next control volume down the channel. Note that mass transport in the porous media in the direction parallel to the gas channels is neglected. The step size along the channel is set such that the molar flow rates for each component change by <2%.

The advantage of the $1 + 2$-D model is its relative simplicity, which translates into reasonably run time. Detailed velocity profiles are not calculated in the channels, and the gas pressure is assumed constant within the gas channels (although through-plane gradients are calculated). This assumption breaks down for cells that have high pressure drop along the gas channel, as is often the case when a serpentine flow geometry is used. Pressure variations affect both relative humidity and velocity profiles significantly (232).

One way to account for this pressure drop is to approximate it from one control volume to the next based on gas flow rates and the channel geometry. Another is to add a full third dimension. The latter approach is taken by Meng and Wang (233). Like the $1 + 2$-D model, their approach is to use the assumptions of single-phase flow and an isothermal cell, but the domain considered includes the full cell sandwich as well as multi-pass anode and cathode channels. 3-D velocity profiles are computed within the channels. A critical disadvantage to the full 3-D approach is in processing time; comparing the two cases results in roughly an order of magnitude longer run time for the 3-D case. This is a considerable price to pay considering that the isothermal assumption adds a significant level of uncertainty to the results of either model.

Thermal management and water management are intricately coupled, and one might expect that, for low-relative-humidity-feeds, thermal effects are significant due to the large water phase change. However, Weber and Newman (52) calculate that the cell is typically more isothermal in-plane than nonisothermal for low relative humidity in the gas channels due to evaporation of water in the gas channels leading to a smaller peak temperature. This analysis is the same as taking the overall heat generation (see equation (7)) and using the enthalpy potential for the lower heating value instead of the higher one (i.e., water is produced as a vapor).

A key boundary condition that is generally used is that the temperature on the back side of each of the bipolar plates is constant. While the temperatures of the cell materials and gases can vary in three dimensions, they are anchored to isothermal boundary planes. For models with relatively small domains, such as that by Djilali and coworkers (234) ($2\,cm^2$) or Ju et al. (201, 235, 236) ($50\,cm^2$), this simplification is probably justified. On the other hand, most commercial cells are considerably larger. As a result, when full-size cells are modeled using an isothermal backplane assumption, as in Van Zee and coworkers ($480\,cm^2$) (237), the results are questionable because in the real system the coolant temperature typically varies by about $10°C$ from inlet to outlet (238, 239).

To address this problem, Wang and Wang (239) have recently extended the model presented in Ju et al. to represent a full-size cell ($200\,cm^2$), including coolant flow channels. By doing so only the inlet conditions for the coolant (flow rate and temperature) need to be specified, and the "backplane" temperature is calculated as a function of position. Results from this type of analysis are shown in Fig. 18. Temperature profiles for the cell at the membrane-cathode interface are shown in Fig. 18a for a cell run at $1.0\,A\,cm^{-2}$ with coolant flow representative of normal operating conditions. The difference between the minimum and the maximum temperature is over $10°C$, illustrating the importance of a nonisothermal backplane

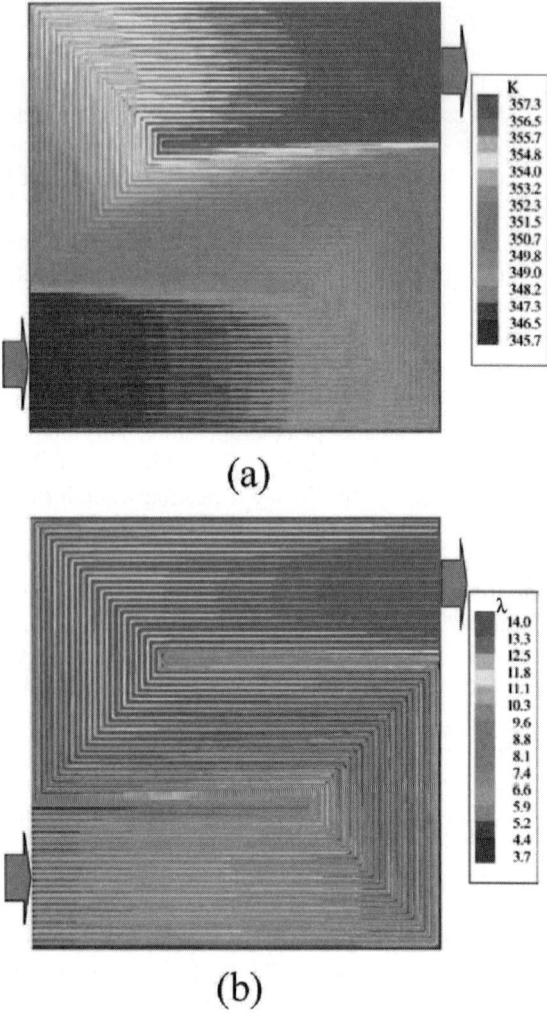

Figure 18. Surface contour plots at membrane-cathode interface for (**a**) temperature and (**b**) water content. (The figure is reproduced from (239) with permission of Elsevier.)

when modeling full-size cells. The corresponding membrane water content at the same location is shown in Fig. 18b. Of course, the drawback to this approach is complexity. The authors report that 23.5 million gridpoint calculations are required, resulting in a 20 h run time on 32 parallel computing nodes.

2. Liquid Water in Gas Channels

Water in the vapor phase can condense within the reactant channels, just as it can in the porous media. If the condensed water agglomerates it can cause significant changes in pressure drop within the cell (possibly forcing flow to adjacent cells (channels) within a stack (cell)) and can also affect mass transfer of the reactants from the channel to the electrodes. Reactant starvation of this type can cause cell power output to drop or fluctuate and may lead to corrosion under some circumstances (48, 240).

To date, simulations of liquid water within the flow fields have been simplified. This is primarily because of the complicated nature of the physical phenomenon, which is unsteady two-phase flow coupled with the complex physics of PEFC operation, and because most modeling activities have been devoted to the PEFC sandwich. At the same time, a lack of reliable experimental observations has made it difficult to understand the level of detail that is sufficient to describe the physical system. Advances in in situ imaging, however, are rapidly negating the latter point, as discussed in Sect. I.1 and shown in Fig. 19.

As Fig. 19 demonstrates, there are several different liquid-water transport mechanisms that can occur in the flow channels. The dominant one depends on the operating conditions (primarily flowrate, temperature, current density or potential) and surface and material properties of the flow-field plate and the GDL. These mechanisms can be classified as mist or fog flow coupled with GDL water-droplet expulsion and detachment (Fig. 19a), (annular) film and corner flow along the flow-field plate (Fig. 19b), and slug flow, where blockage of the channel results (Fig. 19c). The mechanisms portrayed in Fig. 19 can be seen as a progression, where the blockage and slug flow occurs as the film and droplets agglomerate due to liquid-water buildup. In this section, first the movement of liquid water in the flow field is discussed with emphasis on recent models, followed by the coupling of water droplets and the GDL/gas-channel interface.

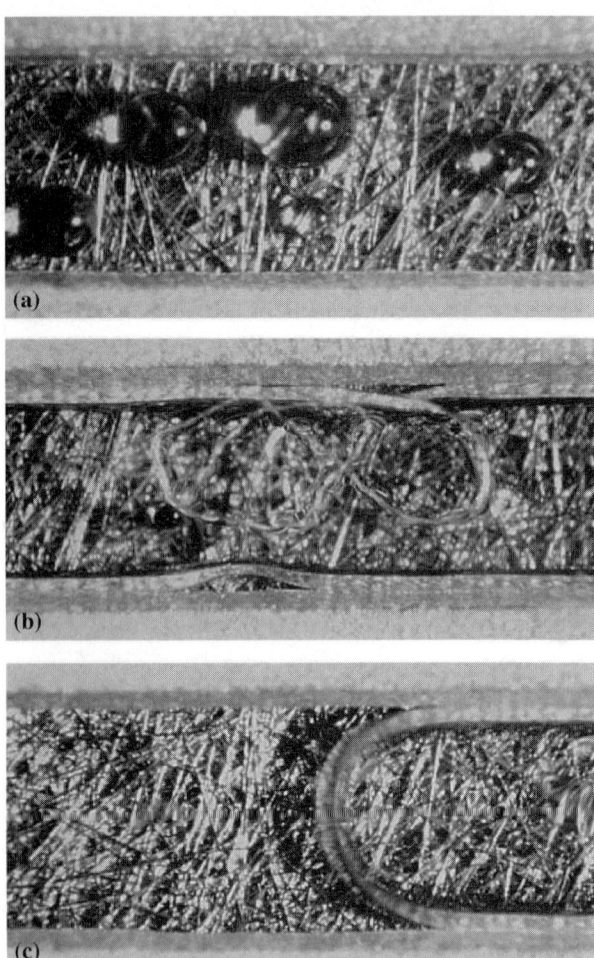

Figure 19. Transparent-cell photographs of liquid-water distributions in an operating PEFC showing (**a**) droplet emergence and flow and (**b**) film and (**c**) slug flow. (The figure is reproduced from (19) with permission of The Electrochemical Society.)

i. Gas-channel analyses

The most common approach taken in multiphase models is to treat the liquid water in the channel as a species that is dispersed in the gas stream (i.e., mist or fog flow) or as a thin liquid film. Either way, its presence is accounted for using the continuity equation, but its volume is considered negligible and does not affect gas transport (i.e., the droplets are ignored) (241, 242). If the model is nonisothermal, the heat of vaporization is included as a source term in the energy equation (52, 241).

One use for these types of models is as a means of predicting at what point along the channel condensation will begin or end as a function of operating conditions and geometry, in other words, where the wet-to-dry and dry-to-wet transitions occur. An example is the study of Lee and Chu (243), who use a 3-D, CFD, straight-channel isothermal model to show the effects of cathode relative humidity on the location of the interface between the water in the vapor and liquid phases. Figure 20 contains the results of one such analysis for a cell operating at a fixed voltage of 0.7 V at 70° C. The anode relative humidity is fixed at 100% at the cell temperature. The lines on the plot correspond to the predicted location of condensation at different cathode inlet relative humidity values. A similar analysis was completed in 1-D by Yi et al (244). In both of these cases, the effect the liquid water has once it is formed in the channel is not rigorously accounted for, but the probable location for liquid-water formation is estimated.

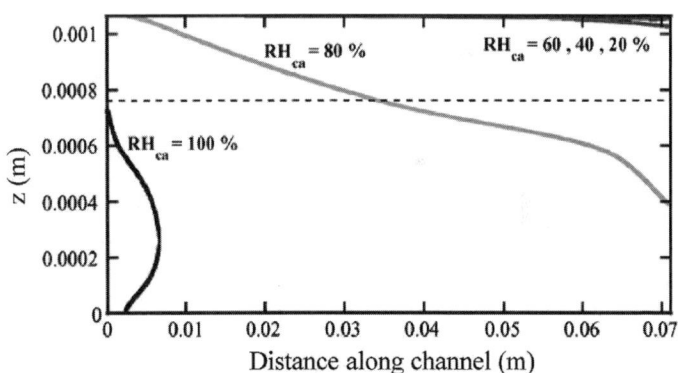

Figure 20. Effects of cathode inlet relative humidity on the vapor–liquid interface location in the channel (above the *dashed line*) and GDL. (The figure is reproduced from (243) with permission of Elsevier.)

While neglecting transport effects due to liquid water in the channel may be appropriate as a first approximation (or under some operating conditions), the imaging studies clearly indicate the presence of agglomerated liquid water in the channels. (19–22, 25, 245) It is highly unlikely that these large droplets have no effect on cell performance. The challenge is that describing the behavior of these droplets rigorously constitutes an unsteady two-phase flow problem that is highly coupled with the operation of the PEFC. While attempts to simulate this phenomenon are discussed in the next section, with regard to the flow field specifically, one of two approaches is taken. Either electrochemical models are used and the presence of water droplets is neglected (as described above), or CFD simulations of droplets in channels are used, and the presence of the rest of the PEFC is largely neglected (i.e., it is just an interface).

Examples of the latter approach are presented by Zhan et al. (246), Jiao et al. (247), and Quan and Lai (248). In all three cases, 3-D CFD modeling is used assuming isothermal conditions and saturated inlet gases (i.e., no phase change), no gas-phase transport through the channel or GDL walls, and no PEFC reactions. Zhan et al. examine the movement of the droplets as a function of the Reynolds, Capillary, and Weber numbers in the channel. They find that straight channels are better than serpentine in discharging water, inertial forces are dominant for gas velocities higher than $4\,\mathrm{m\,s^{-1}}$, and for lower gas velocities, the wettability of the flow-field plate and the GDL determine the ease of water movement; the more hydrophobic the GDL and hydrophilic the plate the better.

Jiao et al. use their model to predict the behavior of liquid water as it travels through a stack of three cells with serpentine channels. The cells are connected by straight inlet and outlet manifolds. The distribution of liquid water within the system is specified as an initial condition. Gas of constant inlet velocity then flows through the system, and the redistribution of water with time is simulated. Examples of initial conditions include suspended droplets in the inlet manifold or a constant-thickness film on a channel wall. Figure 21 shows results based on the initial condition of a 0.2 mm liquid water film distributed along the leeward side of each gas channel. Each subplot shows the predicted location of the water droplets (shaded) at a different time step. After 0.075 s most water has been purged from the system since there is no source for liquid water within the system.

Modeling Water Management in Polymer-Electrolyte Fuel Cells

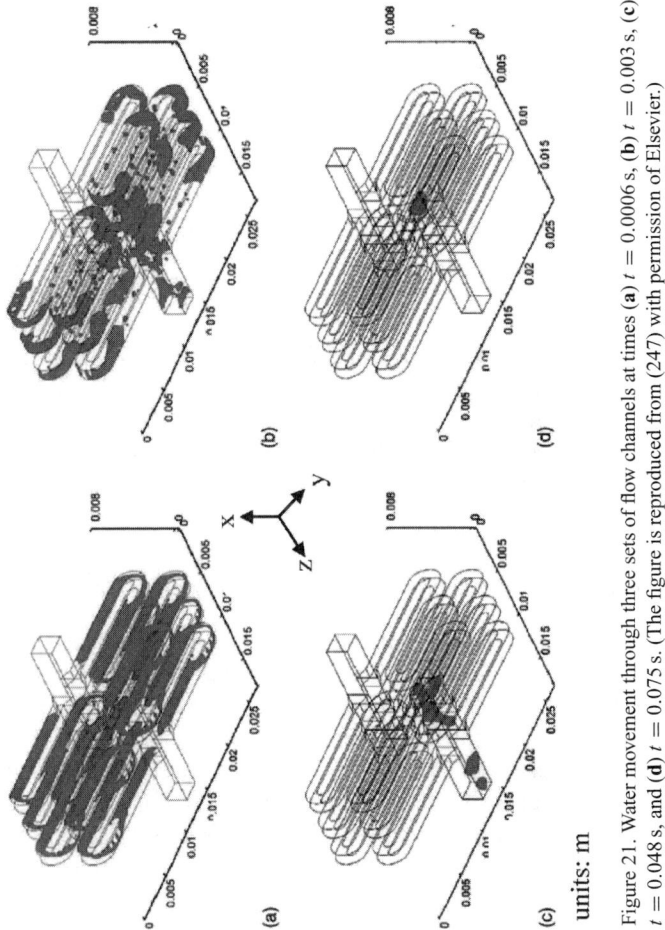

Figure 21. Water movement through three sets of flow channels at times (**a**) $t = 0.0006$ s, (**b**) $t = 0.003$ s, (**c**) $t = 0.048$ s, and (**d**) $t = 0.075$ s. (The figure is reproduced from (247) with permission of Elsevier.)

Quan and Lai consider a single U-shaped flow channel. Rather than specifying an initial distribution of water, a constant flux of water through the face representing the GDL is applied as a boundary condition. A constant-inlet-velocity gas flow is applied, and the simulation runs until the water film thickness arrives at a steady state. A comparison is made between a channel with square corners and one with a radius and no corners. The effect of the interior surface contact angle is also investigated.

The value of these types of models is that they provide insight into the dynamics of liquid water as a function of local features and surface properties. For example, Jaio's model predicts that liquid water tends to collect in films in the turns of the serpentine channels and that water leaves these films primarily in the form of small droplets. Quan and Lai's model predicts that whether the liquid water accumulates in a relatively uniform film on the GDL surface or whether it separates and moves along the corners formed by the intersection of the GDL and channel is dependent upon the contact angle of the surfaces. Using these models to predict actual liquid water distribution within a full-size cell, however, cannot be done accurately because the details of energy, momentum, and mass transport into and out of the gas channel through the PEFC sandwich is neglected.

ii. Droplet models and gas-diffusion-layer/gas-channel interface

From the previous section, one can see the importance of accounting for liquid-water flow in the gas channel rigorously, and the need to connect such models of water movement in the gas channel with that of the PEFC sandwich. As noted, in most models this is accomplished by simultaneous mass and energy balances in the channel along with possible pressure drop and gas-flow equations. The inherent assumption is that of mist flow. Although simplistic, this treatment allows for pseudo dimensional or $(1 + 1\text{-D}$ or $1 + 2\text{-D})$ models where the 1 is the along-the-channel balances (52).

The key issue is how to treat the GDL/gas-channel interface. This boundary condition is extremely important since it determines the water management and saturations inside the PEFC sandwich to a significant amount. This boundary condition is typically a specification of flux or concentration for water vapor and one of saturation or liquid pressure for liquid water (see Sect. II.6). However, if one is using a two-phase model without a residual effective permeability, then setting saturation equal to zero could be problematic in terms of convergence since this condition enforces the fact that all water must leave the GDL in the vapor phase since the effective permeability will go to zero. We believe that it is better to set the liquid pressure or capillary pressure, and ideally this pressure should be associated with the formation and existence of droplets on the GDL surface. However, such droplets emerge from defined locations, and in the absence of a microstructural GDL model, one is required to assume some kind of average value. Meng and Wang (249) do this in their model by assuming a film on the GDL surface that essentially acts as an interfacial liquid-pressure increase, much like a contact resistance. They demonstrate that higher saturations and lower performance are obtained with such a method, although the value of the film thickness requires empirical fitting.

To understand water droplet behavior, emergence, and detachment, and to provide more physical basis for modeling the interface, detailed droplet-specific studies have been accomplished. (19, 245, 250, 251) These studies focus only on single droplets and are not necessarily valid next to a rib, where the hydrophilic-plate interaction can result in annular and corner flow along the plate (see Fig. 19). The four models take the same approach of a force balance for the droplet in the gas channel that is attached to the GDL

$$F_\gamma + F_D = 0, \qquad (64)$$

where F_γ is the surface-tension or adhesion force and F_D is the drag force on the droplet. When the drag force is greater than or equal to the surface-tension force, the droplet becomes unstable and detaches from the GDL surface. For these models, fully developed laminar flow in the channel is assumed. For the geometric analysis, it is known that the droplet exhibits a contact-angle hysteresis in that the advancing angle is typically greater than the receding one. In essence, this gives the drop a deformation from a perfect hemisphere or sphere on the GDL surface, and thus the contact-angle hysteresis can be used as an interaction parameter of the droplet with the GDL surface, as discussed below.

All of the models use a similar form for the surface-tension force, which arises from integrating the force around the droplet (245)

$$F_\gamma = \frac{\pi}{2}\gamma c \left[\frac{\sin(\Delta\theta - \theta_A) - \sin(\theta_A)}{\Delta\theta - \pi} + \frac{\sin(\Delta\theta - \theta_A) - \sin(\theta_A)}{\Delta\theta + \pi} \right], \tag{65}$$

where c is the chord length of the contact area between the droplet and GDL surface, γ is the surface tension, $\Delta\theta$ is the contact angle hysteresis, $\Delta\theta = \theta_A - \theta_R$, and θ_A and θ_R are the advancing and receding contact angles, respectively. The models vary slightly in their geometric analysis and trigonometric identities. (He et al. (251) assume that the chord length is the same as the mean pore size in the GDL, Zhang et al. (19) assume symmetric deviations from the static contact angle, and Chen et al. (250) use $r_d \sin(\theta_A)$ for the wetted area, where r_d is the droplet radius).

For the drag force, He et al. and Zhang et al. both use flow past a sphere

$$F_D = \frac{1}{2} K c_D \rho_G \langle v \rangle^2 A_d, \tag{66}$$

where $\langle v \rangle$ is the average velocity in the rectangular channel for laminar flow (50), A_d is the projected normal area of the droplet to the flow, which is determined from geometric analysis, c_D is the coefficient of drag,

$$F_D = \frac{24}{Re}\left(1 + 0.1925\ Re^{0.63}\right), \tag{67}$$

where Re is the Reynolds number

$$Re = \frac{2\rho_G r_d \langle v \rangle}{\mu_G}. \tag{68}$$

In equation (66), Zhang et al. use K as a fitting function that accounts for the fact that the droplet is not a perfect sphere and also for the assumption of only creeping flow (used for the determination of c_D). He et al. integrate the drag force (without k) over the actual droplet geometry.

The other two droplet models, those of Kumbur et al. and Chen et al., adopt a different strategy for the drag force. They assume that the drag force is made up of a pressure force that acts on the droplet itself and a shear force that acts on the top of the droplet. The expression differs slightly due to the assumed geometry, but are of the form (245)

$$F_D = \frac{12\mu_G \langle v \rangle B h_d^2}{\left(B - \frac{h_d}{2}\right)^2 (1 - \cos(\theta_A))^2} \left(1 + \frac{2B}{\left(B - \frac{h_d}{2}\right)}\right), \quad (69)$$

where h_d is the droplet height and B is the half-width of the channel.

In the above expressions, most of the parameters are known since they are functions of the channel geometry and operating conditions (temperature, flowrate, gas composition, etc.). The ones that are not known can be related to the drop morphology on the GDL surface. In turn, these can be related to the contact-angle hysteresis (assuming a deformed hemispherical drop with the given advancing and receding contact angles). There are various ways to examine such a value. Chen et al. use the contact-angle hysteresis as the independent variable, and then match stability predictions with experimental findings. He et al. assume that the drop behaves with the same form of hysteresis as on Teflon, and they use that function along with the observed static contact angle of water on the GDL. Zhang et al. essentially fold the hysteresis into K, to get an expression for the droplet diameter in terms of the unknown constant, which is then fit to data. Finally, Kumbur et al. perform a linear regression from data for the contact-angle hysteresis, where it is assumed to depend on the channel Reynolds number, and the droplet height and wetted radius or chord. All of the models rely on empirical functions and values for the a priori unknown contact-angle hysteresis, with the last one being the most attractive since it is the most physically reasonable and could be used for model predictions.

The use of the force balance and droplet models provides a means to determine droplet-stability diagrams. Two such diagrams are shown in Fig. 22. In the figure, the independent variable is chosen to be the aspect ratio of the drop, where larger aspect ratios will correspond to more unstable drops, meaning that a lower channel Reynolds number is required to cause drop detachment. The figure clearly shows that spreading of the drop makes it more stable and thus will cause larger mass-transfer limitations for the reactant gases. The figure also displays the impact of hydrophobicity, with more hydrophobic surfaces lowering the drop stability. From Fig. 22, it can be seen that the force model is slightly overpredicting the stability of the drops. This behavior was investigated by Chen et al. (245), who compared the simple force model with a complex 2-D CFD one in which all relevant interactions are accounted. They

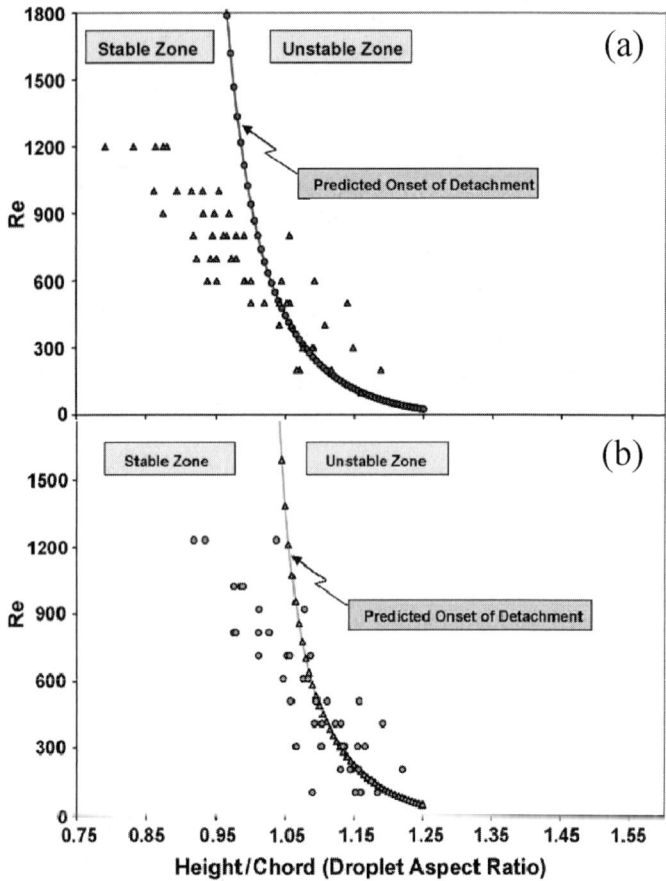

Figure 22. Critical Reynolds number as a function of droplet aspect ratio for a droplet with a GDL contact length and Teflon amount of (**a**) 0.23 cm and 5% and (**b**) 0.19 cm and 20%, respectively. (The figure is reproduced from (245) with permission of Elsevier.)

determined that the reason why the force models overpredict stability is due to the neglect of the inertial effects on the droplet by the flow (i.e., it is not purely creeping flow).

While the above droplet models could be used in full-cell simulations (with some assumption or function for the contact-angle hysteresis), the only study so far to attempt this has been that of

He et al (251). As noted, they use a form of equation (66) for the drag coefficient and assume that the particles are the same diameters as the mean GDL pore size and the contact-angle hysteresis is similar to that on Teflon with the static values fit to Zhang et al (19). Their droplet model is incorporated into the mass-conservation equations in a 2-D, along-the-channel model. Their results show that low surface tension and hydrophobic surfaces are better overall for water removal. In all, the GDL/gas-channel boundary condition is extremely important and complex. The droplet models are a start, but they need to be coupled with the effects of the ribs and flow-field plates as discussed in Sect. IV.2.i, as well as detailed PEFC sandwich models.

3. Water-Management Strategies

Various cell-design strategies have been developed to enhance water management. Each of these has in common the goal of providing optimal cell hydration to as large a fraction of the cell area as possible. Optimal cell hydration occurs when the membrane conductivity is maximized with concurrent reactant-transport-loss minimization. Several of the most common strategies are discussed below, with a focus on the models that have been developed to predict their effects.

i. Gas-flow direction

One method for optimizing cell hydration is to manipulate the orientation of the flow of the reactants and coolant relative to one another. A simple example would be to coflow the cathode gas stream and the coolant. By doing so, the amount of cathode inlet dryout is likely to be reduced relative to a counterflow arrangement due to the lower local cell temperature (238, 239). Of course, at the stack level this approach has the disadvantage that the cathode air leaves the cell with a higher water content, making water balance more of a challenge, and so a trade-off must be made between the durability and performance of the cell and the size and complexity of the system.

Numerous models for cell hydration as a function of fluid-flow orientation have been developed, where it should be noted that counterflow is much harder to simulate than coflow since it requires an extra external iteration loop (i.e., it is a boundary-value instead of an initial-value problem). For this reason, most simulations, especially the 3-D ones, assume coflow. Most of the low-orientation studies focus on reactant flows, assuming either a constant coolant temperature or a predefined temperature gradient. An example is the model by Wilkinson and St-Pierre (252), which is 1-D (down the channel), assumes a uniform current distribution, a linear temperature gradient, and does not account for reactant pressure drop or net water flow through the PEFC sandwich. These simplifications enable a rapid first approximation of reactant relative humidity and concentration as a function of position along the channel for different reactant flow orientations relative to one another as well as relative to the coolant-flow direction. This approach is extended to $1+1$-D by Berg et al. (253), who retain the assumption of a prescribed temperature gradient down the channel while improving accuracy by calculating water transport through the membrane and membrane conductivity. These models demonstrate that counterflow results in more uniform hydration profiles than coflow with low-relative-humidity feeds. Finally, due to manifolding issues, the most appropriate type of reactant-gas flow is probably crossflow. This situation was looked at by Weber and Newman (78), through the use of a 1-D model that is run in a 2-D array of node points, where each point was connected to each other through simultaneous mass and energy balances. They found that crossflow increases the average humidity and provides around a 30% increase in current density for 25% relative-humidity feeds than the coflow case.

CFD models have also been used to study relative flow orientation effects, although most have been isothermal, which limits their applicability towards full-size cells (254–256). While use of a nonisothermal model can improve the simulation accuracy, the complexity of fully-nonisothermal, 3-D, CFD models limits the usefulness of these models for rapid-design iteration (239). A simplified approach is presented by Büchi and coworkers (107,238). Here, a detailed nonisothermal 1-D through-plane model, which includes heat transfer to coolant channels as well as multiphase flow effects, is coupled via a $2+1$-D network. This approach utilizes a plug flow assumption in the channels; detailed velocity profiles are not computed. However, both the through-plane and in-plane temperature gradients are computed for full-size cells with a fraction of the complexity and computational cost of a full CFD model. Figure 23 shows a comparison of the current density profile predicted and measured in a $200 \, \text{cm}^2$ cell as a function of reactant-flow orienta-

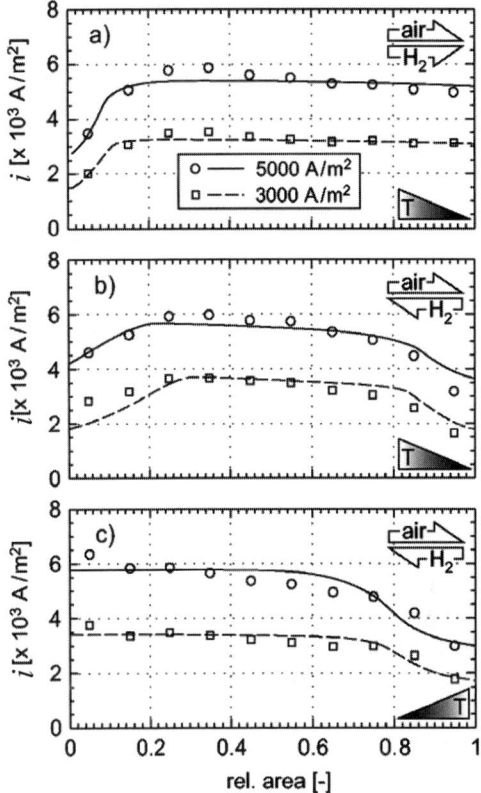

Figure 23. Comparison of measured (*symbols*) and calculated (*lines*) current densities for (**a**) coflow and (**b** and **c**) counterflow cells operated with dry hydrogen and humid air at $0.3\,\text{A}^{-2}$ and $0.5\,\text{A}\,\text{cm}^{-2}$. The temperature drops along the air path from 75 to 65°C in (**a**) and (**b**) while rising along the air path in (**c**). (The figure is reproduced from (238) with permission of The Electrochemical Society.)

tion. The triangular symbol containing a "T" in each plot represents the temperature gradient imposed on the cell by the coolant. Comparing plots "a" and "b" to plots "b" and "c" shows that the effect of the coolant temperature gradient is at least as important as that of reactant-flow orientation.

ii. Interdigitated flow fields

One strategy that is used in an attempt to reduce mass-transport limitations due to water buildup is that of interdigitated flow fields (IDFF) (257–259). In this design, the reactant flow channels are not continuous, and there is flow through the GDL from one channel to the other. The intent is to force reactant transport to the electrode by convection rather than diffusion and to use momentum transfer to reduce liquid water accumulation in the GDLs. Cells with IDFF often do show higher performance at higher current densities (i.e., in the mass-transfer-limited regime). However, this gain comes at the expense of higher reactant pressure drop. As a result, either the parasitic power required for driving the flow increases or the flow channel depth must increase. Whether a net increase in system power density can be expected, therefore, depends on the amount of performance gained per square centimeter relative to the amount of additional pressure drop.

A number of models have been developed to simulate the effect of IDFFs. These have been used in two ways. First, some have been developed to investigate whether this approach truly lowers mass-transport limitations in the manner described above. Second, some are intended as design tools that can be used in the performance-versus-pressure-drop tradeoff study. Before examining IDFF models, it should be mentioned that in-plane convection through the GDL may occur between adjacent channels in cells where an IDFF is not used. This is most likely to occur between adjacent legs of a serpentine flow channel due to the high-pressure drop per unit length that this geometry normally imposes, especially if there is blockage of the bend due to water droplets. Models have been developed to study the relative importance of convection and diffusion as a function of channel geometry (such as rib-to-channel width ratio) and GDL properties. These range in complexity from the simplified but fully-analytical approach presented by Advani and coworkers (260) to the 3-D CFD numerical simulation presented by Park and Li (261). Both approaches show that cross-leakage is a strong function of GDL permeability; for typical values of permeability (ca. 10^{-12} m^2) they predict that the fraction of "cross-leakage" through the GDL will generally be <5% (assuming no blockage of the gas channel).

When IDFF is used on all or part of the cell, 100% of the flow will be through the GDL where the IDFF channels end. Models describing this phenomenon can be divided into two categories. The first is full CFD models (262–265) in which all terms in the Navier–Stokes equations are accounted for and the full cell sandwich is simulated. Models in the second category (258, 259, 266–271) simplify the momentum equation to Darcy's law and restrict the simulated domain to only part of the full cell sandwich, usually the cathode GDL. Some of the earliest models of IDFFs, those of Nguyen and coworkers (258, 259) and Kazim et al. (272) demonstrated the importance of having more gas channels with smaller widths, among other things. An early full 3-D approach for IDFFs was developed by Wang and Liu (262). In this case the full cell sandwich is considered in with coflow in straight reactant channels. For a given side of the cell, the exit of the channel that the reactant gas enters is impermeable (blocked), while the entrance of the adjacent cell is impermeable and the exit is permeable (open). As a result, the gas must flow through the GDL. The model is nonisothermal, although the condensation of water vapor is not considered in the energy equation. Yan et al. (263) extend this approach to include multiple sets of IDFF channels and to compare different $x - y$ plane channel flow patterns (e.g., the Z-type flow pattern). The planform considered, however, is still quite small ($5\,\text{cm}^2$).

A feature of the CFD models is that they estimate both the pressure drop associated with IDFF as well as the performance impact with one set of equations once the geometry is defined. The disadvantage is computational complexity and run time. In the second category of models, simplifications are made in order to reduce the computational load. An example is presented by Yamada et al. (266), where a 2-D channel/rib model is used, considering the GDL only. In addition, Darcy's law, instead of the full Navier–Stokes equation is used to describe the gas flow. The model is isothermal, but, unlike any of the CFD models, the presence of liquid water is explicitly accounted for. Water is allowed to condense in locations where the activity of water is greater than unity, and local liquid water saturation is calculated. Saturation, in turn, affects both gas permeability and gas diffusion coefficients. A similar isothermal 2-D model is presented by Zou et al. (267) although a single-phase assumption is used. Unlike all of the other models reviewed in this section, however, this is a transient model. While these models can calculate cell performance as a function of the specific IDFF geometry, they do not calculate the overall pressure drop (that of the channels plus the substrate), which is also important.

To include the pressure drop, one approach is to calculate performance using a simplified model similar to those described above and to predict pressure drop separately. An example is presented by Arato and Costa (268) (performance model) and Arato and Costa (269) (pressure-drop model). In the pressure-drop model, the system is simplified to two dimensions, x and y. Two adjacent IDFF channels are defined such that the long dimensions align with the x direction. The migration velocity from channel 1 to channel 2 is defined by the variable v, which is a function of x only. The gas pressures in each channel, p_1 and p_2, are also functions of x only. The migration velocity is then given by a simplified version of Darcy's law,

$$v = \frac{f_1 k (p_1 - p_2)}{\mu d}, \tag{70}$$

where f_1 is a shape factor. The gas-phase pressure and velocity for channel j is given by

$$\frac{dp_j}{dz} = -\frac{f_2 32 u_j \mu}{(2B)^2}, \tag{71}$$

$$\frac{du_j}{dz} = -\frac{f_3 v h}{(2B)^2}, \tag{72}$$

where f_2 and f_3 are shape factors, h is the thickness of the GDL, and b is the width of the gas channel. By solving the above three equations, the gas-phase velocity, migration velocity, and pressure as a function of channel position are obtained.

An alternate approach for the pressure-drop calculation is presented by Inoue et al (270). A detailed 2-D channel/rib model of the GDL only is used to calculate oxygen concentration and cell performance based on Darcy flow. However, this model is then integrated with a 2-D in-plane thermal model and a plug-flow, 1-D channel model, which provides boundary conditions for the GDL model. A single-phase approximation is used, and diffusion and heat conduction are neglected in the flow channel. Using this approach, complex flow geometries on the scale of a full-size cell can be simulated.

While the Arato and Inoue approaches are simplified relative to the full CFD models, they offer a design tool to estimate rapidly whether or not IDFF may offer a significant power density advantage and to approximate the impact of different channel-geometry parameters, such as rib-to-channel ratio.

iii. Water-transport plates

Nonuniformities in cell hydration arise either because there is no source for water in a given location or because there is no sink for excess water. Often these conditions coexist on the same cell planform, at the cathode inlets and exits, respectively. A strategy that has been developed to counter this effect is to provide simultaneously a source and a sink for water by using a hydrophilic porous bipolar plate (called a water-transport plate, or WTP), which is filled with water and is maintained at a liquid pressure that is lower than the gas pressure (244, 273). By doing so, the reactants can be internally humidified throughout the entire planform, thereby minimizing dry regions, while at the same time excess water can be removed in the liquid phase through the WTP, thereby minimizing flooded regions.

A detailed analysis of the WTP system is presented by Weber and Darling (274). Here the $1 + 1$-D, multiphase, nonisothermal model developed by the authors to describe solid-plate cells (52) is adapted to porous bipolar plates. Properties of all of the layers in the PEFC sandwich are identical to the solid-plate case with the exception of the plates themselves. Governing equations for the plates include Darcy's law and an energy balance. Boundary conditions at the back side of the WTP are the coolant pressure and temperature. Excess coolant flow is assumed, so that the surface in contact with the coolant is assumed to be at the coolant inlet temperature all along the length of the channel. The model is used to compare WTP and solid-plate performance under low-relative-humidity conditions as well as to explore the performance of the WTP as a function of various parameters, such as gas-to-liquid pressure difference, GDL wettability, and WTP properties. The idea of having a passive or low-power active liquid-water-management strategy similar to WTPs has been investigated with such applications as direct liquid water injection (which is also discussed in the next section) (275), wicking of liquid water through adsorbent wicks or sponges in the flow field (276–278), and electroosmotic pumps within the cell (279). However, those applications have not been modeled extensively. In addition, as discussed in Sect. III.2.iii, advanced GDL designs and the use of MPLs can be used to mitigate water-management concerns.

iv. Alternate cooling approaches

For many applications, liquid or forced-air cooling utilizing dedicated coolant flow channels is the predominant approach to thermal management. Alternatives include evaporative cooling and passive cooling. In evaporative cooling, liquid water is injected into the PEFC, changes phase due to the heat production within the cell, and exits the cell in the vapor phase. Because the heat of vaporization for water is so high, the total liquid flow rate is very small compared with that for conventional cooling. For example, using pure water, to achieve 1 kW of cooling using a 10°C ΔT requires roughly 24 g s^{-1} of coolant flow. With evaporative cooling the flow rate drops to 0.43 g s^{-1}, or 55 times less. As a result of the very small flow rates required, the liquid can either be sprayed into the reactant inlet(s) as a fog or it can be distributed via fine channels or a WTP. The advantage of the former approach is that it can eliminate one of the bipolar plates. The advantage of the latter is greater uniformity across the planform. Modeling of evaporatively cooled cells is limited to date. A systems-level view relating the required stack air exit temperature to the operating pressure is presented by Meyers et al (280)

Passively-cooled cells rely on natural convection to provide both fresh oxidant to the cathode as well as cell cooling. Passive cooling generally is limited to single-cell applications such as portable power. To model these devices, the coupled effects of heat and mass transfer must be considered. An example is the model presented by Djilali and coworkers (281), in which a steady-state 2-D CFD analysis is used to predict oxygen concentration and temperature as a function of position within the cell. Heat transfer is assumed to occur by natural convection only. A similar approach is taken by Hwang et al. (282) and by Litster and Djilali (283), although in the latter case the solution was obtained through a semi-analytical technique rather than CFD. Common to each of these models, however, are the following assumptions that greatly limit their applicability and accuracy. First, there is no net water transport through the membrane and the conductivity of the membrane does not depend on water content. In addition, water is formed and exists only in the vapor phase.

V. TRANSIENT OPERATION AND LOAD CHANGES

Although some (stationary) applications for PEFCs require a relatively constant power output (providing base-load electricity for a building, for example), many target applications, including materials handling, back-up power generation, and transportation, require frequent load changes and transient operation. Several processes affect cell performance during these transients. Of primary importance are the changes in temperature profile, cell hydration, and reactant availability. The amount of waste heat generated by the cell, given by equation (7), will change during the transient. This means that the temperature profile within the cell will change as the cell finds a new steady-state temperature to drive the removal of the waste heat. The steady-state hydration profile will also change, which can result in water within the cell changing phase, leading to two-phase-flow effects in either the PEFC sandwich or the channels. Reactant availability can be an issue if the time constant for transport of hydrogen or oxygen to the catalyst surface is on the same order as the transient time or if a dramatic increase in mass transport resistance occurs due to the presence of liquid water films or droplets.

Since our last review in 2004 (2), there has been a larger focus on simulating transient operation. This section discusses the more recent models for transient operation above 0°C, and Sect. VI.2 discusses those models for transient operation below 0°C (i.e., where freezing affects water management). The models presented in this section are categorized based on their thermal- and water-management strategies (i.e., single-phase or two-phase flow for water and isothermal and nonisothermal, respectively). Before discussing the details of individual models, it is important to contemplate the timescales of the processes involved in a load transient. By doing so, the reader will be in a better position to evaluate the importance of inclusion of each process and therefore the validity of the assumptions made in the models covered below.

1. Relative Timescales

To judge whether or not a process may be rate limiting, it is necessary first of all to understand the desired maximum time for a transient. This will vary with the application. As an example, one can consider automotive targets as put forward by the Department of Energy (284). These targets specify a time of 1 s or less for a step change from 10 to 90% of rated power.

A summary of time scales for the different fuel-cell processes is presented in Table 1, which is a modified version of a similar table in Mueller et al (285). The table is arranged in order of ascending time constant. From this analysis, one can see that the charging and discharging of the double layer and the electrochemical reaction rate are very fast relative to other transients. For this reason they can be neglected without concern (i.e., assume pseudo steady-state for them). On the other hand, species diffusion, membrane equilibration, and heat transfer all occur on time scales that are relevant to the 1 s automotive requirement. An accurate model for this application, therefore, would consider each of these processes. It is also worth noting that, according to this analysis, it will take tens of seconds for the cell to equilibrate completely after a transient has occurred, which is in disagreement with some findings that water rearrangement in the GDL can be on the order of minutes and tens of minutes. (18, 182, 286, 287)

Table 1.

Summary of time-constant analysis

Process	Governing equation	Typical value for time constant (s)	Reference
Charging or discharging of the electrochemical double layer	$\delta_{CL}^2 aC \left(\frac{1}{\kappa} + \frac{1}{\sigma} \right)$	2.0×10^{-7}	(298)
Electrochemical reaction rate	N/A	1.0×10^{-3}	(30)
Species diffusion (gas phase)	$\frac{L_{GDL}^2}{D_{species}}$	0.05	(292)
Heat transfer	$\frac{L_{cell}^2}{\alpha_{cell}}$	2	N/A
Membrane hydration	$\frac{L_{Membrane}^2}{D_{w,mem}}$	10	(292)

2. Single-Phase-Flow Models

As discussed throughout this article, the use of single-phase models is appropriate only if one is dealing with situations where water is not expected to condense. While the neglect of two-phase-flow effects changes the overall water-management results, it does provide for a simpler model and one that is much easier to run for transient conditions. Such models tend to rely heavily on empirical inputs, such as polarization-curve fits, and generally are useful for power-plant-level controls development but have limited utility in the cell design process.

i. Isothermal

The simplest approach to modeling transient behavior is with a 0-D, lumped model in which the cell properties and operational parameters are considered to be independent of position. An example is the model of Haddad et al. (288), which attempted to incorporate hydration effects into a 0-D model. In this case, the conductivity of the membrane was made an explicit function of the reactant humidification. Using this relationship, the change in cell voltage due to fluctuating ionic resistance is predicted. Note that the relationship used in this case is purely empirical and assumes instantaneous equilibration of the membrane with the reactant humidification level. In addition, no distinction is made between anode and cathode humidification.

Considering the effects of membrane hydration more rigorously requires computing membrane water content as a function of position, as by Friede et al. (289) and Yu and Ziegler (290). In their models, λ values are computed as a function of the z-coordinate, and membrane conductivity, water diffusivity, and the electroosmotic coefficient are all modeled as direct functions of $\lambda(z)$. This enables explicit treatment of differences in water content between anode and cathode (due to reactant humidity, product water, etc.). Chen et al. (291) Vorobev et al. (292) use a similar approach, but also account for membrane swelling as a function of λ. Vorobev et al. (292) account for the time required for equilibrium to be reached between liquid water in the membrane phase in the CLs and the surrounding vapor-phase water. Finally, the model of Nazarof and Promislow (293) examines the ignition and extinction behavior of a PEFC system with well mixed gas channels (294). Their analysis agrees with experimental data showing that a minimum membrane water content is necessary for ignition behavior during PEFC startup. This value depends mainly on the feedback between proton conduction and water production.

The 0-D and 1-D models discussed so far can be used to model the effects of membrane conductivity on performance during a transient. However, to predict reactant-transport effects accurately, it is usually necessary to model in more than one dimension due to the fact that neither current density nor reactant concentrations are uniform throughout the cell. As a result, it can be difficult to predict the appropriate boundary conditions for a 1-D model. Along-the-channel models represent one means of addressing this issue. An example is the 2-D model presented by Rao and Rengaswamy (295). In this case, the domain consists of the membrane, cathode CL, GDL, and gas channel. Average current density or cell voltage is provided as an input while the oxygen-partial-pressure profile as a function of time and position down the channel is the primary output. An alternate approach is presented by Yan et al. (296), who formulated a 2-D model to look at the current-density and oxygen-concentration profiles under the channel and rib rather than down the channel. As with Rao, the model considers only the membrane and cathode.

Simplified 2-D models, such as those described above, provide tools with which to evaluate both system-design parameters, such as optimal reactant-feed stoichiometry during a transient, and cell-design parameters, such as acceptable rib to channel ratios. 3-D models combine these two domains (along-the-channel and under-the-rib) to provide a unified solution to the transient problem. The 3-D model of Van Zee and coworkers (194, 297) is an example. Here the domain considered contains both anode and cathode flow channels and ribs, GDLs, and the membrane and CLs are considered to be one layer. The flow channels are oriented in a serpentine fashion, with a total of 20 passes. Membrane hydration is explicitly considered, meaning that the effects of reactant relative humidity can be investigated. Using this model, the oxygen-concentration and current-density profiles at the CL-membrane-CL surface can be observed as a function of time.

Wang and Wang (298, 299) also developed a 3-D, isothermal, transient model. Although the domain considered is a straight channel as opposed to serpentine, the PEFC sandwich is resolved into its

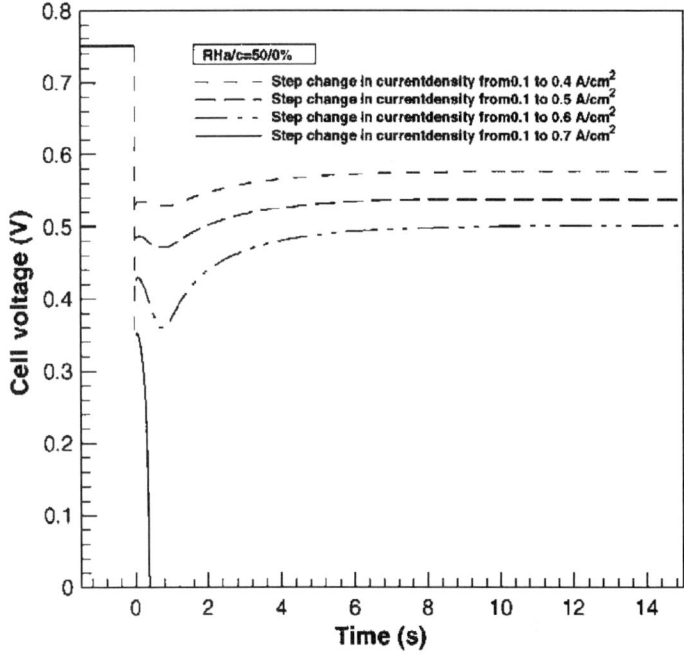

Figure 24. Dynamic response of cell potential to a step change in current density. (The figure is reproduced from (299) with permission of Elsevier.)

constituent layers, providing an additional level of resolution. Results from a sample simulation are provided in Fig. 24. In this case, a step change in current is applied, and the cell-voltage response is simulated. The operating temperature is 80°C, and the inlet relative humidity of the anode and cathode gases is set to be 50 and 0%, respectively, at the cell operating temperature. In the figure, four cases are shown, with each one corresponding to a larger step change in current density. The larger the step change, the greater the "undershoot" in cell voltage – that is, the voltage minimum relative to the steady-state value. Undershoot in this case is caused by temporary anode-side dehydration. When the current changes, the increased electroosmotic flux tends to dry out the anode while the increased water production tends to increase cathode hydration. The undershoot represents the time required for the excess cathode water to back-diffuse, rehydrating the anode and decreasing anode-side ionic resistance. In the most extreme case, for a step change from 0.1 to 0.7 A cm^{-2}, the increase in resistance is so high that the current cannot be supported and the cell voltage drops to zero.

ii. Nonisothermal

An early example of a single-phase, nonisothermal, transient model is that of Amphlett et al (300). Their approach is to couple a 0-D, steady-state, cell-performance model with a transient thermal model. The steady-state performance model is based on an semi-empirical formulation and simplifies to

$$V = V(i, T), \qquad (73)$$

where T is the lumped stack temperature. Fitting parameters based on experimental data are used to determine V. The transient thermal model is

$$\frac{dT}{dt} = \frac{1}{m_{\text{stack}} \hat{C}_P^{\text{stack}}} (Q - Q_{\text{sens}} - Q_{\text{loss}}), \qquad (74)$$

where the energies on the right side of the equation, from left to right, correspond to the theoretical heat production (see equation (7)), the heat transferred to the coolant and reactants (including the heat of vaporization of water), and the heat lost to the environment from the surface of the stack. Each of these terms are included in expanded form in equation (20) with the exception of the last one, which is simply the difference between the lumped stack temperature and the ambient temperature multiplied by a heat transfer coefficient and an effective surface area. By using this model, the authors demonstrate good agreement with stack data. However, the timescale of the transients considered is on the order of minutes, which is roughly the time required to change the bulk temperature of the stack by tens of degrees.

For a higher-frequency load profile, where the current density oscillates rapidly while the stack temperature changes very little, the voltage response caused by rapid changes in hydration will not be captured by this type of model. For example, Chen et al. (301) modeled membrane hydration in 1-D during a transient (considering electroosmotic flow and diffusion effects) and found that the time to reach steady state varied between 1 and 10 s depending on the magnitude of the change in current density.

Shan and Choe (200, 302) present a nonisothermal 1-D model that addresses this issue. In their system, the PEFC sandwich is considered, and membrane hydration effects are included. Reactant transport from the channels to the CL is by diffusion only. An energy balance is completed with respect to each control volume as follows:

$$\sum_i C_{p_i} c_i A_c l_{CV} = \sum \dot{m}_{in} A_c C_{p_j}(T_{in} - T_{CV}) + Q_{conv} A_c + Q_{cond} A_c + Q_{res} A_c + Q A_c, \quad (75)$$

where the terms on the right are for mass flow in, convective heat transfer, conductive heat transfer, heat production due to ohmic losses, and heat production due to the fuel cell reaction (as given by equation (7)), respectively. This 1-D cell model is further integrated into a stack level (41). The 1-D cell stack consists of cells divided by cooling layers (in a 1-to-1 ratio), with this layered entire layered structure bounded on either side by the layers that represent stack end structure. An energy balance is performed, allowing the temperature profile within a given cell as well as throughout the stack to be modeled during a transient.

Multidimensional nonisothermal transient models are also present in the literature. In 2-D, both the down-the-channel case as well as under-the-rib case (303) have been considered. An example of the down-the-channel case is the approach taken by Huang et al (232). In this case, the cell temperature in the z-direction is considered constant, but the temperature is allowed to vary along the channel, in a similar approach to that of Fuller and Newman (304). In this way, the effect of temperature on the current-density distribution and reactant concentrations can be approximated without the added complexity of computing the through-plane thermal profile. Given the fact that the down-the-channel temperature variation is generally large relative to the through-plane variation (238, 274, 305), this is often a reasonable approximation. The model also incorporates pressure-drop effects in the channels (neglecting entrance and exit effects), as well as pressure-driven water flow through the membrane via Darcy's law, enabling analysis of the effect of running the anode and cathode at different pressures.

Despite the fact that the reactant pressure and velocity are not constant down the channel in the model of Huang et al., detailed fluid dynamics is neglected. That is, fully-developed laminar flow is assumed, and the average gas velocity is computed parallel to the channel only, while reactant convection is neglected in the through-plane direction. In contrast, CFD methods can be used to solve the Navier–Stokes equations and predict the velocity field within the channel, as shown in the 2-D, transient, nonisothermal model by Shan et al (306).

To model temperature, reactant-concentration, and current-density profiles during a transient at the full-cell level, 3-D analysis is often required, especially when the reactant gases and/or coolant make multiple passes across the cell planform. Naturally, completing transient calculations with a full 3-D model can be computationally intensive, and methods of simplification with minimal loss in accuracy are desirable. One approach is to use a $2 + 1$-D approach in which species transport is modeled in detail in the through-plane direction, while in-plane the only reactant flow is in the gas channels. This is the technique used by Mueller et al (285). They divide the cell into eight control volumes in the through-plane direction: a coolant channel, anode and cathode solid plates, anode and cathode gas channels, anode and cathode GDLs, and an MEA. Each control volume is characterized by a single lumped temperature, pressure, and set of species mole fractions. In-plane, the cell is discretized into 35 nodes arranged in a grid pattern. This model exhibits good agreement with experimental data for both steady-state and transient, as shown in Fig. 25, operation. The figure again shows the dip in potential as seen in Fig. 24; however, the transient response is now much longer than in that figure and matches experimental data. The long transient demonstrates the importance of thermal management and nonisothermal effects on transient operation.

3. Two-Phase-Flow Models

All of the transient models listed above have neglected the presence of liquid water outside the membrane phase. As discussed throughout this article, this is valid only for some regions of the cell where the activity of water is typically <1 (e.g., at the reactant inlets) or for modeling low-relative-humidity operation. For other conditions, one must consider flow in both the liquid and vapor phases. Furthermore,

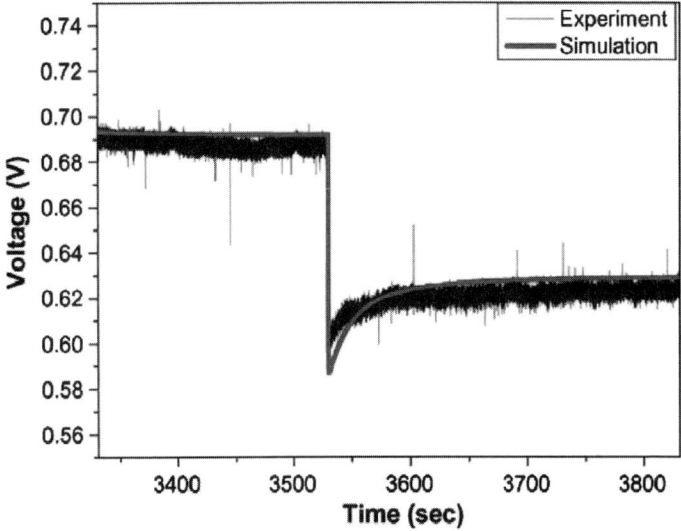

Figure 25. Comparison of simulation and experiment for a change in current density from 0.4 to 0.6 A cm^{-2}. (The figure is reproduced from (285) with permission of Elsevier.)

it is often seen that the rearrangement of liquid water and development of the saturation profiles are the longest time constants in the system (182,287,307). While this is known, transient modeling with two-phase flow becomes very complicated, and there are only a few such models currently.

To consider two-phase-flow effects in a relatively simple manner for transient operation, one could assume that the cell is at a uniform temperature, thereby eliminating the need to have a varying water vapor pressure and computing the complete energy balance. Ziegler et al. (308) present a cathode-side, 1-D model of this type, with the primary application to the study of potential-sweep experiments to understand the potential-current hysteresis in terms of water content changes in the membrane. To calculate two-phase flow, they use the methods described previously with cubic dependence of the permeability on saturation and a GDL model similar to that of Weber et al (85). A similar model to that of Ziegler et al., is the one by Chang and Chu (309), who use an embedded-agglomerate model with a film in the cathode CL instead of only a porous-electrode model. This analysis allowed them to examine in detail the dynamics of water movement, with a focus on the CL and GDL porosities, showing how the approach to steady state (dip in Fig. 20) depends on porosity, with high porosities exhibiting an increase and not a decrease.

A final layer of complexity can be added to the transient models by considering not only multiple phases but also the strongly coupled temperature distribution. This is done by adding the energy equation to the models described above. In Sect. V.2.ii, a similar approach was used to make single-phase models nonisothermal. However, in this case the energy equation must contain a source term for the heat of vaporization to account for energy transferred by the phase change of water.

One-dimensional models of this type have been presented by Song et al. (310) as well as Shah et al (193). The former considers only the cathode GDL while the latter consider the full cell sandwich, including optional microporous layers. These models offer cell designers powerful tools to evaluate the effect of material properties such as porosity, gas and liquid permeability, and contact angle on transient performance. As with single-phase models, however, evaluating reactant-distribution effects most often requires modeling in more than one dimension. This is all the more true in the nonisothermal case because the temperature can vary significantly across the planform.

One approach is to use a 3-D CFD model as by Guilin and Jianren (311) and Van Zee and coworkers (312). In both of these cases the full-cell sandwich is resolved through-plane, and in-plane a serpentine flow geometry is considered. In the channels and GDLs, homogeneous two-phase flow is assumed, meaning that liquid water is assumed to be dispersed within the gas phase and to move at the local gas velocity. This stands in contrast to the 1-D models discussed above, where liquid water is assumed to condense on the surfaces of pores and to move under the influence of a capillary-pressure gradient. To account for the effect of the presence of liquid water on reactant gas transport, Guilin and Jianren use an effective diffusivity, which simply multiplies the diffusion coefficient for a gas species by the gas-phase saturation. Van Zee and coworkers do not use this type of correction; instead they employ a liquid-film resistance on top of the platinum particles in the CL. In Guilin and Jianren's model, the membrane conductivity is not computed as a

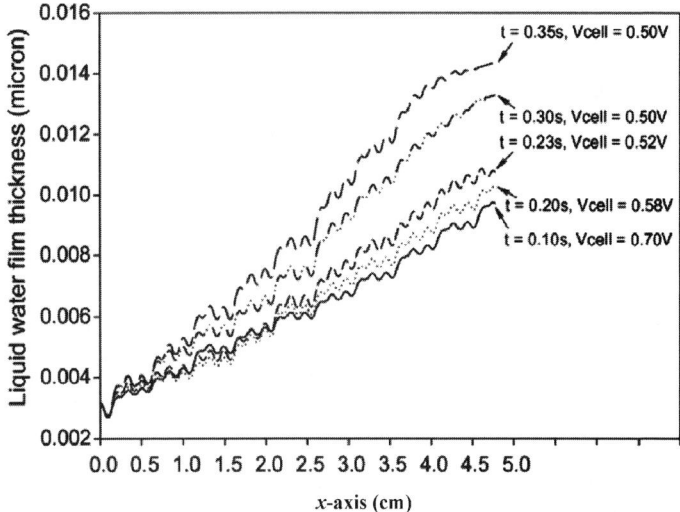

Figure 26. Variation of oxygen mole fraction along the channel width at different times and cell potentials. (The figure is reproduced from (312) with permission of Elsevier.)

function of position; instantaneous equilibration is assumed between membrane water content and the gas phase. For this reason the effects of transients on membrane hydration are highly simplified. Van Zee's model does account for local, time-dependent membrane conductivity.

Oxygen mole fraction as a function of position and time during a transient, as predicted by Van Zee's analysis, is shown in Fig. 26. The x-axis position corresponds to a cross-section of the serpentine cell which is normal to the predominant flow direction of the channels (i.e., it cuts through the channels and ribs). This is the reason for the oscillations as a function of x: the low points correspond to under-the-channel locations while the high points correspond to under-the-rib locations. The higher the value of x, the closer the position is to the exit of the channel. Figure 27 shows values of average cell current density and cell voltage during this transient. For comparison, the current density prediction is also shown for the same voltage transient using the isothermal, single-phase model described in (30), highlighting the importance of accounting for temperature effects.

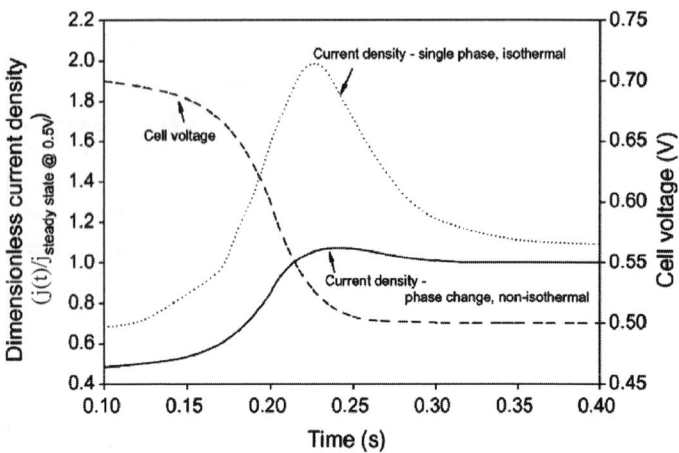

Figure 27. Comparison of predicted transient behavior for a single-phase, isothermal model and a multiphase, nonisothermal model. (The figure is reproduced from (312) with permission of Elsevier.)

VI. FREEZE

Hydrogen fuel cells produce water in addition to electricity and heat, and most proton-exchange-membrane materials require hydration in order to conduct ions. As a result, cold-weather operation of these devices presents unique challenges for their water and thermal management. Residual water in the various PEFC porous media might freeze during shutdown, inhibiting proper functionality on restart while possibly causing irreversible damage. In addition, during the cold-start process, the primary mode of product water removal – that is, in the vapor phase – is unavailable due to the low vapor pressure of water at low temperatures. As a result, if the cell does not heat up fast enough, flooding of the GDLs and CLs will occur, and the PEFC will be unable to continue operating without external heat input.

For automotive applications, several targets relating to cold start have been set by the Department of Energy (313). First, the PEFC must be able to start unassisted from $-40°C$. Second, it must be able to start from $-20°C$ to 50% net power within 30 s. Third, the total amount of energy expended during the start-up *and* shutdown cannot exceed 5 MJ. No requirement is stated by the DOE for the number of freeze start cycles. However, a report by Pesaran et al. (314) at NREL did use historical weather records to estimate how often a PEFC vehicle operated in the US may experience freezing temperatures, concluding that on average between 1961 and 1988, 43 states experienced $-20°C$ at least once per year, while 25 states recorded $-20°C$ over 40 times per year. The expected lifetime of a PEFC system inside a vehicle is 10 years.

One common approach to enable cold start is to dry the PEFC prior to allowing it to freeze, thereby minimizing the risk of damage by ice formation and maximizing the amount of volume available within the porous media to absorb product water on restart. However, while this approach may result in improved start performance, it has significant drawbacks. First, the purge requires energy and time to complete. Second, it reduces the ionic conductivity of the membrane, lowering the amount of power initially available on restart. Third, it exposes the membrane to the stress of a relative-humidity cycle, which can lead to membrane failure (315).

To develop new materials and methodologies that will enable cold start while addressing the issues listed above, modeling is required. However, this subject remains one of the least-explored areas of water modeling in the literature. A number of papers have dealt with stack-level temperature response during shutdown and start-up, but only a handful have attempted modeling water within the PEFC sandwich under these conditions.

1. Shutdown and Freezing

Once the PEFC stack stops operating, its temperature will gradually decay until it reaches ambient temperature, assuming the stack is not restarted. If the ambient temperature is below $0°C$, liquid water present in the PEFC porous media may freeze depending on pore size and wettability. In addition, the thermal gradient that is imposed on the cell during the cool down and freeze can in fact result in the movement of significant amounts of water from one part of the cell to another, first in the vapor phase (above $0°C$), and then in the liquid phase (below $0°C$). Water that has redistributed can then freeze in locations that inhibit proper cell functionality on restart. Understanding the parameters that control this water redistribution, such as the shutdown procedure and the material properties, is critical if system designers are to be able to incorporate mitigation strategies. The mechanisms for the movement of water and the existing models are discussed in this section.

i. Stack-level models

As stated above, it is the presence of thermal gradients within the PEFC sandwich during cooling that result in the redistribution of water. The origin of these thermal gradients comes from the fact that, in most applications, cells are arranged into stacks with lengthscales typically on the order of tens of centimeters. As a result, when a warm stack that is no longer generating heat is exposed to a cool environment, cooling occurs from the outside in. For example, considering only the z-direction, this means that the cells at the end of the stack cool first while the middle cells cool last. As a result, a given cell will experience a change in temperature in the z-direction during the cooling process.

To understand the magnitude of the thermal gradients, stack-level thermal models are generally used. The simplest of these is a 0-D lumped-cell stack model, as presented by Pesaran et al (314). Here the entire cell stack is considered to be a homogeneous mass thermally connected to the environment through a heat-transfer coefficient, h_{stack}. The total cool-down time, t_{cd}, required for the stack to move from the initial temperature, T_0 to a target temperature T_t, with an environmental temperature of T_{ext}, is given by

$$t_{\text{cd}} = \frac{m_{\text{stack}} C_P^{\text{stack}}}{A_{\text{stack}}} \left(\frac{1}{h_{\text{stack}}} + \frac{l_{\text{ins}}}{k_{\text{ins}}} \right) \ln \left(\frac{T_0 - T_{\text{ext}}}{T_t - T_{\text{ext}}} \right), \qquad (76)$$

where C_P^{stack} is the heat capacity of the stack, and A_{stack} is the surface area. The parameter k_{ins} in this case is the thermal conductivity of any insulation used to keep the stack warm.

The lumped stack model gives a first approximation for the cool-down time, and provides a simple framework for evaluating the number of freeze cycles that a stack may see for a given application as well as the impact of insulation thickness, l_{ins}. It should be noted, however, that equation (76) does not contain a term to account for the heat of fusion of ice, which will act to increase t_{cd}. Therefore this model is limited to $T_t > 0°C$. Of course, the missing term could be added, but this would not solve the inherent drawback to this model, which is that, because the entire stack is assumed to be at one temperature, this approach provides no useful information regarding the thermal gradient that a given cell may experience.

To obtain such information, the transient temperature profile within the stack must be determined. A one-dimensional approach to this problem is presented by Bradean et al (316). Here, the heat conduction equation is solved along the length of the stack (z-direction) assuming that the stack is insulated on one end and is cooled from the other. That is,

$$\frac{\partial T}{\partial t} = \alpha_s \frac{\partial^2 T}{\partial z^2} \text{ for } 0 < z < L_{\text{stack}}, \qquad (77)$$

where L_{stack} is the stack length and α_s is the average thermal diffusivity. As in the lumped case, a heat-transfer coefficient is used to connect the stack to the ambient temperature. However, this term removes heat only from the end of the stack, resulting in a temperature gradient in the z-direction during cooling. This boundary condition is given by matching the conductive heat flux out of the system with Newton's law of cooling.

Naturally, the above approach can be extended to 2- or 3-D. Whether or not this is worthwhile largely depends on cell-stack geometry. In many stacks, the "sides" are flat and easily insulated, whereas the fluid, structural, and electrical connections are made through the ends of the stack. These connections are often made of metal and therefore readily conduct heat away from the stack and are difficult to insulate effectively. Therefore it is often the case that most of the cell-stack heat is lost through the ends and the 1-D cooldown model is sufficient.

ii. Cell-level models

Once the thermal gradients within the stack are known, this information may be used to estimate water movement as a function of time during shutdown. There are two primary mechanisms for this water movement that are treated in the literature. The first is vapor-phase movement of water driven by the difference in partial pressure from one side of the cell to the other. Second, there is liquid-phase motion due to changes in capillary pressure that occur while the cell freezes. These will be discussed in turn.

Vapor phase

Modeling transient vapor-phase transport through the cell can be accomplished by applying equation (20), described in Sect. II.1.iii.

This is done by setting the temperature boundary conditions to the temperatures computed using the cool-down model described above, setting the cell current density to zero, and allowing the cell to equilibrate to roughly 5°C, at which point vapor-phase transport becomes insignificant. An alternate, empirical approach is presented by Bradean et al. (316), where the 1-D stack cool-down model described above is used to predict vapor-phase water transport within the cell. The temperature profile is discretized along the stack length, L_{stack}, based on the cell thickness. In other words, the nth cell in the stack is assigned a temperature, $T_n(z, t)$ based on its position within the stack as well as time. Furthermore, $\Delta T_n(z, t)$ represents the approximate temperature difference across the cell, that is $T_{n+1} - T_n$. These variables are then input into an empirically-derived water vapor flux equation of the form

$$f_w(T_n, \Delta T_n) = C_1{}^* \Delta T_n \exp(C_2{}^* T_n) \text{ if } w_{\text{mea}}$$
$$> w_{\text{mea,min}} \, 0 < z < L_{\text{stack}}, \quad (78)$$

$$f_w(T_n, \Delta T_n) = 0 \text{ if } w_{\text{mea}} \leq w_{\text{mea,min}}, \quad (79)$$

where f_w is the water mass flux from the cell sandwich into the cold-side reactant channel, C_1 and C_2 are constants that are dependent on the membrane-electrode-assembly design, and w_{mea} is the mass water content of the cell. The constant $w_{\text{mea,min}}$ represents the minimum cell water content, which is hypothesized to be present primarily in the membrane. Once condensed in the channel, the water can presumably be purged out in the liquid phase prior to freezing. To obtain the water content for a given cell n as a function of time, the flux is integrated and subtracted from the initial water content, $w_{\text{mea},0}$.

Verification of this model was undertaken and a sample result is shown in Fig. 28. In this case, after 12 h of cooling, at which point the cell stack is at a uniform temperature of 24°C, the amount of water present in the PEFC sandwich is measured, and the results compared to the model predictions. Good agreement exists between the simulation and model, and this serves to underscore the importance of vapor-phase water movement during cool-down as well as the importance of cell position. The cells on the end of the stack that was being cooled, where the thermal gradient was the highest,

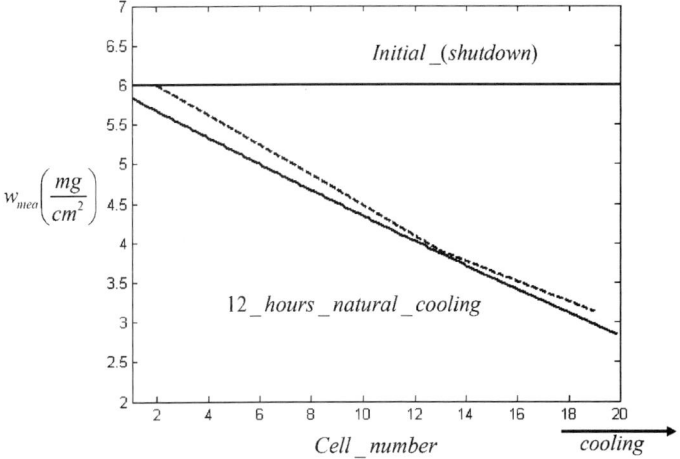

Figure 28. The CL-membrane-CL water content along the length of the stack obtained from the model (*continuous line*) and experiment (*dashed line*) at the end of the stack natural cooling process. (The figure is reproduced from (362) with permission of The Electrochemical Society.)

lost roughly half of their initial water content to the channels. The amount of water movement on the end of the stack, which was adiabatic, was minimal.

Liquid phase.

As the cell continues to cool and the temperature approaches 0°C, vapor-phase transport of water becomes negligible. Once the cell is below 0°C, liquid water transport through the porous layers in the cell can occur, despite the fact that the temperature is below the freezing point of bulk water. This is possible because of the surface energies of the pore network and water droplets, as governed by the Gibbs–Thomson equation (317),

$$T_{\text{FPD}} = \frac{-2\bar{V}_{\text{ice}}\gamma\, T_{\text{m}}}{r\,\Delta H_{\text{f}}}, \tag{80}$$

assuming perfect wetting, where T_{FPD} is the amount of freezing-point depression, $\bar{V}_{\text{ice}}$ is the molar volume of ice, γ is the surface tension of the ice–liquid interface, T_{m} is the melting (freezing) temperature for bulk water, ΔH_f is the heat of fusion of ice, and r is the pore radius. The amount of freezing-point depression in a given pore is primarily a function of pore radius – smaller pores tend to freeze at a lower temperature due to the shift in chemical potential. Because real media have distributions of pore radii, the fraction of unfrozen water versus temperature is generally a continuum below 0°C.

As water freezes within the medium, the average effective pore radius is reduced. Consequently, the liquid pressure is also reduced according to capillary phenomena and assuming hydrophilic pores (see equation (39)). In other words, the freezing of the water lowers the liquid pressure. This change in pressure can drive liquid water flow in two ways. First, within a homogeneous medium, if a temperature gradient exists (the medium is frozen from one side, for example), a pressure gradient will also exist. The liquid water in the cold region will be at a lower pressure than the hot region due to the lower average effective pore radius and to a much lesser extent, the lower density. Consequently, water will flow from hot to cold. Second, for two porous media in fluid contact, if the liquid-pressure versus saturation characteristics are different, a pressure differential will exist between the media as they are cooled. This pressure difference can also drive liquid-water flow.

Under some conditions, the liquid-water flow described above can contribute to a phenomenon known as frost heave. During frost heave, a layer of pure ice known as an ice lens grows within the medium, displacing the surrounding material. Frost heave has been studied extensively in soil science and civil engineering due to its destructive impact on infrastructure. The phenomenon is not related to the volume expansion of water during freezing, as it has long been known to occur also within materials, such as benzene, which contract upon freezing (318). In addition, ice lenses have been shown to grow in both hydrophobic and hydrophilic materials (319).

For an ice lens to form, the forces holding the medium together, namely, the tensile strength of the material and any compressive force supplied externally, must be overcome. The sum of these forces is referred to as the "overburden." The maximum frost heave pressure, p_{max}, must therefore exceed the overburden pressure, p_0. Often the rigid-ice model, first described by Miller (320), is used to determine p_{max}. In this case, it is assumed that the ice within the medium is continuous and at a locally-uniform pressure, p_i. To calculate p_i, the generalized Clapeyron equation (GCE) developed by Loch (321) is used. The GCE is derived from the Gibbs–Duhem equation and relates the ice pressure to the freezing-point depression,

$$\frac{p_i}{\rho_i} - \frac{p_w}{\rho_w} = -\Delta H_f \left(\frac{T_{\text{FPD}}}{T_m} \right). \tag{81}$$

Recent work by Rempel et al. (322) has called into question the validity of the rigid-ice model under conditions where a mixed zone of liquid and ice (called a "frozen fringe") is connected to the ice lens. In this case, the authors maintain, the maximum frost-heave pressure will be less than predicted by the GCE due to the forces acting on the ice within the fringe. Other models of frost heave also exist and have been reviewed by Henry (323).

While frost heave has been investigated for decades, applying the models that have been developed to the PEM environment has only recently been attempted. He and Mench (324, 325) have presented a 1-D, two-phase transient model that seeks to simulate both the movement of water during freeze as well as the growth of ice lenses. In their model, the domain includes half of the cell sandwich, from the membrane through one bipolar plate. The energy equation, which includes a source term for the heat of formation of ice, is used to predict the temperature profile in the domain during the freezing process based on an adiabatic condition on one boundary and heat loss via a heat-transfer coefficient linked to the ambient temperature on the other boundary.

Water is present in the model in membrane, liquid and ice phases; the movement of ice through regelation is neglected. Air is present in the channel, GDL, and CL, but is assumed to be at a constant and uniform pressure. In the membrane, a pure-diffusion model is used to account for liquid-water movement. Within the GDL and CL, liquid water moves under the influence of gradients in capillary pressure that occur once ice begins to form. The local capillary pressure is given by a Leverett J-function based on the effective saturation (see Sect. III.2.i). Gradients in water pressure are related to water flow in the model through the continuity equation and Darcy's law. The effective liquid permeability depends on the effective liquid saturation to the ninth power.

For an ice lens to initiate within the domain, the criteria of $p_i > p_0$ must be met. For growth of the ice lens to occur, in addition to the ice pressure continuing to exceed the overburden pressure, a supply of liquid water must be available to the "hot" side of the lens. That is, the growth rate of the ice lens is given by

$$\rho_i \frac{d\delta_{il}}{dt} = \rho_w(v_{in} - v_{out}), \tag{82}$$

where δ_{il} is the thickness of the ice lens and v_{in} and v_{out} are the velocities of the water coming into and out of the ice lens, respectively. The change in heat transfer due to the presence of the ice lens is also accounted for in the model.

While the model is 1-D, an attempt is made to simulate conditions under the rib as well as under the channel. This is done by adding a domain between the bipolar plate and the GDL which can simulate either an open channel or a rib. In the case of the open channel, liquid water is allowed to be expelled into the channel if the liquid pressure exceeds the air pressure. In the case of the rib, the water is constrained to stay within the porous media and membrane. Ice saturation predictions for both cases (under the rib and under the channel) during a freeze that starts at 5°C at time zero are shown in Fig. 29. In this case, the initial liquid-water saturation was assumed to be 0.6 in the GDL, 0.1 in the CL, and $\lambda = 19$ in the membrane. From the figure, one can clearly see that, as expected, more ice is formed under the rib; furthermore, the ice formation occurs principally from the rib into the GDL, although there is some in the wet membrane.

2. Startup from a Frozen State

Modeling water movement during cell shutdown and freeze is inherently difficult because it is a transient problem with multiple phase transitions that must be considered. It is made considerably simpler, however, by the fact that no electrochemical reaction is taking place. Unfortunately, this is not true during a startup from a frozen condition, making full simulations of this process quite complex.

In a typical startup, the initial condition is that the cell is at steady state at a temperature below 0°C, meaning that the majority of the residual water within the cell is frozen. Reactants are then fed to the stack until open circuit voltage is observed. A load is

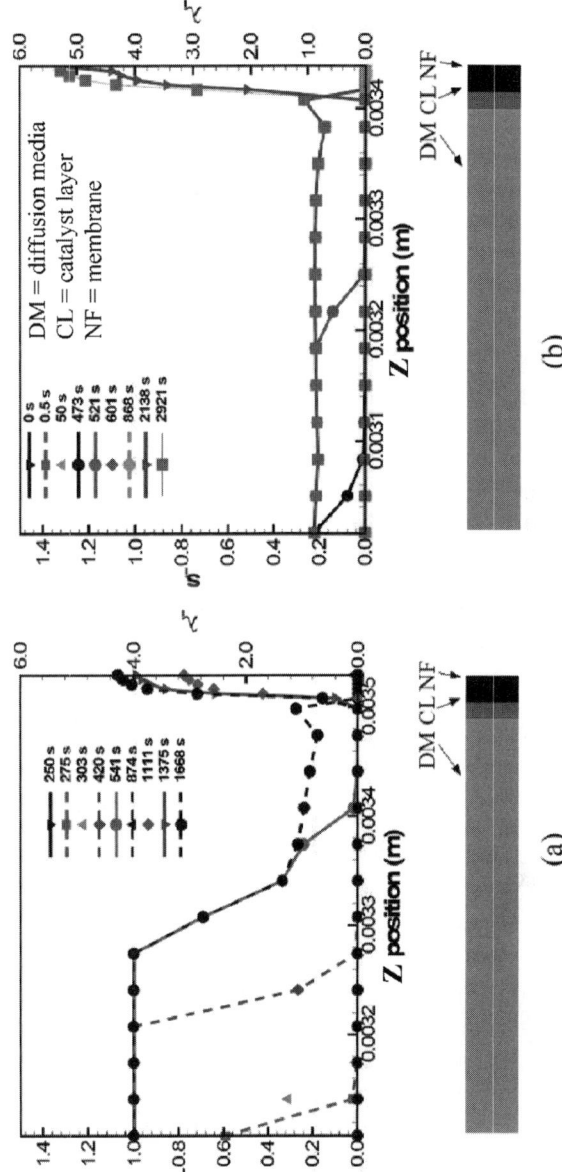

Figure 29. Transient distribution of ice saturation with bipolar plates (**a**) or open channel (**b**) boundary conditions. (The figure is reproduced from (325) with permission of The Electrochemical Society.)

then applied across the stack. This load serves two purposes. First, it provides electricity to the power plant for equipment heating or useful load, depending on the cold-start strategy. Second, the waste heat generated within the stack raises the stack temperature. Eventually the stack warms itself to normal operating temperature, and the system operates as it normally would. Such a procedure is often referred to as a "bootstrap start," because the PEFC is in a sense "pulling itself up by its own bootstraps" since no heat input external to the power plant is utilized.

In practice, achieving a successful bootstrap start (as described above) is quite difficult. First, reactants must have sufficient access to both CLs. This means that, during the shutdown, ice must not have blocked reactant channels or the porous media. Second, assuming that the reactants are initially able to access the CL, the cell must heat up rapidly enough so that product water can be removed before it completely floods the porous media. This is especially troublesome for cells near the ends of the stack because of the high rate at which these cells lose heat to the environment. Of course, the objective of the start-up is not simply to heat the stack but for the power plant to deliver useful power as quickly as possible, adding another challenge to those described above. Available power is reduced during the cold start because all of the principal processes that contribute to cell inefficiency are negatively impacted by the low temperature. Mass-transport limitations increase due to both the presence of ice reducing the available open pore volume and product water build-up. Ionic conductivity is reduced due to the presence of ice in the membrane (326), and the ORR is hindered both by the Arrhenius dependence on temperature (see equation (15)) as well as reduced proton activity (327).

To counter the difficulties of cold start, system designers use both procedural strategies and materials design. Examples of procedural strategies include adjusting the load profile applied during start to optimize performance or circulating coolant through the stack during start to improve thermal uniformity. Materials-design examples include reducing thermal mass to reduce warm-up time or adjusting pore structure to provide a greater reservoir for product water during cold start. However, evaluating these strategies experimentally is time-consuming and expensive. For this reason, having a model available that can be used to assess such strategies rapidly is very valuable. As with shutdown/freeze models, the (relatively few) cold-start models that have been published fall into two broad categories: stack level and cell level. These will be discussed in turn.

i. Stack-level models

The simplest model for startup is 0-D in that it treats the cell stack as a lumped mass with uniform properties. Such an approach was used by Pesaran (314) to estimate the total amount of thermal energy, Q_{tot}, required to raise the temperature of a cell stack from $-20°C$ to $+5°C$. In this case the expression used is

$$Q_{tot} = \left(m_{comp}\hat{C}_{p,comp} + m_i\hat{C}_{p,i}\right)\Delta T_1 + m_i\Delta H_f \\ + (m_{comp}\hat{C}_{p,comp} + m_w\hat{C}_{p,w})\Delta T_2, \quad (83)$$

where the subscript "comp" refers to the components of the stack, ΔT_1 is $0°C$ minus the starting temperature, and ΔT_2 is the final temperature minus $0°C$.

The utility in this simple calculation is that it can be used to estimate start time roughly, assuming that the performance of the stack during startup is known. For example, assuming a stack with $n = 400$ cells, each with an active area of $A = 300\,cm^2$ operating at an average cell voltage of $V_c = 0.750$ V per cell at a current density of $i = 0.1\,A\,cm^{-2}$, the total amount of waste heat per unit time, Q_{stack}, will be 8.8 kW. This is calculated from

$$Q_{stack} = n_c A_c Q, \quad (84)$$

where Q is given by equation (7). Assuming that $Q_{tot} = 5\,MJ$ is required to raise the temperature of the stack from $-20°C$ to $+5°C$, the total heat-up time would be calculated by dividing Q_{tot} by Q_{stack}, yielding roughly 570 s in this case. Note that this represents a lower bound for the actual heat-up time because heat loss during startup is not accounted for.

An alternate approach to that above is given by De Francesco and Arato (328), who define the variable T_{stack} as the lumped stack temperature and present the expression

$$m_{stack}\hat{C}_{p,stack}\frac{\partial T_{stack}}{\partial t} = h_{air}a_{air}(T_{air} - T_{stack}) \\ + h_{ext}a_{ext}(T_{ext} - T_{stack}) + Q, \quad (85)$$

where h_{air} is the heat-transfer coefficient between the cathode air and the stack and a_{air} is the contact area for the cathode air. Similarly, h_{ext} is the heat-transfer coefficient between the stack and the environment and a_{ext} is the external surface area of the stack. This approach has the advantage that the temperature of the stack is explicitly given as a function of time and that the temperature dependence of the cell voltage can be incorporated. It should be noted, however, that this approach does not account for the heat of fusion of ice within the cell.

0-D models can be used to estimate some design parameters for PEFC components. For example, for a given target start time, the maximum allowable thermal mass for the components and a minimum per-cell performance criterion could be specified. However, these would only be preliminary estimates due to nonuniformities in operating temperature, both from cell to cell and within a given cell.

One significant source of cell-to-cell nonuniformity in temperature is heat loss through the stack end structure. One way to address this problem is to use a 1-D model, such as that developed by Sundaresan and Moore (329, 330). This model predicts variations in temperature along the length of the stack, including end structure. To do this, the stack is divided into layers, including two 3-layer end structure assemblies as well as n 7-layer cells. Each layer is assumed to have homogeneous properties, including temperature. To solve for the temperature of a given layer, an energy balance, similar to that in equation (22), is utilized. The amount of heat generated by a given cell will depend on its cell voltage and current, and the cell voltage will in turn depend on the cell temperature; the model incorporates a correction to cell voltage based on the Nernst equation (4).

ii. Cell-level models

Cell performance is a function not only of temperature, but also of the amount and state of water present in the porous media and membrane. These parameters affect both ionic conductivity and mass-transport limitations, and the stack models described above do not account for these effects. To do so, more detailed cell-level models are required.

A semi-empirical approach to this problem is presented by Oszcipok et al (331). The authors performed isothermal, potentiostatic "cold-sweep" experiments using a single cell with a 33 cm^2

active area. Cold-sweep experiments differ from cold starts in that the cell temperature remains fixed at a point below 0°C rather than being allowed to rise under the influence of the cell's waste heat. Cold sweeps may be either potentiostatic or galvanostatic. In either case, the cell power will eventually drop to zero once enough product water is frozen to block reactant access to the catalyst completely.

In this case, during the experiment the current density was observed to rise asymptotically and then decrease rapidly to near zero. The authors attempt to model this behavior using a 1-D isothermal approach that accounts for four effects: the membrane resistance, the contact resistance, the exchange current density, and the oxygen diffusion coefficient. Each of these parameters is related empirically to the cumulative charge transfer. The results of the model predict a decrease in R_m with time during the cold sweep, presumably because the membrane is absorbing water and becoming liquid equilibrated. Other empirical relations are given that predict, with increasing charge transfer, an increasing contact resistance, a decreasing exchange current density, and a decreasing oxygen diffusion coefficient. To predict the overall current density as a function of time, these four parameters are used as inputs into the 1-D model of Springer et al. (154) at each time step. Using this model, the behavior of the current density observed in the experiment is simulated with some success, as shown in Fig. 30. The current density increases as the membrane resistance decreases, but once the cumulative charge transfer reaches a critical threshold, it drops due to increasing contact resistance and GDL flooding.

On the one hand, this type of model provides a framework for explaining a given cell's behavior during a cold sweep. On the other hand, it is difficult to use as a design tool, that is, to use it to predict a priori how a cell will behave based on its material properties. This is because of the fact that, with regards to water management, the various cell layers are lumped together and the properties that one usually specifies in the design process, such as thickness, permeability, and porosity, are not explicitly considered. In addition, although the isothermal nature of the model means that it is fairly simple, it limits its usefulness for two reasons. First, it cannot be used to predict cell heat-up during cold start. It is possible to impose a thermal profile for a heat-up and time-step the model to predict performance based on that profile, but this assumes that one already knows the

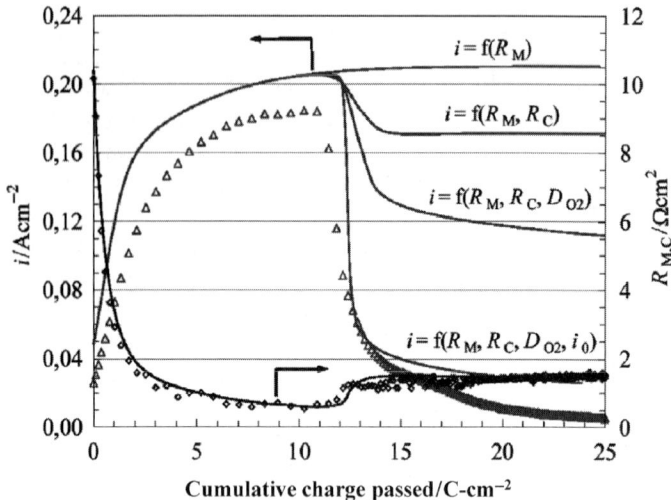

Figure 30. Comparison of simulation and experimental results for current density and ohmic membrane/contact resistance during cold sweep at −8°C. (The figure is reproduced from (331) with permission of Wiley-VCH.)

thermal profile, which implies that the model is being used to explain observed behavior, not to design a cell in advance. Second, the effect of nonuniformities in a given cell's temperature cannot be predicted, which is a significant limitation for modeling full-size cells.

Nonisothermal models can be used to address these problems, as shown by Hishinuma et al (332). In their model, an energy balance is included, meaning that, given adiabatic boundary conditions, the cell temperature depends on the cell performance with time and vice versa. In addition, thermally the model is 3-D. The cell performance in each of the discretized segments is determined using a cathode-performance model in which flooding of the electrode by ice is accounted for by modifying the Butler–Volmer equation (14) to allow the active area to decrease proportionally to the amount of ice frozen in the cell during each time step. Any water produced that is not removed in the gas phase by the reactant gases is assumed to freeze immediately, assuming that the cell temperature is below 0°C.

Using this model, either a cold sweep or a cold start can be evaluated simply by adjusting the thermal boundary conditions. Furthermore, the current-density profile across the cell planform can be predicted. Therefore it can be used in the design process to aid in predicting performance based on startup procedures as well as planform shape (e.g., aspect ratio). However, as with the Oszcipok model, the layers of the PEFC sandwich are not sufficiently distinguishable, and, therefore, the model is not well suited for evaluating proper cell-material specifications. In addition, liquid-phase-water removal from the cathode CL once the cell is above 0°C is not considered.

Mao and Wang (333) present a similar cold-start model that does incorporate a reduced set of material properties for individual layers and therefore may be useful as a preliminary design tool. These properties include CL porosity and ionomer content, GDL porosity, and the heat capacities of each layer. The model is 1-D, and water is considered to exist either in the membrane phase, as vapor, or as ice. An energy balance is included to account for thermal losses to the reactant gases and the surroundings. Because it is a cold-start model, the cell temperature is allowed to change with time, but the cell temperature is taken as a lumped parameter that is constant with position. As with Hishinuma et al., this model assumes that all ice forms in the CL and neglects liquid-water transport out of the CL once the cell is above the freezing point. Unlike Hishinuma et al., removal of product water into the membrane phase is accounted for explicitly.

Mao and Wang (334) have also developed a transient, multiphase, nonisothermal, 3-D model to simulate cold sweep. In the model, liquid water is not allowed, but water does exist in the membrane, vapor, and ice phases. As in their 1-D model, the presence of ice in a given porous medium reduces its permeability and restricts diffusion. In the CL, ice saturation also restricts the electrochemical area. As a result, as time proceeds, more and more product water turns to ice, and eventually the ORR is cut off, and the cell power goes to zero. Similar to the Hishinuma et al. model, a valuable output of this simulation is the current-density distribution down the channel with time. This helps one to understand which parts of the cell flood first. Unlike the Hishinuma et al. model, however, the properties of the individual PEFC layers are modeled explicitly, thereby providing greater insight into where the buildup of water is occurring and providing a tool for investigating the effects of some material properties.

Figure 31 shows results obtained from Mao and Wang given a cold sweep at $-20°C$ and 0.04 A cm^2. Each subplot represents a 2-D representation of ice saturation. The two dimensions are through-plane (z-direction) and in-plane (y-direction, rib and channel cross-section). Each column of subplots represents a different time during the sweep while each row represents the x-position along the channel. In this case, at any point in time, ice is predicted to be the greatest at the channel inlet and to form preferentially under the rib (the top half of the y-axis).

In summary, the most detailed cell models to date have focused on predicting cell performance and flooding during cold-sweep experiments. Due to the fixed-temperature boundary condition and the absence of liquid water in the porous media, this type of analysis is simpler than modeling a true cold start. It is also simpler to verify experimentally. Some models have been constructed for cold start, but to date these treat the cell sandwich as isothermal and do not consider the effects of liquid water once the cell temperature passes through the melting point. Unfortunately, the relationship between the predictions made by the cold-sweep analyses and the behavior of a cell during a cold start remains unclear. In other words, if a cell that is being constructed to start from a frozen condition is designed using a detailed cold-sweep model, it is not clear to what degree the material specifications will be correct. There is at least one basic material property, heat capacity, which plays a significant role in cold start, but which cannot be specified on the basis of a cold-sweep model. Clearly a gap exists in the ability to predict a complete cold start (i.e., from the frozen state all the way to normal operating temperature) and in understanding when such a model should be used as opposed to a (simpler) cold-sweep model. To achieve the ability to optimize fully materials and procedures for cold start, this relationship must be explored further.

VII. HIGHER-TEMPERATURE OPERATION

As the operating temperature of a PEFC is increased, there are several advantages. The most important perhaps is the ability for easier control and thermal (and water) management due to the higher temperature. Automotive manufacturers have a set a target of 120°C for operation, which is akin to the operating temperature of the internal

Modeling Water Management in Polymer-Electrolyte Fuel Cells 393

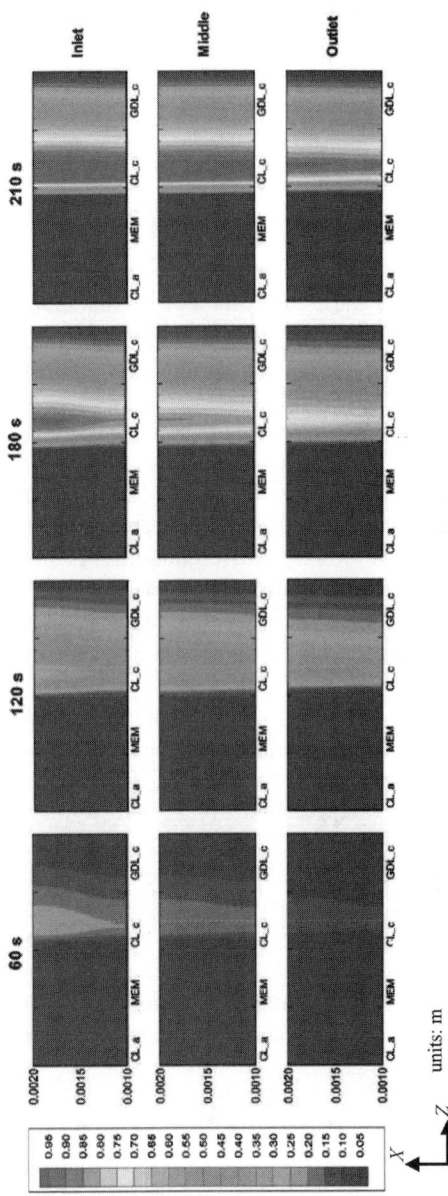

Figure 31. Ice-saturation evolution in the cathode CL. (The figure is reproduced from (333) with permission of The Electrochemical Society.)

combustion engine of today (315, 335). However, operation at that temperature requires the use of novel materials, and specifically, the membrane. The reason is that due to the exponential increase of water vapor pressure with temperature and the need not to pressurize and fully humidify the feed gases, the PEFC must operate at lower humidity to avoid diluting the oxygen too much. Therefore, the membrane must be able to conduct at low relative humidity; furthermore, it must also remain durable, work in the CLs, and conduct in the presence of liquid water that occurs during cool down and startup. The ideal situation is a membrane that conducts with dry feeds. Such a holy grail provides drastic simplifications for the PEFC system; however, there is not such a material currently, although it is an active area of research (222). There are several reviews on the topic of higher-temperature (120–180°C) PEFC material requirements (336–339), and the best and most comprehensive is the recent one by that of de Brujin et al (340).

Before proceeding to discuss the handful of high-temperature PEFC models, the advantages and disadvantages of going to higher temperatures and (hence) lower relative humidities is discussed. In terms of advantages, as mentioned, the most important is system and water-management simplification if the humidity requirement can be removed. If it cannot to a significant degree, then the system is too large and too complex with too many parasitic power losses to be feasible; in addition, while flooding would be avoided, water management becomes a tradeoff between membrane conduction and gas-phase dilution by water vapor. Other advantages of higher-temperature operation include higher impurity tolerance, especially with carbon monoxide (341), and faster kinetics and transport coefficients. However, the latter could be detrimental in terms of increasing the rate of side reactions such as carbon oxidation and platinum dissolution and increasing the gas crossover rates; also, the rate of physiochemical degradation of the various components may increase. Overall, the advantages seem to outweigh the disadvantages, assuming that the necessary material-property targets can be met.

While there has been substantial work experimentally for higher-temperature operation, especially in terms of novel membrane synthesis, there is a dearth in the number of models for these systems (203, 342–347) and overall full-cell results (348–351). This is not too surprising since the materials are still being developed and

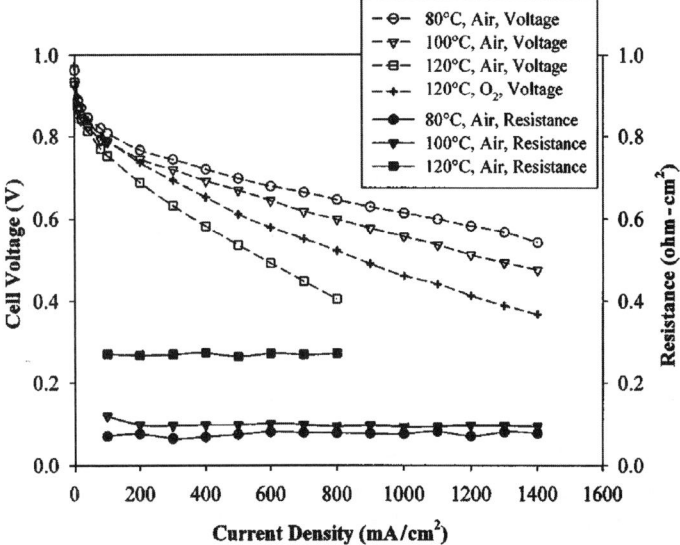

Figure 32. Polarization and resistance curves at various operating temperatures with a fixed water-vapor feed of 100% saturation at 80°C with a Nafion® 112 membrane. The curves therefore correspond to inlet RHs of 100, 70, and 35%, respectively. (The figure is reproduced from (341) with permission of The Electrochemical Society.)

characterized ex situ, and the fabrication of the higher-temperature cells, especially the CLs, are difficult, as noted below. Figure 32 shows the polarization and high-frequency resistance results for a Nafion®-based system as the operating temperature is increased, keeping a more-or-less fixed inlet water partial pressure. These results are similar to the ones observed in Sect. IV.1 and in Fig. 3b. As the temperature increases, the main factor becomes the membrane dehydration as seen in the resistance measurements. However, even if one *IR* corrects the data, there are significant oxygen-dilution mass-transfer effects at the end of the cell, which is one reason that the oxygen gain at 120°C is better than one expects from kinetics alone. As noted, although these data can be collected with Nafion®, the lifetime of the cell in Fig. 32 is extremely short due to the low durability of Nafion® at temperatures >100°C.

In terms of modeling, the higher-temperature systems can be modeled using the approaches and equations discussed throughout this article. In fact, the model is somewhat simpler since liquid water and two-phase flow are no longer significant factors. In terms of the novel membranes, most can be modeled using the same set of material properties as Nafion® (i.e., electroosmotic coefficient, water diffusion coefficient, conductivity) just with a weaker dependence on water content and humidity. An exception to this is the polybenzimidazole (PBI) system since there is a phosphoric-acid electrolyte with both mobile cations and anions (204,352); the modeling of this system is discussed separately below.

The most significant changes in the modeling approach are probably within the CLs, and specifically the cathode CL. These changes arise because of two factors. First, while replacing the membrane as a separator with a novel membrane is somewhat straightforward, placing a high-temperature membrane within a CL is more difficult due to problems of dispersing the ionomer and creating an efficient microstructure (353, 354). It may be that a different ionomer is required in the CLs, such as a low equivalent weight Nafion®-type polymer (355), but this can create dissolution problems when liquid water exists, and there may be interfacial resistances and durability concerns due to the different properties of the membrane separator and CL ionomer (356). The second factor pertains to a change in the rate of the ORR. This rate is now known to be dependent on the local relative humidity, where the reaction rate drops off rapidly below a relative humidity of 60% or so (355, 357, 358). Whether this dependence is due to platinum surface species deactivation, lack of liquid water in the hydrophilic primary pores causing smaller active surface areas, or a decrease in the proton activity, accessibility, and possibly concentration is not known definitively.

Besides the simple models used for the above data analysis including equivalent circuits and the PBI models, the higher-temperature models are from Wang and coworkers (203,346,347). In this set of models, they examine higher temperature (but not 120°C) operation with lower humidity feeds. The models demonstrate the need for water-management strategies and GDL properties to keep the membrane hydrated without diluting the oxygen gas rather than to prevent flooding (in fact, the models are single phase). It is determined that a functionally graded GDL is optimal, where it is more tortuous at the inlet to pressurize the water and hydrate the membrane and less tortuous near the exit to prevent more severe oxygen diffusional losses with the diluted oxygen. None of these models account for the CL effects mentioned above.

One of the most promising and developed membranes for PEFC high-temperature operation is that of PBI (204, 352). This system utilizes a membrane that contains impregnated or possibly tethered phosphoric acid (359). Hence, it is similar to phosphoric acid fuel cells (PAFCs), which operate in the range of 180–200°C (360). PAFCs have been modeled and experimentally explored, and a review of them is outside the scope of this article. Many of the features of PBI cells are similar to those PAFCs including problems of acid leaching and durability concerns of the electrodes and gains of high conductivity and minimal water management. In terms of modeling, a few numerical studies have been conducted with PBI PEFCs (342–345). Of these models, Peng and Lee (342) conclude that thermal effects are dominant in the system, and that a key optimization is the channel to land area ratio. Hu et al. (345) examine durability and degradation concerns with a specific focus on matching experimental data regarding loss of active area and changes within the CLs. Finally, the models of Cheddie and Munroe (343, 344) are perhaps the most detailed and examine such aspects as acid doping level in addition to the more typical analyses. They also show a relatively large influence of the thermal gradients and temperature increase among the cells as well as low catalyst utilization and various limitations in the CL.

VIII. SUMMARY

In this review, we have examined recent modeling efforts to understand and optimize water management in polymer-electrolyte fuel cells (PEFCs) operating with hydrogen. The major focus has been on transport of the various species within the PEFC, and the different facets of water management such as the balance between membrane dehydration and cathode flooding. The basic governing equations and regions of the PEFC were introduced, and the detailed studies involving water-management phenomena discussed. These investigations include design considerations to optimize water management, examination of freeze and subzero effects, accumulation of water and transient effects both within a full cell and a constitutive layer, and detailed models of two-phase flow in the gas-diffusion layers.

Where appropriate, models were compared to one another, but, for the most part, the results of the models were discussed. In addition, the models were broken down into their constitutive parts in terms of describing the phenomena of interest. The reason for this is that model validation occurs with varying sets of experimental data, some of which are cell specific and all of which are somewhat general and tangential to the specific aspect being explored. This is one reason why it is hard to justify one approach over another by just looking at the modeling results, especially when one deals with different levels of model complexity and empiricism. In general, it seems reasonable that the more complex models, which are based on physical arguments, account for several dimensions, and do not contain many fitting parameters, are perhaps closest to reality. Of course, this assumes that they fit the experimental data and observations. For any model, a balance must be struck between the complexity required to describe the physical reality and the additional computational costs of such complexity. In other words, while more complex models more accurately describe the physics of the transport processes, they are more computationally costly and may have so many unknown parameters that their results are not as meaningful. Hopefully, this review has shown and broken-down for the reader the vast complexities and aspects of water management within PEFCs, and the various ways they have been and can be understood better through mathematical modeling.

ACKNOWLEDGEMENTS

The authors would like to acknowledge Dr. Mordechay Schlesinger for his invitation to write this review and Dr. Michael Hickner for providing the neutron-imaging results. This work was supported by UTC Power and the Assistant Secretary for Energy Efficiency and Renewable Energy, Office of Hydrogen, Fuel Cell, and Infrastructure Technologies, of the U.S. Department of Energy under contract number DE-AC02-05CH11231.

NOMENCLATURE

- a_i^α activity of species i in phase α
- $a_{k,p}$ interfacial surface area between phases k and p per unit volume, cm^{-1}
- $a_{1,2}^o$ interfacial area between the electronically conducting and membrane phases with no flooding, cm^{-1}
- A surface area, cm^2
- A_{agg} specific external surface area of the agglomerate, cm^{-1}
- A_c active area, cm^2
- A_d projected normal area, cm^2
- A_{Pt} reactive surface area of platinum, $cm^2\,g^{-1}$
- B channel half-width, cm
- c chord length, cm
- c_D coefficient of drag
- $c_{i,k}$ interstitial concentration of species i in phase k, $mol\,cm^{-3}$
- c_T total solution concentration or molar density, $mol\,cm^{-3}$
- C_j fitting parameter, index j
- $\hat{C}_{p_k}$ heat capacity of phase k, $J\,g$-K^{-1}
- d rib width, cm
- $\mathbf{d}_i$ driving force per unit volume acting on species i in phase k, $J\,cm^{-4}$
- D_{abs} absolute molecular diffusivity, $cm^2\,s^{-1}$
- D_{eff} effective diffusivity, $cm^2\,s^{-1}$
- D_i Fickian diffusion coefficient of species i in a mixture, $cm^2\,s^{-1}$
- D_S capillary diffusivity, $cm^2\,s^{-1}$
- $D_{i,j}$ diffusion coefficient of i in j, $cm^2\,s^{-1}$
- D_{K_i} Knudsen diffusion coefficient of species i, $cm^2\,s^{-1}$
- E effectiveness factor
- f_k shape factor, index number k
- f_w mass flux of water, $g\,cm^{-2}$-s^{-1}
- F Faraday's constant, 96487 C equiv^{-1}
- F_D Drag force, N
- F_γ Surface tension force, N
- $\mathbf{g}$ acceleration due to gravity, $cm\,s^{-2}$
- ΔG_h Gibbs free energy of reaction h, $J\,mol^{-1}$
- $\mathbf{h}$ GDL thickness, cm
- h_d droplet height, cm

$h_{k,p}$	heat-transfer coefficient between phases k and p, J cm^{-2}s-K^{-2}
$\bar{H}_{i,k}$	partial molar enthalpy of species i in phase k, J mol^{-1}
$H_{i,j}$	Henry's constant for species i in component j, mol cm^{-3} kPa^{-1}
ΔH_f	heat of fusion of ice, J mol^{-1} or J g^{-1}
ΔH_l	heat or enthalpy of reaction l, J mol^{-1}
i	superficial current density through the membrane, A cm^{-2}
$\mathbf{i}_k$	current density in phase k, A cm^{-2}
i_{0_h}	exchange current density for reaction h, A cm^{-2}
i_h	transfer current density of reaction h per interfacial area between phases k and p, A cm^{-2}
$i_{\lim}$	limiting current density, A cm^{-2}
$J(S)$	Leverett J-function
$\mathbf{J}_{i,k}$	flux density of species i in phase
k	relative the mass-average velocity of phase k, mol cm^{-2} s^{-1}
k	effective hydraulic permeability, cm^2
k_{T_k}	thermal conductivity of phase k, J cm^{-2} K^{-1}
k_r	relative hydraulic permeability
k_{sat}	saturated hydraulic permeability, cm^2
k_Φ	electrokinetic permeability, cm^2
k	Stokes law fitting function for droplet
$K_{i,j}$	frictional interaction parameters between species i and j
l_k	thickness of phase or element k
L	catalyst layer thickness, cm
m	mass, g
m_{Pt}	loading of platinum, g cm^{-2}
M_i	molecular weight of species i, g mol^{-1}
$M_i^{z_i}$	symbol for the chemical formula of species i in phase k having charge z_i
n_c	number of cells
n_h	number of electrons transferred in electrode reaction h
$\mathbf{N}_{i,k}$	superficial flux density of species i in phase k, mol cm^{-2} s^{-1}
p_0	overburden pressure, kPa
p_i	partial pressure of species i, kPa
p_C	capillary pressure, kPa
p_k	total pressure of phase k, kPa
p_w^{vap}	vapor pressure of water, kPa
$\mathbf{q}_k$	superficial heat flux through phase k, J cm^{-2} s^{-1}

Q	total amount of heat generated, J cm^{-2} s^{-1}
$Q_{k,p}$	heat flux transferred between phases k and p, J cm^{-3} s^{-1}
r	pore radius, cm
r_d	pore radius, cm
r_{evap}	rate of evaporation, mol cm^{-3} s^{-1}
$r_{l,\text{k-p}}$	rate of reaction l per unit of interfacial area between phases k and p, mol cm^{-2} s^{-1}
R	ideal-gas constant, 8.3143 J mol-K^{-1}
R_{agg}	agglomerate radius, cm
$R_{g,k}$	rate of homogenous reaction g in phase k, mol cm^{-3} s^{-1}
$R_{i,j}$	resistance of resistor i,j in Fig. 8 where ct stands for charge-transfer, Ω cm^2
R'	total ohmic resistance, Ω cm^2
Re	Reynolds number
$s_{i,k,l}$	stoichiometric coefficient of species i in phase k participating in reaction l
S	liquid saturation
S_0	specific surface area, cm^2 cm^{-3}
ΔS_h	entropy of reaction h, J mol-K^{-1}
t	time, s
T	absolute temperature, K
T_m	melting point of ice, K
u_i	mobility of species i, cm^2 mol J-s^{-1}
u_j	velocity in channel j, cm s^{-1}
v	velocity, cm s^{-1}
U	reversible cell potential, V
U_θ	standard potential of reaction
U_H	enthalpy potential, V
$\mathbf{v}_k$	superficial velocity of phase k, cm s^{-1}
V	cell potential, V
$\bar{V}_i$	(partial) molar volume of species i, cm^3 mol^{-1}
w_{mea}	water content per unit area of membrane electrode assembly, g cm^{-2}
$W_{O_2}^{\text{diff}}$	molar flow rate of oxygen to the agglomerate, mol cm^{-3} s^{-1}
x	distance across the flow field, cm
$x_{i,k}$	mole fraction of species i in phase k
y	distance along the flow-field channel, cm
z	distance across the cell sandwich, cm
z_i	valence or charge number of species i

Greek

- α_a anodic transfer coefficient
- α_c cathodic transfer coefficient
- α_s average stack thermal diffusivity, cm^2 s^{-1}
- α_w water transport coefficient, mol^2 J-cm-s^{-1}
- β net water flux per proton flux through the membrane
- γ surface tension, N cm^{-1}
- δ_n diffusion length or thickness of region n, cm
- δ_{il} ice lens thickness, cm
- ζ characteristic length, cm
- ϵ_k volume fraction of phase k
- ϵ_o bulk porosity
- ϵ_V bulk strain
- ν_k kinematic viscosity of phase k, cm^2 s^{-1}
- ξ electroosmotic coefficient
- Π_h Peltier coefficient for charge-transfer reaction h, V
- ρ_k density of phase k, g cm^3
- σ_o standard conductivity in the electronically conducting phase, S cm^{-1}
- η_h electrode overpotential of reaction h, V
- η_{s_h} surface overpotential of reaction h, V
- θ contact angle, degrees
- $\Delta\theta$ contact angle hysteresis, degrees
- κ conductivity of the ionically conducting phase, S cm^{-1}
- λ moles of water per mole of sulfonic acid sites
- λ_L relative mobility of the liquid phase
- μ viscosity, Pa-s
- μ_i (electro)chemical potential of species i, J mol^{-1}
- μ_i^α electrochemical potential of species i in phase α, J mol^{-1}
- τ stress tensor, kPa
- τ_k tortuosity of phase k
- ϕ Thiele modulus, defined by eq 51 for the ORR
- Φ_k potential in phase k, V
- ψ_i Permeation coefficient of species i

Subscripts/superscripts

0	initial
1	electronically conducting phase
2	ionically conducting phase
A	advancing
agg	agglomerate
amb	ambient
cd	cool down
CL	catalyst layer
CV	control volume
eff	effective value, corrected for tortuosity and porosity
ext	external to the control volume
f	fixed ionic site in the membrane
film	film covering the agglomerate
FPD	freezing point depression
g	homogeneous reaction number
G	gas phase
h	electron-transfer reaction number
HOR	hydrogen-oxidation reaction
irrev	irreversible
i	generic species, element index, or ice phase
in	into the control volume
ins	insulation
j	generic species
k	generic phase
l	heterogeneous reaction number
L	liquid phase
m	mixture or membrane
m_0	Oxygen partial pressure dependence (see equation (15))
max	maximum
min	minimum
ORR	oxygen-reduction reaction
out	out of the control volume
p	generic phase
R	receding
ref	parameter evaluated at the reference conditions

res   resistive
rev   reversible
s   solid phases
sens   sensible
stack   stack average value
t   target
tot   total
w   water

REFERENCES

1. A. Prasanna, H. Y. Ha, E. A. Cho, S. A. Hong and I. H. Oh, *J. Power Sources* **137** (2004) 1.
2. A. Z. Weber and J. Newman, *Chem. Rev.* **104** (2004) 4679.
3. C. Y. Wang, *Chem. Rev.* **104** (2004) 4727.
4. P. Costamagna and S. Srinivasan, *J. Power Sources* **102** (2001) 242.
5. P. Costamagna and S. Srinivasan, *J. Power Sources* **102** (2001) 253.
6. T. Okada, *J. New Mater. Electrochem. Syst.* **4** (2001) 209.
7. M. A. J. Cropper, S. Geiger and D. M. Jollie, *J. Power Sources* **131** (2004) 57.
8. W. Q. Tao, C. H. Min, X. L. Liu, Y. L. He, B. H. Yin and W. Jiang, *J. Power Sources* **160** (2006) 359.
9. N. Djilali, *Energy* **32** (2007) 269.
10. S. Tsushima, K. Teranishi and S. Hirai, *Electrochem. Solid State Lett.* **7** (2004) A269.
11. S. Tsushima, K. Teranishi and S. Hirai, *Energy* **30** (2005) 235.
12. K. R. Minard, V. V. Viswanathan, P. D. Majors, L. Q. Wang and P. C. Rieke, *J. Power Sources* **161** (2006) 856.
13. K. W. Feindel, S. H. Bergens and R. E. Wasylishen, *Chemphyschem* **7** (2006) 67.
14. J. St-Pierre, *J. Electrochem. Soc.* **154** (2007) B88.
15. M. A. Hickner, N. P. Siegel, K. S. Chen, D. N. McBrayer, D. S. Hussey, D. L. Jacobson and M. Arif, *J. Electrochem. Soc.* **153** (2006) A902.
16. R. Satija, D. L. Jacobson, M. Arif and S. A. Werner, *J. Power Sources* **129** (2004) 238.
17. A. Turhan, K. Heller, J. S. Brenizer and M. M. Mench, *J. Power Sources* **160** (2006) 1195.
18. J. J. Kowal, A. Turhan, K. Heller, J. Brenizer and M. M. Mench, *J. Electrochem. Soc.* **153** (2006) A1971.
19. F. Y. Zhang, X. G. Yang and C. Y. Wang, *J. Electrochem. Soc.* **153** (2006) A225.
20. X. Liu, H. Guo, F. Ye and C. F. Ma, *Electrochim. Acta* **52** (2007) 3607.
21. K. Tuber, D. Pocza and C. Hebling, *J. Power Sources* **124** (2003) 403.
22. X. G. Yang, F. Y. Zhang, A. L. Lubawy and C. Y. Wang, *Electrochem. Solid State Lett.* **7** (2004) A408.
23. S. Litster, D. Sinton and N. Djilali, *J. Power Sources* **154** (2006) 95.

24. P. K. Sinha, P. Halleck and C. Y. Wang, *Electrochem. Solid State Lett.* **9** (2006) A344.
25. I. Manke, C. Hartnig, M. Grunerbel, W. Lehnert, N. Kardjilov, A. Haibel, A. Hilger, J. Banhart and H. Riesemeier, *Appl. Phys. Lett.* **90** (2007) 3.
26. J. Newman and K. E. Thomas-Alyea, *Electrochemical Systems*, Wiley, New York, 2004.
27. A. J. Bard and L. R. Faulkner, *Electrochemical Methods: Fundamentals and Applications*, Wiley, New York, 2001.
28. D. M. Bernardi, E. Pawlikowski and J. Newman, *J. Electrochem. Soc.* **132** (1985) 5.
29. L. Rao and J. Newman, *J. Electrochem. Soc.* **144** (1997) 2697.
30. K. C. Neyerlin, W. B. Gu, J. Jorne and H. A. Gasteiger, *J. Electrochem. Soc.* **153** (2006) A1955.
31. J. X. Wang, T. E. Springer and R. R. Adzic, *J. Electrochem. Soc.* **153** (2006) A1732.
32. A. J. Appleby, *J. Electrochem. Soc.* **117** (1970) 328.
33. K. Kinoshita, *Electrochemical Oxygen Technology*, Wiley, New York, 1992.
34. A. Parthasarathy, B. Dave, S. Srinivasan, A. J. Appleby and C. R. Martin, *J. Electrochem. Soc.* **139** (1992) 1634.
35. P. D. Beattie, V. I. Basura and S. Holdcroft, *J. Electronanal. Chem.* **468** (1999) 180.
36. A. Parthasarathy, S. Srinivasan, A. J. Appleby and C. R. Martin, *J. Electrochem. Soc.* **139** (1992) 2530.
37. A. Parthasarathy, S. Srinivasan, A. J. Appleby and C. R. Martin, *J. Electrochem. Soc.* **139** (1992) 2856.
38. Y. W. Rho, O. A. Velev and S. Srinivasan, *J. Electrochem. Soc.* **141** (1994) 2084.
39. A. Parthasarathy, S. Srinivasan, A. J. Appleby and C. R. Martin, *J. Electronanal. Chem.* **339** (1992) 101.
40. J. Perez, E. R. Gonzalez and E. A. Ticianelli, *Electrochim. Acta* **44** (1998) 1329.
41. F. A. Uribe, T. E. Springer and S. Gottesfeld, *J. Electrochem. Soc.* **139** (1992) 765.
42. K. C. Neyerlin, H. A. Gasteiger, C. K. Mittelsteadt, J. Jorne and W. B. Gu, *J. Electrochem. Soc.* **152** (2005) A1073.
43. W. Liu and D. Zuckerbrod, *J. Electrochem. Soc.* **152** (2005) A1165.
44. U. A. Paulus, T. J. Schmidt, H. A. Gasteiger and R. J. Behm, *J. Electronanal. Chem.* **495** (2001) 134.
45. S. F. Burlatsky, V. Atrazhev, N. Cipollini, D. Condit and N. Erikhman, *ECS Trans.* **1** (2006) 239.
46. R. M. Darling and J. P. Meyers, *J. Electrochem. Soc.* **150** (2003) A1523.
47. R. M. Darling and J. P. Meyers, *J. Electrochem. Soc.* **152** (2005) A242.
48. C. A. Reiser, L. Bregoli, T. W. Patterson, J. S. Yi, J. D. L. Yang, M. L. Perry and T. D. Jarvi, *Electrochem. Solid State Lett.* **8** (2005) A273.
49. J. P. Meyers and R. M. Darling, *J. Electrochem. Soc.* **153** (2006) A1432.
50. R. B. Bird, W. E. Stewart and E. N. Lightfoot, *Transport Phenomena*, Wiley, New York, 2002.
51. J. J. Hwang, C. H. Chao, C. L. Chang, W. Y. Ho and D. Y. Wang, *Int. J. Hydrog. Energy* **32** (2007) 405.
52. A. Z. Weber and J. Newman, *J. Electrochem. Soc.* **153** (2006) A2205.

53. L. M. Onishi, J. M. Prausnitz and J. Newman, *J. Phys. Chem. B* **111** (2007) 10166.
54. A. Z. Weber and J. Newman, *J. Electrochem. Soc.* **150** (2003) A1008.
55. R. B. Moore and C. R. Martin, *Macromolecules* **21** (1988) 1334.
56. M. A. Hickner, H. Ghassemi, Y. S. Kim, B. R. Einsla and J. E. McGrath, *Chem. Rev.* **104** (2004) 4587.
57. K. D. Kreuer, S. J. Paddison, E. Spohr and M. Schuster, *Chem. Rev.* **104** (2004) 4637.
58. A. Z. Weber and J. Newman, in *Advances in Fuel Cells, Vol. 1*, T. S. Zhao, K.-D. Kreuer and T. V. Nguyen Editors, Elsevier, Amsterdam (2007).
59. J. Fimrite, H. Struchtrup and N. Djilali, *J. Electrochem. Soc.* **152** (2005) A1804.
60. B. Carnes and N. Djilali, *Electrochem. Acta* **52** (2006) 1038.
61. J. Fimrite, B. Carnes, H. Struchtrup and N. Djilali, *J. Electrochem. Soc.* **152** (2005) A1815.
62. T. Thampan, S. Malhotra, H. Tang and R. Datta, *J. Electrochem. Soc.* **147** (2000) 3242.
63. M. Wohr, K. Bolwin, W. Schnurnberger, M. Fischer, W. Neubrand and G. Eigenberger, *Int. J. Hydrog. Energy* **23** (1998) 213.
64. A. Z. Weber and J. Newman, *J. Electrochem. Soc.* **151** (2004) A311.
65. P. N. Pintauro and D. N. Bennion, *Ind. Eng. Chem. Fundam.* **23** (1984) 230.
66. T. F. Fuller, Solid-polymer-electrolyte Fuel Cells, University of California, Berkeley, CA (1992).
67. B. S. Pivovar, *Polymer* **47** (2006) 4194.
68. M. W. Verbrugge and R. F. Hill, *J. Electrochem. Soc.* **137** (1990) 886.
69. R. Schlögl, *Zeitschrift für physikalische Chemie, Neue Folge* **3** (1955) 73.
70. T. Okada, *J. Electronanal. Chem.* **465** (1999) 1.
71. T. Okada, G. Xie and M. Meeg, *Electrochim. Acta* **43** (1998) 2141.
72. J. Divisek, M. Eikerling, V. Mazin, H. Schmitz, U. Stimming and Y. M. Volfkovich, *J. Electrochem. Soc.* **145** (1998) 2677.
73. M. Eikerling, A. A. Kornyshev and U. Stimming, *J. Phys. Chem. B* **101** (1997) 10807.
74. P. Choi, N. H. Jalani and R. Datta, *J. Electrochem. Soc.* **152** (2005) A1548.
75. P. Choi, N. H. Jalani and R. Datta, *J. Electrochem. Soc.* **152** (2005) E84.
76. A. Z. Weber and J. Newman, *AIChE J.* **50** (2004) 3215.
77. I. Nazarov and K. Promislow, *J. Electrochem. Soc.* **154** (2007) B623.
78. A. Z. Weber and J. Newman, *J. Electrochem. Soc.* **154** (2007) B405.
79. S. S. Kocha, J. D. L. Yang and J. S. Yi, *AIChE J.* **52** (2006) 1916.
80. A. Taniguchi, T. Akita, K. Yasuda and Y. Miyazaki, *J. Power Sources* **130** (2004) 42.
81. X. Cheng, Z. Shi, N. Glass, L. Zhang, J. J. Zhang, D. T. Song, Z. S. Liu, H. J. Wang and J. Shen, *J. Power Sources* **165** (2007) 739.
82. M. S. Mikkola, T. Rockward, F. A. Uribe and B. S. Pivovar, *Fuel Cells* **7** (2007) 153.
83. R. C. Makkus, A. H. H. Janssen, F. A. de Bruijn and R. Mallant, *J. Power Sources* **86** (2000) 274.
84. K. Promislow, J. Stockie and B. Wetton, *Proc. R. Soc. London, A* **462** (2006) 789.

85. A. Z. Weber, R. M. Darling and J. Newman, *J. Electrochem. Soc.* **151** (2004) A1715.
86. J. H. Nam and M. Kaviany, *Int. J. Heat Mass Transfer* **46** (2003) 4595.
87. J. T. Gostick, M. W. Fowler, M. A. Ioannidis, M. D. Pritzker, Y. M. Volfkovich and A. Sakars, *J. Power Sources* **156** (2006) 375.
88. J. Benziger, J. Nehlsen, D. Blackwell, T. Brennan and J. Itescu, *J. Membr. Sci.* **261** (2005) 98.
89. J. J. Hwang, *J. Electrochem. Soc.* **153** (2006) A216.
90. A. Z. Weber and J. Newman, *ECS Trans.* **1 (16)** (2005) 61.
91. R. E. De La Rue and C. W. Tobias, *J. Electrochem. Soc.* **106** (1959) 827.
92. D. A. G. Bruggeman, *Ann. Physik.* **24** (1935) 636.
93. G. Li and P. P. Pickup, *J. Electrochem. Soc.* **150** (2003) C745.
94. D. M. Bernardi, Mathematical Modeling of Lithium(alloy) IronSulfide Cells and the Electrochemical Precipitation of Nickel Hydroxide, University of California, Berkeley (1986).
95. M. Knudsen, *The Kinetic Theory of Gases*, Methuen, London, 1934.
96. L. B. Rothfeld, *AIChE J.* **9** (1963) 19.
97. A. Z. Weber and J. Newman, *Int. Commun. Heat Mass Transfer* **32** (2005) 855.
98. E. A. Mason and A. P. Malinauskas, *Gas Transport in Porous Media: The Dusty-Gas Model*, Elsevier, Amsterdam, 1983.
99. C. T. Miller, G. Christakos, P. T. Imhoff, J. F. McBride, J. A. Pedit and J. A. Trangenstein, *Adv. Water Resour.* **21** (1998) 77.
100. F. A. L. Dullien, *Porous Media: Fluid Transport and Pore Structure*, Academic Press, New York, 1992.
101. J. Bear, *Dynamics of Fluids in Porous Media*, Dover, New York, 1988.
102. W. O. Smith, *Physics* **4** (1933) 425.
103. M. C. Leverett, *Pet. Div. Trans. Am. Inst. Min. Metall. Engineers* **142** (1941) 152.
104. C. Y. Wang and P. Cheng, *Adv. Heat Trans.* **30** (1997) 93.
105. C. Y. Wang and P. Cheng, *Int. J. Heat Mass Transfer* **39** (1996) 3607.
106. Z. H. Wang, C. Y. Wang and K. S. Chen, *J. Power Sources* **94** (2001) 40.
107. S. A. Freunberger, M. Santis, I. A. Schneider, A. Wokaun and F. N. Buchi, *J. Electrochem. Soc.* **153** (2006) A396.
108. F. N. Buchi, A. B. Geiger and R. P. Neto, *J. Power Sources* **145** (2005) 62.
109. H. Meng and C. Y. Wang, *J. Electrochem. Soc.* **151** (2004) A358.
110. A. Fischer, J. Jindra and H. Wendt, *J. Appl. Electrochem.* **28** (1998) 277.
111. E. A. Ticianelli, J. G. Beery and S. Srinivasan, *J. Appl. Electrochem.* **21** (1991) 597.
112. X. Cheng, B. Yi, M. Han, J. Zhang, Y. Qiao and J. Yu, *J. Power Sources* **79** (1999) 75.
113. E. Passalacqua, F. Lufrano, G. Squadrito, A. Patti and L. Giorgi, *Electrochim. Acta* **46** (2001) 799.
114. M. Uchida, Y. Aoyama, N. Eda and A. Ohta, *J. Electrochem. Soc.* **142** (1995) 4143.
115. P. Berg, A. Novruzi and K. Promislow, *Chem. Eng. Sci.* **61** (2006) 4316.
116. G. Y. Lin, W. S. He and T. Van Nguyen, *J. Electrochem. Soc.* **151** (2004) A1999.
117. W. Sun, B. A. Peppley and K. Karan, *Electrochim. Acta* **50** (2005) 3359.
118. K. M. Yin, *J. Electrochem. Soc.* **152** (2005) A583.

119. G. Q. Wang, P. P. Mukherjee and C. Y. Wang, *Electrochim. Acta* **51** (2006) 3139.
120. G. Q. Wang, P. P. Mukherjee and C. Y. Wang, *Electrochim. Acta* **51** (2006) 3151.
121. Z. N. Farhat, *J. Power Sources* **138** (2004) 68.
122. J. Xie, D. L. Wood, K. L. More, P. Atanassov and R. L. Borup, *J. Electrochem. Soc.* **152** (2005) A1011.
123. Y. Bultel, P. Ozil and R. Durand, *J. Appl. Electrochem.* **30** (2000) 1369.
124. Y. Bultel, P. Ozil and R. Durand, *Electrochim. Acta* **43** (1998) 1077.
125. Y. Bultel, P. Ozil and R. Durand, *J. Appl. Electrochem.* **28** (1998) 269.
126. Y. Bultel, P. Ozil and R. Durand, *J. Appl. Electrochem.* **29** (1999) 1025.
127. O. Antoine, Y. Bultel, R. Durand and P. Ozil, *Electrochim. Acta* **43** (1998) 3681.
128. J. Euler and W. Nonnenmacher, *Electrochim. Acta* **2** (1960) 268.
129. J. Newman and C. W. Tobias, *J. Electrochem. Soc.* **109** (1962) 1183.
130. J. Newman and W. Tiedemann, *AIChE J.* **21** (1975) 25.
131. J. O. M. Bockris and S. Srinivasan, *Fuel Cells: Their Electrochemistry*, McGraw-Hill, New York, 1969.
132. P. DeVidts and R. E. White, *J. Electrochem. Soc.* **144** (1997) 1343.
133. A. A. Shah, G. S. Kim, W. Gervais, A. Young, K. Promislow, J. Li and S. Ye, *J. Power Sources* **160** (2006) 1251.
134. H. S. Fogler, *Elements of Chemical Reaction Engineering*, Prentice-Hall, Upper Saddle River, NJ, 1992.
135. E. W. Thiele, *Ind. Eng. Chem.* **31** (1939) 916.
136. M. Eikerling, A. S. Ioselevich and A. A. Kornyshev, *Fuel Cells* **4** (2004) 131.
137. M. Secanell, K. Karan, A. Suleman and N. Djilali, *Electrochim. Acta* **52** (2007) 6318.
138. Q. P. Wang, M. Eikerling, D. T. Song, Z. S. Liu, T. Navessin, Z. Xie and S. Holdcroft, *J. Electrochem. Soc.* **151** (2004) A950.
139. D. T. Song, Q. P. Wang, Z. S. Liu, M. Eikerling, Z. Xie, T. Navessin and S. Holdcroft, *Electrochim. Acta* **50** (2005) 3347.
140. M. Eikerling, *J. Electrochem. Soc.* **153** (2006) E58.
141. C. Y. Du, T. Yang, R. F. Shi, G. R. Yin and X. Q. Cheng, *Electrochim. Acta* **51** (2006) 4934.
142. M. Eikerling and A. A. Kornyshev, *J. Electronanal. Chem.* **475** (1999) 107.
143. N. Wagner, *J. Appl. Electrochem.* **32** (2002) 859.
144. C. Y. Yuh and J. R. Selman, *AIChE J.* **34** (1988) 1949.
145. I. D. Raistrick, *Electrochim. Acta* **35** (1990) 1579.
146. R. Makharia, M. F. Mathias and D. R. Baker, *J. Electrochem. Soc.* **152** (2005) A970.
147. O. Himanen and T. Hottinen, *Electrochim. Acta* **52** (2006) 581.
148. J. M. Le Canut, R. M. Abouatallah and D. A. Harrington, *J. Electrochem. Soc.* **153** (2006) A857.
149. M. Schulze, N. Wagner, T. Kaz and K. A. Friedrich, *Electrochim. Acta* **52** (2007) 2328.
150. F. Jaouen and G. Lindbergh, *J. Electrochem. Soc.* **150** (2003) A1699.
151. T. E. Springer, T. A. Zawodzinski, M. S. Wilson and S. Gottesfeld, *J. Electrochem. Soc.* **143** (1996) 587.
152. Q. Z. Guo and R. E. White, *J. Electrochem. Soc.* **151** (2004) E133.

153. K. Wiezell, P. Gode and G. Lindbergh, *J. Electrochem. Soc.* **153** (2006) A749.
154. T. E. Springer, T. A. Zawodzinski and S. Gottesfeld, *J. Electrochem. Soc.* **138** (1991) 2334.
155. S. Patankar, *Numerical Heat Transfer and Fluid Flow*, Hemisphere Publishing Corporation, Washington, DC, 1980.
156. K. S. Udell, *Int. J. Heat Mass Transfer* **28** (1985) 485.
157. J. P. Feser, A. K. Prasad and S. G. Advani, *J. Power Sources* **162** (2006) 1226.
158. E. C. Kumbur, K. V. Sharp and M. M. Mench, *J. Power Sources* **168** (2007) 356.
159. U. Pasaogullari and C. Y. Wang, *J. Electrochem. Soc.* **152** (2005) A380.
160. C. H. Chao and A. J. J. Hwang, *J. Power Sources* **160** (2006) 1122.
161. V. Gurau, M. J. Bluemle, E. S. De Castro, Y. M. Tsou, T. A. Zawodzinski and J. A. Mann, *J. Power Sources* **165** (2007) 793.
162. X. W. Shan and H. D. Chen, *Phys. Rev. E* **47** (1993) 1815.
163. X. W. Shan and H. D. Chen, *Phys. Rev. E* **49** (1994) 2941.
164. C. Pan, M. Hilpert and C. T. Miller, *Water Resour. Res.* **40** (2004).
165. C. R. Ethier, *AIChE J.* **37** (1991) 1227.
166. K. E. Thompson, *AIChE J.* **48** (2002) 1369.
167. H. J. Vogel, J. Tolke, V. P. Schulz, M. Krafczyk and K. Roth, *Vadose Zone J.* **4** (2005) 380.
168. V. P. Schulz, J. Becker, A. Wiegmann, P. P. Mukherjee and C. Y. Wang, *J. Electrochem. Soc.* **154** (2007) B419.
169. P. K. Sinha and C. Y. Wang, *ECS Trans.* **3** (2006) 387.
170. M. S. Valavanides and A. C. Payatakes, *Adv. Water Resour.* **24** (2001) 385.
171. M. Prat, *Int. J. Heat Mass Transfer* **50** (2007) 1455.
172. B. Markicevic and N. Djilali, *Phys. Fluids* **18** (2006).
173. R. D. Hazlett, *Transp Porous Media* **20** (1995) 21.
174. V. Sygouni, C. D. Tsakiroglou and A. C. Payatakes, *Phys. Fluids* **18** (2006).
175. Y. Shi, J. S. Xiao, M. Pan and R. Z. Yuan, *J. Power Sources* **160** (2006) 277.
176. G. L. He, Z. C. Zhao, P. W. Ming, A. Abuliti and C. Y. Yin, *J. Power Sources* **163** (2007) 846.
177. U. Pasaogullari, C. Y. Wang and K. S. Chen, *J. Electrochem. Soc.* **152** (2005) A1574.
178. A. T. Corey, *Producer's Mon.* **18** (1954) 38.
179. R. H. Brooks and A. T. Corey, in *Hydrology Papers*, Colorado State University, Fort Collins (1964).
180. M. T. Vangenuchten, *Soil Sci. Soc. Am. J.* **44** (1980) 892.
181. A. Z. Weber and J. Newman, *J. Electrochem. Soc.* **151** (2004) A311.
182. D. Natarajan and T. V. Nguyen, *J. Electrochem. Soc.* **148** (2001) A1324.
183. X. L. Wang, H. M. Zhang, J. L. Zhang, H. F. Xu, Z. Q. Tian, J. Chen, H. X. Zhong, Y. M. Liang and B. L. Yi, *Electrochim. Acta* **51** (2006) 4909.
184. M. S. Wilson, J. A. Valerio and S. Gottesfeld, *Electrochim. Acta* **40** (1995) 355.
185. N. Hara, K. Tsurumi and M. Watanabe, *J. Electroanal. Chem.* **413** (1996) 81.
186. E. Passalacqua, G. Squadrito, F. Lufrano, A. Patti and L. Giorgi, *J. Appl. Electrochem.* **31** (2001) 449.
187. K. Karan, H. Atiyeh, A. Phoenix, E. Halliop, J. Pharoah and B. Peppley, *Electrochem. Solid State Lett.* **10** (2007) B34.
188. Z. G. Qi and A. Kaufman, *J. Power Sources* **109** (2002) 38.

189. Z. G. Zhan, J. S. Xiao, D. Y. Li, M. Pan and R. Z. Yuan, *J. Power Sources* **160** (2006) 1041.
190. G. J. M. Janssen and M. L. J. Overvelde, *J. Power Sources* **101** (2001) 117.
191. V. A. Paganin, E. A. Ticianelli and E. R. Gonzalez, *J. Appl. Electrochem.* **26** (1996) 297.
192. C. S. Kong, D.-Y. Kim, H.-K. Lee, Y.-G. Shul and T.-H. Lee, *J. Power Sources* **108** (2002) 185.
193. A. A. Shah, G. S. Kim, P. C. Sui and D. Harvey, *J. Power Sources* **163** (2007) 793.
194. S. Shimpalee, U. Beuscher and J. W. Van Zee, *J. Power Sources* **163** (2006) 480.
195. J. Chen, T. Matsuura and M. Hori, *J. Power Sources* **131** (2004) 155.
196. U. Pasaogullari and C. Y. Wang, *Electrochim. Acta* **49** (2004) 4359.
197. A. Z. Weber and J. Newman, *J. Electrochem. Soc.* **152** (2005) A677.
198. E. Birgersson, M. Noponen and M. Vynnycky, *J. Electrochem. Soc.* **152** (2005) A1021.
199. J. Ramousse, J. Deseure, O. Lottin, S. Didierjean and D. Maillet, *J. Power Sources* **145** (2005) 416.
200. Y. Y. Shan and S. Y. Choe, *J. Power Sources* **145** (2005) 30.
201. H. Ju, H. Meng and C. Y. Wang, *Int. J. Heat Mass Transfer* **48** (2005) 1303.
202. G. L. Hu and J. R. Fan, *Energy Fuels* **20** (2006) 738.
203. H. C. Ju, C. Y. Wang, S. Cleghorn and U. Beuscher, *J. Electrochem. Soc.* **153** (2006) A249.
204. Y. Wang and C. Y. Wang, *J. Electrochem. Soc.* **153** (2006) A1193.
205. O. N. Scholes, S. A. Clayton, A. F. A. Hoadley and C. Tiu, *Transp. in Porous Media* **68** (2007) 365.
206. J. G. Pharoah, K. Karan and W. Sun, *J. Power Sources* **161** (2006) 214.
207. U. Pasaogullari, P. P. Mukherjee, C. Y. Wang and K. S. Chen, *J. Electrochem. Soc.* **154** (2007) B823.
208. M. M. Tomadakis and S. V. Sotirchos, *AIChE J.* **37** (1991) 74.
209. J. T. Gostick, M. W. Fowler, M. D. Pritzker, M. A. Ioannidis and L. M. Behra, *J. Power Sources* **162** (2006) 228.
210. T. I. I. Toray, Carbon Fiber Paper "TGP-H" Property Sheet.
211. H. Meng, *J. Power Sources* **161** (2006) 466.
212. M. Khandelwal and M. M. Mench, *J. Power Sources* **161** (2006) 1106.
213. J. T. Gostick, M. W. Fowler, M. D. Pritzker, M. A. Ioannidis and L. M. Behra, *J. Power Sources* **162** (2006) 228.
214. G. W. Jackson and D. F. James, *Can. J. Chem. Eng.* **64** (1986) 364.
215. J. G. Pharoah, *J. Power Sources* **144** (2005) 77.
216. M. V. Williams, H. R. Kunz and J. M. Fenton, *J. Electrochem. Soc.* **151** (2004) A1617.
217. W. K. Lee, C. H. Ho, J. W. Van Zee and M. Murthy, *J. Power Sources* **84** (1999) 45.
218. J. Ihonen, M. Mikkola and G. Lindbergh, *J. Electrochem. Soc.* **151** (2004) A1152.
219. P. Zhou, C. W. Wu and G. J. Ma, *J. Power Sources* **163** (2007) 874.
220. P. Zhou and C. W. Wu, *J. Power Sources* **170** (2007) 93.
221. P. C. Sui and N. Djilali, *J. Power Sources* **161** (2006) 294.
222. J. B. Ge, A. Higier and H. T. Liu, *J. Power Sources* **159** (2006) 922.

223. W. R. Chang, J. J. Hwang, F. B. Weng and S. H. Chan, *J. Power Sources* **166** (2007) 149.
224. T. Hottinen and O. Himanen, *Electrochem. Commun.* **9** (2007) 1047.
225. M. Mathias, J. Roth, J. Fleming and W. Lehnert, in *Handbook of Fuel Cells: Fundamentals, Technology, and Applications, Vol. 3*, W. Vielstich, A. Lamm and H. A. Gasteiger Editors, Wiley, New York (2003).
226. Y. L. Tang, A. M. Karlsson, M. H. Santare, M. Gilbert, S. Cleghorn and W. B. Johnson, *Mater. Sci. Eng. a-Struct. Mater. Prop. Microstruct. Process.* **425** (2006) 297.
227. O. C. Zienkiewicz and R. L. Taylor, *The Finite Element Method*, Butterworth-Heinemann, Oxford, UK, 2000.
228. A. Bazylak, D. Sinton, Z. S. Liu and N. Djilali, *J. Power Sources* **163** (2007) 784.
229. E. Endoh, S. Terazono, H. Widjaja and Y. Takimoto, *Electrochem. Solid State Lett.* **7** (2004) A209.
230. G. H. Guvelioglu and H. G. Stenger, *J. Power Sources* **147** (2005) 95.
231. G. H. Guvelioglu and H. G. Stenger, *J. Power Sources* **163** (2007) 882.
232. W. Huang, B. Zhou and A. Sobiesiak, *J. Electrochem. Soc.* **153** (2006) A1945.
233. H. Meng and C. Y. Wang, *Chem. Eng. Sci.* **59** (2004) 3331.
234. P. T. Nguyen, T. Berning and N. Djilali, *J. Power Sources* **130** (2004) 149.
235. H. C. Ju, C. Y. Wang, S. Cleghorn and U. Beuscher, *J. Electrochem. Soc.* **152** (2005) A1645.
236. H. Ju, G. Luo and C. Y. Wang, *J. Electrochem. Soc.* **154** (2007) B218.
237. S. Shimpalee, S. Greenway, D. Spuckler and J. W. Van Zee, *J. Power Sources* **135** (2004) 79.
238. S. A. Freunberger, A. Wokaun and F. N. Buchi, *J. Electrochem. Soc.* **153** (2006) A909.
239. Y. Wang and C. Y. Wang, *J. Power Sources* **153** (2006) 130.
240. T. W. Patterson and R. M. Darling, *Electrochem. Solid State Lett.* **9** (2006) A183.
241. P. W. Li, L. Schaefer, Q. M. Wang, T. Zhang and M. K. Chyu, *J. Power Sources* **115** (2003) 90.
242. Y. Zong, B. Zhou and A. Sobiesiak, *J. Power Sources* **161** (2006) 143.
243. C. I. Lee and H. S. Chu, *J. Power Sources* **161** (2006) 949.
244. J. S. Yi, J. D. L. Yang and C. King, *AIChE J.* **50** (2004) 2594.
245. E. C. Kumbur, K. V. Sharp and M. M. Mench, *J. Power Sources* **161** (2006) 333.
246. Z. G. Zhan, J. S. Xiao, M. Pan and R. Z. Yuan, *J. Power Sources* **160** (2006) 1.
247. K. Jiao, B. Zhou and P. Quan, *J. Power Sources* **154** (2006) 124.
248. P. Quan and M. C. Lai, *J. Power Sources* **164** (2007) 222.
249. H. Meng and C. Y. Wang, *J. Electrochem. Soc.* **152** (2005) A1733.
250. K. S. Chen, M. A. Hickner and D. R. Noble, *Int. J. Energy Res.* **29** (2005) 1113.
251. G. L. He, P. W. Ming, Z. C. Zhao, A. Abudula and Y. Xiao, *J. Power Sources* **163** (2007) 864.
252. D. P. Wilkinson and J. St-Pierre, *J. Power Sources* **113** (2003) 101.
253. P. Berg, K. Promislow, J. St Pierre, J. Stumper and B. Wetton, *J. Electrochem. Soc.* **151** (2004) A341.
254. E. Birgersson and M. Vynnycky, *J. Power Sources* **153** (2006) 76.

255. S. Um and C. Y. Wang, *J. Power Sources* **156** (2006) 211.
256. Y. Wang and C. Y. Wang, *J. Power Sources* **147** (2005) 148.
257. M. S. Wilson, Fuel cell with interdigitated porous flow-field, The Regents of the University of California Office of Technology Transfer, U.S. (1995).
258. W. He, J. S. Yi and T. V. Nguyen, *AIChE J.* **46** (2000) 2053.
259. J. S. Yi and T. V. Nguyen, *J. Electrochem. Soc.* **146** (1999) 38.
260. J. P. Feser, A. K. Prasad and S. G. Advani, *J. Power Sources* **161** (2006) 404.
261. J. Park and X. G. Li, *J. Power Sources* **163** (2007) 853.
262. L. Wang and H. T. Liu, *J. Power Sources* **134** (2004) 185.
263. W. M. Yan, H. Y. Li and W. C. Tsai, *J. Electrochem. Soc.* **153** (2006) A1984.
264. J. J. Hwang, C. H. Chao, W. Y. Ho, C. L. Chang and D. Y. Wang, *J. Power Sources* **157** (2006) 85.
265. S. Um and C. Y. Wang, *J. Power Sources* **125** (2004) 40.
266. H. Yamada, T. Hatanaka, H. Murata and Y. Morimoto, *J. Electrochem. Soc.* **153** (2006) A1748.
267. J. Zou, X. F. Peng and W. M. Yan, *J. Power Sources* **159** (2006) 514.
268. E. Arato, M. Pinna and P. Costa, *J. Power Sources* **158** (2006) 206.
269. E. Arato and P. Costa, *J. Power Sources* **158** (2006) 200.
270. G. Inoue, Y. Matsukuma and M. Minemoto, *J. Power Sources* **157** (2006) 136.
271. K. V. Zhukovsky, *AIChE J.* **49** (2003) 3029.
272. A. Kazim, H. T. Liu and P. Forges, *J. Appl. Electrochem.* **29** (1999) 1409.
273. C. Reiser, Ion Exchange Membrane Fuel Cell Power Plant with Water Management Pressure Differentials, UTC Fuel Cells, United States (1997).
274. A. Z. Weber and R. M. Darling, *J. Power Sources* **168** (2007) 191.
275. D. L. Wood, Y. S. Yi and T. V. Nguyen, *Electrochim. Acta* **43** (1998) 3795.
276. T. Yang and P. Shi, *J. Electrochem. Soc.* **153** (2006) A1518.
277. S. H. Ge, X. G. Li and I. M. Hsing, *J. Electrochem. Soc.* **151** (2004) B523.
278. S. H. Ge, X. G. Li and I. M. Hsing, *Electrochim. Acta* **50** (2005) 1909.
279. C. R. Buie, J. D. Posner, T. Fabian, C. A. Suk-Won, D. Kim, F. B. Prinz, J. K. Eaton and J. G. Santiago, *J. Power Sources* **161** (2006) 191.
280. J. P. Meyers, R. M. Darling, C. Evans, R. Balliet and M. L. Perry, *ECS Trans.* **3** (2006) 1207.
281. S. Litster, J. G. Pharoah, G. McLean and N. Djilali, *J. Power Sources* **156** (2006) 334.
282. J. J. Hwang, S. D. Wu, R. G. Pen, P. Y. Chen and C. H. Chao, *J. Power Sources* **160** (2006) 18.
283. S. Litster and N. Djilali, *Electrochim. Acta* **52** (2007) 3849.
284. Hydrogen, fuel cells & infrastructure technologies program; multi-year research, development and demonstration plan, Tables 3.4.2 and 3.4.3., p. 14, U.S. Department of Energy (2007).
285. F. Mueller, J. Brouwer, S. G. Kang, H. S. Kim and K. D. Min, *J. Power Sources* **163** (2007) 814.
286. W. S. He, G. Y. Lin and T. V. Nguyen, *AIChE J.* **49** (2003) 3221.
287. R. Mosdale, G. Gebel and M. Pineri, *J. Membr. Sci.* **118** (1996) 269.
288. A. Haddad, R. Bouyekhf, A. El Moudni and M. Wack, *J. Power Sources* **163** (2006) 420.
289. W. Friede, S. Rael and B. Davat, *IEEE Trans. Power Electron.* **19** (2004) 1234.
290. H. M. Yu and C. Ziegler, *J. Electrochem. Soc.* **153** (2006) A570.

291. F. L. Chen, M. H. Chang and C. F. Fang, *J. Power Sources* **164** (2007) 649.
292. A. Vorobev, O. Zikanov and T. Shamim, *J. Power Sources* **166** (2007) 92.
293. I. Nazarov and K. Promislow, *Chem. Eng. Sci.* **61** (2006) 3198.
294. J. Benziger, E. Chia, J. F. Moxley and I. G. Kevrekidis, *Chem. Eng. Sci.* **60** (2005) 1743.
295. R. M. Rao and R. Rengaswamy, *Chem. Eng. Sci.* **61** (2006) 7393.
296. W. M. Yan, C. Y. Soong, F. L. Chen and H. S. Chu, *J. Power Sources* **143** (2005) 48.
297. S. Shimpalee, W. K. Lee, J. W. Van Zee and H. Naseri-Neshat, *J. Power Sources* **156** (2006) 355.
298. Y. Wang and C. Y. Wang, *Electrochim. Acta* **50** (2005) 1307.
299. Y. Wang and C. Y. Wang, *Electrochim. Acta* **51** (2006) 3924.
300. J. C. Amphlett, R. F. Mann, B. A. Peppley, P. R. Roberge and A. Rodrigues, *J. Power Sources* **61** (1996) 183.
301. F. L. Chen, Y. G. Su, C. Y. Soong, W. M. Yan and H. S. Chu, *J. Electronanal. Chem.* **566** (2004) 85.
302. Y. Shan and S. Y. Choe, *J. Power Sources* **158** (2006) 274.
303. H. Wu, P. Berg and X. G. Li, *J. Power Sources* **165** (2007) 232.
304. T. F. Fuller and J. Newman, *J. Electrochem. Soc.* **140** (1993) 1218.
305. Y. Wang and C. Y. Wang, *J. Electrochem. Soc.* **153** (2006) A1193.
306. Y. Y. Shan, S. Y. Choe and S. H. Choi, *J. Power Sources* **165** (2007) 196.
307. D. Natarajan and T. V. Nguyen, *J. Power Sources* **115** (2003) 66.
308. C. Ziegler, H. M. Yu and J. O. Schumacher, *J. Electrochem. Soc.* **152** (2005) A1555.
309. S. M. Chang and H. S. Chu, *J. Power Sources* **161** (2006) 1161.
310. D. T. Song, Q. P. Wang, Z. S. Liu and C. Huang, *J. Power Sources* **159** (2006) 928.
311. H. Guilin and F. Jianren, *J. Power Sources* **165** (2007) 171.
312. S. Shimpalee, D. Spuckler and J. W. Van Zee, *J. Power Sources* **167** (2007) 130.
313. Hydrogen, fuel cells & infrastructure technologies program; multi-year research, development and demonstration plan, Tables 3.4.2 and 3.4.3., in, p. 14, U.S. Department of Energy (2007).
314. A. A. Pesaran, G.-H. Kim and J. D. Gonder, *NREL* **MP-540-38760** (2005).
315. M. F. Mathias, R. Makharia, H. A. Gasteiger, Jason J. Conley, T. J. Fuller, C. J. Gittleman, S. S. Kocha, D. P. Miller, C. K. Mittelsteadt, T. Xie, S. G. Yan and P. T. Yu, *Electrochem. Soc. Interface* **14** (2005) 24.
316. R. Bradean, H. Haas, K. Eggen, C. Richards and T. Vrba, *ECS Trans.* **3** (2006) 1159.
317. M. Modell and R. C. Reid, *Thermodynamics and its Applications*, Prentice-Hall, Englewood Cliffs, NJ 1974.
318. S. Taber, *J. Geol.* **38** (1930) 303.
319. J. D. Sage and M. Porebska, *J. Cold Reg. Eng.* **7** (1993) 99.
320. R. D. Miller, *Proceedings of the Third International Conference on Permafrost* (1978) 707.
321. J. P. G. Loch, *Soil Science* **126** (1978) 77.
322. A. W. Rempel, J. S. Wettlaufer and M. G. Worster, *J. Fluid Mech.* **498** (2004) 227.

323. K. S. Henry, A review of the thermodynamics of frost heave, in, H. US Army Cold Regions Research and Engineering Laboratory, NH, Report TR-00-16 Editor, p. 1 (2000).
324. S. H. He and M. M. Mench, *J. Electrochem. Soc.* **153** (2006) A1724.
325. S. He and M. M. Mench, *ECS Trans.* **3** (2006) 897.
326. E. L. Thompson, T. W. Capehart, T. J. Fuller and J. Jorne, *J. Electrochem. Soc.* **153** (2006) A2351.
327. E. L. Thompson, J. Jorne and H. A. Gasteiger, *J. Electrochem. Soc.* **154** (2007) B783.
328. M. De Francesco and E. Arato, *J. Power Sources* **108** (2002) 41.
329. M. Sundaresan and R. M. Moore, *Fuel Cells* **5** (2005) 476.
330. M. Sundaresan and R. M. Moore, *J. Power Sources* **145** (2005) 534.
331. M. Oszcipok, A. Hakenjos, D. Riemann and C. Hebling, *Fuel Cells* **7** (2007) 135.
332. Y. Hishinuma, T. Chikahisa, F. Kagami and T. Ogawa, *Jsme Int. J. Ser. B-Fluids and Thermal Eng.* **47** (2004) 235.
333. L. Mao and C. Y. Wang, *J. Electrochem. Soc.* **154** (2007) B139.
334. L. Mao and C. Y. Wang, *J. Electrochem. Soc.* **154** (2007) B341.
335. H. A. Gasteiger and M. F. Mathias, in *Proton Conducting Membrane Fuel Cells III*, J. W. Van Zee, T. F. Fuller, S. Gottesfeld and M. Murthy Editors, The Electrochemical Society Proceeding Series, Pennington, NJ (2002).
336. J. L. Zhang, Z. Xie, J. J. Zhang, Y. H. Tanga, C. J. Song, T. Navessin, Z. Q. Shi, D. T. Song, H. J. Wang, D. P. Wilkinson, Z. S. Liu and S. Holdcroft, *J. Power Sources* **160** (2006) 872.
337. C. Yang, P. Costamagna, S. Srinivasan, J. Benziger and A. B. Bocarsly, *J. Power Sources* **103** (2001) 1.
338. Y. Y. Shao, G. P. Yin, Z. B. Wang and Y. Z. Gao, *J. Power Sources* **167** (2007) 235.
339. T. M. Thampan, N. H. Jalani, P. Choi and R. Datta, *J. Electrochem. Soc.* **152** (2005) A316.
340. F. A. de Bruijn, R. C. Makkus, R. K. A. M. Mallant and G. J. M. Janssen, in *Advances in Fuel Cells, Vol. 1*, T. S. Zhao, K.-D. Kreuer and T. V. Nguyen Editors, Elsevier, Amsterdam (2007).
341. R. C. Jiang, H. R. Kunz and J. M. Fenton, *J. Electrochem. Soc.* **152** (2005) A1329.
342. J. Peng and S. J. Lee, *J. Power Sources* **162** (2006) 1182.
343. D. F. Cheddie and N. D. H. Munroe, *J. Power Sources* **160** (2006) 215.
344. D. F. Cheddie and N. D. H. Munroe, *Int. J. Hydrog. Energy* **32** (2007) 832.
345. J. W. Hu, H. M. Zhang, Y. F. Zhai, G. Liu, J. Hu and B. L. Yi, *Electrochim. Acta* **52** (2006) 394.
346. P. K. Sinha, C. Y. Wang and U. Beuscher, *J. Electrochem. Soc.* **154** (2007) B106.
347. P. K. Sinha, C. Y. Wang and A. Su, *Int. J. Hydrog. Energy* **32** (2007) 886.
348. J. L. Zhang, Y. H. Tang, C. J. Song, X. A. Cheng, J. J. Zhang and H. J. Wang, *Electrochim. Acta* **52** (2007) 5095.
349. H. Xu, H. R. Kunz and J. M. Fenton, *Electrochim. Acta* **52** (2007) 3525.
350. Y. Song, J. M. Fenton, H. R. Kunz, L. J. Bonville and M. V. Williams, *J. Electrochem. Soc.* **152** (2005) A539.

351. Y. H. Tang, J. J. Zhang, C. J. Song, H. Liu, J. L. Zhang, H. J. Wang, S. Mackinnon, T. Peckham, J. Li, S. McDermid and P. Kozak, *J. Electrochem. Soc.* **153** (2006) A2036.
352. J. S. Wainright, J. T. Wang, D. Weng, R. F. Savinell and M. Litt, *J. Electrochem. Soc.* **142** (1995) L121.
353. S. von Kraemer, M. Puchner, P. Jannasch, A. Lundblad and G. Lindbergh, *J. Electrochem. Soc.* **153** (2006) A2077.
354. L. J. Bonville, H. R. Kunz, Y. Song, A. Mientek, M. Williams, A. Ching and J. M. Fenton, *J. Power Sources* **144** (2005) 107.
355. K. C. Neyerlin, H. A. Gasteiger, C. K. Mittelsteadt, J. Jorne and W. B. Gu, *J. Electrochem. Soc.* **152** (2005) A1073.
356. B. S. Pivovar and Y. S. Kim, *J. Electrochem. Soc.* **154** (2007) B739.
357. H. Xu, Y. Song, H. R. Kunz and J. M. Fenton, *J. Electrochem. Soc.* **152** (2005) A1828.
358. C. J. Song, Y. H. Tang, J. L. Zhang, J. J. Zhang, H. J. Wang, J. Shen, S. McDermid, J. Li and P. Kozak, *Electrochim. Acta* **52** (2007) 2552.
359. L. Xiao, H. Zhang, T. Jana, E. Scanlon, R. Chen, E. W. Choe, L. S. Ramanathan, S. Yu and B. C. Benicewicz, *Fuel Cells* **5** (2005) 287.
360. S. Srinivasan, *Fuel Cells. From Fundamentals to Applications*, Springer, New York, Berlin, Heidelberg, 2006.
361. P. Zhou, C. W. Wu and G. J. Ma, *J. Power Sources* **163** (2007) 874.
362. R. Bradean, H. Haas, K. Eggen, C. Richards and T. Vrba, *ECS Trans.* **3** (2006) 1159.

8

Adaptive Characterization and Modeling of Electrochemical Energy Storage Devices for Hybrid Electric Vehicle Applications

Mark W. Verbrugge

General Motors Research and Development, Warren, MI 48090-9055, USA

Abstract The control and adaptive characterization of batteries or supercapacitors, which are central to the construction of electrochemical energy storage devices operating in hybrid electric vehicles, require that the state estimator forming the basis of the control system be informed by the underlying electrochemistry. In this chapter, we overview tools and methods useful for the integration of batteries and supercapacitors into HEV systems, with a focus on the construction of state estimators. The approach discussed may be viewed as combining electroanalytical procedures with formal controls methods.

I. INTRODUCTION

For efficient energy management of a system employing batteries or supercapacitors, an adaptive algorithm that can characterize the state of the electrochemical energy-storage system (EESS) is required. Inputs to the algorithm include the system current, voltage, and temperature, and outputs characterize the energy content (state of charge, or SOC), predicted power capability (state of power, or

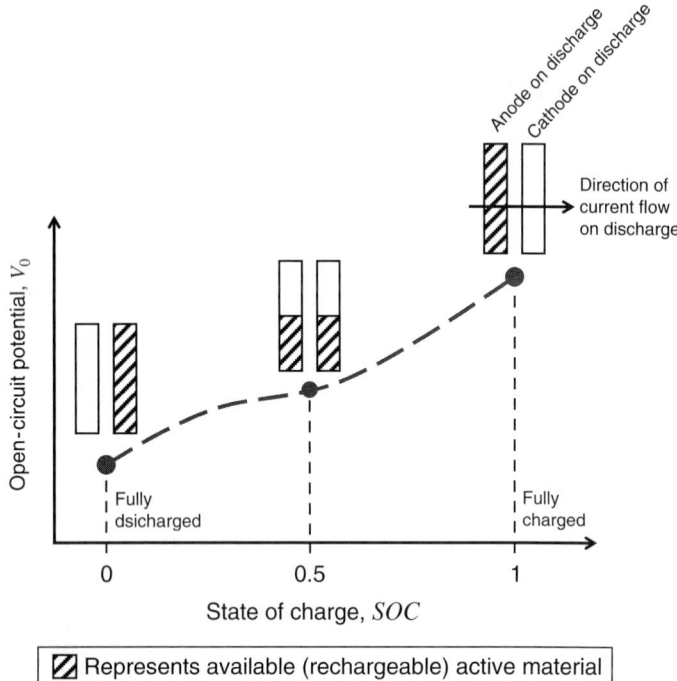

Figure 1. Schematic representation of state of charge, SOC. For each of the paired rectangle, the anode (on discharge) is left of the cathode. The SOC reflects the potential energy available to do work, while $1 - SOC$ represents the energy that can be absorbed upon cell charging under equilibrium conditions (Side reactions and other complications are neglected in this simplified perspective.).

SOP), and performance relative to the new and end-of-life condition (state of health, or SOH). The SOC corresponds to the stored charge available to do work relative to that which is available after the battery has been fully charged. A schematic representation of the SOC is provided in Fig. 1, and the SOH is described qualitative in Fig. 2. The SOH schematic reflects the common observation that with cycling and aging, EESS's lose coulombic capacity and increase in impedance. The impedance is readily extracted by state estimators, as shall be described later. Hence, we assume that we can assess the SOH if we have a method to identify the impedance spectrum for the battery system over the frequency range of interest

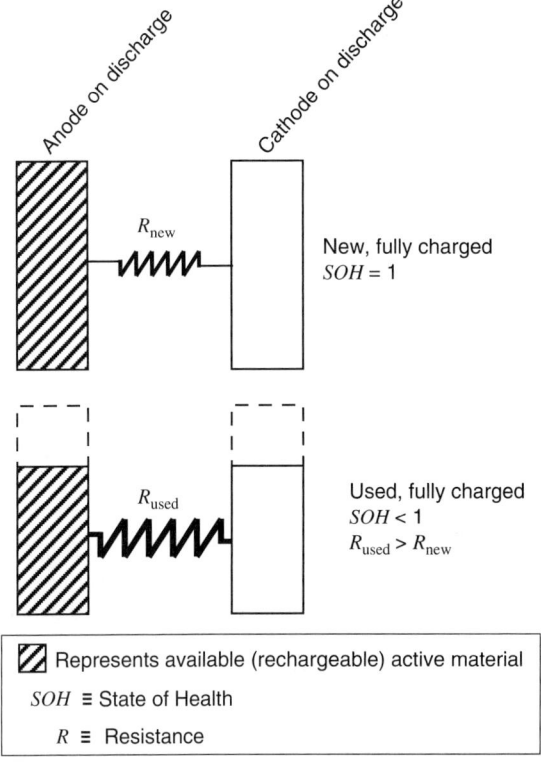

Figure 2. Schematic representation of state of health, SOH. As electrochemical energy storage devices are cycled or aged over prolonged durations, the common observation is that the coulombic capacity decreases and the resistance increases. State of power, SOP, reflects the ability of the system to deliver discharge power or accept charge power, which is a function of the available active material (capacity times SOC) and the resistance. A more complete description replaces the resistance depicted above with the system impedance, which is frequency dependent.

in an on-line (or adaptive) manner. In this review, we do propose a specific definition of SOH. Thus the thermodynamics allow one to assess the potential energy (SOC) of the system, and irreversible losses as well as the SOH can be assessed once the impedance spectrum is identified. In addition to SOC and SOH, the state of power capability (SOP) is of critical importance in the context of hybrid electric vehicles or any other device in which two power sources are available, as control of the composite power system requires knowledge of the available (charge or discharge) power for each energy source. For automotive applications, the conversion of input information to outputs must be fast and not require substantial amounts of computer storage, consistent with embedded-controller and serial-data-transfer capabilities. Generally these two limitations mandate that algorithms be fully recursive, wherein all information utilized by the algorithm stems from previous time-step values and measurements that are immediately available (Fig. 3).

Before delving further into the topic of state estimators and SOC, SOH, and SOP algorithms, it is helpful to understand the context within which EESS's operate and the utility of these quantities. First, we are often interested in power systems that comprise more than one power source, as is depicted by the schematic in Fig. 4. For such systems, it is important to know the state of each power system, thus enabling efficient operation. The propulsion system architecture of hybrid electric vehicles (HEV's) corresponds to a composite power system, as shown in the more detailed schematic of Fig. 4; a parallel HEV configuration is displayed, but the ensuing discussion is germane to all HEV architectures. A representative

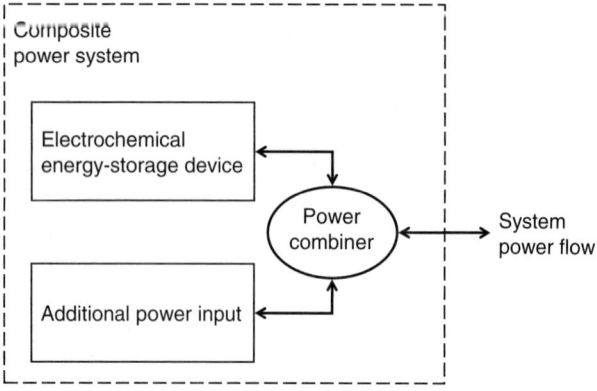

Figure 3. Schematic of a composite power system. For effective system operation, a real-time state estimator for electrochemical energy storage device is required (as well as for the additional power input device).

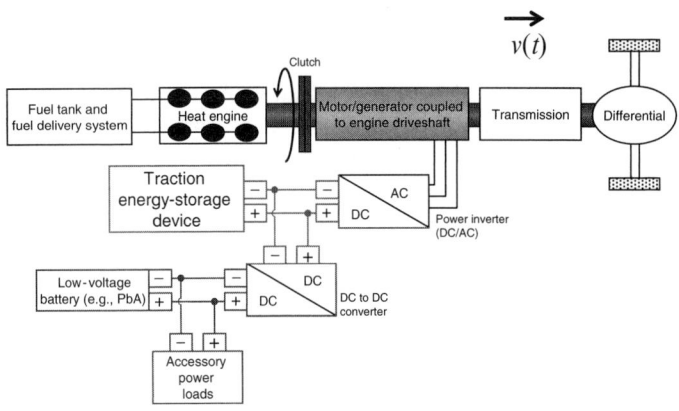

Figure 4. Schematic illustration of a parallel hybrid propulsion system. This schematic is a specific and more detailed example of a composite power system (cf. Fig. 3).

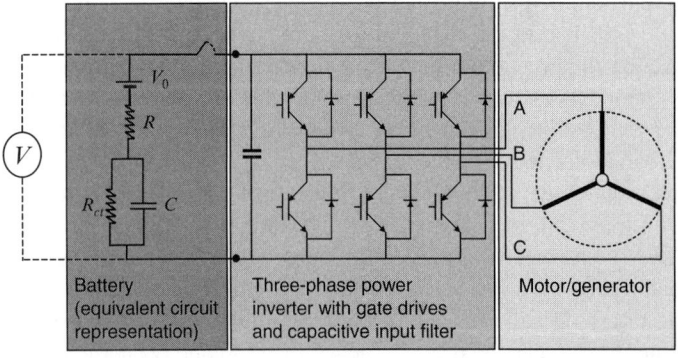

Figure 5. Representative three-phase electric-traction system. This schematic corresponds to the (1) traction energy storage device, (2) DC/AC power inverter, and (3) motor/generator depicted in Fig. 4.

HEV electric-traction system is depicted in Fig. 5, which corresponds to an expanded view of the traction energy storage device, DC/AC power inverter, and motor/generator depicted in Fig. 4. On discharge, the battery current flows into the three-phase power inverter module (PIM). Upon control of the gate drives (often powered by the low-voltage, 12 V electrical system) in the PIM, the alternating current drives the three-phase electric motor. For regenerative braking, the voltage V rises above that of the battery, and charging current flows into the battery. A switch (contactor) is shown in the upper right of the battery system, which allows for the immediate isolation of the high-voltage battery. An example of a vehicle architecture comprising a battery pack is provided in Fig. 6. We note in closing this discussion of composite power systems that even when a power system utilizes a single electrochemical source, as is the case for a battery electric vehicle, knowledge of the battery SOC, SOP, and SOH is still required for effective vehicle operation.

Obvious questions arise in the context of HEV operation that underscore the need to ascertain the state of the EESS. Should the heat engine be turned off when the vehicle comes to rest? To answer this question, it is key to know the SOP. For a charge-sustaining HEV's,[1] if the EESS's *discharge* power capability is too low, then the heat engine cannot be started upon a subsequent request by the driver and the customer is stranded, as the vehicle fails to provide propulsion. In a similar vein, when the vehicle must be slowed, can the EESS accept the regenerative braking energy delivered by the traction motor through the power inverter module? To answer this question, the state estimator must be able to forecast the *charge* power capability of the EESS. We can also outline scenarios that demand knowledge of the EESS energy content. If, for example, it is a requirement to run the vehicle electrical accessory power loads for 5 min with the heat engine off when the vehicle comes to a stop (e.g., run an electrical air-conditioning system), then the state estimator must track the SOC, and the SOC must be maintained above the threshold value to (1) enable the 5-min accessory power load requirement once the heat engine is turned off and (2) maintain sufficient SOC to restart the engine after the 5-min. In the context of

---

[1] Hybrid electric vehicles with small batteries relative to the energy content of the on-board fuel (e.g., gasoline) tank are run in a charge-sustaining mode, versus a vehicle that can charge off the electrical grid, often termed a plug-in hybrid or an extended-range electric vehicle (EREV). Charge-sustaining hybrids are more common, as the costs of today's batteries as well as those of the electric motors and power electronics are reduced relative to plug-in HEVs and EREVs. For a charge-sustaining HEV, the EESS oscillates about a specified SOC and all charging of the EESS is delivered by regenerative braking or by the heat engine through a generator (cf. Fig. 4). In contrast, for plug-in HEVs and EREVs, the SOC declines throughout driving to a lower threshold, after which the charge-sustaining mode is realized.

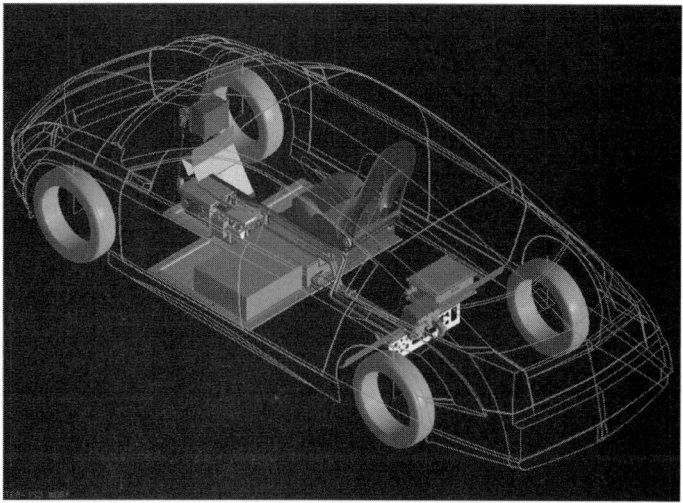

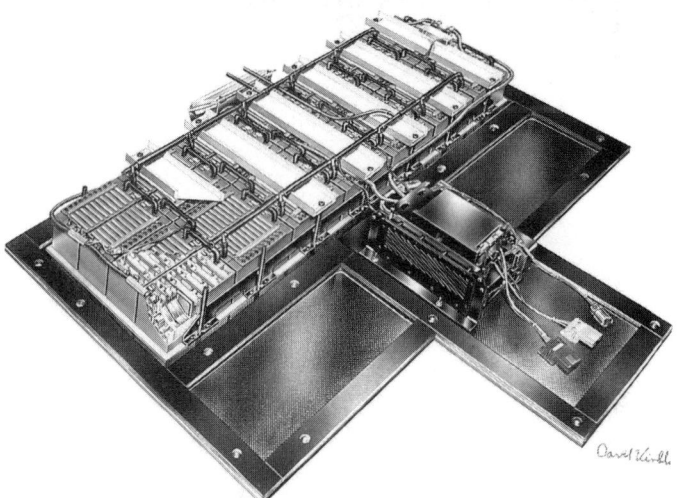

Figure 6. *Top*: Rendering of the Precept hybrid electric vehicle. The battery system is located under the front seats. The front axel has a coaxial electric traction system, and the rear axel receives torque through a traction system comprising an electric motor, a compression ignition heat engine, and a manual transmission that is automatically shifted. Regenerative braking takes place at both axels. *Bottom*: Battery pack system. The 28 battery modules are liquid cooled; the coolant enters and exits at the rear of the pack. Controllers, high-voltage devices (contactors, relays, etc.) are located in front and behind the modules. The vehicle achieved 80 mpg over the FTP cycle; battery data over the cycle is analyzed later in this paper.

Fig. 4, the implication is that the EESS would discharge through the dc/dc converter to power the air-conditioning system. HEV architectures exist wherein larger accessory power loads operate at the same voltage as the EESS (with improved energy efficiency). Last, for consistent vehicle operation, it is necessary to reflect the EESS SOH; when the SOH falls below a threshold value, then vehicle functionality will be noticeably impaired, the customer must be alerted (e.g., by a telltale light on the vehicle display panel), and the EESS must be replaced.

To construct a state estimator for the SOC, SOH, and SOP, a model reference adaptive systems can be employed. Early work in this area came out of the University of Aachen and was focused on lead-acid batteries in starting, lighting, and ignition (SLI) systems (1, 2). An overview of how a model reference adaptive system is constructed, relative to more traditional proportional-integral-differential (PID) control systems is provided by the block diagram of Fig. 7. A block diagram representing conventional control

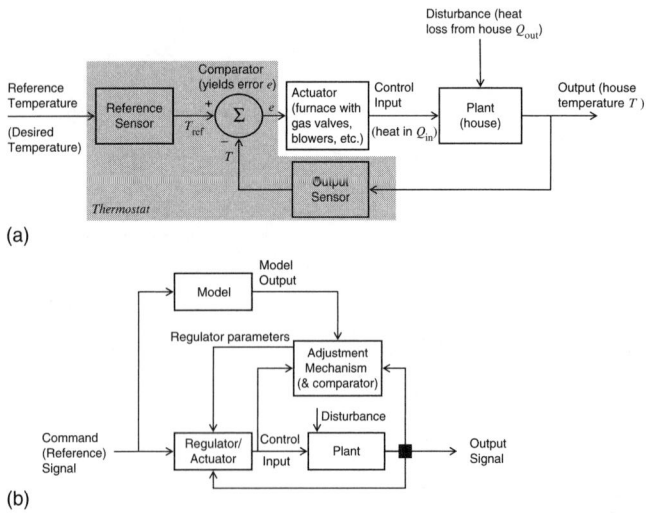

Figure 7. Block diagrams representing control system architectures. (**a**): Conventional control system commensurate with proportional-integral-differential (PID) schemes, where the error e is used to determine actuation. A heating system for a building is depicted. (**b**): Model Reference Adaptive System (MRAS). A model of the system is employed to better inform the control system relative to optimal actuation.

system commensurate with proportional-integral-differential (PID) schemes, wherein the error is used to determine actuation, is displayed in Fig. 7a. The methodology dates back to at least the 1600s, with examples including Drebbel's incubator in 1620 and speed control of wind-driven flour mills in the early 1700s leading to James Watt's use of the fly-ball (centrifugal) governor to control shaft rotation speeds. A model reference adaptive system, by contrast, is shown in Fig. 7b. In this case, a model of the system is employed to better inform the control system relative to optimal actuation. The parameters of the model are regressed adaptively, motivating the development of optimal parameter estimation. For hybrid electric vehicles, the plant may be viewed as the HEV traction system, including the battery, power inverter module, electric motor(s), internal combustion engine, and transmission. Early work on control theory within the automotive industry can be found in (3), and more recent accounts of work in the area of battery and HEV control systems are provided in Wiegman's (4) and Tate's theses (5), respectively. Reference (6) overviews recent progress in the development and deployment of hybrid electric vehicles worldwide.

For approaches associated with optimal estimation methods, the subject of this review, a model of the plant (e.g., the EESS) is constructed, and the parameters appearing in the model are regressed from the available measurements. For example, using an equivalent electrical circuit or network, one can construct a mathematical expression to correlate the behavior of an EESS, and the values of the circuit elements can be regressed from the available current, voltage, and temperature data during vehicle operation by employing a system identification scheme. The need to regress the value of all parameters that impact the impedance spectrum motivates the formulation of a generalized algorithm for the system identification problem that can treat an arbitrary number of parameters. While many optimal estimation methods have been employed for EESS's (1,2,7–24), the method of weighted recursive least squares (WRLS) with exponential forgetting has proven to be a pragmatic approach for parameter regression when model reference adaptive systems are employed (14,16,21,23). The time weighting of data is damped exponentially with this approach; hence, new data has a preferential impact in determining the value of regressed parameters and thus the state of the system. The focus of this review will be on the application of WRLS methods for constructing state estimators for EESS's (batteries and supercapacitors) consistent with the needs identified in the discussion of Figs. 4 and 5. A review of early work on state estimators for batteries can be found in the paper by Pillar et al. (7). A discussion of more recent developments of battery state estimators can be found in the thesis by Smith (20) and the paper by Santhanagopalan and White (24).

Because (1) equivalent circuits based on electrical circuit elements have been shown to mimic the behavior of EESS's and (2) equivalent circuits can be used to construct adaptive filters, the use of electrical circuit elements provides a convenient foundation upon with to base state estimators for EESS's. The mathematical manipulation of electrical circuit networks has a long history. Carlin and Giodano (25), Belevitch (26), and Scanlan and Levy (27) provide general texts on the topic, replete with mathematical methods and example problems involving *RLC* circuits. Van Valkenburg's text (28) is particularly helpful in the context of formulating initial conditions for circuit excitation problems, examining transmission lines, and construting Nyquist plots (frequency response plots). Van Valkenburg's analysis of a tunnel diode is quite similar to the use of an equivalent circuit for constructing a state estimator for an EESS.

The use of network analysis to determine the output signal in response to an input excitation falls under the topic areas of signal processing and filter design (cf. Fig. 8a). The many applications of adaptive filters is reflected by the large number of monographs devoted to this topic area; refs. (29–39) provide a subset of texts that are relevant to the class of adaptive filters overviewed in this work. Antoniou (32), in addition to providing a general overview of digital signal processing, makes extensive use of block diagrams to represent filters, giving a geometric representation to the topic area. The focus by Ghausi and Laker (34) on filters derived from *RC* circuits is particularly relevant to the circuits of most interest for the filter discussed in this work, much like Van Valkenburg's network analysis text mentioned previously. Of the texts cited in this paper, the derivation of equations for adaptive filters provided in Chap. 11 of Bellanger (35) and Chap. 8 of Widrow and Stearns (37) are most similar to the methods used in this work. A short and informative overview of adaptive filters based on least squares estimation is included in Haykin's text (36). While much of the literature devoted to this topic was prompted by the development of transistors

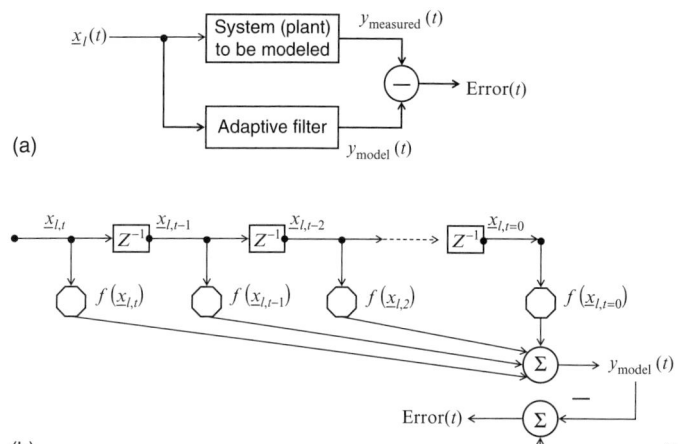

Figure 8. (**a**): Adaptive filter. The measured temperature and current is represented by $\underline{x}_l$. The measured voltage corresponds to y_{measured}; y_{model} is the modeled voltage. The measured and known parameters contained in the vector $\underline{x}_l$ influence both the plant and the adaptive filter. (**b**): Expanded view of an adaptive transversal filter. The operator z^{-1} retrieves the previous time-step value of the variable. For a least-squares filter, the square of the error is minimized by adjusting the parameters contained in $f(\underline{x}_l)$.

and communications systems in the late 1960s and throughout the 1970s, the development of least squares estimation can be traced back to Gauss in 1809 (40). Recursive least squares (RLS), as employed in this work, was formalized in 1950 by Plackett (41). An illustration of a least squares filter is depicted in Fig. 8b. While conventional least squares analyses utilizes all data points collected over time, a recursive relation removes the need to sequentially go back in time beyond the previous time step in order to obtain data for state estimation, as will be shown later. The weight factors (as in weighted recursive least squares) and other terms are carried in $f(\underline{x}_l)$. Adaptive filters based on RLS are closely related to variants based on the Kalman filter (42), as is discussed by Tretter (30) and Haykin (36).

The previous two paragraphs have provided synopses of network analysis and signal processing, respectively. One can view the field of optimal estimation (43–50) as a bridge between these to topics. System specification and requirements generation (51, 52) play key roles as well in the overall construction of algorithms and software. In many cases, it is difficult to characterize a work as belonging to either the field of optimal estimation or that of signal processing and adaptive filtering. Ljnug and Söderström (48) survey the most widely used recursive identification methods; Sect. 2.7 provides a general survey of existing methods, and Sect. 5.6 provides a step-by-step comparison of weighted recursive least squares (WRLS) implemented with exponential forgetting and the Kalman filter. (See Figs. 5 and 6 for specific graphical comparisons on a selected problem.) It can be inferred from the authors' text that if sufficient knowledge is known about the time evolution of the parameters characterizing the system state as well as the measurement and process noise, then the Kalman filter represents the optimal algorithm for recursive identification. Similar conclusions are suggested by Åström and Wittenmark (49); both survey texts suggest that in the absence of such complete information, the WRLS method with the time weighting corresponding to exponential forgetting constitutes a pragmatic approach, which explains continued research on this method for adaptive control (53–59). In this review, we focus on the implementation of WRLS methods so as to regress all relevant model parameters for the construction of a state estimator for an EESS. In closing this section, it is noted that the correlation of equivalent-circuit parameters for EESS's with SOC, a closely related topic, is the subject of (60–66). The efficacy of such methods for state estimators has been problematic, however, and no further commentary on this topic is provided in this review.

A plot of the open-circuit potentials for electrochemical cells of interest for HEV applications is provided in Fig. 9a. Various metal oxides (e.g., CoO_2, NiO_2, Mn_2O_4, physical mixtures of these materials as well as substantial substitution of the metal centers, as in $Co_{1/3}Mn_{1/3}Ni_{1/3}O_2$ and $Al_{0.05}Co_{0.15}Ni_{0.8}O_2$ and related variants) are of immediate interest as well as the spinel lithium manganese oxide used to reflect conventional lithium ion technology in Fig. 9b. The voltages shown in Fig. 9b are approximate and intended to be representative. Figure 9a contains actual cell data.

A recent trend in HEV battery technology has been to utilize reactions resulting in lower voltage systems that are more abuse tolerant and provide higher specific power (kW kg^{-1}) at the expense of specific energy (kWh kg^{-1}). Lower voltage (e.g., phosphate based)

Characterization and Modeling of Electrochemical Energy

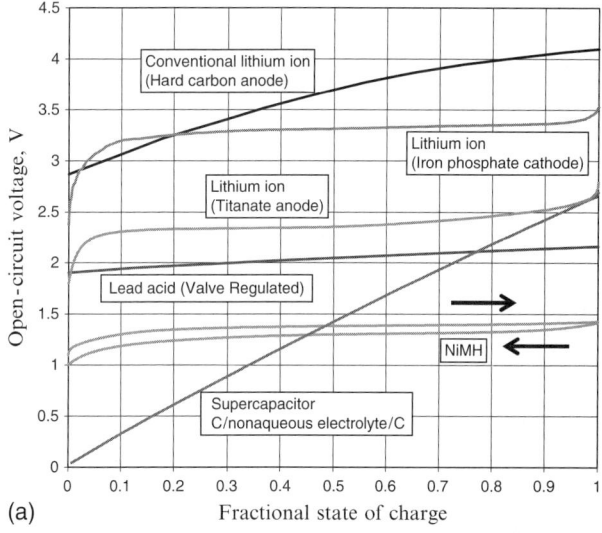

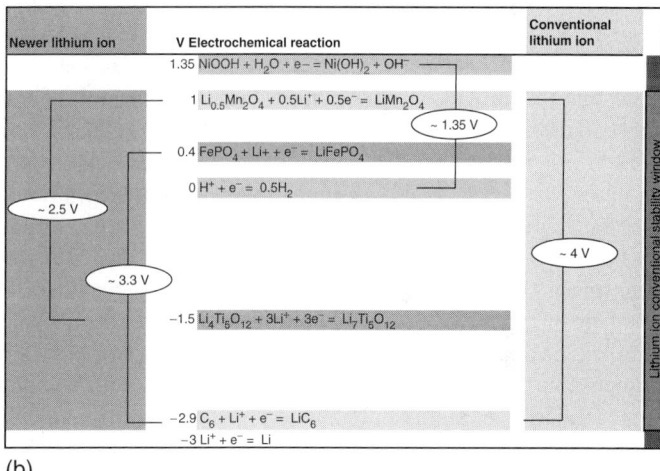

Figure 9. (**a**): Equilibrium voltage of representative electrochemical energy storage devices. The systems shown are of interest for traction applications. (**b**): Advanced battery technologies and associated cell reactions.

cathodes (67–70) and higher voltage (e.g., titanate based) anodes (71–75) reflect this trend. Because the voltage part of state estimator algorithms employs the extracted open-circuit potential to deduce the SOC, those energy storage devices that exhibit substantial and linear voltage change with SOC exhibit greater signal to noise in the determination of the SOC. The rather flat voltage vs. SOC relations for these newer systems implies that the optimal estimation of SOC, SOP, and SOH will be more difficult than for the of conventional lithium ion systems.

While the discussion associated with Fig. 9 focused on batteries, supercapacitors are also of immediate interest for traction applications for mild hybrid architectures, and much has been written on the behavior of these devices (76, 80–90, 133–147). If it is desired to have the engine (or fuel cell) off upon deceleration or when the vehicle is stopped, and one is willing to forego substantial downsizing of the primary power plant, then capacitors based on high-surface-area carbon electrodes and acetonitrile solvents represent a promising low-cost, high-performance option over a broad range of temperatures. There is concern, however, over the abuse tolerance of acetonitrile-based systems, particularly in the context of cell overcharge, and this provides motivation for clarifying the accuracy of control algorithms. It is also worth noting that for plug-in HEVs designed to provide zero-emission-vehicle (ZEV) range, the use of a high-energy battery coupled to a high power capacitor is a promising architecture, as the power to energy ratio of the battery can be reduced, and performance and cycle life for the tandem system can be improved. To summarize automotive interests in supercapacitors of the type investigated here, there is strong interest in these devices for mild HEVs and plug-in HEVs, but less interest in the devices for a conventional charge-sustaining strong HEV, as a single energy storage device (e.g., a lithium ion battery) would likely provide the most effective energy storage configuration. Asymmetric capacitors that utilize a robust battery electrode along with an activated carbon (capacitor) anode (148, 149) may provide a means to deliver high power density and sufficient energy density for charge-sustaining strong HEVs.

II. ELECTROCHEMICAL MODELING

The first task in the construction of an optimal estimator is to develop and validate a suitable model of the EESS. By identifying a simple circuit model that can describe the salient features of both transient and pseudo-steady state experimental data, we can develop a useful conceptual scheme. Central to this realization is that if an equivalent circuit is to be constructed to match experimental data, then the circuit should contain the fewest number of circuit elements, thereby removing extraneous information and simplifying the problem. The utility (and the pitfalls) of equivalent circuits for the representation of electrochemical systems has received much attention (77–79). The transmission line circuits shown Figs. 10–12 were adapted from Ong and Newman (76). By reference to semi-infinite transmission line representations shown, it is possible to glean the origin of the simplified circuits depicted in the lowermost schematics. For a supercapacitor (Fig. 10), the transmission line representation corresponds to a capacitor dominated by ohmic drop and comprising ideal capacitive behavior at the electrode-electrolyte interface. For a battery (Fig. 11), an additional resistance is added to comprehend interfacial charge-transfer resistance. An incremental improvement in the generality of the equivalent circuits would be to add a capacitor in series with the interfacial resistors (C_S in Fig. 12) or a frequency-dependent Warburg impedance. Such additions are intended to approximate mass-transfer impedance (77, 79). It is not clear that such additions are necessary to simulate the data of this work, however, and we shall not discuss adaptive filters beyond the complexity of the lowermost circuit depicted in Fig. 11. In this section, we develop and implement models based on the simplified equivalent circuits of Figs. 10 and 11 to represent a supercapacitor's and a battery's current-voltage relationship while knowing the state of charge. The inverse is desired for an imbedded controller in an electric or hybrid-electric vehicle application; in these situations, the current-voltage history is known, and the battery's SOC and SOH must be determined. To the extent that this work is successful in representing experimental data with the simple circuits, it should be possible to examine efficiently the inverse problem; that is, to develop a sufficiently accurate, real-time state-of-charge algorithm for an imbedded controller.

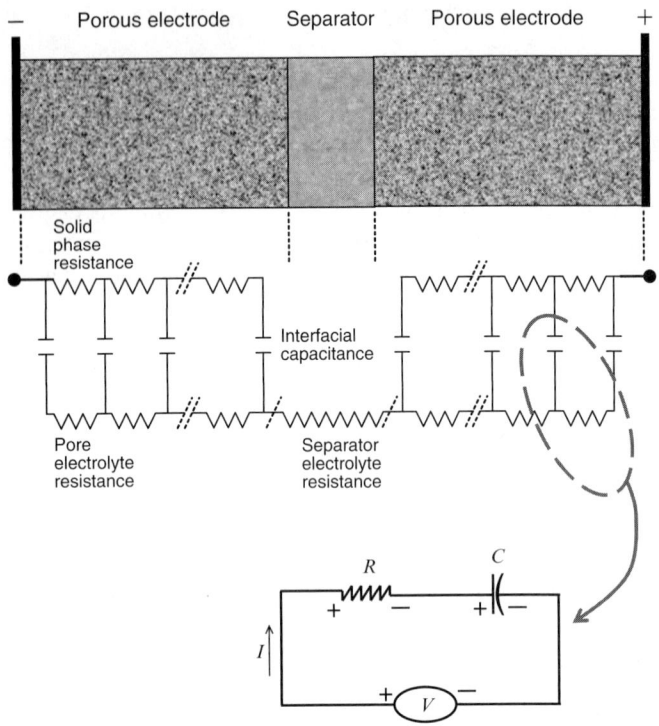

Figure 10. Equivalent circuit representation for a symmetric supercapacitor. The simplified equivalent circuit shown in the lowermost figure is used to construct an adaptive filter.

For modeling purposes, we develop the equations for the lowermost circuit depicted in Fig. 11, with the recognition that R_{ct} goes to infinity corresponds to the supercapacitor equivalent circuit (lowermost circuit in Fig. 10). For batteries, the open-circuit potential V_0 will vary with SOC, which ranges from 0 (fully discharged) to 1 (fully charged). By fully discharged and charged, we intend that the cell's equilibrium potential V_0 corresponds to the lowermost allowed voltage or uppermost allowed voltage, respectively (cf. Fig. 1). The state of charge (SOC) is related to current passage,

$$\text{SOC}(t) = \text{SOC}(t_0) + \frac{1}{Q} \int_{t_0}^{t} I \, dt, \tag{1}$$

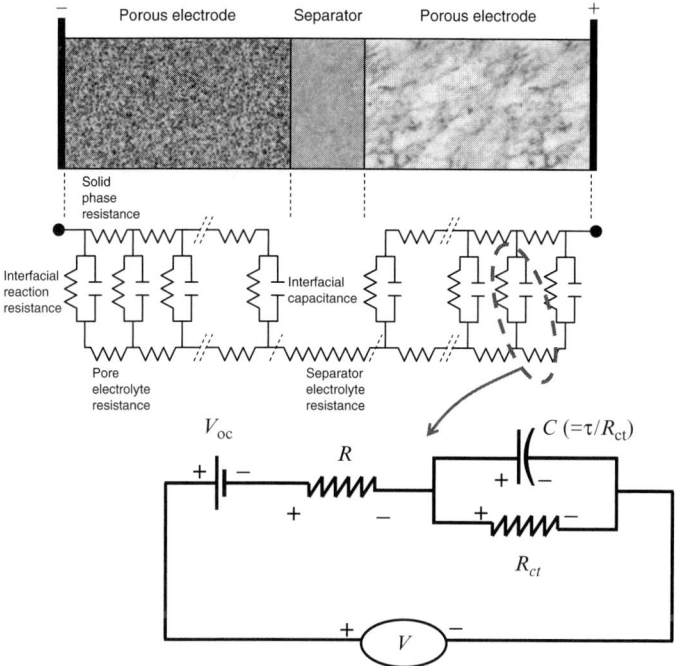

Figure 11. Equivalent circuit representation for a battery. The simplified equivalent circuit shown in the lowermost figure is used to construct an adaptive.

where Q is the Coulombic capacity of the battery, t denotes time, and I refers to current devoted to the desired electrochemical reaction. We shall specify the V_0 (SOC) relation for each battery modeled in this work.

The equations governing the circuit depicted in the lowermost circuit of Fig. 11 can be written as

$$I = I_2 + I_3$$
$$V = V_0 + IR + I_3 R_{ct}$$
$$V = V_0 + IR + \frac{q}{C}. \quad (2)$$
$$I_2 = \frac{dq}{dt}$$

A current balance yields the first equation, summing the voltage about loops including the interfacial charge-transfer resistance R_{ct}

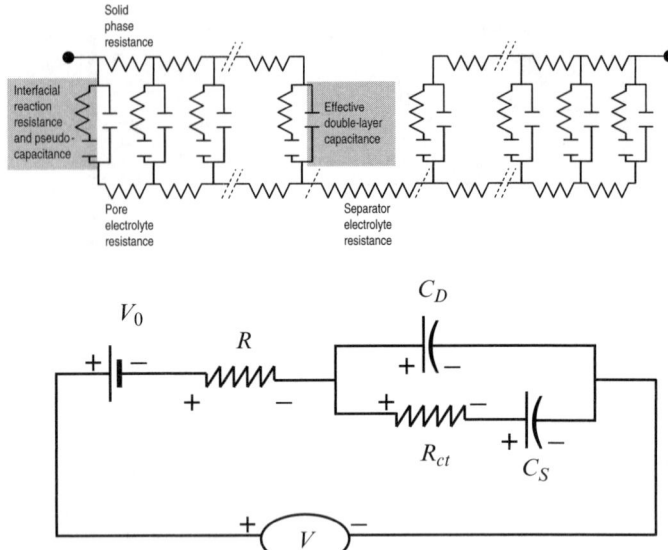

Figure 12. Equivalent circuit representation for a battery with additional pseudo-capacitance (C_S). The simplified equivalent circuit shown in the lowermost figure is used to construct an adaptive.

and capacitance C provides the second and third equations (respectively), and the last equation relates the charge q on the capacitor to the current passing through it. Substitution can be used to eliminate the currents I_2 and I_3 and the charge q, which yields a single equation relating the cell current I and potential V:

$$R\frac{dI}{dt} + \frac{1}{C}\left(1 + \frac{R}{R_{ct}}\right)I = \frac{dV}{dt} + \frac{1}{R_{ct}C}(V - V_0). \qquad (3)$$

We have assumed that the open-circuit voltage changes slowly with time; hence, $dV_0/dt = 0$. For a supercapacitor (lowermost circuit of Fig. 10, $R_{ct} \to \infty$), this equation reduces to

$$\frac{dV}{dt} = \frac{I}{C} + R\frac{dI}{dt}. \qquad (4)$$

More complete models utilizing nonlinear, partial differential equations based on the underlying transport phenomena, interfacial kinetics, and thermodynamics can be found in (76, 80–89) for supercapacitors and (84, 91–98) for batteries.

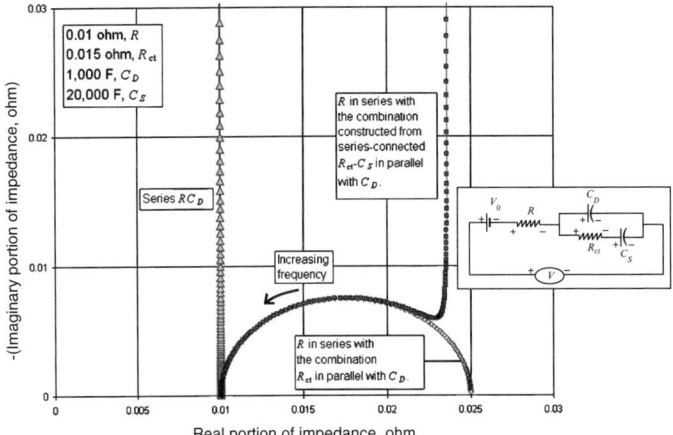

Figure 13. *Triangles*: R_{ct} infinite (simplified supercapacitor model). *Diamonds*: C_S infinite (simplified battery model). *Squares*: Parameter values displayed in upper left box, with the full circuit depicted to the right of the plot.

Cyclic voltammetry. Cyclic voltammetry (79), wherein the cell potential is changed linearly with time, renders an easily implemented and interpreted procedure for analysis. By controlling the cell potential (and enforcing current limits), one can ensure that the cell is not abused during the test conditions. Second, we note that one might attempt to ascertain the complete behavior of a capacitor through a series of potential steps. A more pragmatic approach is to sweep the potential linearly with time (potential sweep rate given by $\nu = dV/dt$, cf. upper plot in Fig. 14) and monitor the current, thereby clarifying the voltage-current-time relationship. Hence, voltammetry provides a convenient method to control the state of the electrochemical system. For the battery analysis, we utilize a small amplitude variation in potential about V_0 (± 15 mV for the experimental data examined in this work), and we make the approximation that V_0 is constant. Introducing the nondimensionalization

$$\tau = \frac{t}{\theta}, \ \tilde{V} = \frac{V - V_0}{V_{\max} - V_0}, \ \tilde{I} = \frac{I}{\nu C}, \ \alpha = \frac{RC}{\theta}, r = \frac{R}{R_{ct}},$$

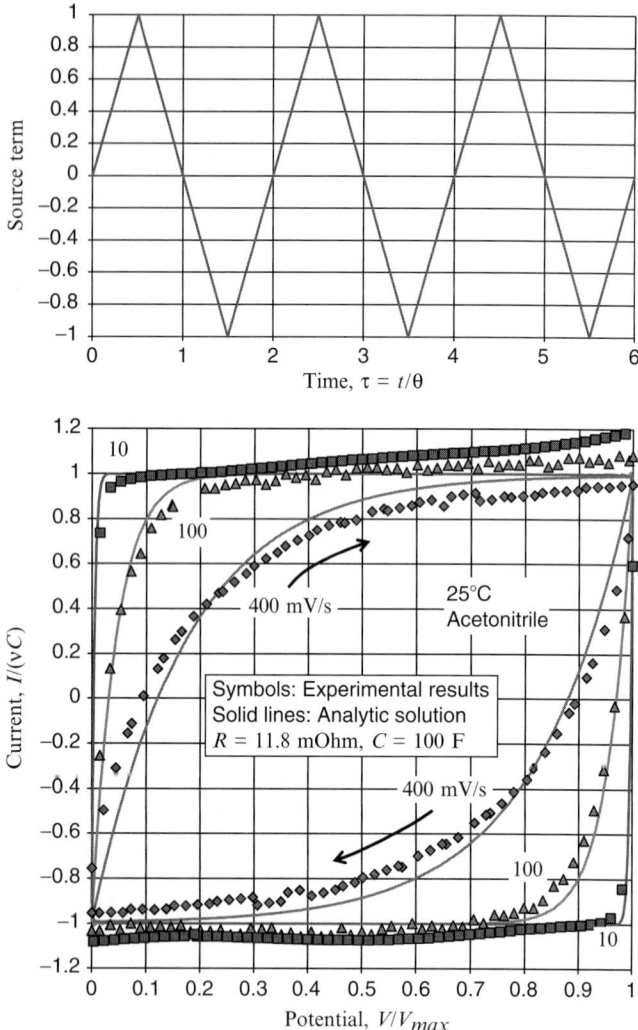

Figure 14. *Top*: Dimensionless voltage source ($\tilde{V}$). *Bottom*: Experiment-theory comparison for the acetonitrile-based supercapacitor in the uniform and sustained steady state (8).

we can recast (3) as

$$\alpha \frac{d\tilde{I}}{d\tau} + (1+r)\tilde{I} = \frac{1}{2}\left(\frac{r}{\alpha}\tilde{V} + \frac{d\tilde{V}}{d\tau}\right) \quad (5)$$

with the potential source give by

$$\tilde{V} = -\frac{8}{\pi^2} \sum_{n=1,3,5,\ldots}^{\infty} \frac{1}{n^2} \cos\left[n\pi\left(\tau + \frac{1}{2}\right)\right]. \quad (6)$$

The full solution to (5) for supercapacitors and batteries are provided in (90,99), respectively; the same references contain solutions for triangular current source excitations. For the purposes of this review, we employ the uniform and sustained periodic state, denoted by subscript ∞, which is of interest in terms of comparing experiment and theory:

$$\tilde{I}_\infty = 4 \sum_{n=1,3,5,\ldots}^{\infty} \frac{\alpha n\pi \sin\left[n\pi\left(\tau+\frac{1}{2}\right)\right] - (\alpha^2 n^2 \pi^2 + r + r^2)\cos\left[n\pi\left(\tau+\frac{1}{2}\right)\right]}{\alpha n^2 \pi^2 [\alpha^2 n^2 \pi^2 + (1+r)^2]}. \quad (7)$$

For a supercapacitor, we take $R_{ct} \to \infty$ and $r \to 0$, which yields

$$\tilde{I}_\infty = 4 \sum_{n=1,3,5,\ldots}^{\infty} \frac{\sin(n\pi\tau) - \alpha n\pi \cos(n\pi\tau)}{n\pi + \alpha^2(n\pi)^3}. \quad (8)$$

A comparison of experimental data with (8) is shown in the lower plot of Fig. 14 for a supercapacitor. Analogous plots of experimental data versus calculations from (7) for the lithium ion cell (a lithium titanate, lithium cobalt oxide system) are provided in the four plots (corresponding to 20, 40, 60, and 80% state of charge, respectively) of Figs. 15 and 16.

The parameter values extracted from the supercapacitor analysis are displayed in Fig. 14. Both the open-circuit voltage and the parameters R, R_{ct}, and C are provided in Fig. 17 for the lithium titanate, lithium cobalt oxide cell investigated in this work. A more complete description of equilibrium relations based on molecular

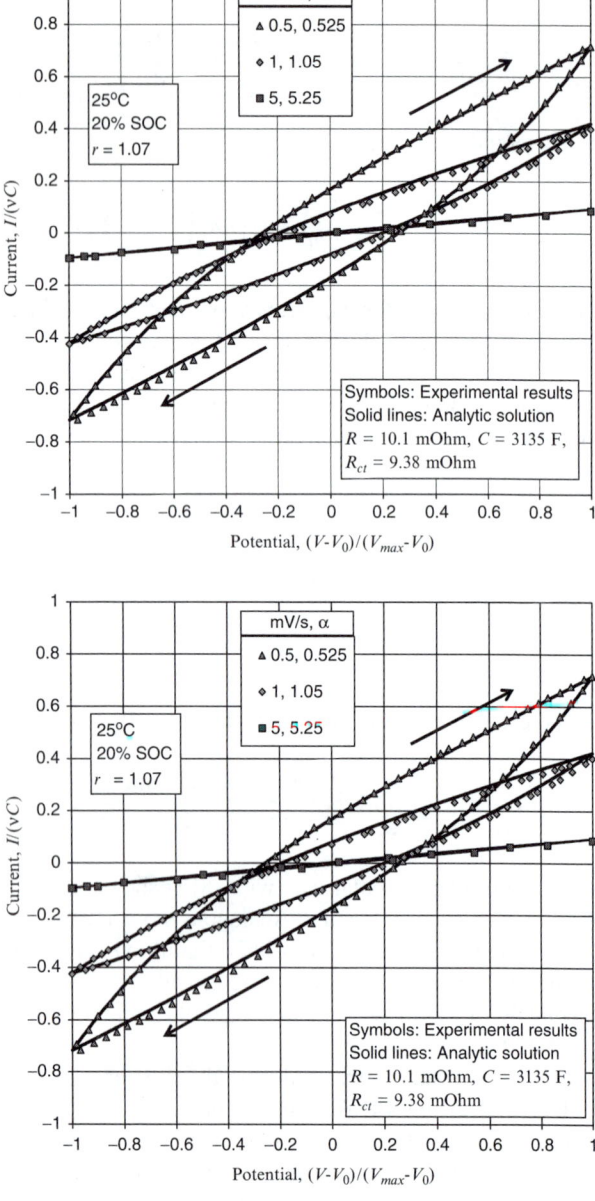

Figure 15. Experiment-theory comparison at 20% (*top*) and 40% (*bottom*) percent state of charge. The model response corresponds to (7).

Characterization and Modeling of Electrochemical Energy

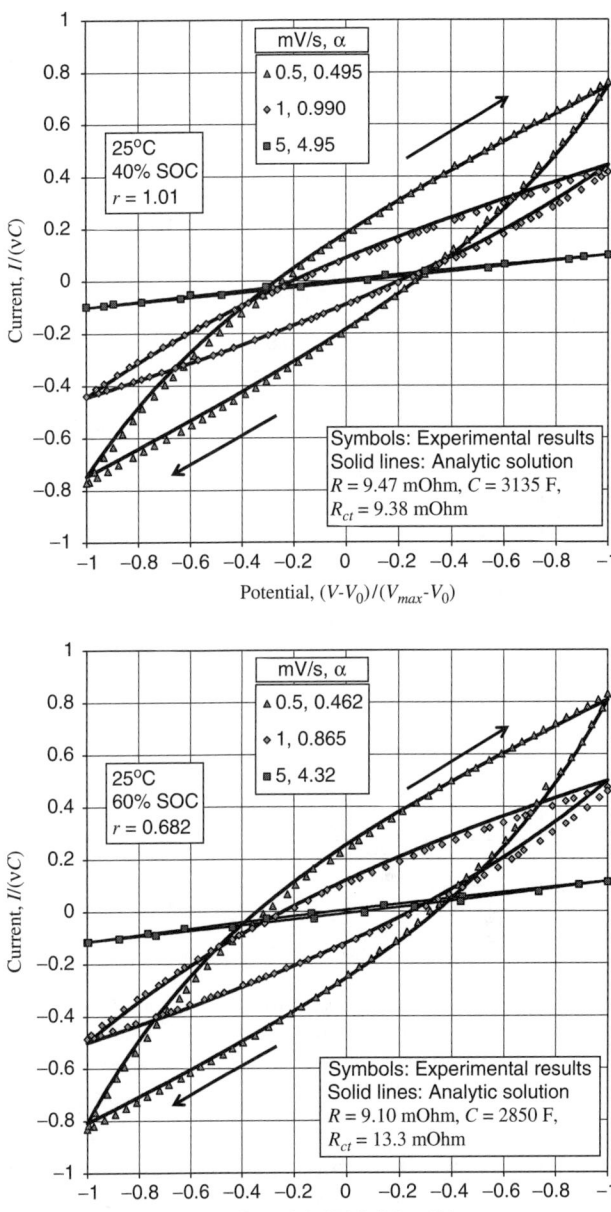

Figure 16. Experiment-theory comparison at 60% (*top*) and 80% (*bottom*) state of charge. The model response corresponds to (7).

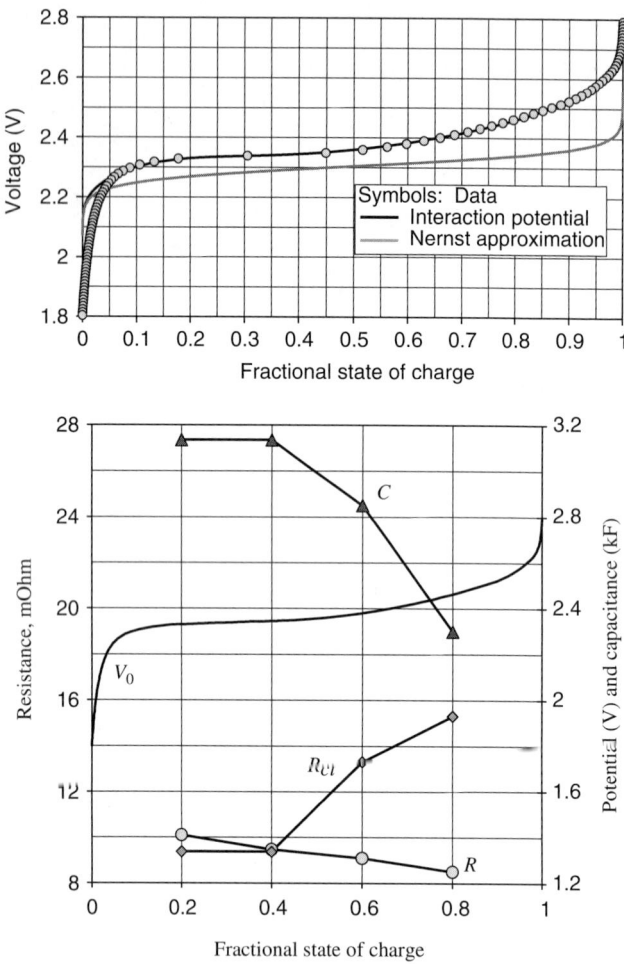

Figure 17. *Top*: Open-circuit potential of the lithium titanate, lithium cobalt oxide cell. The Interaction Potential is provided by (9), and (10) corresponds to the depicted Nernst equation. *Bottom*: Open-circuit voltage and parameters used in the cyclic voltammetry simulations.

thermodynamics for intercalation electrodes can be found in (101–103). By considering repulsion between intercalate species within the capacity-limiting intercalation electrode, we obtain the following relation for the cell potential (102),

$$V_0(\text{SOC}) = U^0 - \frac{RT}{F} \ln \frac{1-\text{SOC}}{\text{SOC}} + \sum_{j=1,\infty} a_j \text{SOC}^j, \quad (9)$$

where $j+1$ denotes the number of intercalates involved in the intercalate-intercalate interaction and SOC reflects the fractional state of charge. For our purposes, we truncate the series at $j=6$. The Nernst approximation results from neglecting the summation term used to address lithium–lithium interactions:

$$V_0(\text{SOC}) = U^0 - \frac{RT}{F} \ln \frac{1-\text{SOC}}{\text{SOC}}. \quad (10)$$

The coefficients a_j used to generate the Interaction Potential depicted in Fig. 17 are provided in Table 1. The uppermost plot of Fig. 17 indicates that interactions between lithium guests in the capacity-limiting lithium titanate electrode are quite weak, as there is less than 1.2 kcal mol^{-1} difference between the Nernst expression and the data at 50% SOC. The parameters governing the cell impedance are constant over the constant-voltage portion of the cell potential, and vary monotonically throughout the sloping portion of the open-circuit potential (i.e., above approximately 50% state of charge).

For both the supercapacitor and lithium ion battery voltammetry analyses, the favorable comparison of the analytic solutions with complementary experimental data implies that control algorithms

Table 1. Interaction coefficients and other values used in (9) and (10). The temperature corresponds to 25°C

Quantity	Value, V
RT/F	0.02568
U^0	2.3044
a_1	0.8889
a_2	−4.4742
a_3	9.4101
a_4	−8.7417
a_5	3.1588
a_6	2.3354×10^{-6}

based on the simple equivalent circuits employed are promising for constructing state estimators.

Constant power operation. For constant power operation, $P = IV$, where P denotes power. We shall cast the cell potential as the dependent variable of interest, thereby transforming (3) to

$$-\frac{PR}{V^2}\frac{dV}{dt} + \frac{1}{C}\left(1 + \frac{R}{R_{ct}}\right)\frac{P}{V} = \frac{dV}{dt} + \frac{1}{R_{ct}C}(V - V_0),$$

or

$$\frac{dV}{dt} = \frac{PV(R + R_{ct}) - V^2(V - V_0)}{R_{ct}C(PR + V^2)}, \quad (11)$$

which can be restated as an integral equation,

$$\int_{v(0)}^{v} \frac{R_{ct}C(PR + V^2)}{PV(R + R_{ct}) - V^2(V - V_0)} dV = t. \quad (12)$$

For this investigation, the system is initially at rest (equilibrated at zero current). Immediately after the power excitation, the current flows through the resistor R, and the capacitor element C offers no impedance. Thus the initial voltage is given by

$$V(0) = \frac{V_0 + \sqrt{V_0^2 + 4PR}}{2}, \quad (13)$$

and this provides the lower limit to the integral. This equation can also be used to estimate the resistance R from the initial voltage change from V_0 to $V(0)$.

The solution to (12) for constant-power operation of a supercapacitor ($R_{ct} \to \infty$) is (90)

$$\frac{1}{2}\left(V^2 - V_i^2\right) + RP \ln\frac{V}{V_i} = \frac{P}{C}t.$$

This equation can be recast so as to isolate the cell voltage V:

$$V = \sqrt{RPW\left(\frac{V_i^2}{RP}\exp\left[\frac{V_i^2}{RP} + \frac{2t}{RC}\right]\right)}. \quad (14)$$

Note that the cell current is obtained from $I = P/V$. $W(z)$ is the Lambert W function,

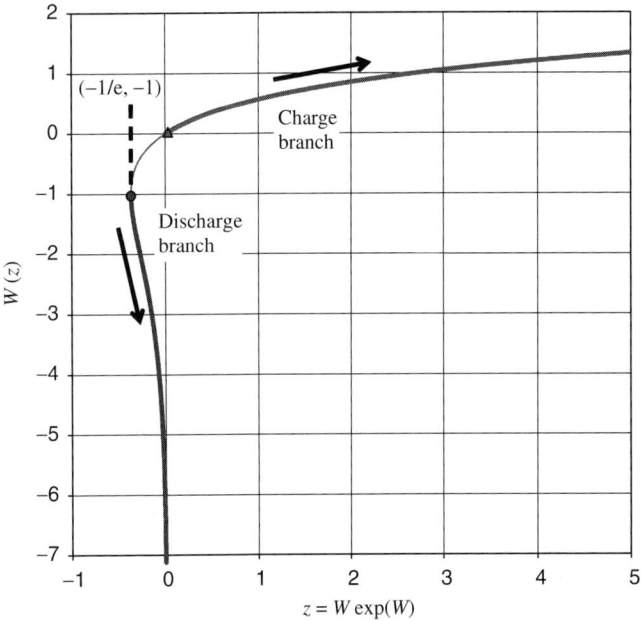

Figure 18. Lambert W function (104–107) and relevant solution branches for the implementation of (14).

$$z = W(z)e^{W(z)},$$

which frequently arises in the solution of problems in the physical sciences (104–107). As depicted in Fig. 18, for charge ($P > 0$) the branch corresponding to $W > 0$ applies, whereas for discharge ($P < 0$), the branch corresponding to $W < -1$ applies. Much has been written on the Lambert W function, and symbolic manipulators such as Macsyma, Maple, and Mathematica include this function in their libraries (starting in the early 1980s). Correlations are available to provide values of W as a function of z, and this fact enables one to use (14) as a closed-form analytic solution. The cited papers can be used to select correlations and methods to calculate $W(z)$.

We compare the analytic expression (14) with experimental data for the current and voltage histories at 20 W cell^{-1} in the upper plot of Fig. 19. Similar to the good agreement observed in the voltammetry analyses, the analytic expressions for constant-power operation

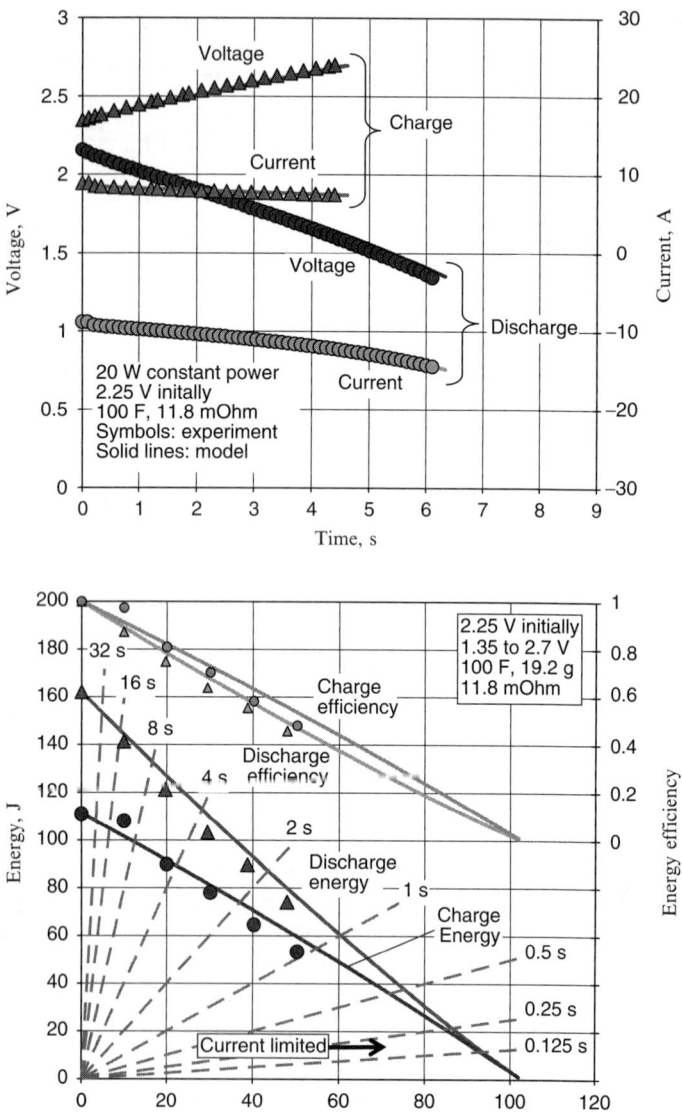

Figure 19. *Top*: Voltage and current histories for 20 W constant-power operation. After equilibrating at 2.25 V, the capacitor was charged (respectively discharged) at 20 W to 2.7 V (respectively 1.35 V). *Bottom*: Ragone plot. *Symbols*: experimental data. *Curves*: model calculations.

exhibit good agreement when compared with experimental data. The difference between the model and experiment at 20 W is so small that it is difficult to resolve the two as plotted. It is important to note that the same values extracted from the voltammetry analysis (Fig. 14) were used to generate the constant-power calculations.

To summarize the constant-power operation, application engineers are most interested in Ragone plots, which provide the power versus energy relationship subject to voltage limits; such a plot is shown in the lower portion of Fig. 19 for the acetonitrile-based capacitor at room temperature. The symbols in Fig. 19 correspond to experimental results. The dashed lines emanating from the origin correspond to the duration of the constant-power operation. For example, the capacitor can be discharged at 40 W from an equilibrated state of 2.25 V for just over 2 s, and about 90 J are delivered before 1.35 V is obtained. When the capacitor is charged at 40 W from an equilibrated state of 2.25 V, it takes just under 2 s for the capacitor to reach the 2.7 V upper voltage limit. The rationale behind choosing a set point voltage of 2.25 V for HEV applications is discussed in (90). Current limitations are provided to avoid cell abuse, thus limiting the charge and discharge power to about 50 W cell^{-1}. At this power level, the energy efficiency, which can be calculated from the above equations (90), is approximately 50%, and considerable internal heating of the cell takes place over the short duration of the constant-power operation.

Table 2.
Coefficients for open-circuit potential at 25°C for the lithium ion cell corresponding to the data shown in Fig. 20. The coefficient are applied in the formula

$$V_0(SOC) = \sum_{j=0,3} a_j SOC^j$$

Quantity	Value, V
a_0	2.7077
a_1	2.5025
a_2	−1.6617
a_3	0.5358

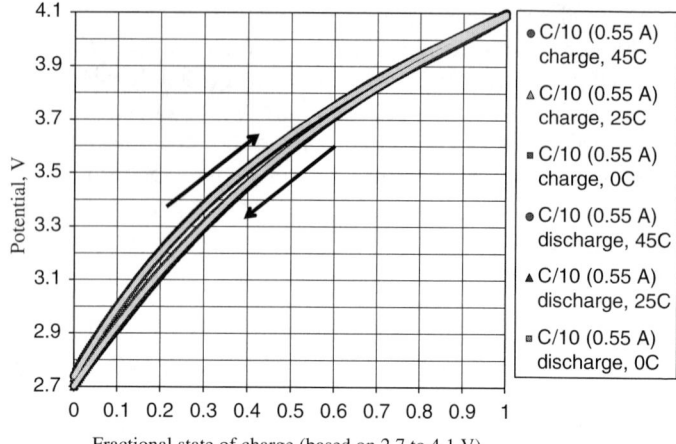

Figure 20. Low-rate (0.55 A) discharge and charge responses for the lithium ion cell, which comprised a hard carbon anode and a $Co_x Mn_y Ni_z O_2$ cathode ($x + y + z = 1$). The arrows denote the scan direction. Little difference is seen between the scans for temperatures ranging form 0 to 45°C, as the symbols representing each temperature are blocked visually by the line passing through the data. The average of the forward and reverse scans yields the coefficients shown in Table 2.

Numerical integration of (12) is straightforward and can be used to analyze constant-power operation of battery systems. We present results here for a 5.5 Ah cell (100) comprising a hard carbon anode and a $Co_x Mn_y Ni_z O_2$ cathode ($x + y + z = 1$). The formula used for the open-circuit potential and the necessary coefficients are provided in Table 2, and the associated (low current) plots are displayed in Fig. 20.

It is difficult to distinguish the response of the model equations from that of the experimental cell for temperatures between 25 and 45°C. Potential and current traces versus time for 25°C are plotted in Fig. 21. By plotting time on a logarithmic scale, differences between model calculations and experimental data are more easily observed; the abscissa in Fig. 22 is thus logarithmic for the 25°C experiment-model comparison. Even with this expanded view, it is difficult to distinguish differences between the model calculations and the experimental results. The same is true for the 45°C comparison.

Characterization and Modeling of Electrochemical Energy

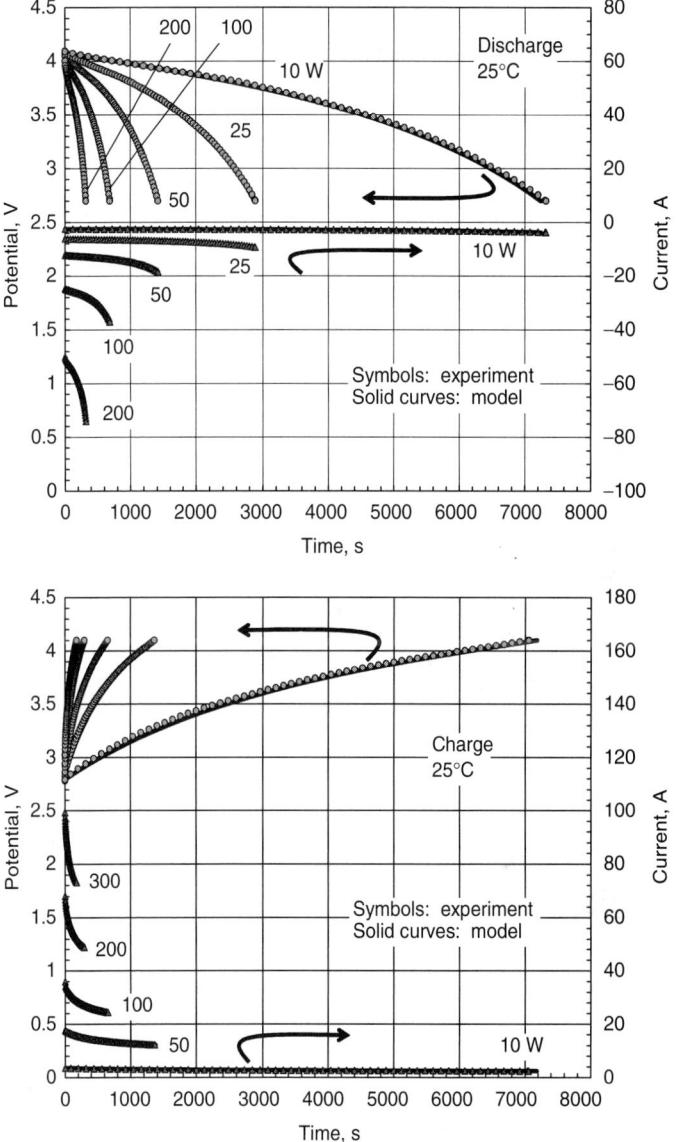

Figure 21. Constant-power discharge and charge at 25°C.

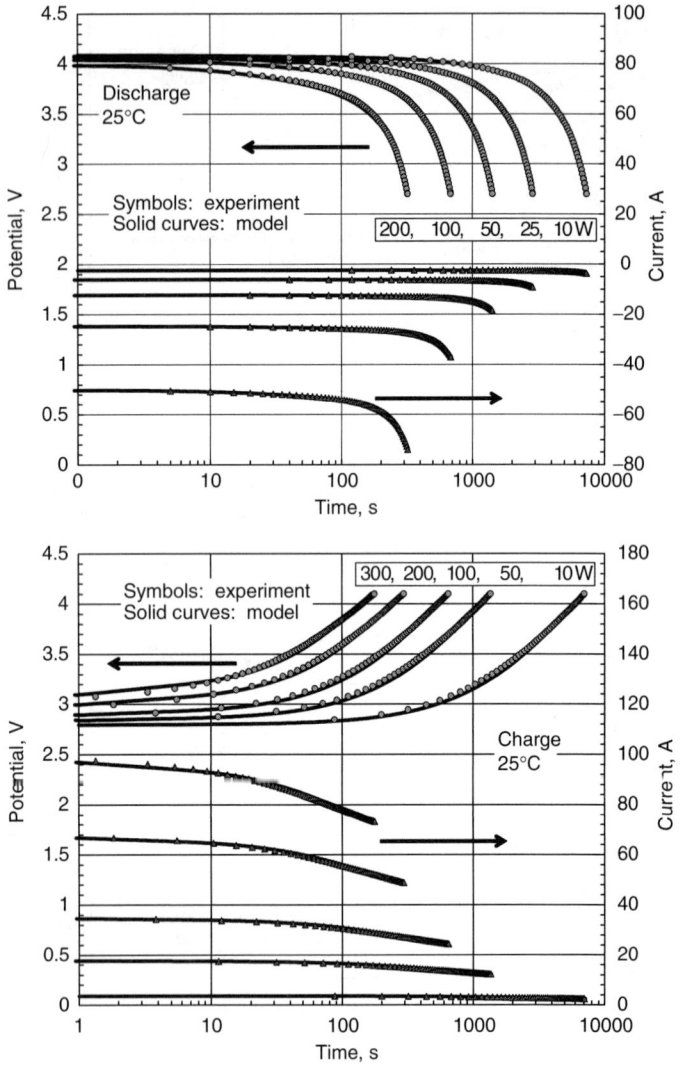

Figure 22. Constant-power discharge and charge. A logarithmic scale has been employed for the abscissa to facilitate viewing relative to Fig. 21.

Below 0°C, however, differences between the model calculations and the experiment data become increasingly evident. The generally good agreement between model and experiment at 0°C is shown in Fig. 23 for constant-power charge and discharge. Qualitative agreement is depicted in Fig. 24 for constant-power discharge at −20°C, and the anomalous behavior in the experiment relative to the model forwarded here is clearly evident in Fig. 24 for the −30°C discharge. Phenomena not captured in the equivalent circuit representation intrude on the system response below 0°C; in particular, it is postulated that the cell is heating substantially during current passage due to the high resistance at low temperatures. Upon heating, the resistance can subsequently decrease, and increases in the voltage during constant-power discharge emerge, consistent with the −30°C discharge response.

One might speculate that the proposed model is useful if the extracted parameters follow a logical variation with temperature (e.g., an Arrhenius dependence for the resistances R and R_{ct}) or are otherwise largely invariant with temperature (e.g., for capacitance C and coulombic capacity Q). The variations in the cell capacity (Ah) and capacitance (F) with temperature are displayed in Fig. 25. The coulombic capacity is obtained from the low rate ($C/10$, 0.55 A) discharge. The effective interfacial capacitance C was obtained by regressing the model parameters to the higher rate discharge and charge data. The fall in measured coulombic capacity with declining temperature is due in part to the increasing impedance observed at lower temperatures, making it more difficult to transport lithium ions in the electrolyte phase and lithium guests in the intercalation electrodes. We do not have a concise mechanistic explanation as to why the capacitance C drops at higher temperatures on discharge but is not affected on charge, and why the capacitance drops at lower temperatures on charge but is not affected on discharge. Individual electrode data (relative to a reference electrode) would be helpful in elucidating the causes of such behavior.

The variation is the cell resistances R and R_{ct} are in agreement with what one normally sees in lithium ion systems; i.e, as depicted in Fig. 26, the resistance decreases as temperature is increased consistent with an Arrhenius relationship,

$$\ln \frac{R}{R(25°C)} = -\frac{\Delta E}{R_{gas}} \left(\frac{1}{T} - \frac{1}{298} \right),$$

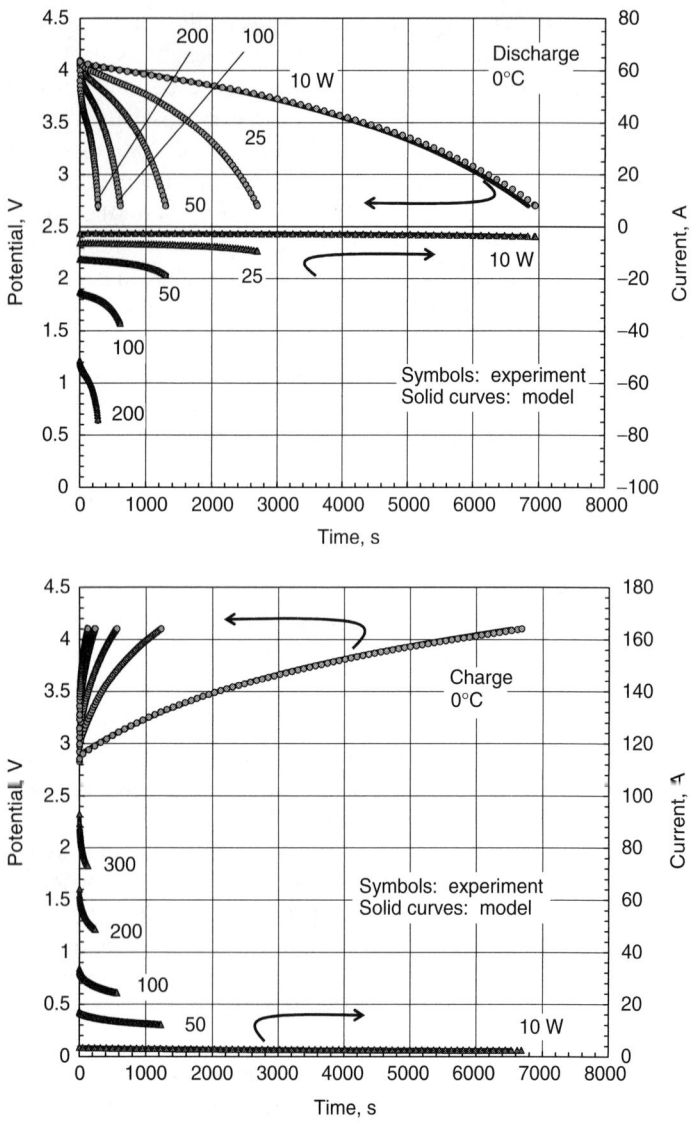

Figure 23. Constant-power discharge and charge at 0°C.

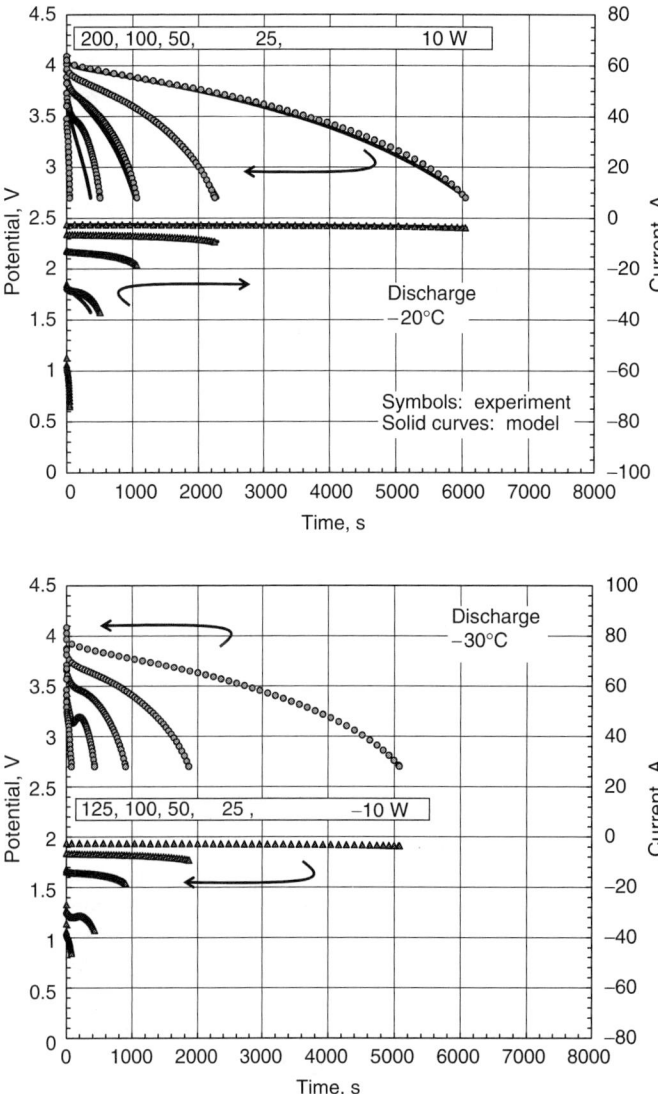

Figure 24. *Upper plot*: −20°C constant-power discharge. *Lower plot*: −30°C constant-power discharge.

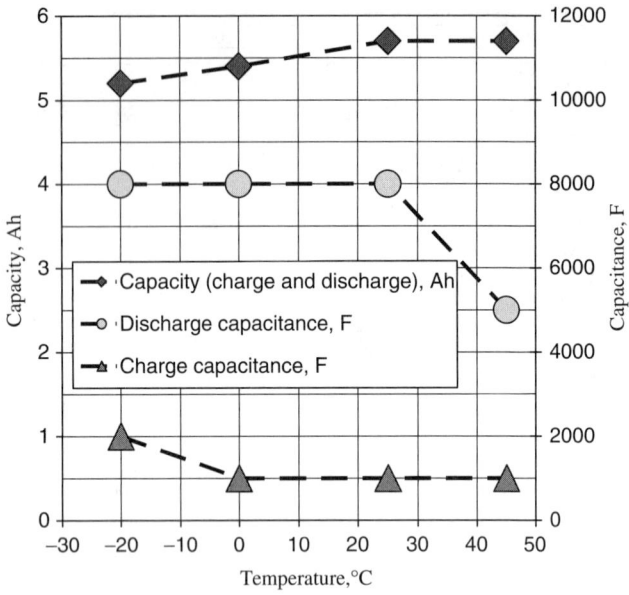

Figure 25. Coulombic capacity Q and capacitance C.

where R_{gas} is the universal gas constant and temperature T is in Kelvin. The activation energy ΔE for the high frequency resistance R is nearly identical for charge and discharge (about -4.65 kcal mol^{-1}). To summarize the resistance trends, a consistent Arrhenius dependence for the resistances R and R_{ct} is observed between 0 and 45°C. It should be noted that while the parameters R, R_{ct}, and C vary from charge to discharge and with temperature, the parameters have been kept constant over the full SOC range.

The ability of the model to represent the time for charge and discharge is depicted in Fig. 27; such a plot is similar to a Ragone plot (cf. Fig. 19 for the supercapacitor), as the discharge time multiplied onto the (constant) power yields energy. The upper plot in Fig. 27 corresponds to discharge, and the charge analysis is provided in the lower plot.

While we shall not provide the mathematical details here, a similar analysis can be done for constant current operation. Results of such an analysis are depicted in Fig. 28 for a high energy density NiMH module. The same parameters extracted from the

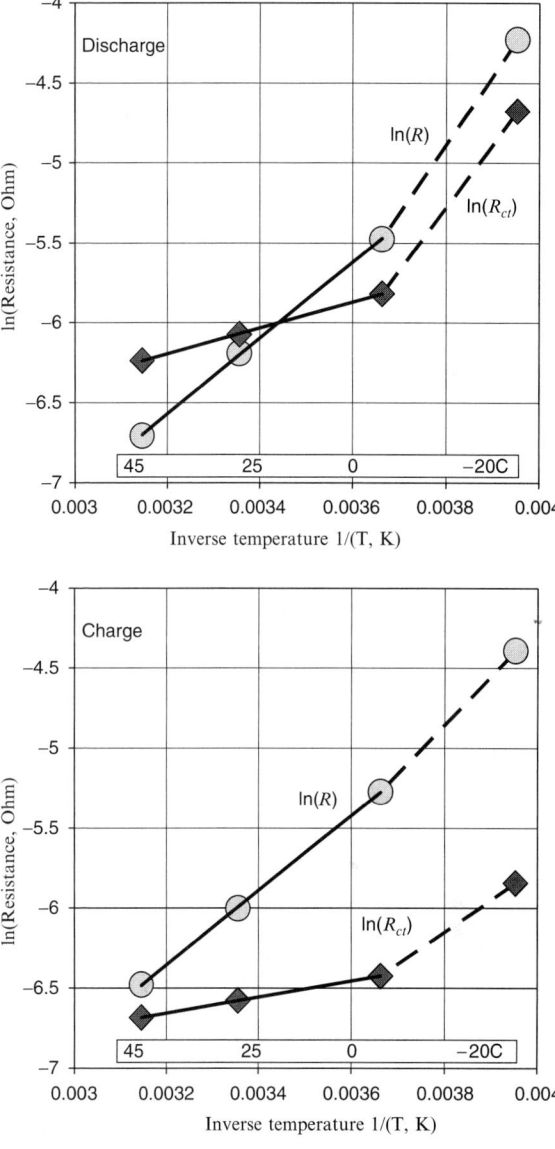

Figure 26. Arrhenius plots for the resistance on discharge (*upper plot*) and charge (*lower plot*). The corresponding (0–45°C) activation energies for R on discharge and charge are −4.7 and −4.6 kcal mol^{-1}, respectively, and 1.6 and 1 kcal mol^{-1} for R_{ct} on discharge and charge, respectively.

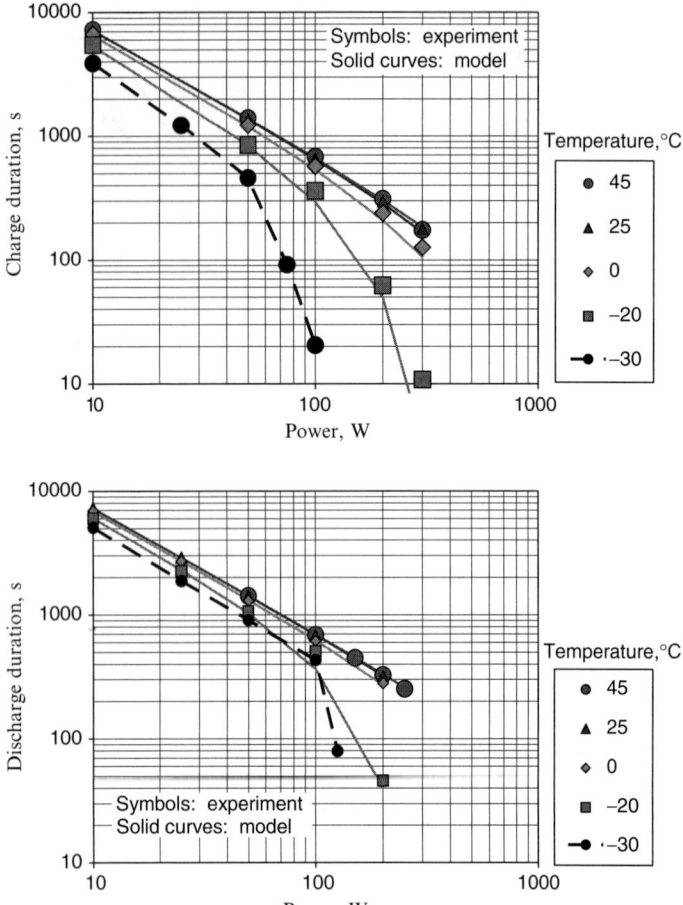

Figure 27. Discharge (*upper plot*, 4.1–2.7 V) and charge (*lower plot*, 2.7–4.1 V) as a function of the test-chamber temperature. For the $-30°C$ condition, dashed lines are used to facilitate viewing and are not calculations.

experiment-model comparison of Fig. 28 were used to generate the model calculations displayed in the drive cycle data of Fig. 29. The details of this work can be found in (102).

As was observed in the voltammetry modeling, the favorable comparison of the analytic solutions with complementary experimental data during constant current and power operation implies that

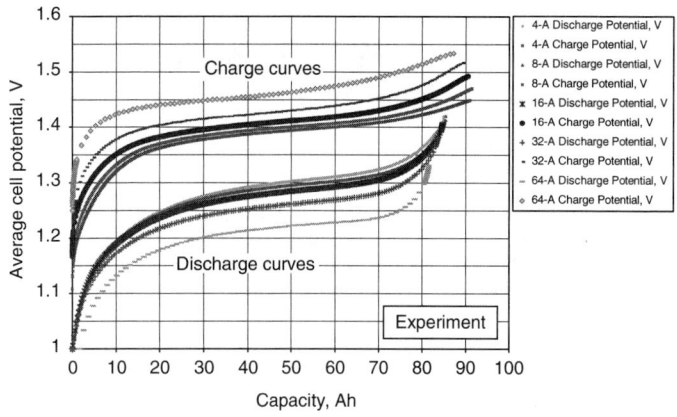

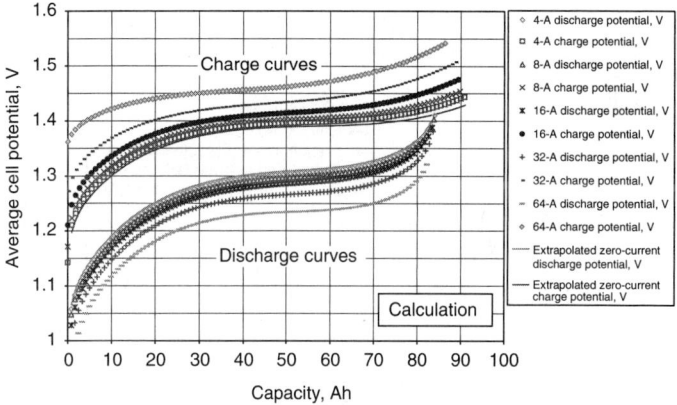

Figure 28. *Upper plot*: experimental charge and discharge potential responses for the NiMH module. *Lower plot*: calculated charge and discharge potential responses.

we can construct state estimators based on the simple equivalent circuit employed for supercapacitors, lithium ion, and NiMH batteries. The same conclusion holds for lead acid batteries operating at cell potentials below that which excessive hydrogen and oxygen gases evolve, as will be discussed below.

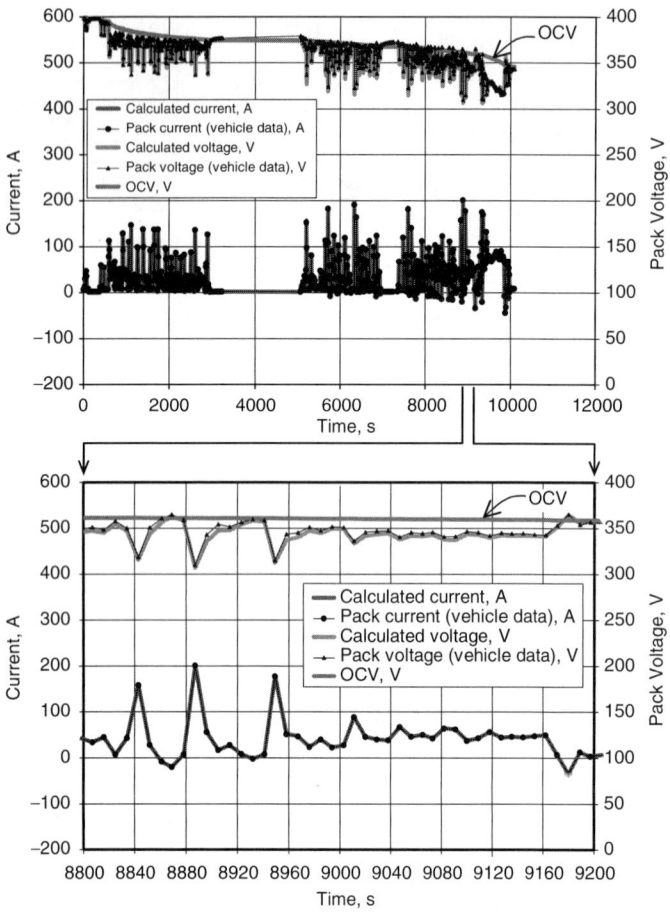

Figure 29. S10E vehicle data. The same model used to characterize the module is applied to the vehicle-level data. The corresponding drive is characteristic of a city schedule. The lower plot provides an expanded view of the upper plot. The battery pack comprises 26 modules.

III. STATE ESTIMATORS

Method of Least Squares. A very simple but useful example of constructing a state estimator for an EESS employs the method of least-squares with a fixed frame of data points, and only two parameters to be regressed. Much of the general framework and insight into the construction of more multivariable, recursive state estimators to be described can be understood by this introductory example. To enable discussion, we shall investigate a NiMH battery, modeled by the equivalent circuit depicted in the lowermost graphic of Fig. 11 and regress the high-frequency resistance and the open-circuit voltage. As will be discussed in detail, the NiMH system is complicated by hysteresis in the open-circuit voltage, which must also be regressed adaptively.

Before analyzing the specific application of the algorithm, an overview of the data shown in Fig. 30 provides a compelling reason as to why this approach is promising. A drive schedule similar to that used to determine the fuel economy values for the Precept HEV (108) yielded the data points depicted in Fig. 30. Illustrations of the vehicle and battery pack are depicted in Fig. 6. Note that the points are substantially co-linear, and a simple ohmic-battery description

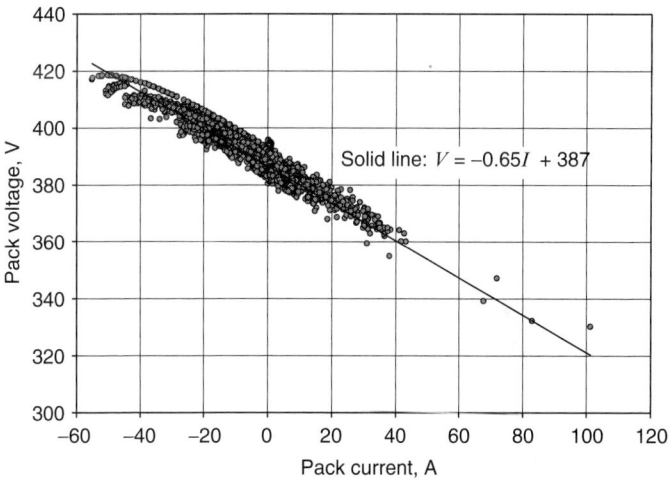

Figure 30. Battery system current-voltage relation over a representative drive cycle. Discharge currents are positive. The pack power was allocated based on vehicle modeling for the Precept vehicle (Fig. 6) over a drive cycle of the following characteristics: 1,384 s city schedule (FUDS), 10 min rest, 505 s of city schedule (FUDS), and two highway schedules (HWYFET). The collected current and voltage data for the described schedules correspond to the plotted data points. The fit line indicates an open-circuit voltage of 387 V, reflecting about 50% SOC.

($V=V_0+IR$) fit to the data captures the basic trend in the data. This result is partially enabled by the small variation in the SOC throughout the test; if large variations in SOC are manifest, then V_0 would not be substantially constant and the ohmic-battery description would be less insightful.

The algorithm used for the state estimator consists of two parts. There is an electrical circuit model (lowermost circuit in Fig. 11) that is used to describe the relationship between the currents and voltages observed at the terminals of the battery and a coulomb-accumulation model. We shall ignore self-discharge and current inefficiency on charge. The two SOC values are referred to as SOC_C (for the coulomb-accumulation contribution) and SOC_V (for the voltage-based contribution). Both contributions yield useful information regarding the SOC; thus a composite SOC is calculated:

$$SOC = w(SOC_C) + (1-w)(SOC_V), \qquad (15)$$

where w is a weighting factor. For SOC_C,

$$SOC_C(t) = SOC(t-\Delta t) + \int_{t-\Delta t}^{t} 100 \frac{I}{Ah_{nominal}} \frac{dt}{3{,}600}. \qquad (16)$$

It is important to note that the first term on the right is SOC, not SOC_C; the use of the composite SOC as the first term on the right strongly couples the voltage-based and coulomb-accumulation models. For the module of this NiMH example, the nominal capacity $Ah_{nominal}$ is 12.5 Ah (35°C, $C/3$ discharge rate); this quantity will have a mild temperature dependence over the range of application. The factor 3,600 has units of s/h, and the factor 100 is employed to keep a consistent percent basis.

To extract the voltage-based SOC_V, the integral solution for the voltage associated with (2), for an arbitrary current source, is employed:

$$V = V_0 + IR - A \int_{\zeta=t}^{\zeta=0} I(\zeta)\exp[-B(t-\zeta)]d\zeta. \qquad (17)$$

The first two terms on the right side give rise to an ohmic description of the battery, as the battery voltage V is related to the open-circuit potential V_0 reduced by the ohmic drop IR, where R is the battery resistance. The last term on the right side corresponds to a superposition integral, through which past currents influence the system potential. (Because of the exponential weighting function, the influence of early currents is less that that of more recent currents.) The inverse of the capacitance C^{-1} is given by A and $B = 1/R_{ct}C$. We shall apply (17) at the pack level. Thus the pack resistance is given by the number of modules multiplied by R_{module}, as other resistances (e.g., harnesses, etc.) have been allocated to the module resistance. For the pack utilized in this example, 28 modules were employed.

The open-circuit voltage is described by a modified Nernst equation (U immediately below) and an empirical expression to capture the salient features associated with voltage hysteresis (cf. (9) for a similar treatment of an insertion-electrode based battery):

$$V_0 = \text{Function}(T, \text{SOC}, V_H)$$
$$= U + V_H \qquad (18)$$
$$= U^0 + \frac{R_g T}{F} \ln \frac{\text{SOC} - \Pi}{100 - \text{SOC}} - (\gamma)(SOC) + V_H.$$

Faraday's constant (96,487 C mol^{-1}) corresponds to F; for consistent units, a value of 8.314 J mol-K^{-1} should be used for R_g. To construct the open-circuit voltage curve shown in Fig. 31, $U^0 = 1.37$ V, $\Pi = 8$ (taken as dimensionless, as is the percent state of charge), and $\gamma = 0.04$ V. The number of cells in the battery pack for this example is given by $N = 280$; thus the pack open-circuit voltage is given by $V_0 = N V_{0,cell}$. A plot of (18) against experimental data is provided in Fig. 31 for $V_H = 0$ (i.e., the function U is plotted).

For the hysteresis contribution, which plays a significant role in NiMH systems (11, 102, 109–113), we construct the following first-order differential equation to formulate a cell hysteresis voltage V_H:

$$\frac{\partial V_H}{\partial t} = \beta I [V_{H,\max} + \text{sign}(I) V_H]. \qquad (19)$$

This equation constructs a varying hysteresis voltage; for this work, the hysteresis voltage is set up so that for prolonged charge currents, or short but very large charge currents, the hysteresis voltage tends to 50 mV cell^{-1} (for 280 cells/pack, this corresponds to 11.2 V); thus $V_{H,max} = 50$ mV cell^{-1}. The exact opposite holds for discharge (positive) currents. The ±50 mV cell^{-1} corresponds to the 100 mV cell^{-1} difference between the upper and lower open-circuit voltage curves displayed in Fig. 31.

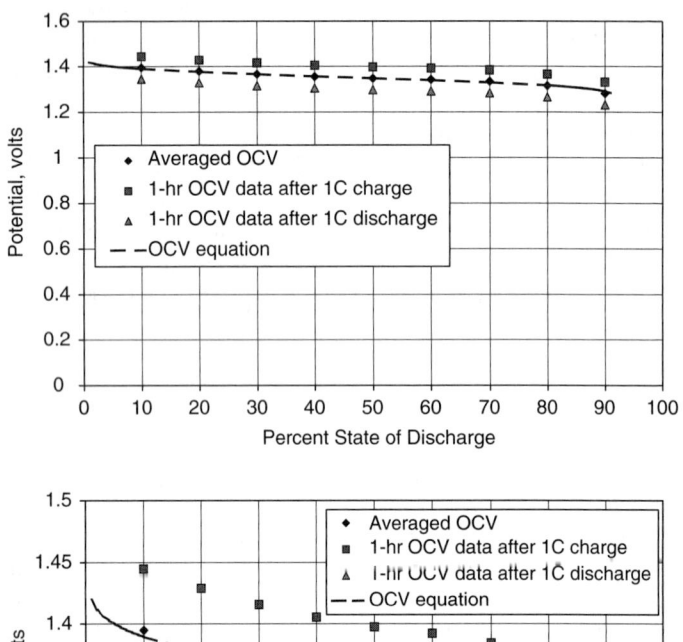

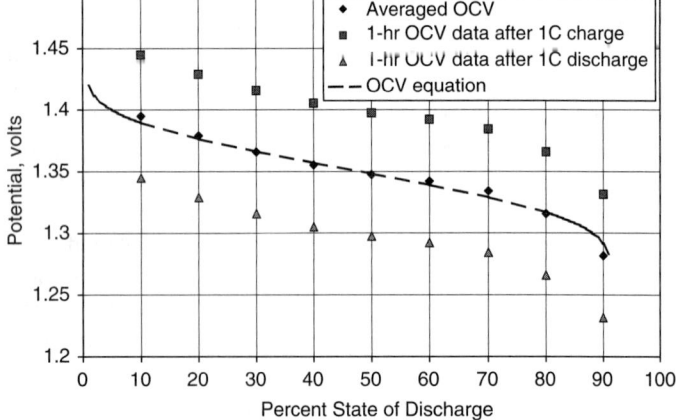

Figure 31. Open-circuit voltages. The upper panel illustrates the flatness of the open-circuit voltage versus SOC. The lower plot utilizes an expanded ordinate.

The description of the governing equations for the state estimator is now complete. The next step is to recast the equations in discrete form. First, the coulomb-integration expression is written as

$$\text{SOC}_C(t) = \text{SOC}(t - \Delta t) + \frac{I_{t-1} + I_t}{2} \frac{100}{Ah_{\text{nominal}}} \frac{\Delta t}{3{,}600}.$$

The difference between the present time and the last recorded time is given by Δt. Next, (17) is recast as a recursion relation:

$$V = (V_0 + IR)_t + A\left(\frac{I_{t-\Delta t} + I_t}{2}\right)\Delta t + e^{-B\Delta t}(V - V_0 - IR)_{t-\Delta t}, \tag{20}$$

where the subscripts t and $t - \Delta t$ denote the time at which the quantities are to be evaluated. This equation is a particularly simple recursion relation in that only variables calculated at the previous time step are required to calculate the voltage at time t. The equation is derived in the Appendix (48). To implement (20), one replaces the battery voltages with measured values:

$$\begin{aligned} V^{\text{to be used in regression}} = \hat{V} &= V^{\text{measured}}\Big|_t \\ &\quad - A\left(\frac{I_{t-\Delta t} + I_t}{2}\right)\Delta t \\ &\quad - e^{-B\Delta t}(V^{\text{measured}} - V_0 - IR)_{t-\Delta t} \\ &= (V_0 + IR)_t \end{aligned} \tag{21}$$

Thus the regression analysis to determine the open-circuit potential and resistance is based on the voltage $\hat{V}$, the *regression voltage*, and a least-squares analysis of the corrected voltage data should yield a good approximation for the ohmic resistance R and open-circuit potential V_0. For all results shown in this work, $A = 0.0229 \, \text{F}^{-1}$ and $B = 0.0366 \, \text{s}^{-1}$.

The open-circuit voltage V_0 and the resistance R are found using a standard least-squares approach, applied to data corresponding to a specified time interval, as is commonly done for fitting two-parameter models to data (114–116). The following definitions are applied:

$$s_I = \frac{1}{N}\sum_{j=1}^{N} I_j$$

$$s_{II} = \frac{1}{N}\sum_{j=1}^{N} I_j^2$$

$$s_V = \frac{1}{N}\sum_{j=1}^{N} \hat{V}_j$$

$$s_{IV} = \frac{1}{N}\sum_{j=1}^{N} I_j \hat{V}_j,$$

where N represents the number of recorded current-potential data points to be included in the extraction of the open-circuit voltage V_0 and the resistance R. For all analyses presented in this work, the time step is 1 s and $N = 91$. The summations are made effectively recursive by recognizing that a first-in, first-out method keeps the memory full of the most recent N data points and does not require a full summation over all components within the sum at each time step. Using these expressions, one obtains the following,

$$R = -\frac{S_{IV} - S_I S_V}{S_{II} - (S_I)^2}$$

and

$$V_0 = \frac{S_{II} S_V - S_{IV} S_I}{S_{II} - (S_I)^2}.$$

These equations fail when the variance $s_{II} - (s_I)^2 = 0$, or when this quantity is nearly zero. In addition, the equations can fail to provide a reasonable result when many of the recorded currents used in the least-squares analysis are of similar value with the exception of one or two data points; thus in application it is important to not update (regress) R and V_0 when the variance is small or there is a large skewness (118) in the data.

The hysteresis contribution is now addressed. The following formulation is applied:

$$V_H(t) = w_H(V_H + \beta(\Delta t)I[V_{H,\max} + \text{sign}(I)V_H])_{t-\Delta t} \\ + (1 - w_H)(V_0 - U)_{t-\Delta t}. \tag{22}$$

The subscript on the outer parentheses indicates that values to the right of this equation can be evaluated at the previous time step. This equation is not a straight-forward (numerical) time integration of (19) unless the weighting factor w_H is set to unity. Thus the term $(1 - w_H)(V_0 - U)_{t-\Delta t}$ allows for a correction to the extraction of the hysteresis voltage through the recognition that the previous time step value for the SOC can be used to calculate an open-circuit voltage. This back-calculated open-circuit voltage provides an adaptive routine for deducing V_H.

Both the combined SOC and the hysteresis voltage expressions utilize weighting factors (w and w_H, respectively), and the influence of the time step size needs to be addressed. That is, if very small time steps are employed, then the weighting factors should be altered so that time-dependent quantities are not lost from the calculation and instead are allowed to evolve in accordance with their particular time constants. The following approach has been found to work well:

$$w = w_{\max} - \alpha_w(\Delta t) \text{ and } w_H = w_{H,\max} - \alpha_H(\Delta t).$$

The weighting factors are bounded between 0 and 1 here. The factor α_H is taken as 0.005 s^{-1} for all of the plots in this work. For short times (i.e., before the regression analysis allows for accurate fitting of the resistance and open-circuit potential, 91 s in this work), w is set to unity, and the SOC is calculated for this short time based solely on coulomb integration from the previously stored SOC. The factor α_w is taken as 0.001 s^{-1}. Reference (11) provides more discussion around the selection of these two weight factors for NiMH systems.

Figure 32 can be viewed as the base case for this work, as it shows the results for the drive data plotted in Fig. 30. The battery was first charged to 50% SOC, then the battery was stimulated using a power signal that approximated the operation of the battery when the vehicle is driven on the FTP driving schedule. The measured current and voltage are shown in the upper plot, along with the SOC deduced by purely coulomb counting and the open circuit voltage extracted from the data using the full algorithm. For the relatively short times (less than a few hours) of power cycling, pure coulomb counting allows one to determine the SOC. For longer durations (e.g., days), this would not be the case. The lower plot in Fig. 32 provides the error relative to the pure coulomb counting

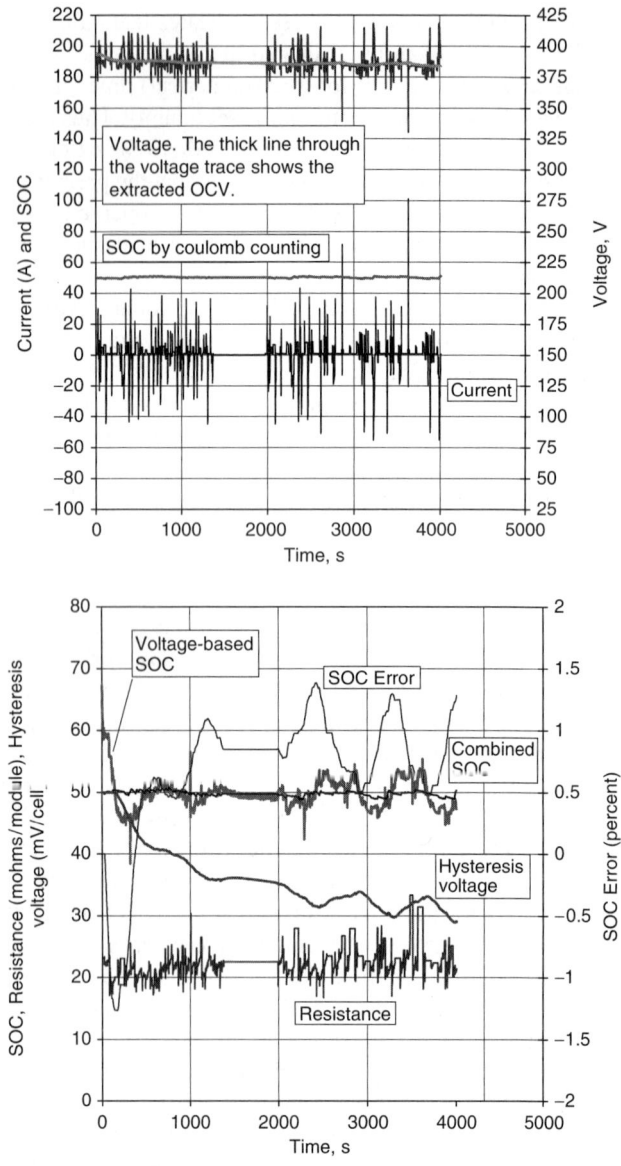

Figure 32. Least squares algorithm applied to data shown in Fig. 30. The voltage and current correspond to pack data; with the exception of the percent error, all other results are outputs of the algorithm.

SOC. The percent error is taken as $100(SOC_{\text{pure coulomb counting}} - SOC)/SOC_{\text{pure coulomb counting}}$. Other quantities plotted in the lower portion of Fig. 32 include the extracted module resistance (about 23 mohm module^{-1} throughout this work, corresponding to a pack resistance of 650 mohms as shown in Fig. 30), the regressed combined state of charge (SOC), and the regressed voltage-based state of charge (SOC_V). The hysteresis voltage declines from nearly 50 mV/cell initially, reflecting a battery that had just been charged, to nearly 0 V, indicating that the operation of the battery system is largely charge sustaining throughout the remainder of the test.

A test of the algorithm rate of convergence and stability is depicted in Fig. 33, where the initial SOC for the algorithm is set to 10%, even though the actual initial SOC is 50%. In order to speed convergence of the algorithm to less than a couple minutes,

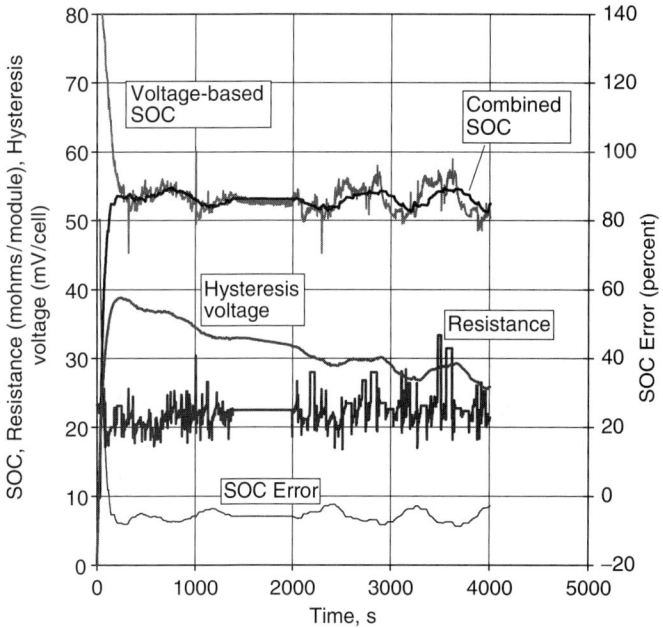

Figure 33. Robustness test. The hysteresis voltage was set to zero initially. In addition, the beginning SOC used to start the algorithm was set to 10% (instead of the actual 50%). The error in the regressed SOC is less than 10% within 100 s.

the weighting of the voltage-based portion of the algorithm is increased; $\alpha_w = 0.01$ s^{-1} (more adaptive than that of Fig. 32, wherein $\alpha_w = 0.001$ s^{-1}). However, the combined SOC shows more variation (albeit still acceptably stable), and an expected tradeoff between stability and adaptability must be comprehended in application.

Generalized Weighted Recursive Least Squares. The previous section used a fixed number of data points and standard least squares methodology to regress two parameters (R and V_0), along with a separate and less rigorous adaptive procedure was found to be useful in determining the voltage hystersis contribution. In this section, we derive and implement an algorithm that can accommodate an arbitrary number of model parameters, thereby allowing for more complicated battery models to be employed in formulating model reference adaptive systems as part of energy management scheme for systems employing batteries.

The instantaneous error ε in the desired voltage response is defined as

$$\varepsilon(t) = [y - (m_1 x_1 + m_2 x_2 + \cdots + m_L x_L + b)],$$

where y represents the experimentally obtained voltage at time t (i.e., $y = V^{\text{measured}}$) and the values $x_1, x_2, \ldots, x_L$ represent the measured quantities on which the parameters $m_1, m_2, \ldots, m_L$ multiply, respectively, to complete the linear model once the parameter b is included.[2] The relationship between $x_1, x_2, \ldots, x_L, m_1, m_2, \ldots, m_L$, and b relative to the quantities in governing the EESS shall be made explicit after the WRLS formulation is complete. For weighted least squares analysis (weight factor w), we seek to minimize the error term

$$\begin{aligned}
\varepsilon &= \sum_{j=1}^{j=N} w_j [y_j - (b + m_{1} x_{1,j} + m_{2} x_{2,j} + \cdots + m_{L,j} x_{L,j})]^2 \\
&= \sum_{j=1}^{j=N} w_j \left[y_j - b - \sum_{l=1}^{l=L} m_l x_{l,j} \right]^2,
\end{aligned} \tag{23}$$

---

[2] We employ y, m, x, and b simply because the formula $y = mx + b$ is commonly used in exposition to describe a line, and this is in fact the explicit basis for the two-parameter model, which we generalize to a multi-parameter algorithm in this section, leading to the use of vectors **Y** and **m** and the matrix **X**.

where the subscript j denotes the time step. An important decision has been made in using (23) for the loss (error) term that is often overlooked. One could, for example, choose to weight the error by the standard deviations of the data points so as to diminish the influence of data that fall outside expected confidence boundaries, as is done in Chi-Square fitting. That is, not all measurement errors are independent and normally distributed with identical standard deviations, and least-squares analyses can be skewed by anomalous data points. A helpful discussion on robust estimation and related matters appears in Chap. 14 of Press et al. (116).

By setting the partial derivatives of ε with respect to each of the parameters $m_1, m_2, \ldots, m_L$ and b to zero, we obtain the following $L+1$ equations:

$$b = \frac{1}{\sum_{j=1}^{j=N} w_j} \left(\sum_{j=1}^{j=N} w_j y_j - m_1 \sum_{j=1}^{j=N} w_j x_{1,j} \right.$$

$$\left. - m_2 \sum_{j=1}^{j=N} w_j x_{2,j} - \cdots - m_L \sum_{j=1}^{j=N} w_j x_{L,j} \right),$$

$$0 = \sum_{j=1}^{j=N} w_j y_j x_{1,j} - b \sum_{j=1}^{j=N} w_j x_{1,j} - m_1 \sum_{j=1}^{j=N} w_j x_{1,j} x_{1,j} - m_2$$

$$\times \sum_{j=1}^{j=N} w_j x_{2,j} x_{1,j} - \cdots - m_L \sum_{j=1}^{j=N} w_j x_{L,j} x_{1,j},$$

$$0 = \sum_{j=1}^{j=N} w_j y_j x_{2,j} - b \sum_{j=1}^{j=N} w_j x_{2,j} - m_1 \sum_{j=1}^{j=N} w_j x_{1,j} x_{2,j} - m_2$$

$$\times \sum_{j=1}^{j=N} w_j x_{2,j} x_{2,j} - \cdots - m_L \sum_{j=1}^{j=N} w_j x_{L,j} x_{2,j},$$

$$\vdots$$

$$0 = \sum_{j=1}^{j=N} w_j y_j x_{L,j} - b \sum_{j=1}^{j=N} w_j x_{L,j} - m_1 \sum_{j=1}^{j=N} w_j x_{1,j} x_{L,j} - m_2$$
$$\times \sum_{j=1}^{j=N} w_j x_{2,j} x_{L,j} - \cdots - m_L \sum_{j=1}^{j=N} w_j x_{L,j} x_{L,j}.$$

The above L equations can be more compactly expressed as

$$0 = \sum_{j=1}^{j=N} w_j y_j x_{\ell,j} - b \sum_{j=1}^{j=N} w_j x_{\ell,j}$$
$$- \sum_{l=1}^{l=L} m_l \sum_{j=1}^{j=N} w_j x_{l,j} x_{\ell,j} \quad (L \text{ equations}; \ell = 1, L).$$

There are two reasons why one would want to weight data sets differently through the value of w_j. First, some observations may be subject to greater disturbance; here, a disturbance refers to a phenomenon not accounted for in the system model. For example, the onset of secondary reactions during battery charging (hydrogen and oxygen evolution (119–122)) may lead one to discount the charge observations relative to those of discharge, as the impact of the secondary reactions is not accounted for explicitly in the model equations. Second, newer observations are generally more important than older observations in determining the state of the system and therefore should be given a larger weight factor relative to older observations. For these two reasons, we decompose the weight factor into a time-weighting factor λ and the factor γ; the latter can be used to weight discharge events over those of charge, for example. Hence,

$$w_j = \gamma_j \lambda^{N-j}.$$

It can be shown that the use of λ^{N-j} yields an exponential decay in the influence of past data points on the determination of the current value of m_l and b:

$$\lambda^{N-j} = e^{\ln \lambda^{N-j}} = e^{(N-j)\ln \lambda} \approx e^{-(N-j)(1-\lambda)} \quad \text{for } \lambda \to 1.$$

Thus $\Delta t/(1-\lambda)$ reflects the time constant over which past points impact the regression analysis. For example, if Δt is kept near 1 s and $\lambda = 0.99$, then $(1 \text{ s})/(1 - 0.99) = 100 \text{ s}$. Hence, a data point taken 100 s prior to the current point has less than 40% of the impact on the regression analysis relative to that of the current data point. A data point taken 500 s prior to the current point has less than 1% of the impact on the regression relative to that of the current data point. More complete discussions on exponential forgetting can be found in Sect. 5.6.1 of Ljnug and Söderström (48), Sect. 6.2 of Anderson and Moore (45), and Sects. 5.3 and 5.4 of Kulhavý (50).

The summations displayed in the $L + 1$ equations are made recursive with the following definitions:

$$s_w |_N = \gamma_N + \lambda(s_w |_{N-1}),$$
$$s_u |_N = \gamma_N u_N + \lambda(s_u |_{N-1}), \text{ and}$$
$$s_{u,v} |_N = \gamma_N u_N v_N + \lambda(s_{u,v} |_{N-1}).$$

In these sums, u and v refer to y or x_l; e.g., $s_{x_1} = \sum_{j=1}^{j=N} w_j x_{1,j}$, $s_{x_1,x_2} = \sum_{j=1}^{j=N} w_j x_{1,j} x_{2,j}$, and $s_{y,x_1} = \sum_{j=1}^{j=N} w_j y x_{1,j}$. Note that $s_{u,v} = s_{v,u}$.

Initially,

$$s_w |_1 = \gamma_1,$$
$$s_u |_1 = \gamma_1 u_1, \text{ and}$$
$$s_{u,v} |_1 = \gamma_1 u_1 v_1.$$

We utilize the following covariance expressions:

$$V_{uv} |_N = \left(s_{u,v} |_N - \frac{s_u |_N \; s_v |_N}{s_w |_N} \right) \frac{1}{s_w |_N}$$

(where $V_{uv} = V_{vu}$). We can now formulate the following (symmetric) matrix system of equations

$$m_1 V_{1,1} + m_2 V_{1,2} + m_3 V_{1,3} + \cdots + m_L V_{1,L} = V_{y,1},$$
$$m_1 V_{1,2} + m_2 V_{2,2} + m_3 V_{2,3} + \cdots + m_L V_{2,L} = V_{y,2},$$
$$m_1 V_{1,3} + m_2 V_{2,3} + m_3 V_{3,3} + \cdots + m_L V_{3,L} = V_{y,3},$$
$$\vdots$$
$$m_1 V_{1,L} + m_2 V_{2,L} + m_3 V_{3,L} + \cdots + m_L V_{L,L} = V_{y,L},$$

or

$$\sum_{l=1}^{l=L} m_l V_{\ell,l} = V_{y,\ell} \quad (L \text{ equations}; \ell = 1, L; V_{\ell,l} = V_{l,\ell}).$$

This symmetric system can be expressed as

$$\mathbf{Xm} = \mathbf{Y},$$

where

$$\mathbf{Y} = \begin{bmatrix} V_{y,1} \\ V_{y,2} \\ V_{y,3} \\ \vdots \\ V_{y,L} \end{bmatrix}, \mathbf{m} = \begin{bmatrix} m_1 \\ m_2 \\ m_3 \\ \vdots \\ m_L \end{bmatrix}, \text{ and } \mathbf{X} = \begin{bmatrix} V_{1,1} & V_{1,2} & V_{1,3} & \cdots & V_{1,L} \\ V_{1,2} & V_{2,2} & V_{2,3} & \cdots & V_{2,L} \\ V_{1,3} & V_{2,3} & V_{3,3} & \cdots & V_{3,L} \\ \vdots & \vdots & \vdots & \ddots & \vdots \\ V_{1,L} & V_{2,L} & V_{3,L} & \cdots & V_{L,L} \end{bmatrix}.$$

Upon solution of the matrix system for the parameters m_l, the parameter b can be extracted for the following (recursive) expression:

$$b = \frac{1}{[\gamma_N + \lambda(S_w|_{N-1})]} \left\{ \gamma_N y_N + \lambda(S_y|_{N-1}) - \sum_{l=1}^{l=L} m_l \left[\gamma_N X_{l,N} + \lambda(S_{r_l}|_{N-1}) \right] \right\} \tag{24}$$

The benefit of removing b from the matrix system $\mathbf{Xm} = \mathbf{Y}$ is related to the fact that b is not a vector or tensor and therefore need not be included in operations associated with solving the matrix system. (Although it would be less efficient, one could place b in the $\mathbf{m}$ vector, with the corresponding changes to the entries in $\mathbf{X}$ that multiply onto b.)

In summary, we have provided the equations that will be used to implement the WRLS algorithm. We require the solution to

$$\mathbf{m} = \mathbf{X}^{-1}\mathbf{Y}.$$

This expression is equivalent to

$$\tilde{\mathbf{m}} = (\tilde{\mathbf{x}}^T \mathbf{w} \tilde{\mathbf{x}})^{-1} \tilde{\mathbf{x}}^T \mathbf{w} \mathbf{y},$$

which is often seen in the literature.[3] Here $\tilde{\mathbf{x}}$ denotes the coefficient matrix comprising the x_j entries, the $\mathbf{y}$ vector contains the y_j entries, and $\mathbf{w}$ comprises the w_j weighting elements (including the forgetting factor); as formulated in this work, $\mathbf{w}$ is a diagonal matrix with entries $w_1, w_2, w_3, \ldots, w_N$. The parameter vector is given by $\tilde{\mathbf{m}}$. The tilde over the vector $\tilde{\mathbf{m}}$ and matrix $\tilde{\mathbf{x}}$ is used to indicate that the parameter b is now included in the vector $\tilde{\mathbf{m}}$, with corresponding changes to elements in $\tilde{\mathbf{x}}$ that multiply onto b. It is more difficult to find recursive formulations in the literature that are presented in a form similar to $\mathbf{m} = \mathbf{X}^{-1}\mathbf{Y}$; although mathematically equivalent formulations exist, they are constructed with intermediate calculations involving covariance and gain matrixes, both of which contain useful analytical information. We do not utilize such formulations, however, as they are less direct with respect to the implementation of symbolic manipulators (117).

Because of the recursive nature of the algorithm, it is important to seed the initial values for the parameters m_l; this can be done by using previously measured values to provide $m_{l,}|_{t=0}$. Prior measurements of the battery parameters (cf. Table 4) are also needed in order to provide reasonable bounds on the variation of the regressed parameter values relative to their nominal values.

We are now in a position to use the derived WRLS algorithm to formulate a state estimator for an EESS. Our voltage expression is for the system voltage at time t is obtained by substituting (22) into (18) ($V_0 = U + V_H$), and then substituting the result into (20):

$$V_t = U + w_H(V_H)_{t-\Delta t} + (1 - w_H)(V_0 - U)_{t-\Delta t}$$
$$+ \beta w_H \Delta t I_{t-\Delta t}[V_{H,\max} - \text{sign}(I)V_H]_{t-\Delta t} + I_t R$$
$$+ \left(\frac{I_{t-\Delta t} + I_t}{2}\right) A_d r_C \Delta t + E(V - V_0 - IR)_{t-\Delta t}, \quad (25)$$

where $E = \exp(-\Delta t/\tau)$ and $\tau = 1/R_{\text{ct}}C$. The quantity A_d is the inverse of the capacitance on discharge, and r_C is the ratio of A for charge to that of discharge; i.e.,

$$r_C(T, \text{SOC}) = A_c/A_d = C_{D,\text{discharge}}/C_{D,\text{charge}}.$$

---

[3] For examples, see Gelb (46) and Conte and de Boor (117). Transpose is denoted by superscript T, and a superscript -1 represents inverse.

Table 3.
Quantities, values, and units unless otherwise specified for the NiMH HEV state estimator

Value	Units	Quantity
50	mV	Half the maximum cell hysteresis, $V_{H,max}$
12.5	Ah	Nominal capacity
0.0229	F^{-1}	$A = 1/C$, inverse capacitance
0.0366	s^{-1}	$B = 1/(R_{ct}C)$, inverse time constant
10		Cells per module (280 cells/pack)
28		Modules per pack
2.47×10^{-5}	C^{-1}	β for charge-augmented operation
3.70×10^{-5}	C^{-1}	β for charge-depleted operation
0.650	Ohms	Default high-frequency pack resistance R
0.001	s^{-1}	α_W at 50% SOC
0.005	s^{-1}	α_H

Table 4.
Lead acid and lithium ion cell characteristics

Lead acid cell	Lithium ion cell	Quantity, units
0.7800	1.637	R, mohm
0.7	0.905	E
2.500×10^{-4}	8.000×10^{-5}	A_d, F^{-1}
1.122×10^{-4}	1.122×10^{-4}	β, C^{-1}
6.8	0.75	$r = C_{D,dis}/C_{D,chg}$
4,000	12,500	$C_{D,dis}$, F
588.2	16,667	$C_{D,chg}$, F
2.81	5.00	τ, s
0.702	0.4	$R_{ct,dis}$, mohm
4.773	0.3	$R_{ct,chg}$, mohm
16.5	20	$V_{H,max}$, mV
55	4.2	Ah
1	1×10^{-10}	Det_cal
1	1	w_H

During discharge, $r_C=1$. The use of r_C eliminates the need to regress separate values for $A = 1/C$ on charge and discharge. Because A will change with surface area in a consistent manner on both charge and discharge, we expect the ratio r_C to remain constant, and we regress A_d only.

For a two-parameter system, $y = m_1 x_1 + b$, and

$$m_1 = \frac{V_{y,1}}{V_{1,1}}.$$

The parameter b is always to be calculated using (24). This two-parameter approach is useful for regressing the open-circuit potential and the resistance, as described for NiMH previously and will be discussed below for a lead-acid cell. In this case, the following assignments are made, consistent with regressing the two values V_0 and R in (21).

$$\begin{aligned}
y &= V^{\text{measured}}|_t \\
x_1 &= I_t \\
m_1 &= R|_t \\
b &= U|_t + (1 - w_{\text{H}})(V_0 - U)_{t-\Delta t} + w_{\text{H}}(V_{\text{H}})_{t-\Delta t} \\
&\quad + E(V^{\text{measured}} - V_0 - IR)_{t-\Delta t} \\
&\quad + A_d \left(\frac{I_{t-\Delta t} + I_t}{2}\right) \left[r_{\text{C},(I_t + I_{t-\Delta t})/2}\right] \Delta t \\
&\quad + \beta w_{\text{H}} \Delta t I_{t-\Delta t} [V_{\text{H,max}} - \text{sign}(I) V_{\text{H}}]_{t-\Delta t}.
\end{aligned} \quad (26)$$

After the parameter b is regressed, $U|_t$ is isolated by subtracting the other (constant or measured, but not regressed) quantities from the value of b.

For three-parameter models, $y = m_1 x_1 + m_2 x_2 + b$, and

$$\begin{aligned}
\text{Det} &= V_{1,1} V_{2,2} - V_{1,2}^2, \\
m_1 &= \frac{1}{\text{Det}} [V_{y,1} V_{2,2} - V_{y,2} V_{1,2}], \text{ and} \\
m_2 &= \frac{1}{\text{Det}} [V_{y,2} V_{1,1} - V_{y,1} V_{1,2}],
\end{aligned} \quad (27)$$

where Det represents the determinant of **X**. This approach can be used with the following assignments allowing one to extract the high-frequency resistance R, the exponential of the time constant τ, and the open-circuit potential V_0:

$$y = V^{\text{measured}}|_t$$
$$x_1 = I_t$$
$$x_2 = (V^{\text{measured}} - V_0 - IR)_{t-\Delta t}$$
$$m_1 = R|_t$$
$$m_2 = \exp(-B\Delta t) = e^{-\Delta t/\tau} = E|_t$$
$$b = U|_t + (1-w_H)(V_0 - U)_{t-\Delta t} + w_H(V_H)_{t-\Delta t}$$
$$+ A_d \left(\frac{I_{t-\Delta t} + I_t}{2}\right) \left[r_{C,(I_t + I_{t-\Delta t})/2}\right] \Delta t$$
$$+ \beta w_H \Delta t I_{t-\Delta t} [V_{H,\max} - \text{sign}(I) V_H]_{t-\Delta t}.$$

For four-parameter models, $y = m_1 x_1 + m_2 x_2 + m_3 x_3 + b$, and

$$\text{Det} = V_{1,3}^2 V_{2,2} - 2 V_{1,2} V_{1,3} V_{2,3} + V_{1,1} V_{2,3}^2 + V_{1,2}^2 V_{3,3}$$
$$- V_{1,1} V_{2,2} V_{3,3},$$
$$m_1 = \frac{1}{\text{Det}} [V_{2,3}^2 V_{y,1} - V_{2,2} V_{3,3} V_{y,1} + V_{1,2} V_{3,3} V_{y,2}$$
$$+ V_{1,3} V_{2,2} V_{y,3} - V_{2,3} (V_{1,3} V_{y,2} + V_{1,2} V_{y,3})],$$
$$m_2 = \frac{1}{\text{Det}} [V_{1,3}^2 V_{y,2} - V_{1,1} V_{3,3} V_{y,2} + V_{1,2} V_{3,3} V_{y,1}$$
$$+ V_{1,1} V_{2,3} V_{y,3} - V_{1,3} (V_{2,3} V_{y,1} + V_{1,2} V_{y,3})], \text{ and}$$
$$m_3 = \frac{1}{\text{Det}} [V_{1,3} V_{2,2} V_{y,1} - V_{1,2} V_{2,3} V_{y,1} - V_{1,2} V_{1,3} V_{y,2}$$
$$+ V_{1,1} V_{2,3} V_{y,2} + V_{1,2}^2 V_{y,3} - V_{1,1} V_{2,2} V_{y,3}].$$

This approach works well for systems wherein hysteresis does not play a significant role (e.g., lead acid and lithium ion batteries), as all parameters with the exception of β are regressed adaptively:

$$y = V^{\text{measured}}|_t$$
$$x_1 = I_t$$
$$x_2 = (V^{\text{measured}} - V_0 - IR)_{t-\Delta t}$$
$$x_3 = \left(\frac{I_{t-\Delta t} + I_t}{2}\right) \left[r_{C,(I_t + I_{t-\Delta t})/2}\right] \Delta t$$

$$m_1 = R|_t$$
$$m_2 = \exp(-B\,\Delta t) = e^{-\Delta t/\tau} = E|_t$$
$$m_3 = A_d|_t$$
$$b = U|_t + (1 - w_\text{H})(V_0 - U)_{t-\Delta t} + w_\text{H}(V_\text{H})_{t-\Delta t}$$
$$\quad + \beta w_\text{H} \Delta t\, I_{t-\Delta t}[V_{\text{H,max}} - \text{sign}(I)V_\text{H}]_{t-\Delta t}.$$

Last, for the five-parameter model, all relevant parameters are regressed, and the system identification scheme is substantially complete:

$$y = V^{\text{measured}}|_t$$
$$x_1 = I_t$$
$$x_2 = (V^{\text{measured}} - V_0 - IR)_{t-\Delta t}$$
$$x_3 = \left(\frac{I_{t-\Delta t} + I_t}{2}\right)\left[r_{C,(I_t + I_{t-\Delta t})/2}\right]\Delta t$$
$$x_4 = w_\text{H}\Delta t\, I_{t-\Delta t}\left[V_{\text{H,max}} - \text{sign}(I)V_\text{H}\right]_{t-\Delta t} \qquad (28)$$
$$m_1 = R|_t$$
$$m_2 = \exp(-B\,\Delta t) = e^{-\Delta t/\tau} = E|_t$$
$$m_3 = A_d|_t$$
$$m_4 = \beta|_t$$
$$b = U|_t + (1 - w_\text{H})(V_0 - U)_{t-\Delta t} + w_\text{H}(V_\text{H})_{t-\Delta t}.$$

Expressions for m_l for the five-parameter regression are presented in second portion of the Appendix, along with a discussion explaining how to derive the expressions using a symbolic manipulator.

Statistical tests for the regression analysis. Expressions for the regressed parameters all contain the determinant[4] of the relevant square matrix, a single-valued function, which provides a convenient means to assess the suitability of the collected data for parameter regression. The first statistical test pertains to the value of the determinant. Specifically,

---

[4] For the solution to the linear system $\mathbf{Xm} = \mathbf{Y}$, $\mathbf{X}$ should be invertible in order that the system has exactly one solution. The matrix $\mathbf{X}$ is invertible if and only if the determinant of $\mathbf{X}$ is nonzero, $\det(\mathbf{X}) \neq 0$. From a practical point of view, we require that the determinant of $\mathbf{X}$ be of magnitude greater than a specified (calibration) value.

$$\text{Det\_test} = \begin{cases} 0 & \text{if abs(Det)} < \text{Det\_cal} \\ 1 & \text{if abs(Det)} \geq \text{Det\_cal} \end{cases}.$$

Thus if the magnitude of the determinant is too small to ensure numerical accuracy in the evaluation of the matrix equations, the value of Det_test is set to zero, and the analytic expressions used to extract the parameter values m_l are not to be employed; the appropriate actions under these conditions are discussed below. (The value of Det_cal represents a calibration; i.e., a constant that is set prior to algorithm operation.)

The next test concerns the nature of the variation in the current excitation source. For our purposes, we define skewness as (118),

$$\text{skewness} = \left| \frac{1}{N\sigma^3} \sum_{j=1}^{j=N} (x_j - \bar{x})^3 \right|,$$

where $\bar{x}$ is the average of the x-values and $\sigma^2 = \frac{1}{N} \sum_{j=1}^{j=N} (x_j - \bar{x})^2$ is the variance. We restrict the skewness test to the actual current-time (excitation source) values $I(t)$ and do not incorporate the charge-discharge weighting. Following the same logic used previously, recursive relations are employed, with the subscript s is added to indicate quantities associated with the skewness calculation:

$$S_{w,s}|_N = \sum_{j=1}^{N} \lambda^{N-j} = 1 + \lambda \sum_{j=1}^{N-1} j\lambda^{N-1-j} = 1 + \lambda \left(S_{w,s}|_{N-1} \right)$$

$$S_{I,s}|_N = \frac{1}{\sum_{j=1}^{N} \lambda^{N-j}} \sum_{j=1}^{N} \lambda^{N-j} I_j = \frac{I_N + \lambda \left(S_{I,s}|_{N-1} \right) \left(S_{w,s}|_{N-1} \right)}{\left(S_{w,s}|_N \right)}$$

$$S_{II,s}|_N = \frac{1}{\sum_{j=1}^{N} \lambda^{N-j}} \sum_{j=1}^{N} \lambda^{N-j} I_j^2 = \frac{I_N^2 + \lambda \left(S_{II,s}|_{N-1} \right) \left(S_{w,s}|_{N-1} \right)}{\left(S_{w,s}|_N \right)}$$

$$\text{skewness}|_N = \begin{cases} \text{skew\_cal if } Det\_test \\ \left| \frac{(I_N - S_{I,s}|_N)^3}{\left[S_{II,s}|_N - (S_{I,s}|_N)^2 \right]^{3/2}} \right| \frac{1}{N} \\ + (\text{skewness}|_{N-1}) \left(1 - \frac{1}{N} \right) \quad \text{if Det\_test} = 1. \end{cases}$$

The value of skew_cal represents a calibration. To start the recursive calculations for skewness, the following conditions are used.

$$S_{w,s}|_1 = 1$$
$$S_{I,s}|_1 = I_1$$
$$S_{II,s}|_1 = I_1^2.$$

For the first time step, $\text{Det\_test}|_{t=0} = 0$ and $\text{skewness}|_{t=0}$ is set to skew_cal. Analogous to Det_test, define

$$\text{skew\_test} = \begin{cases} 0 & \text{if skewness} \geq \text{skew\_cal}, \\ 1 & \text{if skewness} < \text{skew\_cal}. \end{cases}$$

As with Det_test, the regression analysis is not employed if the skew test is not passed. The final result for our statistical tests corresponds to

$$\text{if Det\_test or skew\_test} = 0, \text{ then } \begin{cases} (m_l)_t = (m_l)_{t-\Delta t}, \text{ and} \\ b_t = y_t - \sum_{l=1}^{l=L}(m_l)_{t-\Delta t}(x_l)_t. \end{cases}$$

In summary, the parameter vector **m** is not updated if Det_test or skew_test = 0; however, the approach allows one to calculate the parameter b under all conditions, which implies that the open-circuit voltage and SOC_V can always be extracted

Power capability, SOP. The state of power (SOP), requires knowledge of the maximum discharge power, which can be expressed as

$$P_{\max,\text{discharge}} = IV = IV_{\min}.$$

That is, when the battery voltage obtains its lowest acceptable value, the maximum discharge power results. First, consider an ohmic battery, wherein the superposition integral (low frequency impedance) can be ignored. For the ohmic battery, $V = V_0 + IR$, and

$$P_{\max,\text{discharge}} = IV_{\min} = \frac{(V_{\min} - V_0)}{R} V_{\min}. \tag{29}$$

Similarly, for the maximum charge power for the ohmic battery is given by

$$P_{\max,\text{charge}} = IV_{\max} = \frac{(V_{\max} - V_0)}{R} V_{\max}. \tag{30}$$

For the maximum ohmic resistance, obtained at long times, R is replaced by $R + R_{ct}$.

The ohmic battery does not address transient effects such as those correlated by the superposition integral. To improve the estimate, for times that are greater than $\sim 0.1 R_{ct} C$, we include the superposition integral and calculate the maximum charge and discharge powers available for the time interval Δt, with $\Delta t \ll R_{ct} C$:

$$I|_t = -\frac{(V_0 - V)_t + (A I_{t-\Delta t} \Delta t/2) + \exp(-B \Delta t)[V - (V_0 + IR)]_{t-\Delta t}}{R + (A_d r_C \Delta t/2)}, \quad (31)$$

$P_{\max,\text{discharge}}(\Delta t)$
$= I V_{\min}$
$= \left[-\frac{(V_0 - V_{\min})_t + (A_d I_{t-\Delta t} \Delta t/2) + \exp(-B \Delta t)[V - (V_0 + IR)]_{t-\Delta t}}{R + (A_d r_C I_{t-\Delta t} \Delta t/2)} \right] V_{\min},$

and

$P_{\max,\text{charge}}(\Delta t)$
$= I V_{\max}$
$= \left[-\frac{(V_0 - V_{\max})_t + (A_c I_{t-\Delta t} \Delta t/2) + \exp(-B \Delta t)[V - (V_0 + IR)]_{t-\Delta t}}{R + (A_d r_C I_{t-\Delta t} \Delta t/2)} \right] V_{\max},$

where it is recognized that $r_C = 1$ on discharge. To implement these equations, the respective powers are calculated immediately after the algorithm has been employed to finish the SOC determination at time t. In this case, quantities calculated or measured at time t are then stored in the variables listed in the respective power expressions at time $t - \Delta t$. Then one must state the duration corresponding to the desired estimate for power. For example, if we want to know the power estimates 3 s from "now," then the measured and extracted values are placed in the $t - \Delta t$ quantities, t and Δt are set to 3 s, and the right sides of the above equations yield the desired power estimates.

While (31) is used in this work for the transient power calculations, inaccuracies can give rise to errors in the expressions for power capability, particularly for larger values of Δt. For longer times, we can use the constant-voltage solutions to approximate the power capability. In this case, $dV/dt = 0$. Power P will be our dependent variable, with $P = IV$. We define the following:

$$\tau_V = \left(\frac{R + R_{ct}}{R R_{ct} C} \right) t \text{ and } p_V = \left[\frac{R + R_{ct}}{V(V - V_0)} \right] P.$$

The subscript V is used to denote constant-voltage operation. Our governing equations (2) can be recast as

$$\frac{\mathrm{d}p_V}{\mathrm{d}\tau_V} + p_V = 1.$$

The solution to this equation is

$$\frac{p_V - 1}{p_V(0) - 1} = \mathrm{e}^{-\tau_V}$$

or

$$P(t) = \frac{V(V - V_0)}{R + R_{\mathrm{ct}}} + \left[P(0) - \frac{V(V - V_0)}{R + R_{\mathrm{ct}}} \right] \exp\left[-\left(\frac{R + R_{\mathrm{ct}}}{R R_{\mathrm{ct}} C} \right) t \right].$$

For long times, p_V tends to 1, and

$$P(\infty) = \frac{V(V - V_0)}{R + R_{\mathrm{ct}}},$$

indicating that all of the cell current flows through the two resistors, and the capacitor C is fully charged. To identify $p_V(0)$, we recognize that immediately prior to the voltage being changed to its new, constant value,

$$V_{\mathrm{init}} = V_0 + I_{\mathrm{init}} R + \frac{q}{C}.$$

Immediately after the voltage has been changed to its new, constant value, the same loop of voltage components yields

$$\begin{aligned} V &= V_0 + I(0)R + \frac{q}{C} \\ &= V_0 + I(0)R + V_{\mathrm{init}} - V_0 - I_{\mathrm{init}} R, \\ &= V_{\mathrm{init}} + I(0)R - I_{\mathrm{init}} R \end{aligned}$$

which provides an expression for the current immediately after the voltage has been changed to its new, constant value:

$$I(0) = \frac{V - V_{\mathrm{init}}}{R} + I_{\mathrm{init}}.$$

Last, the initial power, $P(0)$, can now be expressed as

$$P(0) = I(0)V = \left(\frac{V - V_{\mathrm{init}}}{R} + I_{\mathrm{init}} \right) V,$$

where V is the constant-voltage value. The final equation for power becomes

$$P(t) = \frac{V(V - V_0)}{R + R_{ct}} + \left[V \left(\frac{V - V_{init}}{R} + I_{init} \right) - \frac{V(V - V_0)}{R + R_{ct}} \right] \\ \times \exp\left[-\left(\frac{R + R_{ct}}{R R_{ct} C} \right) t \right]. \quad (32)$$

Two-parameter application: lead-acid cell. This section is analogous to that devoted to the NiMH two-parameter analysis above; whereas the NiMH example utilized a fixed number of data points ($N = 91$), the analysis in this section utilizes the two-parameter WRLS method derived in the previous section (26). Following this section, we shall examine the full five-parameter method on a lithium ion cell, thereby providing a full description of the WRLS method and analysis of lead acid, NiMH, and lithium ion cells. We investigate 12-V Panasonic HV1255 VRLA module of 55 Ah rated capacity. The module and the algorithm have been employed in GM's 42V Parallel Hybrid Truck, as described in (16). The open-circuit potential is depicted in Fig. 34 and is described by

$$V_0(\text{volts/cell}, 25\text{--}45^\circ\text{C}) = U + V_H \\ = 1.9214 + 0.2949 \frac{\text{SOC}}{100} + V_H.$$

The hysteresis contribution V_H is minimal. For w_H, because the lead-acid battery does not exhibit significant hysteresis effects, we set $w_H = 1$ throughout; that is, the hysteresis voltage is calculated and included in the results, but it is not adapted (cf. (26)). The value of $V_{H,max}$ is 16 mV. Parameters characterizing the lead acid cells are presented in Table 4.

As illustrated in the lower plot of Fig. 34 and consistent with the two-parameter WRLS algorithm provided by (26), the state estimator regresses the open-circuit potential V_0 and the high-frequency resistance R. From the value of $U = V_0 - V_H$, the voltage-based SOC_V is obtained, which, when combined with coulomb counting yields the extracted SOC (16). For the rather short test durations of this work, pure coulomb counting can be used to access the SOC, unlike in actual continuous HEV operation, and the accuracy of the regressed SOC is reflected by the agreement between regressed SOC and the curve labeled $\text{SOC}_{\text{coulomb counting}}$. A C/3 discharge of the

Characterization and Modeling of Electrochemical Energy

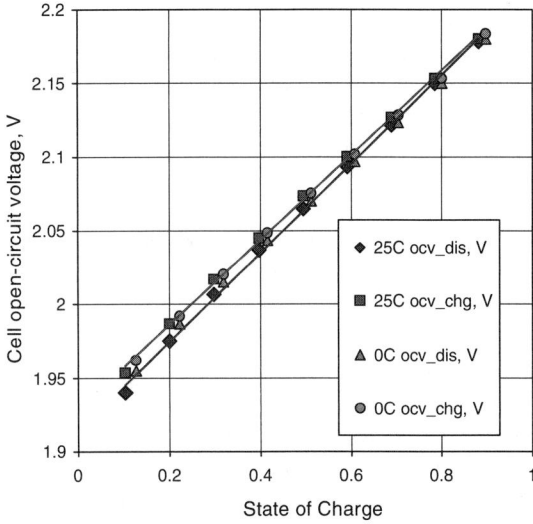

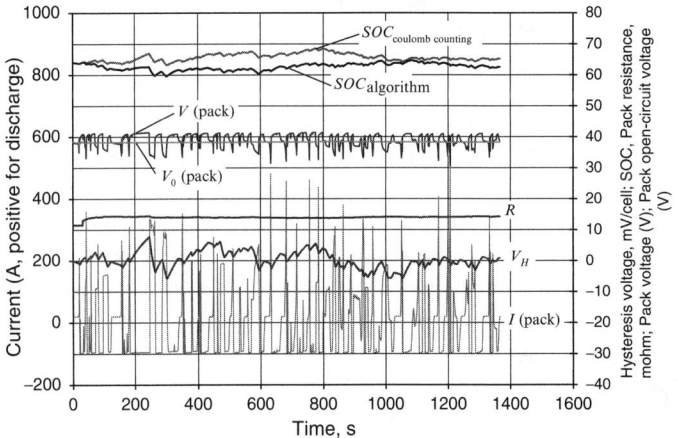

Figure 34. Lead acid cell. *Top*: Open circuit potential. These data correspond to the measured open circuit voltage (ocv) collected 15 min after a sustained discharge (dis) or charge (chg). For temperatures up to about 45°C, the ocv curves do not differ significantly from those for 25°C. *Bottom*: Measured pack voltage and current and SOC (by purely coulomb counting). Also shown are the calculated hysteresis voltage and the regressed SOC, open-circuit voltage, and high-frequency pack resistance.

battery at the end of the experiment yielded an SOC of 62%, in close agreement with the regressed SOC. Generally similar results were obtained when the SOC was cycled about an average SOC value ranging from 40 to 70% (16).

The high-frequency resistance may be useful in assessing the health of the battery. This assertion is based on two observations. First, as electrodes degrade upon cycling, intimate contact between conductive particles necessary for battery operation is slowly lost, resulting in larger values of R. Similarly, loss of acid due to sulfation (growth of inert lead sulfate crystals, removing sulfate ions from the electrolyte) leads to an increased ohmic resistance due to the loss of acid and a concomitant drop in electrolyte conductivity along with pore blockage attributable to the covering of otherwise active electrode reaction sites by inert lead sulfate crystals (122). Degradation phenomena take place in NiMH (123, 124) and lithium ion batteries as well, and upon cycling the high-frequency resistance increases.

Just as Fig. 30 facilitates an immediate understanding as to why least-square methods should be helpful in analyzing batteries, Fig. 35 is very helpful in explaining why the two-parameter regressions work well. Shown in the upper plot of Fig. 35 is the error in representing the measured voltage using the regressed parameters is shown in the lowermost plot of Fig. 35, corresponding to $y_j - (m_1 x_{1,j} + b)$ in (26). The time per point is 1 s for these data, and the power prediction error refers to the power predicted at time t when the parameters regressed from the previous time step $t - \Delta t$ are used to calculate the battery voltage using the measured current at time t and (31). The error on charge and discharge was generally less than $100\,\text{mV}$ module^{-1}, yielding less than 1% for both the voltage and power capability error under these conditions. The immediate increase in the skewness of the current source near 245 s (lower plot) results from a nearly constant charge power excitation (cf. Fig. 34) followed by an immediate change to discharge. The upper plot corresponds to rather random power excitation around 1,021 s and low skewness in the current excitation source.

Thus we see form the upper plot in Fig. 35 that when the parallel resistor-capacitor contribution to the impedance is removed from the measured voltage so as to form the regression voltage (cf. (26)), a substantially linear relation holds for the resulting regression voltage versus current, and it is not surprising that the two-parameter

Characterization and Modeling of Electrochemical Energy

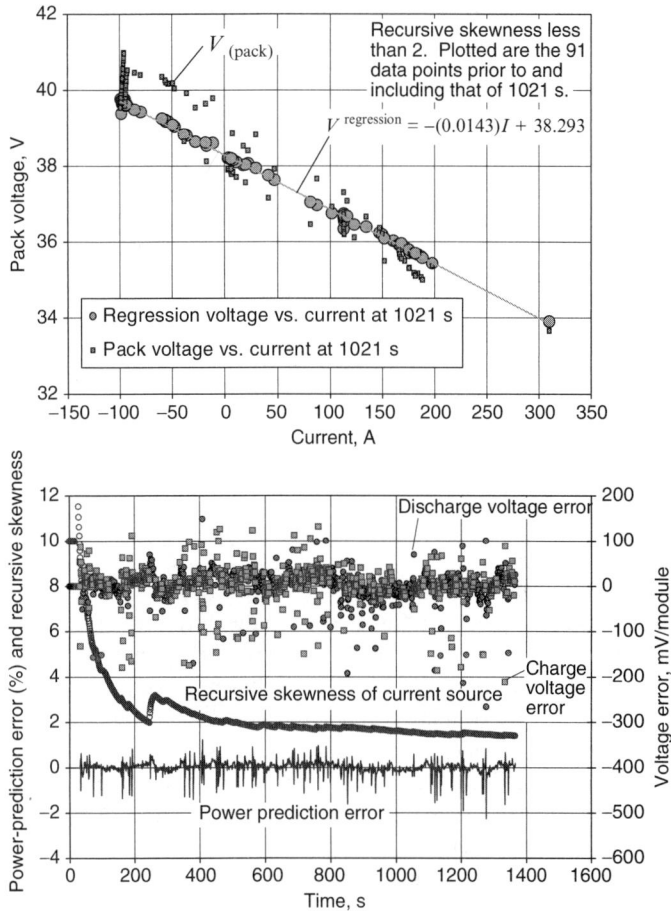

Figure 35. *Top*: Statistics for weighted recursive least square regression associated with the profiles shown in Fig. 34. Discharge currents are positive. The power prediction error results from using the regressed values at time step $t - \Delta t$ to predict the power at time t given the current at time t versus the actual power (current multiplied onto voltage) at time t. *Bottom*: Recursive skewness analysis. More random current histories give rise to smaller skewness values and good regression statistics, as evidenced by the nearly straight line for the regression voltage.

linear regression represents the data with little error. Note that the 91 points used to generate the upper plot in Fig. 35 corresponds to 91 s, which is on the order of the duration of influence when $\lambda = 0.99$, the base-case value for the forgetting factor λ (cf. Fig. 36). The influence

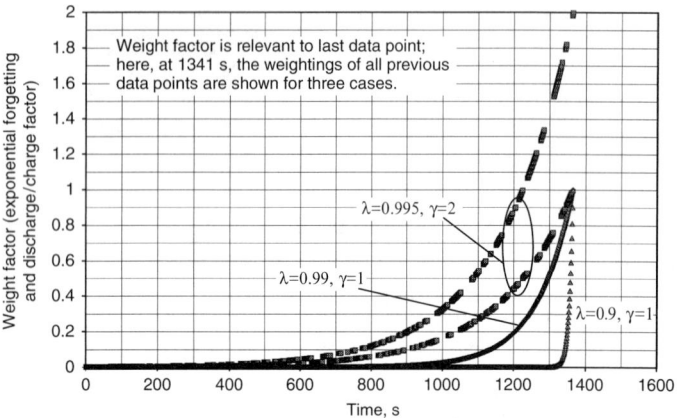

Figure 36. Weight factor components for weighted recursive least squares formulation. For the variable γ curves (*box symbols*), on discharge $\gamma = 2$ and on charge $\gamma = 1$. For all subsequent lead-acid analyses described in this work, $\lambda = 0.99$ and $\gamma = 1$.

of past points on the WRLS analysis and the attributes of selecting γ are further examined in Fig. 36. As described previously in the WRLS derivation, low values of λ are shown to weight preferentially the influence of more recent data. Two curves are seen when $\gamma \neq 1$ in Fig. 36; one for discharge data points ($\gamma = 2$, reflecting a higher weigthing of discharge data and more confidence in the model to represent discharge phenomena) and one for charge data points ($\gamma = 1$). When $\gamma > 1$ (respectively $\gamma < 1$), the curve for the discharge weight factor is above (respectively below) the charge curve.

The convergence of the algorithm is highly dependent on value of the weight factor w_{SOC}. The voltage based portion of the SOC calculation, SOC_V, converges rapidly, but it is highly sensitively to variations in the battery voltage. In addition to actual SOC variations, both measurement and model errors can give rise to unwanted instabilities in SOC_V. Conversely, the coulomb-counting-based portion of the SOC calculation, SOC_C, is a much more stable integral quantity, but it can steadily become inaccurate due to current efficiencies that are not known with enough certainty to be included in the model reference adaptive system. Thus SOC_V contains the more adaptive portion of the algorithm, whereas SOC_C

provides stability. For the base-case conditions chosen, the algorithm converges smoothly from intentionally erroneous starting values for the SOC in about 200 s, as is indicated in Fig. 37. The initial value of SOC was 64% (cf. Fig. 34, lower plot, $t = 0$); the initial value for SOC was arbitrarily set to 90 and 30% for the upper and

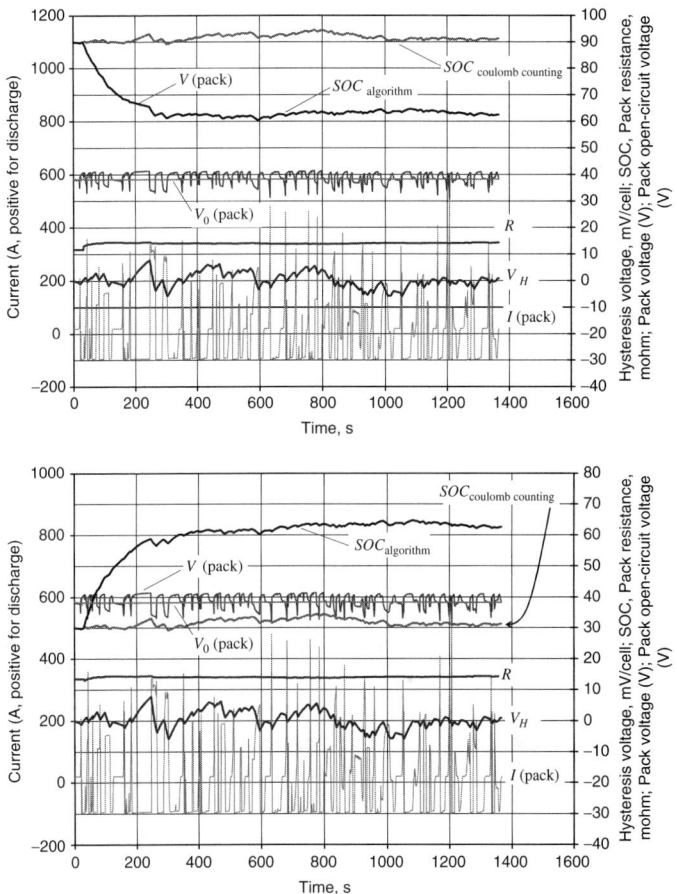

Figure 37. Algorithm convergence test, lead acid cell. For the upper plot, the initial SOC was set to 90% (the correct value being 65% as in Fig. 34, lower plot, at $t = 0$). For the lower plot, the initial SOC was set to 30%. A comparison with Fig. 34 shows that the algorithm converges to the correct SOC within about 200 s. Note that the regressed value for the high frequency resistance is not affected substantially by the choice of the initial SOC.

lower plots of Fig. 37, respectively. The results of Fig. 37 help to explain why a coulomb-counting SOC method cannot be used in actual vehicle applications, as any error in the SOC is not removed adaptively, and the pure coulomb-counting value for the SOC remains close to the erroneous initial starting value in each case. Note also that power-capability calculations are quite sensitive to the value of the SOC, as the driving force for discharge (respectively charge) power is the difference between the minimum (respectively maximum) voltage allowed and V_0, and V_0 is obtained from the value of the regressed SOC. Last, we note that the regressed high-frequency resistance R is not sensitive to erroneous values for the initial SOC, in agreement with the previous NiMH analysis, which bodes well for the use of an adapted high-frequency resistance as a robust diagnostic for a battery state of health.

A 25°C discharge power test of the algorithm for the GM Parallel Hybrid Truck (PHT) application is shown in Fig. 38. First, the power versus time trace for the PHT over a drive similar to that of the Federal Test Procedure (FTP) was recorded. Then, in a separate laboratory test employing the recorded drive power-versus-time trace, six different max discharge power pulses were enforced at various times throughout the FTP-like drive cycle by setting the pack voltage to 30 V (10 V module^{-1}), indicated by the arrows in Fig. 38. Generally the predicted 2-s and instantaneous power calculations bounded the immediate discharge power available (cf. ovals), and the available power always exceeded the 2 s predicted power capability. Similar results were obtained at -20 and $0°C$ (16).

Five-parameter application: lithium-ion cell. To comprehend the full capability of the WRLS algorithm, we now employ the equation system 28 to a lithium ion cell, with the characteristics provide in Table 4. The cell comprises a graphite anode and an $Al_{0.05}Co_{0.15}Ni_{0.8}O_2$ cathode (19, 23). The open-circuit potential V_0 is a function of temperature, SOC, and the hysteresis function:

$$V_0 = U(T, SOC) + V_H.$$

In this case, a look-up table is used to determine the SOC_V once the value of U is obtained (i.e., values of U can be placed in a table such that U is a function of both temperature T and SOC; thus knowing U and T allows one to determine the SOC). This approach is pragmatic and lends no insight into the phenomena governing the

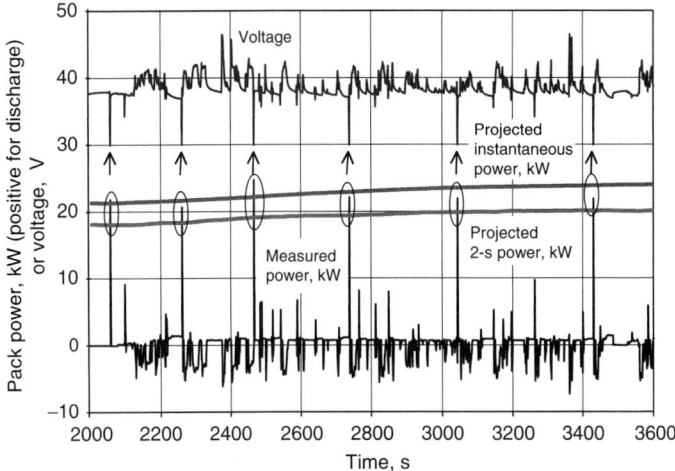

Figure 38. Discharge power test. Over an FTP-representative cycle recorded from a (GM) Parallel Hybrid Truck drive, the pack voltage was set abruptly to 30 V (10 V module^{-1}) at six different times, as indicated by the arrows. This corresponds to a maximum discharge power request. Generally the predicted 2-s and instantaneous power calculations bounded the immediate discharge power available (cf. ovals), and the available power always exceeded the 2-s predicted power capability.

open circuit behavior. The open circuit voltage is shown in the upper plot of Fig. 39. For the lithium ion system examined in this section, voltage hysteresis does not impact the analysis appreciably.

The variation in the current excitation source depicted in the lower plot of Fig. 39 yields a robust application of the WRLS algorithm, and the difference between the model and measured voltages is less that 0.5% for times greater than 3,000 s, as indicated the statistics plot in Fig. 40. Analysis of the forgetting factor (cf. Fig. 36) indicates that the exponential character of the weighting function allows only the last ∼100 s of data receive sufficient weighting to impact the extracted cell parameters. Along with the current source, displayed in Fig. 40 are the SOC, voltage, hysteresis voltage (which is negligible), and open-circuit potential. As with the lead acid and NiMH analyses, discharge of the battery at the completion of the indicated experiment yielded a capacity that was within ±2.5% of the final SOC. The ability of the algorithm to adapt to an erroneous initial SOC is examined in Fig. 41; the upper plot shows the effect

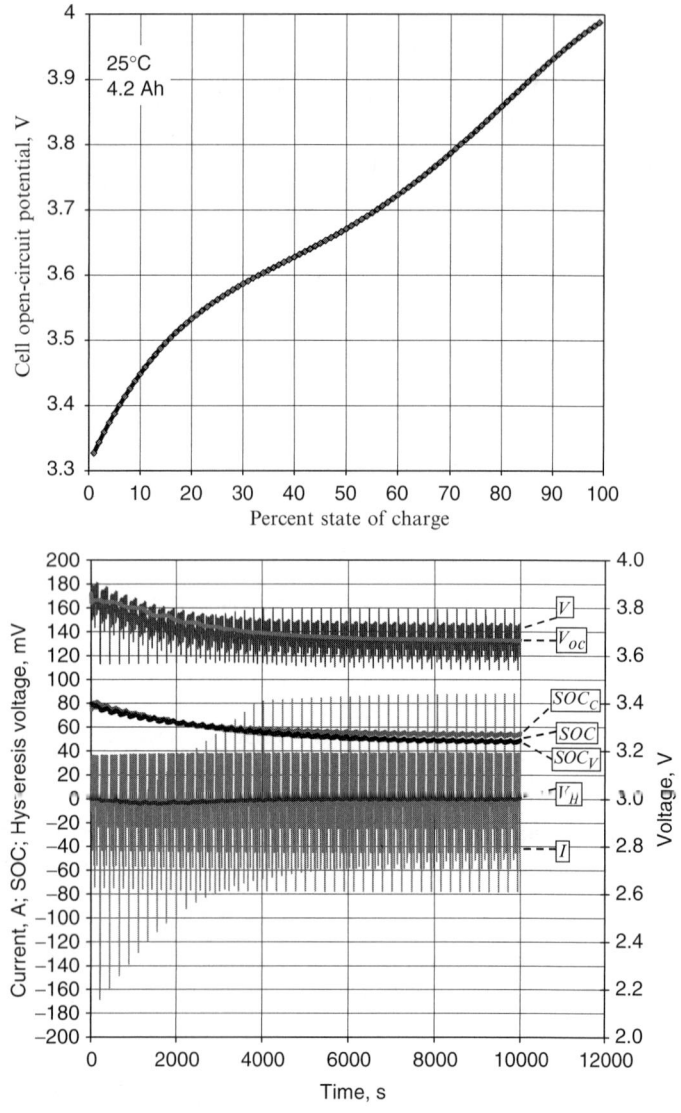

Figure 39. Lithium ion cell. *Top*: Open circuit potential. Potentials above 4 V and below 3.3 V are treated as 100% SOC and 0% SOC, respectively. *Bottom*: Five-parameter algorithm results including SOC and related values for the lithium-ion cell. Pulses to the maximum and minimum voltages are contained in the current source every 225 s so as to test the power capability projections of the algorithm. As plotted, it is difficult to resolve the differences between SOC_V and SOC, as they are nearly identical.

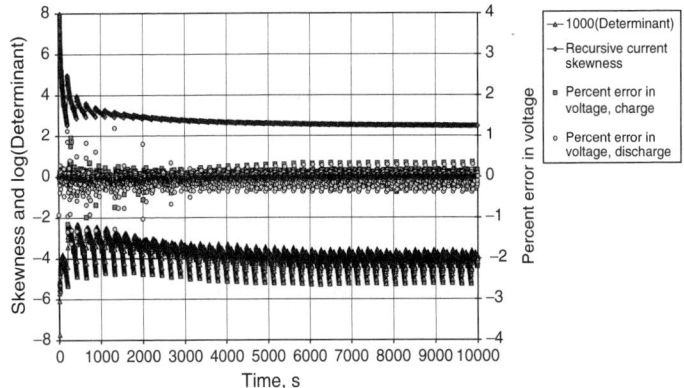

Figure 40. Statistics for the five-parameter algorithm applied to the lithium-ion cell. The skewness (recursive), determinant, and voltage error are plotted for the analysis of Fig. 39.

of incorrectly specifying a 10% initial SOC, and 100% initial SOC was specified for the lower plot. Convergence takes place within about 100 s. Because the initial SOC was incorrectly specified, the coulomb-counting SOC_C remains inaccurate, whereas the voltage-based SOC_V corresponds closely to that of the composite SOC.

To determine the state of an EESS, it is helpful to obtain knowledge of the electrochemical parameters adaptively. The regressed values for the electrochemical parameters of the lithium-ion battery are shown in Fig. 42 (corresponding to the results plotted in Figs. 39 and 40), and the utility of using a 300-s moving average of the values displayed in the upper plot is depicted in the lower plot. Overall, the regressed values for the electrochemical parameters are close to those measured independently for the lithium ion battery (cf. Table 4, uppermost five rows of the lithium ion column); similar findings are reported for the application of the five-parameter model to lead acid and NiMH cells (19).

The most important output for energy management algorithm associated with EESS operation of a charge-sustaining HEV is the power capability, as the EESS provides the power, and gasoline provides the energy. Power capability projections for the lithium-ion cell are shown in Fig. 43 for both charge and discharge. The abscissa in the lower plot is expanded about 4,565 s to allow for a closer

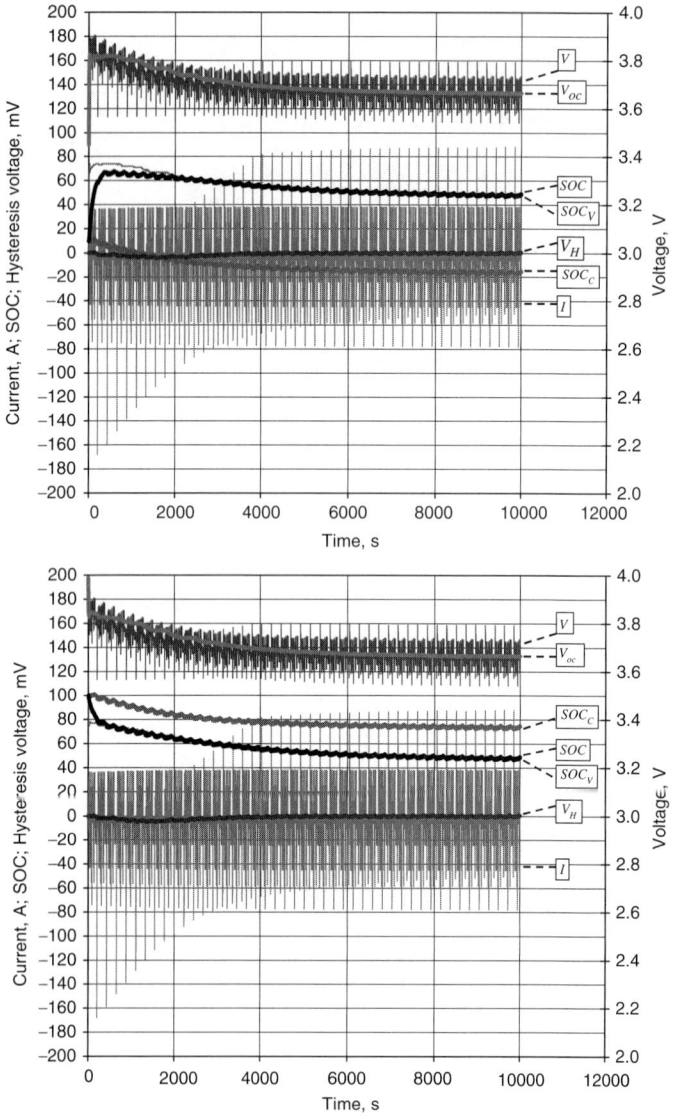

Figure 41. Algorithm convergence test, lithium ion cell. Effect of incorrectly specifying a 10% initial SOC (*top*) and 100% SOC (*bottom*). The ability of the algorithm to adapt to the correct SOC within about 100 s is demonstrated.

Characterization and Modeling of Electrochemical Energy

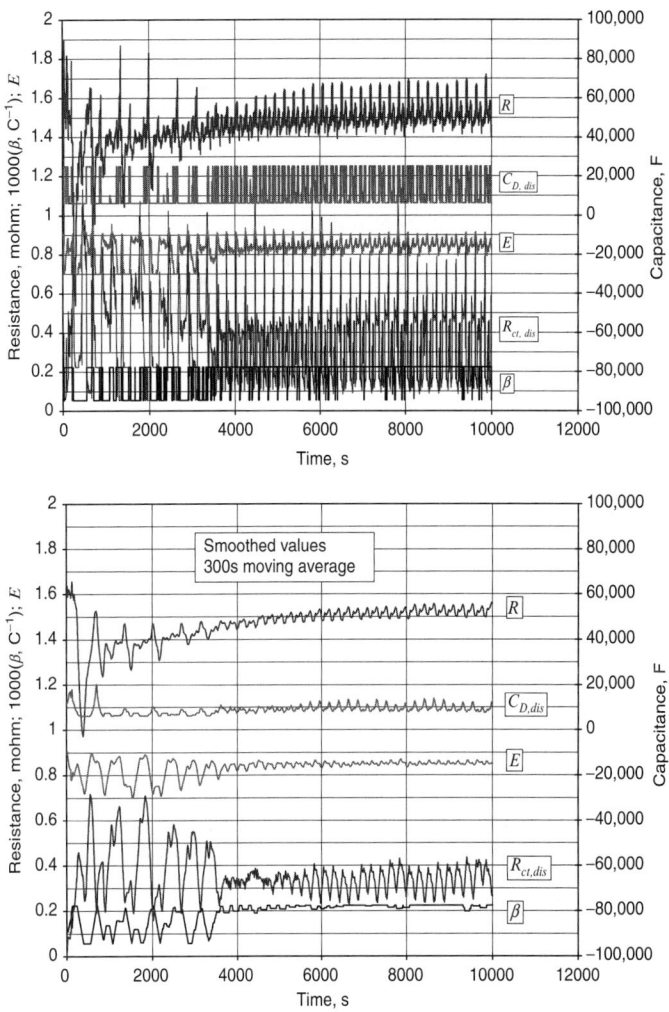

Figure 42. Electrochemical parameters extracted for the lithium-ion battery. Note that $C_{dis}/C_{chg} = 0.75$, and this is also ratio for $R_{ct,chg}/R_{ct,dis}$. The lower plot provides a 300-s moving average of the values displayed in the upper plot (i.e., the data 300 s prior to the time value are averaged), which can facilitate the assessment of the cell state of health SOH.

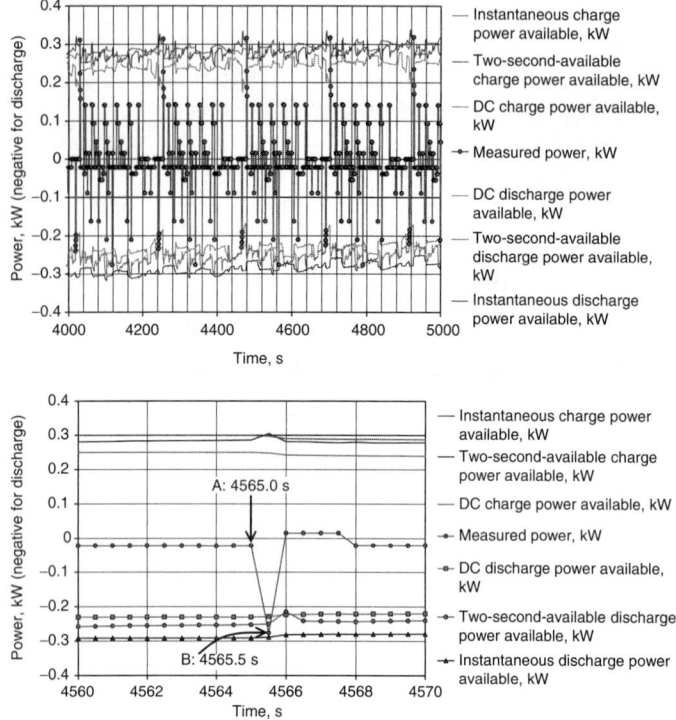

Figure 43. Power capability projections for the lithium-ion cell. The maximum amplitudes in the measured power correspond to the voltage being set to the maximum or minimum value, 3.57 and 3.80 V, respectively, for these plots. The abscissa is expanded about 4,565 s in the lower plot, wherein the discrete nature of the excitation source is exhibited. The order of the description in the legend corresponds to the order of the curves at the far right of each plot.

inspection of how the algorithm operates and the accuracy of the algorithm projections. We first discuss the lower plot. Immediately after 4,565 s, the algorithm has completed calculations corresponding the outputs depicted in Figs. 39, 40, and 42; the power capability is then projected for the case of enforcing minimum and maximum voltages, 3.57 V (for maximum discharge power) and 3.80 V (for maximum charge power), respectively, at 4,565.5 s. Thus the value for the power capability projections associated with the three uppermost and three lowermost curves are obtained prior to that of the measured power, and we see that the algorithm bounds the subsequently measured maximum discharge power at 4,565.5 s between the high-frequency and dc power projections. Inspection of the upper plot in Fig. 43 indicates that the power capability is always greater in magnitude than that of the dc power projection (corresponding to (29) and (30) with R replaced by $R + R_{ct}$). For charge, the measured power after 0.5 s corresponds closely to that of the high-frequency resistance (cf. (31)). In summary, for the lithium ion cell, the dc power capability can be used to conservatively estimate the power capability of the battery.

Variable forgetting factors. The last algorithm research topic we address is the use of variable forgetting factors; that is associating a forgetting factor unique to each parameter to be regressed. Two issues arise in the standard implementation of WRLS. First, one would like to assign a time-weighting (forgetting) factor for each of the parameters extracted instead of single forgetting factor as is commonly employed (44–50); this is an active area of research within the controls community (125–131). It will be shown in this section that a substantially more accurate state estimator for SOC, SOH, and SOP is obtained when individual forgetting factors are incorporated. Second, it is desirable to select the value of the forgetting factors using an optimization function. Here we (1) propose a method based on WRLS such that forgetting factors can be assigned to each individual parameter to be regressed and (2) provide a means to determine the optimal values of the forgetting factors. The approach is applied to the same high-power-density lithium ion battery analyzed in the previous section.

We begin with the instantaneous error ε, or loss term, (23):

$$\varepsilon(t) = [y - (m_1 x_1 + m_2 x_2 + \cdots + m_L x_L + b)],$$

where y represents the experimentally obtained dependent variable at time t (i.e., $y = V^{\text{measured}}$, the measured voltage for the energy storage system) and the values $x_1, x_2, \ldots, x_L$ represent the measured quantities on which the L parameters $m_1, m_2, \ldots, m_L$ multiply, respectively, to complete the linear model once the parameter b (resulting from the regressed open-circuit potential in the case of energy storage devices) is included. Unlike the previous development, because we shall formulate an iterative scheme that does not require matrix inversion, it is expedient to fold b into the parameter vector $\mathbf{m} = [m_1, m_2, \ldots, m_L]^T$, recognizing that the corresponding value of x associated with b is unity, as will be made clear below. The weighted square of the error term summed over N data points can be expressed as

$$\varepsilon = \sum_{j=1,N} \gamma_j \lambda^{N-j} \left[y_j - (m_1 x_{1,j} + m_2 x_{2,j} + \cdots + m_{L,j} x_{L,j}) \right]^2. \quad (33)$$

For a system wherein only one of the L parameters changes with time, designated as m_l, and all others correspond to fixed values, the weighted square of the error associated with the single parameter l is

$$\varepsilon_l = \sum_{j=1,N} \gamma_{l,j} \lambda_l^{N-j} \left[y_j - m_l x_{l,j} - \sum_{\substack{k=1,L \\ k \neq l}} m_k x_{k,j} \right]^2. \quad (34)$$

A clarification must be provided for (33) and (34). Recall that all of the L parameters may vary with time. Our approach is to let only one parameter (denoted by subscript l) change relative to its previously calculated value, and the remaining $L-1$ parameters are fixed at their values obtained from the previous time step (i.e., at their regressed values corresponding to time $t - \Delta t$ and the integer time index $j = N - 1$). Thus (34) specifies that the exponential forgetting factor λ (and the factor γ) is to be associated with a parameter l; up to this point our formulation is similar to that of Vahidi et al. (cf. (13) of Ref. (130)).

Consistent with (33) and (34), we shall term the total error as the square of the sum of the L individual errors,

$$\varepsilon = \sum_{l=1,L} \varepsilon_l = \sum_{l=1,L} \sum_{j=1,N} \gamma_{l,j} \lambda_l^{N-j} \left[y_j(t) - m_l(t) x_{l,j}(t) - \sum_{\substack{k=1,L \\ k \neq l}} m_k(t - \Delta t) x_{k,j}(t) \right]^2. \quad (35)$$

By minimizing the total error ε with respect to m_l at time step N [employing (35) to determine $\partial \varepsilon / \partial m_l(t) = 0$], we obtain an equation for the l'th parameter m_l:

$$m_{l,N} = \frac{1}{\sum_{j=1,N} \gamma_{l,j} \lambda_l^{N-j} x_{l,j}^2}$$

$$\times \left(\sum_{j=1,N} \gamma_{l,j} \lambda_l^{N-j} y_j x_{l,j} - \sum_{\substack{k=1,L \\ k \neq l}} m_{k,N-1} \sum_{j=1,N} \gamma_{l,j} \lambda_l^{N-j} x_{k,j} x_{l,j} \right). \quad (36)$$

One can view this approach as the sequential minimization of the total error term with respect to individual parameters. This relation can be used to regress individually each of the L parameters at time step N, and we now have an expression reflecting a weight factor λ_l for each of the L parameters m_l. Equation (36) is implemented L times at each time step, with the l ranging from 1 to L. Thus there are no matrix equations to solve in this approach, and the method can be viewed as iterative. We do not address parameter convergence (126), which remains an open question.

For illustrative purposes, it is helpful once again to view a two-parameter model.

$$y = m_1 x_1 + m_2 x_2. \quad (37)$$

This example corresponds to our previous "ohmic battery," wherein the hysteresis and parallel resistor-capacitor contributions to the equivalent circuit in are removed:

$$\begin{aligned} V &= V_0 + IR \\ y &= V^{\text{measured}} \\ m_1 &= V_0 \\ m_2 &= R \\ x_1 &= 1 \\ x_2 &= I \end{aligned} \quad (38)$$

and $V = V_0 + IR = m_1 x_1 + m_2 x_2$. The total error is written as

$$\varepsilon = \sum_{j=1}^{N} \gamma_{1,j} \lambda_1^{N-j} \left[y_j(t) - m_1(t) x_{1,j}(t) - m_2(t - \Delta t) x_{2,j}(t) \right]^2$$

$$+ \gamma_{2,j} \lambda_2^{N-j} \left[y_j(t) - m_2(t) x_{2,j}(t) - m_1(t - \Delta t) x_{1,j}(t) \right]^2. \quad (39)$$

The parameters (R and V_0) can be regressed using the following:

$$m_{1,N} = \frac{1}{\sum_{j=1,N} \gamma_{1,j} \lambda_1^{N-j} x_{1,j}^2}$$
$$\times \sum_{j=1,N} \left[\gamma_{1,j} \lambda_1^{N-j} y_j x_{1,j} - m_{2,N-1} \gamma_{1,j} \lambda_1^{N-j} x_{2,j} x_{1,j} \right]$$

$$m_{2,N} = \frac{1}{\sum_{j=1,N} \gamma_{2,j} \lambda_2^{N-j} x_{2,j}^2}$$
$$\times \sum_{j=1,N} \left[\gamma_{2,j} \lambda_2^{N-j} y_j x_{2,j} - m_{1,N-1} \gamma_{2,j} \lambda_2^{N-j} x_{1,j} x_{2,j} \right], \quad (40)$$

with λ_1 and λ_2 being the forgetting factors for m_1 and m_2, respectively.

For the general case of L parameters, we can make (36) fully recursive with the following definitions:

$$s_{x,l}|_N = \sum_{j=1,N} \gamma_{l,j} \lambda_l^{N-j} x_{l,j}^2 = \gamma_{l,N} x_{l,N}^2 + \lambda_l (s_{x,l}|_{N-1})$$
$$s_{yx,l}|_N = \sum_{j=1,N} \gamma_{l,j} \lambda_l^{N-j} y_j x_{l,j} = \gamma_{l,N} y_N x_{l,N} + \lambda_l (s_{yx,l}|_{N-1}).$$
$$s_{xx,l}|_N = \sum_{j=1,N} \gamma_{l,j} \lambda_l^{N-j} x_{k,j} x_{l,j} = \gamma_{l,N} x_{k,N} x_{l,N} + \lambda_l (s_{xx,l}|_{N-1})$$
$$(41)$$

These recursive expressions can be used to recast (36) as

$$m_{l,N} = \frac{1}{s_{x,l}|_N} \left[s_{yx,l}|_N - \sum_{\substack{k=1,L \\ k \neq l}} m_{k,N-1}(s_{xx,l}|_N) \right]. \quad (42)$$

This expression is used for each of the L parameters m_l at each time step. As noted previously, no matrix inversion is required.

For the equivalent circuit employed to represent the battery system (Fig. 11), (28) relates the x and m values, with the exception of eliminating b. Hence,

$$y = V^{\text{measured}}|_t$$

$$x_1 = I_t$$

$$x_2 = (V^{\text{measured}} - V_0 - IR)_{t-\Delta t}$$

$$x_3 = \left(\frac{I_{t-\Delta t} + I_t}{2}\right)\left[r_{C,(I_t + I_{t-\Delta t})/2}\right]\Delta t$$

$$x_4 = \Delta t I [V_{\text{H,max}} - \text{sign}(I)V_{\text{H}}]_{t-\Delta t} \quad (43)$$

$$x_5 = 1$$

$$m_1 = R$$

$$m_2 = \exp(-B\Delta t) = e^{-\Delta t/\tau} = E$$

$$m_3 = A_d$$

$$m_4 = \beta$$

$$m_5 = U + (V_{\text{H}})_{t-\Delta t}$$

We shall apply the algorithm to the same lithium ion data of the last section. For the case of fixed and variable forgetting factors, the error for the entire data set of N data points was minimized to find the optimal forgetting-factor values. We define the unweighted total error to be minimized as:

$$\begin{aligned}\varepsilon_{\text{opt}} &= \sum_{j=1,N}\left[V_j^{\text{measured}} - V_j^{\text{model}}\right]^2 \\ &= \sum_{j=1,N}\left[V_j^{\text{measured}} - \sum_{k=1,L} m_{k,j} x_{k,j}\right]^2. \end{aligned} \quad (44)$$

Newton's method (115) was employed to optimize the forgetting factors:

$$\lambda^{(n+1)} = \lambda^{(n)} - \frac{\varepsilon_{\text{opt}}^{(n)}}{\varepsilon_{\text{opt}}'^{(n)}}, \quad (45)$$

where $\varepsilon_{\text{opt}}'(\lambda)$ is the Jacobian matrix of the unweighted total error term that we minimize by determining the optimal values of the forgetting factor vector λ for the entire data set; the superscript (n) refers to the step in the Newton iteration. For this work, we found convergence [abs$(\lambda_l^{(n+1)} - \lambda_l^{(n)})/\lambda_l^{(n)} < 10^{-6}$] was obtained in about six iterations. For a fixed forgetting factor, the optimal value of λ is 0.9847, as depicted in Fig. 44 (lower plot). Note that for the case of

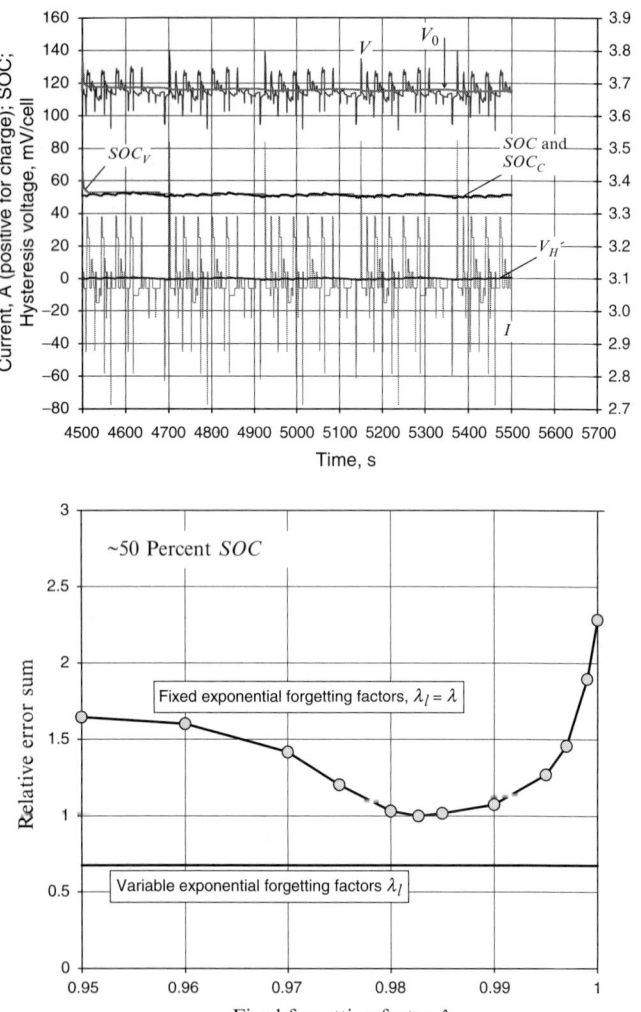

Figure 44. *Top*: State of charge and measured voltage, equilibrium and hysteresis voltages employing variable forgetting factors shown in the inset table of Fig. 45. *Bottom*: Relative error sum for the optimization of forgetting factors. The error is increased by 53% in going to a fixed forgetting factor relative to variable forgetting factors.

a constant forgetting factor, $\lambda = \lambda$, a single-valued scalar quantity. The optimal values for the individual forgetting factors employed in the lower plot were close to values shown in the inset table of Fig. 45, which is discussed below. In experimenting with the variable forgetting factors, we learned that while a larger forgetting factor is appropriate for some parameters, the forgetting factor for V_0 must be smaller in order to capture SOC variations with current.

As noted in the Introduction of this paper, for charge-sustaining HEV's, the battery is cycled about a set point SOC, generally near 50% SOC; we shall restrict much of our analysis to this condition. Analogous to the lower plot in Fig. 39, the state of charge and measured (V), equilibrium (V_0) and hysteresis (V_H) voltages are depicted in Fig. 44; the algorithm was started at 4,500 s (cf. Fig. 39), facilitating the analysis of algorithm operation about 50% SOC.

To appreciate the influence of the forgetting factor on the unweighted total error ε_{opt} ((44)), we turn to the lower plot of Fig. 44. The ordinate values correspond to the unweighted total error ε_{opt} normalized by that which is obtained for the optimal fixed forgetting factor. The unweighted total error is increased by 53% in going from variable forgetting factors to a fixed forgetting factor. Hence, one can expect that employing a variable forgetting factor for a charge-sustaining hybrid utilizing a lithium ion battery of the type investigated here will increase the accuracy of the algorithm by about 50%. In support of the optimal fixed forgetting factors of 0.9827 depicted in Fig. 44 (lower plot, upper curve), a single (fixed) value of 0.99 was used in the previous sections for lead acid, NiMH, and lithium ion cells after we had gained some familiarity and experience with those algorithms. The optimization process employed in this work provides a quantitative basis for why a value near 0.99 worked well for the case of fixed forgetting factors in the previous work. That is, while 0.99 was chosen without any quantitative insight, we find here the total error is minimized with a constant forgetting factor quite close to 0.99.

The optimized values for the variable forgetting factors and the associated parameter values m_l are shown in Fig. 45 for the analysis shown in Fig. 44. Four of the extracted parameters are displayed in the lower plot, and the fifth (V_0) is shown in Fig. 44. The lower plot in Fig. 45 shows that the parameters effectively converge within about 50 s (see the R and E curves, in particular). The high-frequency resistance R is seen to be quite stable, and maintains

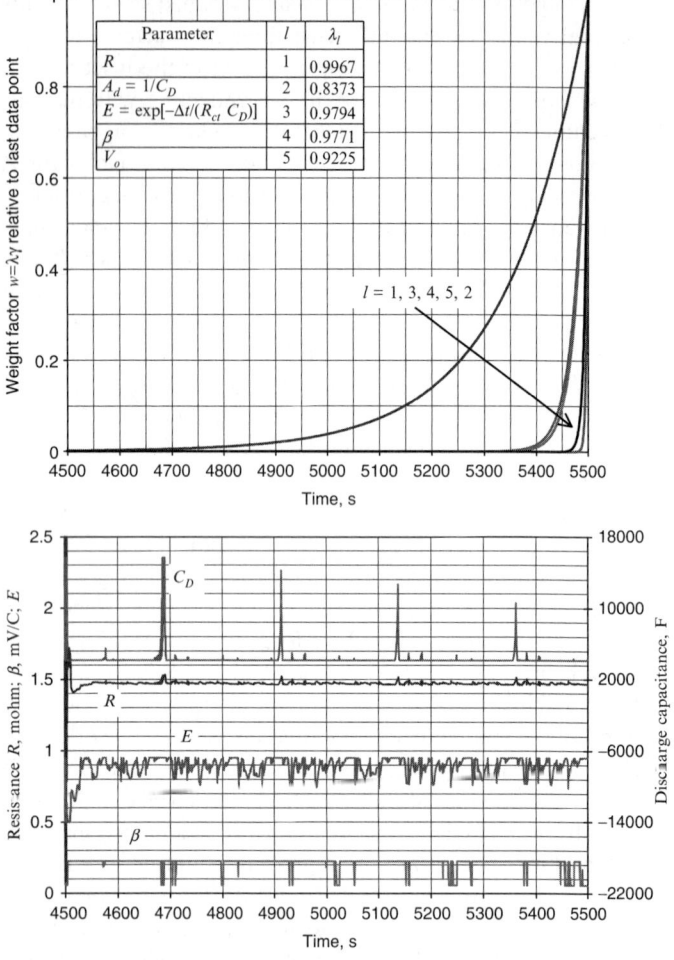

Figure 45. Optimized (variable) forgetting factors and parameter profiles. The high-frequency resistance is seen to be quite stable, and a large forgetting factor, reflecting time average over a longer duration, results from the optimization. Conversely, more rapid changes in the open-circuit potential are required for the high-power cycling regime, resulting in a smaller optimized forgetting factor for V_0.

a larger forgetting factor, reflecting time averaging over a longer duration. Conversely, more rapid changes in the open-circuit potential are required for the high-power cycling regime, resulting in a smaller forgetting factor for V_0. Lithium ion and NiMH batteries are both insertion systems wherein the average concentration of the ions in the entire electrolyte phase should not change on charge and discharge. For lithium ion batteries on discharge, lithium ions are ejected from the carbon anode and inserted into the metal oxide cathode, and there is no net change in the number of ions within the electrolyte phase. The same conclusion holds for charge, wherein lithium ions are discharged from the metal oxide cathode and inserted into the carbon anode. While local concentration gradients will influence the cell potential (84, 91–93, 97), to a first approximation we might expect the high-frequency resistance R to be rather constant over a drive profile, consistent with the secondary current distribution for the cell (93, 94, 98) and a constant number of charge carriers in the electrolyte phase. The same arguments hold for protons (versus lithium ions) in NiMH batteries. The fact that the algorithm yields a stable value for R is also important in the context of SOH, as it is likely that the definition

$$\text{SOH} \equiv R_{\text{nominal}}(T, \text{SOC})/R(T, \text{SOC}) \qquad (46)$$

will provide a means to rationalize the term state of health, as mentioned previously. In this relation, the nominal resistance for a (new and healthy) battery is R_{nominal}, which can be a tabulated quantity within an embedded controller so as to be a function of temperature and SOC. By the definition provided in (46), we would expect new batteries to have an SOH value near unity, and the SOH would decline as the battery ages. (A short-circuit within a cell would lead to an abnormally high value of SOH, significantly greater than unity, and would imply failure of the system.)

The remainder of our discussion is concerned with power projections provided by the algorithm. Equations (29–31) provide the necessary relations. Plots of the power projections provided by the algorithm along with the actual measured power are provided in Fig. 46, with an expanded view of power projections is displayed in Fig. 47. In addition to the traces shown in Fig. 46, the low-frequency discharge-power capability (resistance corresponding to $R + R_{\text{ct}}$) is included. The 0.5-s power projection (large circles, $\Delta t = 0.5$ s

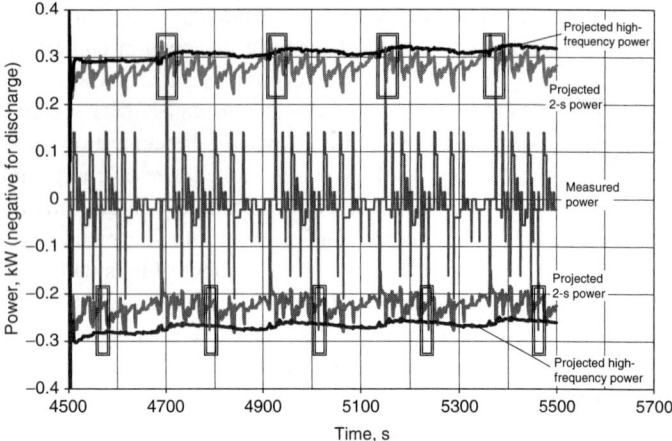

Figure 46. Power-capability validation test. The current source for these data is shown in Fig. 39 (*lower plot*). The cell power is the known (recorded) excitation source, and all other quantities plotted have been calculated using the regressed parameters values.

for the implementation of (31)) is shown to accurately predict the measured power; that is, using past information and the voltage set point taken to be that which is 0.5 s into the future, the algorithm is able to predict the measured power with high accuracy. Due to charging and discharging of the capacitor circuit element C, the 0.5-s power-projection magnitudes can exceed those of the high-frequency projection. We see that conservative battery operation is accomplished by employing the 2-s maximum power projection as the system's maximum power capability for the next 0.5 s; i.e., the risk of the voltage exceeding or dropping below the maximum or minimum voltage, respectively, is very low when the 2-s maximum power projection is employed to represent battery's maximum power capability for the next 0.5 s.

Algorithm verification and validation. We have now covered the implementation of WRLS methods for the formulation of a state estimator for an EESS. Independent of the approach used to determine the SOC, SOH and SOP, it is important to verify that the algorithm is behaving as designed and to examine the efficacy of the algorithm.

In order to verify and validate the state estimator, it is helpful to place the energy storage device in an experimental set up that closely

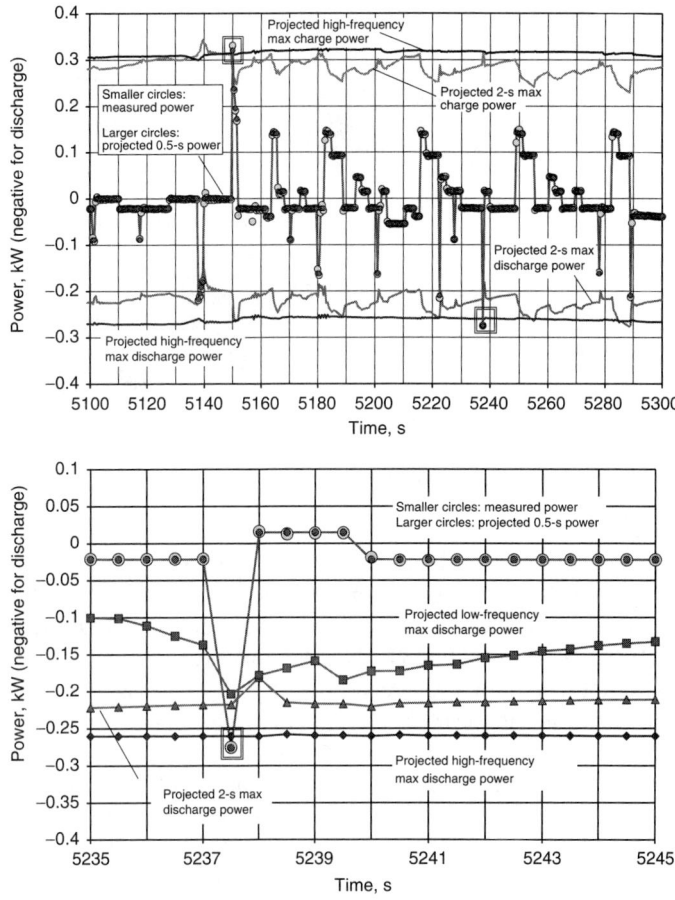

Figure 47. Expanded view of power projections. The 0.5-s power projections (*large circles*) is shown to accurately predict the measured power, consistent with the time per point for the measured power being 0.5 s.

mimics the intended application. In (132), a so-called hardware-in-the-loop (HWIL) system is described wherein an electrochemical energy storage device is placed within the set up, and the state estimator can be analyzed. Here we describe the set up in the context of an electric double layer capacitor cell; a schematic of the set up is provided in Fig. 48.

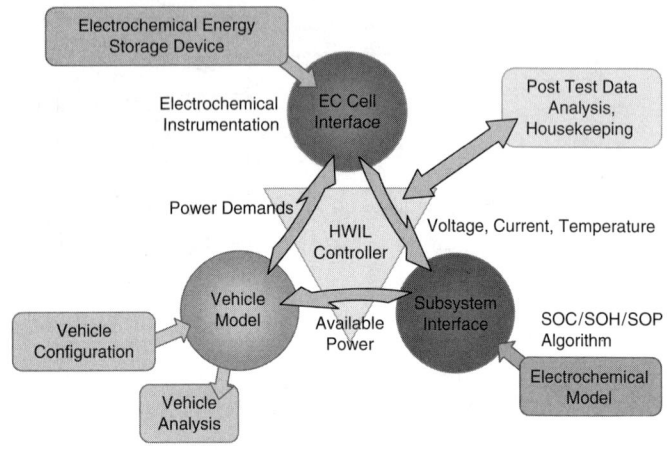

Figure 48. Hardware-in-the-Loop system (HWIL).

A supercapacitor cell is chosen here as it is the most straightforward electrochemical energy storage device to analyze with a state estimator, and thus we can better clarify the character and utility of the HWIL set up. In addition to the nearly linear equilibrium voltage versus SOC curve for a supercapactior, the actual measured voltage of the supercapacitor is much closer to that of the open-circuit value in the case of supercapacitors because of their low impedance (i.e., low sheet resistance, ohm-cm^2). The Introduction of this paper outlined automotive interest in supercapacitors. Because of the concern over the abuse tolerance of acetonitrile-based systems, particularly in the context of cell overcharge, there is additional motivation for clarifying the accuracy of control algorithms at the cell level versus the pack level, as much less energy is associated with a single cell than a pack.

The HWIL system comprises three modules (cf. Fig. 48): the Electrochemical Cell Interface (ECI), the Subsystem Interface, and the Vehicle Model. The modules are linked together via the HWIL controller, which passes data and commands between the modules. The ECI module provides electrical and thermal environment for the electrochemical cell. Based on the measured response of the cell through this interface, the HWIL controller simulates the vehicle energy storage system.

A single-channel testing system and microcontroller (Arbin Instruments) is employed. The tester (model BT-2000) is configured for large electrochemical energy storage devices and provides up to ±5 kW at potentials between 0.6 and 5 V. For increased accuracy, the linear bipolar channel is configured with two current ranges, 1,000 and 20 A. Rise times range from 0.5 to 1 ms. The tester is controlled by a microcontroller with the Motorola Coldfire processor. It includes a generic high-level command set based on the TCP/IP communications protocol, which allows for a direct hardware interface from the HWIL controller, minimizing the system response time.

The electrochemical cell is housed within a thermal chamber (Model 105A Test Equity, Thousand Oaks, CA). This chamber has a programmable range from -40 to $130°C$, controllable within $\pm 1°C$. For all of the data reported in this review, the electrochemical cell is an electric double layer capacitor acquired from Saft America. The nominal capacitance is 3,500 F, and the maximum voltage is 2.8 V.

The Subsystem Interface module includes the State Estimator algorithm rendered in the Matlab/Simulink tool. The Subsystem Interface module supplies the SOC and SOP estimates for the capacitor deduced from the history of voltage and current measurements.

The vehicle model is based on the Hybrid Powertrain Simulation Program (HPSP), a modeling tool developed by GM and based on the backward-driven simulation approach (150, 151). The analysis is initiated by an instantaneous road load requirement specified by a driving cycle (vehicle speed as a function of time). Empirical quasi-steady state models using efficiency maps and system specific input parameters such as inertias and gear ratios represent the components of the powertrain. The speed and torque requirements of each component are tracked backwards from the road load through the driveline components eventually determining the engine speed and torque operating points. The current and voltage requirements for the electric drive are determined from the torque, speed, and acceleration requirements of the electrical components, which include their electrical and mechanical losses. An electric accessory load is also accounted for in the model. This approach is ideal for following a driving schedule to determine the engine operating regions under optimum controls of the powertrain or based on specialized control and energy management strategies. Within the HWIL system, HPSP performs its calculations and generates system requests at a regular time intervals (typically 0.1–0.25 s). These calculations are used to continuously update the vehicle performance parameters (fuel consumption, available supercapacitor power, etc.). HPSP subsequently records all powertrain related activity over the test profile and formats them for post processing.

The primary function of the HWIL controller is to interface the multiple application environments associated with the modules. Because these modules are intended to be flexible, the responsibility for overall abuse tolerance, robustness and performance is handled by the HWIL controller. It is also the control application for the cell tester, relaying power commands from the vehicle model, recording cell current, voltage and temperature and transferring data to the Subsystem Interface module. The HWIL controller oversees the test cell performance parameters, providing them to the vehicle model, and stopping or limiting a test if these parameters are exceeded.

By summing the voltage around the lowermost circuit shown in Fig. 10, we can express the capacitor voltage as

$$V = \frac{q(0)}{C} + IR + \frac{1}{C}\int_{\zeta=0}^{\zeta=t} I|_\zeta \, d\zeta.$$

Consistent with the integration of (4). Analogous to the procedure outline in the Appendix, this equation can be recast as the following fully recursive relation for time step N:

$$V|_N = (t_N - t_{N-1})\frac{I|_N + I|_{N-1}}{2}\frac{1}{C} + I|_N R + (V|_{N-1} - I|_{N-1}R).$$

Hence, two parameters are to be adapted, R and C, from a history of currents and voltages, which provide the basis for state estimation of the capacitor.

The following definitions stem directly from the two parameter version of the WRLS algorithm previously described:

$$x_1 = I_t - I_{t-\Delta t} \qquad x_2 = (I_t + I_{t-\Delta t})\frac{\Delta t}{2}$$

$$m_1 = R \qquad m_2 = \frac{1}{C}$$

$$y = V_{\text{meas},t} - V_{\text{meas},t-\Delta t}$$

These definition yield the linear model for the voltage-based portion of the supercapacitor state estimator:

$$y = m_1 x_1 + m_2 x_2.$$

The remainder to the WRLS algorithm is similar to that previously described for the two-parameter systems previously (cf. (27) with $b = 0$). However, to construct the composite SOC and the power expressions, minor variations on the previous discussion are useful for the supercapacitor, and we shall describe these items next.

For the voltage-based SOC_V calculation, the charge on the capacitor depicted in can be expressed as

$$q = q(0) + \int_{\zeta=0}^{\zeta=t} I|_\zeta \, d\zeta = C(V - IR),$$

where q may be viewed as the magnitude of the charge on the (symmetric) capacitor electrodes. We define SOC_V, in terms of the minimum charge on the capacitor $q_{\min}$ relative to its maximum value $q_{\max}$ allowed under normal operating conditions,

$$\begin{aligned}
\text{SOC}_V &= \frac{q - q_{\min}}{q_{\max} - q_{\min}} \\
&= \frac{C(V - IR) - q_{\min}}{CV_{\max}|_{I=0} - q_{\min}} \\
&= \frac{V - IR - V_{\min}|_{I=0}}{V_{\max}|_{I=0} - V_{\min}|_{I=0}},
\end{aligned}$$

where $q_{\min}$ and $q_{\max}$ correspond to the minimum and maximum voltages ($V_{\min}$ and $V_{\max}$, respectively) under zero current conditions. In vehicle applications, the power electronics (i.e., the power inverter) used to convert the dc current associated with the energy storage device (e.g., a battery, fuel cell, or capacitor) to ac current for the electric machines require that $V_{\min} \geq 0.5 V_{\max}$ in order that the power inverter maintains an acceptable efficiency. Thus, for the figures presented herein, $V_{\min} = 0$ in formulating the SOC, and we investigate SOC's ranging from 50 to 100% in the power capability analyses ($1.4 \leq V_0 \leq 2.8$ V).

We can also calculate an SOC based on coulomb counting so as to construct a current-based SOC, SOC_C, in a manner analogous to that described for batteries (16):

$$\text{SOC}_C(t) = \text{SOC}(t - \Delta t) + \left(\frac{I_t + I_{t-\Delta t}}{q_{\max} - q_{\min}}\right)\frac{\Delta t}{2}$$
$$= \text{SOC}(t - \Delta t) + \frac{1}{C}\left(\frac{I_t + I_{t-\Delta t}}{V_{\max}|_{I=0} - V_{\min}|_{I=0}}\right)\frac{\Delta t}{2}.$$

As noted for batteries, both the voltage and current based SOC's contain useful information, and a weighted average is thus rendered as the final composite SOC described in 15, with the weight factor w chosen to be closer to 1 for enhanced stability and closer to 0 for increased responsiveness. For supercapacitor analysis, $w = 0.99$ was used. Note that in the formulation of SOC_C, we increment from the previous value of SOC (not SOC_C), linking explicitly the current-based SOC_C to the voltage-based SOC_V.

As with batteries, the maximum discharge power can be expressed as:
$$P_{\max,\text{discharge}} = IV = IV_{\min}.$$

That is, when the capacitor voltage obtains its lowest acceptable value, the max discharge power results. Consider the available instantaneous power; i.e., the power available before the charge on the capacitor is depleted significantly by the discharge event. In this case, $V = V(0) + IR$, where $V(0)$ is the system voltage at zero current immediately prior to the discharge, and
$$P_{\max,\text{discharge}} = IV_{\min} = \frac{[V_{\min} - V(0)]}{R}V_{\min}.$$

Similarly, the instantaneous charge power for the ohmic battery is given by
$$P_{\max,\text{charge}} = IV_{\max} = \frac{[V_{\max} - V(0)]}{R}V_{\max}.$$

This ohmic description does not address transient effects, which are important for times that are greater than $\sim 0.01 RC$. Analogous to (32), with R_{ct} infinite and operation from an equilibrated state, we obtain the following transient expressions for the $P(t)$:

$$P_{\max,\text{discharge}} = IV_{\min} = \frac{[V_{\min} - V(0)]}{R}V_{\min}e^{-t/(RC)}$$
$$P_{\max,\text{charge}} = IV_{\max} = \frac{[V_{\max} - V(0)]}{R}V_{\max}e^{-t/(RC)}.$$

As $t \to 0$, the instantaneous power expressions are obtained.

The test protocol employed to analyze the supercapacitor is now described. For all of the tests reported herein, the capacitor was initially set to 2.3 V (i.e., time $t = 0$ in the upper plot of Fig. 49). The state estimator was provided default values for the resistance R and capacitance C; the state estimator converged to the substantially constant values for R and C after about 500 s, as noted in the lower plot of Fig. 49. The first 600 s of the test correspond to 10 min of the US EPA Urban Cycle (defined by the velocity versus time relationship in Fig. 49). While this forcing of the state estimator to converge at the start of a drive event is a severe test relative to initialization methods employed in actual vehicle applications, it does allow one to assess the convergence properties of the state estimator. During the first 600 s, the HWIL system randomly selects which test will be run once 600 s is reached and when during the next 770 s the test will be conducted. Three different tests can be selected: (1) SOC test, (2) discharge power test, or (3) charge power test. For the results presented in this work, the number of SOC, discharge power, and charge power tests correspond to 67, 70, and 63. For verification and validation of an algorithm, it is important to not introduce systematic aberrations into the analysis, thus explaining the random selection of both the test time and type. In addition, while not further addressed in this work, a myriad of different drive schedules and events must be investigated. (The HWIL system described here is fully automated, allowing an extensive collection of data for analysis.)

For the SOC tests, at predetermined test time, the projected composite SOC rendered by the state estimator is recorded, the current is set to zero for 180 s, after which the cell potential is recorded in order to acquire a voltage-based SOC,

$$\text{SOC}_V = \frac{V - V_{\min}|_{I=0}}{V_{\max}|_{I=0} - V_{\min}|_{I=0}} \qquad (47)$$

Per Eq. (47) under equilibrium conditions. The cell is then discharged at low current (6 A) to measure the coulombic capacity and ascertain a second measure of the cell SOC per SOC $= w\text{SOC}_I + (1-w)\text{SOC}_V$.

For the power tests, at predetermined test time, the projected power 0.5 and 1 s into the future is rendered by the state estimator and recorded, the voltage is set to its minimum or maximum value

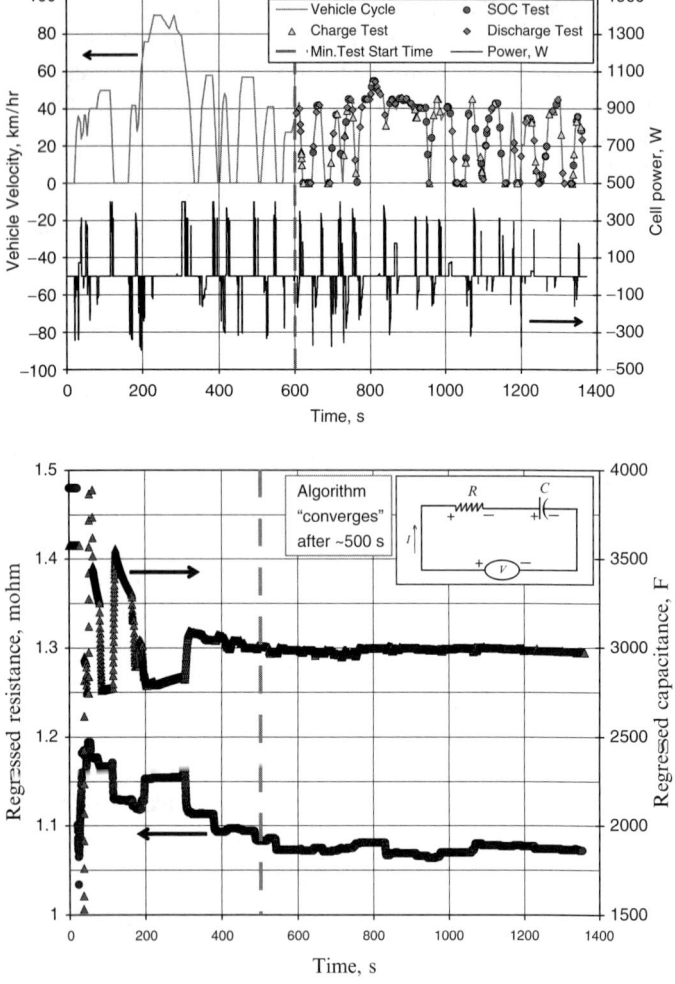

Figure 49. *Top*: Vehicle velocity, power profile of the electric double layer capacitor, and (randomly selected) times power and SOC tests. *Bottom*: Regressed capacitance and resistance for the equivalent circuit representation over the drive cycle depicted in the upper plot.

for the discharge or charge power test, respectively, and the measured power after 0.5 and 1 s is recorded.

The results for the SOC and power-projection tests are presented in Fig. 50. We find that the voltage-based projection (47) is in good agreement with the measured SOC and that the coulombic-based projection over-estimates the SOC, giving rise to an overestimation rendered by the composite SOC. A plot of the projected maximum (discharge and charge) power capability versus the measure maximum power is provided in the lower plot of Fig. 50. For times less than 0.5 s into the future, the state estimator provides a very accurate estimate of the power capability. For longer times, the state estimator under predicts the maximum power capability for both charge and discharge.

IV. OPEN QUESTIONS

Low temperature performance remains a challenging problem for battery state estimators. As mentioned briefly in the discussion of Figs. 24–27, the simple model proposed does not represent lithium ion batteries well below $0°C$, and the same is true for lead acid and NiMH batteries. Because these systems can be expected to operate at temperatures below $-30°C$, this is an important unresolved problem.

Another important and general question deserving of more work is clarifying the merits of various SOH approaches. Here we propose a figure of merit based on the high-frequency resistance, (46), but there is no general agreement within the battery community regarding an optimal SOH definition.

A third problem common to the state estimator approaches covered in this work is a procedure to bound the regressed parameters. For example, for the lithium ion state estimations, the high-frequency resistance R was allowed to vary between 0.1 and 10 times the nominal value listed in Table 4. The parameter E was allowed to vary between 0.7 and 0.95 the nominal value, and the parameter A_d was allowed to vary between 0.5 and 2 times the nominal. At this time we do not have a specific procedure for determining appropriate parameters bounds. A related question is the issue of parameter convergence, which was mentioned briefly in the context of the variable forgetting factor, wherein the iterative nature of that

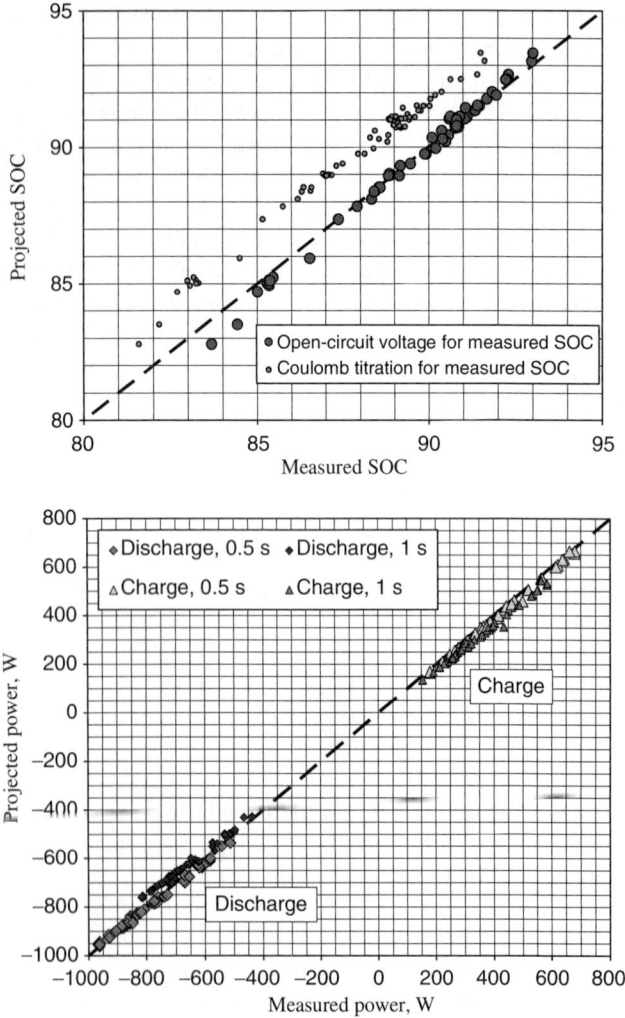

Figure 50. *Top*: Projected and measured state of charge. The (randomly selected) times at which the SOC was measured are indicated by the circular symbols in Fig. 49 (*upper plot*). *Bottom*: Projected and measured maximum charge and discharge power. The (randomly selected) times at which the power tests were conducted are indicated by the diamond (discharge) and triangular (charge) symbols in Fig. 49 (*upper plot*).

algorithm raises concern. For all of the WRLS methods discussed in this review, while one can look at voltage errors as well as errors in power and SOC predictions, we are not aware of a robust method that allows one to state categorically that the parameters have converged and the algorithm is behaving well. This problem is made difficult by the fact that the model upon which the state estimator is designed is approximate, which in turn leads one to expect temporal variations in the model parameters when implemented. What variations are acceptable, however, is an open question.

The remaining problematic topic specifically related to the algorithms discussed in this work concerns adaptive hysteresis. The open questions concerning the state of understanding for the NiMH battery are discussed briefly in this review and are addressed in (102) as well. The approach used for adaptive hysteresis outlined here is primitive (cf. (19) and (22)) and we expect better formulations to follow. Similar hysteresis problems may arise in cells that exhibit little variation in open-circuit potential with SOC, as in the titanate (71–73) and iron phosphate (67–70) cells.

V. SUMMARY

The initial portion of this chapter is devoted to short description of hybrid electric vehicles, their propulsion systems, and the role of the electrochemical energy storage system (EESS). Integration of the EESS into the propulsion system requires real-time information on the state of the EESS while in operation. Specifically, given the current, voltage, and temperature of the EESS, the on-board controller must provide the energy content (state of charge), predicted power capability for both charge and discharge (state of power), and performance relative to the new and end-of-life condition (state of health). This task is best accomplished with a state estimator.

Most of the state estimators for EESSs, including those used within GM, rely on equivalent circuits for describing the electrochemical phenomena, versus the more complicated system of coupled, nonlinear, partial differential equations that describe the governing thermodynamics, transport phenomena, and interfacial kinetics. Network analysis plays an important role in developing equation systems for the equivalent circuits, and relevant aspects of this topic are reviewed, along with adaptive filters and optimal estimators.

Before constructing a state estimator based on a physical model of the EESS, it is best to first ascertain whether the physical model provides a robust description of the EESS. The use of an equivalent circuit to represent various batteries and carbon–carbon supercapacitors is examined for controlled current, potential, and power excitation sources. The generally good agreement between the equivalent circuit model and experimental data supports the state-estimator implementation described afterwards for lead acid, nickel metal hydride, and lithium ion battery systems as well as for a carbon–carbon supercapacitor.

The last topic addressed in this review is the verification and validation of state estimators in a hardware-in-the-loop system. With this closing topic, we complete the process in going from the system electrochemistry to prototyping algorithms for vehicle applications. The review culminates with a description of pressing open questions that have not been fully comprehended and are generally difficult topics to address.

ACKNOWLEDGMENTS

I have had the pleasure to work with and learn from a number of researchers and engineers over the past few years in the development of state estimators and electrochemical energy storage systems. From GM: Daniel Baker, Robert Conell, Damon Frisch, Michael Gielniak, Scott Jorgensen, Brian Koch, Kevin Liu, Joe LoGrasso, Mutasim Salman, Lance Turner, Trudy Weber, and Ramona Ying. From HRL Labs: Arthur Bekaryan, Ping Liu, Cameron Massey, and Souren Soukiazian. This paper benefited from numerous comments and helpful suggestions offered by Alan Taub of GM R&D. Lastly, I want to acknowledge the permission granted to me by Elsevier, Springer, and The Electrochemical Society to use figures in this chapter from the referenced publications of their respective journals.

VI. APPENDIX

In this first section of the Appendix, we derive the voltage recursion relation. We begin with the superposition integral (17),

$$V = V_0 + IR - A \int_{\zeta=t}^{\zeta=0} I(\zeta) \exp[-B(t - \zeta)]d\zeta.$$

At time zero,
$$(V - V_0 - IR)_{t=0} = 0.$$

For the first time step,

$$(V - V_0 - IR)_{t=t_1} \approx e^{-Bt_1} A \left(\frac{I_{t=0} + I_{t=t_1}}{2} \right) e^{Bt_1} \Delta t_1, \text{ or}$$

$$(V - V_0 - IR)_{t=t_1} e^{Bt_1} \approx A \left(\frac{I_{t=0} + I_{t=t_1}}{2} \right) e^{Bt_1} \Delta t_1,$$

where the current in the integrand has been taken as an average of the current at the beginning and end of the time step. For the second time step,

$$(V - V_0 - IR)_{t=t_2}$$
$$\approx e^{-Bt_2} \left[A \left(\frac{I_{t=t_1} + I_{t=t_2}}{2} \right) e^{Bt_2} \Delta t_2 + A \left(\frac{I_{t=0} + I_{t=t_1}}{2} \right) e^{Bt_1} \Delta t_1 \right], \text{ or}$$
$$(V - V_0 - IR)_{t=t_2} \approx A \left(\frac{I_{t=t_1} + I_{t=t_2}}{2} \right) \Delta t_2 + e^{-B \Delta t_2} (V - V_0 - IR)_{t=t_1}.$$

Following this procedure, similar recursion relations result for subsequent time steps, with the general expression corresponding to

$$V = (V_0 + IR)_t + A \left(\frac{I_{t-\Delta t} + I_t}{2} \right) \Delta t + e^{-B \Delta t} (V - V_0 - IR)_{t-\Delta t}. \tag{48}$$

The utility of the recursion relation stems from the fact that time integration of past currents need not be used to determine the voltage; only the present and previous time-step values are required. It should be recognized that there are other methods to approximate the integral that are equivalent from an implementation point of view when it comes to constructing a recursive voltage-current relation for SOC and SOH extraction. For example, instead of averaging the current over the time step, giving rise to the factor $\left(\frac{I_{t-\Delta t}+I_t}{2}\right)$ in the above equations, a forward-difference formulation could be used, wherein the current is simply set to its final value at the end of the time interval, which leads to I_t in place of $\left(\frac{I_{t-\Delta t}+I_t}{2}\right)$. Likewise, a backwards difference approach could be formulated (leading to a $I_{t-\Delta t}$ factor only), or higher-order approximations could be employed. Since the expression is placed in an adaptive routine, and the time steps in practice are to be taken as small relative the characteristic time $\tau = 1/B$, all of the aforementioned approaches yield effectively identical results.

Five-parameter expression for m_l. The purpose of this section is to show how one can obtain direct, analytic expressions for regression parameters through the use of a symbolic manipulator. To make this statement concrete, we must chose and discuss a specific symbolic manipulator and provide exact commands. We shall provide results for the five-parameter model ($y = m_1 x_1 + m_2 x_2 + m_3 x_3 + m_4 x_4 + b$). The governing matrix equation is given by

$$\mathbf{X} = \begin{bmatrix} V_{1,1} & V_{1,2} & V_{1,3} & V_{1,4} \\ V_{1,2} & V_{2,2} & V_{2,3} & V_{2,4} \\ V_{1,3} & V_{2,3} & V_{3,3} & V_{3,4} \\ V_{1,4} & V_{2,4} & V_{3,4} & V_{4,4} \end{bmatrix}, \mathbf{m} = \begin{bmatrix} m_1 \\ m_2 \\ m_3 \\ m_4 \end{bmatrix}, \text{ and } \mathbf{Y} = \begin{bmatrix} V_{y,1} \\ V_{y,2} \\ V_{y,3} \\ V_{y,4} \end{bmatrix}.$$

The variances $V_{u,v}$ require the recursive evaluations of $s_w, s_{x_1}, s_{x_2}, s_{x_3}, s_{x_4}, s_y, s_{x_1,x_1}, s_{x_1,x_2}, s_{x_1,x_3}, s_{x_1,x_4}, s_{x_1,y}, s_{x_2,x_2}, s_{x_2,x_3}, s_{x_2,x_4}, s_{x_2,y}, s_{x_3,x_3}, s_{x_3,x_4}, s_{x_3,y}, s_{x_4,x_4}$, and $s_{x_4,y}$. These quantities are then used to construct the covariances of the matrix $\mathbf{X}$. The final output expressions are obtained directly from the Mathematica environment. For the five-parameter model, the covariance matrix $\mathbf{X}$ is entered within Mathematica as

$m = \{\{V_{11}, V_{12}, V_{13}, V_{14}\}, \{V_{12}, V_{22}, V_{23}, V_{24}\},$
$\{V_{13}, V_{23}, V_{33}, V_{34}\}, \{V_{14}, V_{24}, V_{34}, V_{44}\}\}$

where "m" denotes the matrix function in Mathematica (i.e., $m = \mathbf{X}$; unfortunately, the parameter vector in this work corresponds to $\mathbf{m}$, while m refers to the matrix function in Mathematica). The

Characterization and Modeling of Electrochemical Energy

nomenclature relative to that of the main text corresponds to $V\text{uv} = V_{u,v}$. The matrix equation system $\mathbf{Xm}=\mathbf{Y}$, with $L + 1 = 5$, is then entered as

$$m.\{m_1, m_2, m_3, m_4\} == \{V\text{y}_1, V\text{y}_2, V\text{y}_3, V\text{y}_4\}.$$

The left side is $\mathbf{Xm}$ (thus $[m_1\ m_2\ m_3\ m_4] = \mathbf{m}^T$) and the right side is $\mathbf{Y}$ (with $[V\text{y}_1\ V\text{y}_2\ V\text{y}_3\ V\text{y}_4] = \mathbf{Y}^T$). The solution is obtained with the command

$$\text{Simplify}[\text{Solve}[m.\{m_1, m_2, m_3, m_4\} == \{V\text{y}_1, V\text{y}_2, V\text{y}_3, V\text{y}_4\}]].$$

For the parameters m_1 to m_4, one can associate the denominators of these quantities with the determinant of $\mathbf{X}$ by calculating the determinant itself within Mathematica; i.e.,

$$\text{Det}[\{\{V_{11}, V_{12}, V_{13}, V_{14}\}, \{V_{12}, V_{22}, V_{23}, V_{24}\},$$
$$\{V_{13}, V_{23}, V_{33}, V_{34}\}, \{V_{14}, V_{24}, V_{34}, V_{44}\}\}].$$

The following five expressions are obtained:

$$\begin{aligned}
\text{Det} = &-(V^*_{11} V^{**}_{24} 2^* V_{33}) + V^{**}_{14} 2^* (V^{**}_{23} 2 - V^*_{22} V_{33}) + \\
& 2^* V^*_{11} V^*_{23} V^*_{24} V_{34} + V^{**}_{12} 2^* V^{**}_{34} 2 - V^*_{11} V^*_{22} V^{**}_{34} 2 - \\
& 2^* V^*_{14} (V^*_{13} V^*_{23} V_{24} - V^*_{12} V^*_{24} V_{33} - V^*_{13} V^*_{22} V_{34} + \\
& V^*_{12} V^*_{23} V_{34}) - V^*_{11} V^{**}_{23} 2^* V_{44} - V^{**}_{12} 2^* V^*_{33} V_{44} + \\
& V^*_{11} V^*_{22} V^*_{33} V_{44} + V^{**}_{13} 2^* (V^{**}_{24} 2 - V^*_{22} V_{44}) + V^*_{13} (- \\
& 2^* V^*_{12} V^*_{24} V_{34} + 2^* V^*_{12} V^*_{23} V_{44})
\end{aligned}$$

$$\begin{aligned}
m_1 = &(V^{**}_{23} 2^* V^*_{44} V\text{y}_1 + V^*_{14} V^*_{23} V^*_{34} V\text{y}_2 - V^*_{12} V^{**}_{34} 2^* V\text{y}_2 - \\
& V^*_{13} V^*_{23} V^*_{44} V\text{y}_2 + V^*_{12} V^*_{33} V^*_{44} V\text{y}_2 - V^*_{12} V^*_{23} V^*_{44} V\text{y}_3 + \\
& V^{**}_{24} 2^* (V^*_{33} V\text{y}_1 - V^*_{13} V\text{y}_3) - V^*_{14} V^{**}_{23} 2^* V\text{y}_4 + \\
& V^*_{12} V^*_{23} V^*_{34} V\text{y}_4 + V^*_{22}(V^{**}_{34} 2^* V\text{y}_1 - V^*_{33} V^*_{44} V\text{y}_1 - \\
& V^*_{14} V^*_{34} V\text{y}_3 + V^*_{13} V^*_{44} V\text{y}_3 + V^*_{14} V^*_{33} V\text{y}_4 - V^*_{13} V^*_{34} V\text{y}_4) + \\
& V^*_{24}(-(V^*_{14} V^*_{33} V\text{y}_2) + V^*_{13} V^*_{34} V\text{y}_2 + V^*_{12} V^*_{34} V\text{y}_3 - \\
& V^*_{12} V^*_{33} V\text{y}_4 + V^*_{23}(-2^* V^*_{34} V\text{y}_1 + V^*_{14} V\text{y}_3 + V^*_{13} V\text{y}_4)))/(-\text{Det})
\end{aligned}$$

$$\begin{aligned}
m_2 = &(-(V^*_{12} V^{**}_{34} 2^* V\text{y}_1) + V^*_{12} V^*_{33} V^*_{44} V\text{y}_1 + V^*_{11} V^{**}_{34} 2^* V\text{y}_2 - \\
& V^*_{11} V^*_{33} V^*_{44} V\text{y}_2 - V^*_{11} V^*_{24} V^*_{34} V\text{y}_3 + V^*_{11} V^*_{23} V^*_{44} V\text{y}_3 +
\end{aligned}$$

$V_{14}^{**}2^*(V_{33}^*Vy_2 - V_{23}^*Vy_3) + V_{11}^*V_{24}^*V_{33}^*Vy_4 -$
$V_{11}^*V_{23}^*V_{34}^*Vy_4 + V_{13}^{**}2^*(V_{44}^*Vy_2 - V_{24}^*Vy_4) + V_{14}^*(-$
$(V_{24}^*V_{33}^*Vy_1) + V_{23}^*V_{34}^*Vy_1 - 2^*V_{13}^*V_{34}^*Vy_2 + V_{13}^*V_{24}^*Vy_3 +$
$V_{12}^*V_{34}^*Vy_3 + V_{13}^*V_{23}^*Vy_4 - V_{12}^*V_{33}^*Vy_4) + V_{13}^*(V_{24}^*V_{34}^*Vy_1$
$- V_{23}^*V_{44}^*Vy_1 - V_{12}^*V_{44}^*Vy_3 + V_{12}^*V_{34}^*Vy_4))/(-\mathrm{Det})$

$m_3 = (V_{12}^*V_{24}^*V_{34}^*Vy_1 - V_{12}^*V_{23}^*V_{44}^*Vy_1 - V_{11}^*V_{24}^*V_{34}^*Vy_2 +$
$V_{11}^*V_{23}^*V_{44}^*Vy_2 + V_{11}^*V_{24}^{**}2^*Vy_3 + V_{12}^{**}2^*V_{44}^*Vy_3 -$
$V_{11}^*V_{22}^*V_{44}^*Vy_3 + V_{14}^{**}2^*(-(V_{23}^*Vy_2) + V_{22}^*Vy_3) -$
$V_{11}^*V_{23}^*V_{24}^*Vy_4 - V_{12}^{**}2^*V_{34}^*Vy_4 + V_{11}^*V_{22}^*V_{34}^*Vy_4 +$
$V_{14}^*(V_{23}^*V_{24}^*Vy_1 - V_{22}^*V_{34}^*Vy_1 + V_{13}^*V_{24}^*Vy_2 + V_{12}^*V_{34}^*Vy_2$
$-2^*V_{12}^*V_{24}^*Vy_3 - V_{13}^*V_{22}^*Vy_4 + V_{12}^*V_{23}^*Vy_4) + V_{13}^*(-$
$(V_{24}^{**}2^*Vy_1) + V_{22}^*V_{44}^*Vy_1 - V_{12}^*V_{44}^*Vy_2 + V_{12}^*V_{24}^*Vy_4))/(-\mathrm{Det})$

$m_4 = (-(V_{12}^*V_{24}^*V_{33}^*Vy_1) + V_{12}^*V_{23}^*V_{34}^*Vy_1 + V_{11}^*V_{24}^*V_{33}^*Vy_2$
$-V_{11}^*V_{23}^*V_{34}^*Vy_2 - V_{11}^*V_{23}^*V_{24}^*Vy_3 - V_{12}^{**}2^*V_{34}^*Vy_3 +$
$V_{11}^*V_{22}^*V_{34}^*Vy_3 + V_{14}^*(-(V_{23}^{**}2^*Vy_1) + V_{22}^*V_{33}^*Vy_1 +$
$V_{13}^*V_{23}^*Vy_2 - V_{12}^*V_{33}^*Vy_2 - V_{13}^*V_{22}^*Vy_3 + V_{12}^*V_{23}^*Vy_3) +$
$V_{11}^*V_{23}^{**}2^*Vy_4 + V_{12}^{**}2^*V_{33}^*Vy_4 - V_{11}^*V_{22}^*V_{33}^*Vy_4 +$
$V_{13}^{**}2^*(-(V_{24}^*Vy_2) + V_{22}^*Vy_4) + V_{13}^*(V_{23}^*V_{24}^*Vy_1 -$
$V_{22}^*V_{34}^*Vy_1 + V_{12}^*V_{34}^*Vy_2 + V_{12}^*V_{24}^*Vy_3 -$
$2^*V_{12}^*V_{23}^*Vy_4))/(-\mathrm{Det}).$

Clearly the risk of error is reduced when the results from a symbolic manipulation can be copied directly to the source code.

LIST OF SYMBOLS

Ah Coulombic capacity, C-h s^{-1}
A   C^{-1}, F^{-1}
b   Regressed intercept
B $1/(R_{ct}C)$, s^{-1}
C   Capacitance, F
I   Current, A
L   Number of parameters *m* to be regressed adaptively
m   Parameter to be regressed

N	Number of time steps (data points) in the regression
P	Power, W
q	Charge on the capacitor C of the equivalent circuit
r_C	Capacitance ratio, $C_{discharge}/C_{charge}$
R	High-frequency resistance, ohm
R_{ct}	Effective interfacial resistance, ohm
SOC	State of charge
SOH	State of health
SOP	State of power
s	Sum
t	Time, s
V	System voltage, V
V_H	Hysteresis voltage, V
V_0	Open-circuit voltage, V
x	Time-dependent values multiplying onto parameters m
y	Dependent variable
w_{SOC}	Weighting factor for SOC calculation
β	Hysteresis parameter, C^{-1}
ε	Error or loss term
ε_{opt}	Unweighted total error as defined by (44)
λ	Forgetting factor
γ	Parameter for selective weighting of data beyond that of the forgetting factor
σ	Variance

REFERENCES

1. W. Steffens, "Verfahren zur Schätzung der inneren Größen von Starterbatterien," Ph.D. Thesis, Technical University of Aachen, Germany (1987).
2. H. P. Schoener, "Über die Auswertung des elektrischen Verhaltens von Bleibatterien beim Entladen und Laden", Ph.D. Thesis, Technical University of Aachen, Germany (1988).
3. M. A. Dorgham, R. D. Fruechte (eds.), *Application of Control Theory in the Automotive Industry*, Inderscience, olneg, UK., 1983.
4. H. L. N. Wiegman, "Battery State Estimation and Control for Power Buffering Applications," Ph.D. Thesis, University of Wisonsin, Madison (1999).
5. E. D. Tate, Jr. "Techniques for Hybrid Electric Vehicle Controller Synthesis," Ph.D. Thesis, University of Michigan, Ann Arbor (2006).
6. M. van Walwijk, C. Saricks (eds.), *Hybrid and Electric Vehicles*, International Energy Agency, Implementing Agreement on Hybrid and Electric Vehicle Technologies and Programmes annual report of the Executive Committee and Anne I over 2006, February 2007.

7. S. Pillar, M. Perrin, A. Jossen, *J. Power Sourc.*, 96(2001)113.
8. A. Tenno, R. Tenno, T. Suntio, *J. Power Sourc.*, 103(2001)42.
9. O. Barbarisi, R. Canaletti, L. Glielmo, M. Gosso, F. Vasca, *Proceedings of the 41st IEEE Conference on Decision and Control*, Las Vegas, NV, paper WeM05-5, (2002) 1739.
10. E. Meissner, G. Richter, *J. Power Sourc.*, 116(2003)79.
11. M. W. Verbrugge, E. D. Tate, *J. Power Sourc.*, 126(2004)236.
12. G. L. Plett, *J. Power Sourc.*, 134(2004)252, 262, 277.
13. G. L Plett, *IEEE Trans. Veh. Technol.*, 53(2004)1586.
14. M. W. Verbrugge, P. Liu, S. Soukiazian, *J. Power Sourc.*, 141(2005)369.
15. V. Pop, H. J. Bergveld, P. H. L. Notten, P. P. L. Regtien, *Meas. Sci. Technol.*, 16(2005)R93.
16. M. W. Verbrugge, D. Frisch, B. Koch, *J. Electrochem. Soc.*, 152(2005)A333.
17. H. Ashizawa, H. Nakamura, D. Yumoto, H. Asai, Y. Ochi, *SAE*, paper 05P-319 (2004).
18. H. Asai, H. Ashizawa, D. Yumoto, H. Nakamura, Y. Ochi, *SAE*, paper 2005-01-0807 (2005).
19. O. Barbarisi, F. Vasca, L. Glielmo, *Control Eng. Practice*, 14(2006)267.
20. K. A. Smith, "Electrochemical Modeling, Estimation, and Control of Lithium Ion Batteries," Ph.D. Thesis, Pennsylvania State University, (2006).
21. M. W. Verbrugge, B. J. Koch, *J. Electrochem. Soc.*, 153(2006)A187.
22. S. Santhanagopalan, R. E. White, *J. Power Sourc.*, 161(2006)1346.
23. M. W. Verbrugge, *J. Appl. Electrochem.*, 37(2007)605.
24. S. Santhanagopalan, Q. Guo, R. E. White, *J. Electrochem. Soc.*, 154(2007) A198.
25. H. J. Carlin, A. B. Giordano, *Network Theory*, Prentice-Hall, Englewood Cliffs, NJ (1964).
26. V. Belevitch, *Classical Network Theory*, Holden-Day, San Francisco, CA (1968).
27. J. O. Scanlan, R. Levy, *Circuit Theory*, Oliver and Boyd, Edinburgh, Great Britain (1970).
28. M. E. Van Valkenburg, *Network Analysis*, 3rd edition, Prentice-Hall, Englewood Cliffs, NJ (1974).
29. L. R. Rabiner, B. Gold, *Theory and Application of Digital Signal Processing*, Prentice-Hall, Englewood Cliffs, NJ (1975).
30. S. A. Tretter, *Introduction to Discrete-Time Signal Processing*, Wiley, New York, NY (1976).
31. A. Peled, B. Liu, *Digital Signal Processing*, Wiley, New York, NY (1976).
32. A. Antoniou, *Digital Filters: Analysis and Design*, McGraw-Hill, New York, NY (1979).
33. L. P. Huelsman, P. E. Allen, *Introduction to the Theory and Design of Active Filters*, McGraw-Hill, New York, NY (1980).
34. M. S. Ghausi, K. R. Laker, *Modern Filter Design: Active RC and Switched Capacitor*, Prentice-Hall, Englewood Cliffs, NJ (1981).
35. M. Bellanger, *Digital Processing of Signals*, Wiley, New York, NY (1984). (Originally published as *Traitement Numérique Du Signal—Théorie Et Pratique* by M. Bellanger, Masson, Paris, 1980).
36. S. Haykin, *Introduction to Adaptive Filters*, MacMillan, New York, NY (1984).

37. B. Widrow, S. D. Stearns, *Adaptive Signal Processing*, Prentice-Hall, Englewood Cliffs, NJ (1985).
38. M. G. Bellanger, *Adaptive Digital Filters and Signal Analysis*, Marcel Dekker, New York, NY (1987).
39. H. Baher, *Analog and Digital Signal Processing*, Wiley, New York, NY (1990).
40. C. F. Gauss, *Theoria Motus Corporum Coelestium in Sectionibus Conicus Solem Ambientum*, Hamburg, (1809), (translation: Dover, 1963).
41. R. L. Plackett, *Biometrika*, 37(1950)149.
42. R. E. Kalman, *Trans. ASME. J. Basic Eng.*, 82D(1960)35.
43. A. H. Jazwinski, *Stochastic Processes and Filtering Theory*, Academic Press, New York, NY (1970).
44. A. Gelb (ed.), *Applied Optimal Estimation*, edited by M.I.T. Press, Cambridge, MA (1974).
45. B. D. O. Anderson, J. B. Moore, *Optimal Filtering*, Prentice-Hall, Englewood Cliffs, NJ (1979).
46. P. S. Maybeck, *Stochastic Models, Estimation and Control*, volume 141-1 of *Mathematics in Science and Engineering*, Academic Press, UK (1979).
47. W. L. Brogan, *Modern Control Theory*, 2nd edition, Prentice-Hall, Englewood Cliffs, NJ (1985).
48. L. Ljnug, T. Söderström, *Theory and Practice of Recursive Identification*, M.I.T Press, MA (1986).
49. K. J. Åström, B. Wittenmark, *Adaptive Control*, Addison-Wesley, MA, USA, (1989), (second edition: 1995).
50. R. Kulhavý, *Recursive Nonlinear Estimation. A Geometric Approach*, Springer, Berlin Hiedelberg New York (1996).
51. T. Demarco, *Structured Analysis and System Specification*, Prentice-Hall, Englewood Cliffs, NJ (1979).
52. D. H. Hatley, I. A. Pirbhai, *Strategies for Real-Time System Specification*, Dorset House, New York, NY (1988).
53. S. Bittanti, P. Bolzern, M. Campi, E. Coletti, *Proceedings of the American Control Conference, IEEE*, Austin, Texas, December 1988, pp. 1530–1531.
54. L. Ljung, S. Gunnarsson, *Automatica*, 26(1990)7.
55. J. E. Parkum, N. K. Poulsen, J. Holst, *Int. J. Control*, 55(1992)109.
56. R. Kulhavý, *Int. J. Control*, 58 (1993)905.
57. A. Vahidi, M. Druzhinina, A. Stefanopoulou, H. Peng, *Proceedings of the American Control Conference, IEEE*, Denver, Colorado, June 2003, pp. 4951–4956.
58. Y. Zheng, Z. Lin, *IEEE Trans. Circuits Syst.—II: Analog. Digital Signal Process.*, 50(2003)602.
59. C. S. Ludovico, J. C. M. Bermudez, *IEEE ICASSP*, 2(2004) II-673.
60. S. Sathyanarayana, S. Venugopalan, M. L. Gopikanth, *J. Appl. Electrochem.*, 9(1979)369.
61. M. L. Gopikanth, S. Sathyanarayana, *J. Appl. Electrochem.*, 9(1979)369.
62. M. S. Suresh, S. Sathyanarayana, *J. Power Sourc.*, 37(1992)335.
63. S. Rodrigues, N. Munichandraiah, A. K. Shukla, *J. Power Sourc.*, 87(2000)12.
64. V. V. Viswanathan, A. J. Salkind, J. J. Kelley, J. B. Ockerman, *J. Appl. Electrochem.*, 25(1995)716.

65. V. V. Viswanathan, A. J. Salkind, J. J. Kelley, J. B. Ockerman, *J. Appl. Electrochem.*, 25(1995)729.
66. F. Huet, *J. Power Sourc.*, 70(1998)59.
67. A. K. Padhi, K. S. Najundaswamy, J. B. Goodenough, *J. Electrochem. Soc.*, 144(1997)1188.
68. A. K. Padhi, K. S. Najundaswamy, C. Masquelier, S. Okada, J. B. Goodenough, *J. Electrochem. Soc.*, 144(1997)1609.
69. S. Y. Chung, J. T. Bloking, Y. M. Chiang, *Nat. Mater.*, 1(2002) 123, (www.nature.com/naturematerials).
70. W. F. Howard, R. F. Spotnitz, *J. Power Sourc.*, 165(2007)887.
71. K. M. Colbow, J. R. Dahn, R. R. Haering, *J. Power Sourc.*, 26(1989)397.
72. T. Ohzuku, A. Ueda, N. Yamamoto, *J. Electrochem. Soc.*, 142(1995)1431.
73. Y. H. Rho, K. Kanamura, J. Solid State Chem., 177(2004)2094.
74. R. K. B. Gover, J. R. Tolchard, H. Tukamoto, T. Murai, J. T. S. Irvine, *J. Electrochem. Soc.*, 146(1999)4348.
75. Y. H. Rho, K. Kanamura, *J. Solid State Chem.*, 177(2004)2094.
76. I. J. Ong, J. Newman, *J. Electrochem. Soc.*, 146(1999)4360.
77. P. Delahay, *Double Layer and Electrode Kinetics*, Interscience Publishers, New York, NY (1965).
78. D. D. Macdonald, *Transient Techniques in Electrochemistry*, Plenum, New York, NY, 1977.
79. A. J. Bard, L. R. Faulkner, *Electrochemical Methods: Fundamentals and Applications*, Wiley, New York, NY (1980).
80. B. Pillay, "Design of Electrochemical Capacitors for Energy Storage," Ph.D. Thesis, University of California, Berkeley, CA (1996).
81. B. Pillay, J. Newman, *J. Electrochem. Soc.*, 143(1996)1806.
82. V. Srinivasan, J. W. Weidner, *J. Electrochem. Soc.*, 146(1999)1650.
83. D. Dunn, J. Newman, *J. Electrochem. Soc.*, 147(2000)820.
84. K. E. Thomas, R. M. Darling, J. Newman, "Mathematical Modeling of Lithium Batteries," in: *Advances in Lithium-Ion Batteries*, W. van Schalkwaijk, D. Scrosati (eds.), Kluwer, Dordrecht (2002), Chap. 12.
85. C. Lin, J. A. Ritter, B. N. Popov, R. E. White, *J. Electrochem. Soc.*, 146(1999)3168.
86. C. Lin, B. N. Popov, H. J. Ploehn, *J. Electrochem. Soc.*, 149(2002)A167.
87. H. Kim, B. N. Popov, *J. Electrochem. Soc.*, 150(2003)A1153.
88. S. Devan, V. R. Subramanian, R. E. White, *J. Electrochem. Soc.*, 151(2004) A905.
89. M. W. Verbrugge, P. Liu, *J. Electrochem. Soc.*, 152(2005)D79.
90. M. W. Verbrugge, P. Liu, *J. Electrochem. Soc.*, 153(2006)A1237.
91. J. S. Dunning, "Analysis of Porous Electrodes with Sparingly Soluble Reactants," Ph.D. Thesis, University of California, Los Angeles (1971).
92. J. Newman, W. Tiedemann, *AIChE J.*, 21(1975)25.
93. J. Newman, *Electrochemical Systems*, second edition, Prentice-Hall, Englewood Cliffs, NJ (1991).
94. M. W. Verbrugge, *J. Electrostatics*, 34(1995)61.
95. M. W. Verbrugge, *AIChE J.*, 41(1995)1550.
96. M. W. Verbrugge, D. W. Glander, D. R. Baker, *J. Cryst. Growth*, 81(1995)155.
97. C. Y. Wang, W. B. Gu, B. Y. Liaw, *J. Electrochem. Soc.*, 145(1998)3407.

98. D. R. Baker, M. W. Verbrugge, *J. Electrochem. Soc.*, 146(1999)2413.
99. M. W. Verbrugge, P. Liu, *J. Power Sourc.* (Hybrid Vehicles Special Battery Issue), 174(2007)2.
100. M. W. Verbrugge, R. Y. Ying, *J. Electrochem. Soc.*, 154(2007)A949.
101. M. W. Verbrugge, B. J. Koch, *J. Electrochem. Soc.*, 146(1999)833.
102. M. W. Verbrugge, R. S. Conell, *J. Electrochem. Soc.*, 149(2002)A45.
103. M. W. Verbrugge, B. J. Koch, *J. Electrochem. Soc.*, 143(1996)600.
104. R. M. Corless, G. H. Gonnet, D. E. G. Hare, D. J. Jeffrey, *Maple Tech. Newsletter*, 9(1993)12.
105. R. M. Corless, G. H. Gonnet, D. E. G. Hare, D. J. Jeffrey, D. E. Knuth, *Adv. Comput. Math.*, 5(1996)329.
106. F. Chapeau-Blondeau, A. Monir, *IEEE Trans. Signal Process.*, 50(2002)2160.
107. B. Hayes, *Am. Sci.*, 93 (2005) 104.
108. M. W. Verbrugge, "Application of a Simplified Model for the Analysis of a Novel Battery Used in General Motors' Precept Hybrid Electric Vehicle," *Proceedings Volume (CD-ROM) from the 17th International Electric Vehicle Symposium*, Montreal, Canada, October 2000.
109. P. Milner, U. Thomas, *Advanceds in Electrochemistry and Electrochemical Engineering*, C. W. Tobias (ed.), Interscience Publishers, New York, NY (1967).
110. X. G. Yang, B. Y. Liaw, *J. Electrochem. Soc.*, 148(2001)A1023.
111. K. P. Ta, J. Newman, *J. Electrochem. Soc.*, 146(1999)2769.
112. W. B. Gu, C. Y. Wang, *J. Electrochem. Soc.*, 147(2000)2910.
113. V. Srinivasan, J. W. Weidner, J. Newman, *J. Electrochem. Soc.*, 148(2001)A969.
114. R. M. Felder, R. W. Rousseau, *Elementary Principles of Chemical Processes*, Wiley, New York, NY (1978), pp. 501–503.
115. S. D. Conte, C. de Boor, *Elementary Numerical Analysis*, 3rd edition, McGraw-Hill, New York, NY (1980), Chap. 6.
116. W. H. Press, B. P. Flannery, S. A. Teukolsky, W. T. Vetterling, *Numerical Recipes*, Cambridge University Press, Cambridge, Great Britain (1989), Chap. 14.
117. S. Wolfram, *The Mathematica Book*, 3rd edition, Wolfram Media, Champaign, IL (1996).
118. W. H. Beyer (ed.), *Standard Mathematical Tables*, 24th edition, CRC Press, Cleveland, OH (1976), pp. 476–477.
119. D. Berndt, *Maintenance-Free Batteries*, Research Studies Press, Taunton, U.K., (1993).
120. D. M. Bernardi, M. K. Carpenter, *J. Electrochem. Soc.*, 142(1995)2631.
121. Y. Guo, J. Wu, L. Song, M. Perrin, H. Doering, J. Garche, *J. Electrochem. Soc.*, 148(2001)A1287.
122. H. Bode, translated by R. J. Brodd, Karl V. Kordesch, *Lead-Acid Batteries*, New York, Wiley (1977).
123. P. Leblanc, C. Jordy, B. Knosp, P. Blanchard, *J. Electrochem. Soc.*, 145(1998)860.
124. P. Bernard, *J. Electrochem. Soc.*, 145(1998)456.
125. T. R. Fortescue, L. S. Kershenbaum, B. E. Ydstie, *Automatica*, 17(1981)831.
126. S. Bittanti, P. Bolzern, M. Campi, E. Coletti, *Proceedings of the American Control Conference, IEEE*, Austin, Texas, December 1988, pp. 1530–1531.
127. L. Ljung, S. Gunnarsson, *Automatica*, 26(1990)7.

128. J. E. Parkum, N. K. Poulsen, J. Holst, *Int. J. Control*, 55(1992)109.
129. R. Kulhavý, *Int. J. Control*, 58 (1993)905.
130. A. Vahidi, M. Druzhinina, A. Stefanopoulou, H. Peng, *Proceedings of the American Control Conference, IEEE*, Denver, Colorado, June 2003, pp. 4951–4956.
131. Y. Zheng, Z. Lin, *IEEE Trans. Circuits Syst.—II Analog. Digital Signal Process.*, 50(2003)602.
132. C. Massey, A. Bekaryan, P. Liu, L. Turner, D. Frisch, T. Weber, M. Verbrugge, *SAE* paper O5CV-137 (2005).
133. B. E. Conway, *Electrochemical supercapacitors: scientific fundamentals and technological applications*, Kluwer, Dordrecht, 1999.
134. R. Kötz, M. Carlen, *Electrochim. Acta*, 45(2000)2483.
135. A. Burke, *J. Power Sourc.*, 91(2000)37.
136. S. Buller, E. Karden, D. Kok, R. W. DeDoncker, *IEEE Trans. Ind. Appl.*, 38(2002)1622.
137. A. Chu, P. Braatz, *J. Power Sourc.*, 112(2002)236.
138. R. B. Wright, D. K. Jamison, T. Q. Duong, Abstract 237, *204th Meeting of The Electrochemical Society*, Orlando, October 2003.
139. D. Y. Jung, Y. H. Kim, S. W. Kim, S-H. Lee, *J. Power Sourc.*, 114(2003)366.
140. M. W. Verbrugge, "Supercapacitors and Automotive Applications," *Proceedings Volume for the World Summit on Advanced Capacitors*, Washington, DC, August 2003.
141. L. Li, "Effects of Activated Carbon Surface Chemistry and Pore Structure on the Adsorption of Trace Organic Contaminants from Aqueous Solution," Ph.D. Dissertation, North Caroline State University, Raleigh, NC, 2002.
142. G. Sikha, R. E. White, B. N. Popov, *J. Electrochem. Soc.*, 152(2005)A1682.
143. R. S. Prabaharan, R. Vimala, Z. Zainal, *J. Power Sourc.*, 161(2006)730.
144. M. W. Verbrugge, P. Liu, "On the merits of supercapacitors for vehicle propulsion systems and open questions," *Proceedings from the Advanced Capacitors World Summit*, San Diego, CA, USA, July 17–19, 2006.
145. T. Wang, M. Fujita, M. Inagaki, *Electrochim. Acta*, 51(2006)4096.
146. J. R. Miller, *Electrochim. Acta*, 52(2006)1703.
147. M. Sevilla, S. Álvarez, T. A. Centeno, A. B. Fuertes, F. Stoeckli, Electrochim. Acta, 52(2007)3207.
148. X. Wang, J. P. Zhenga, *J. Electrochem. Soc.*, 151(2004)A1683.
149. S. A. Kazaryan, S. N. Razumov, S. V. Litvinenko, G. G. Kharisov, V. I. Koganb, *J. Electrochem. Soc.*, 153(2006)A1655.
150. Part 2 of "Well-to-Wheel Energy Use and Greenhouse Gas Emissions of Advanced Fuel/Vehicle Systems – North American Analysis, Executive Summary Report", Argonne National Laboratory, Center of Transportation Research, 2001. (Posted at http://www.transportation.anl.gov/publications/index.html.)
151. T. Weber, "Vehicle System Modeling in the Automotive Industry," *ARO/ERC Engine Modeling Symposium*, University of Wisconsin, Madison, June 2003.

Index

A
Accad, Y., 45
Adsorption time scales, 153–154
Advani, S.G., 360
Advective timescale
 large Peclet number, 150–151
 small Peclet number, 149–150
Amphlett, J.C., 370
Anderson, B.D.O, 469
Antoniou, A, 426
Araki, H., 56
Arato, E., 362, 387
Åström, K.J., 428
Atomic units
 Bohr radius, 37
 correction factor, 38

B
Baker, J.D., 49
Balakrishnan, A., 60
Baldwin, K.G.H., 60
Belevitch, V, 426
Bellanger, M.G., 426
Bergeson, S.D., 60
Berg, P., 358
Bethe, H.A., 44
Bhatia, A.K., 53
Bingel, W.A., 43

Bockris, J.O.M., 312
Bohr–Sommerfeld quantum theory, 44
Bradean, R., 379, 380
Brinkman equation, 300
Bruggeman expression, 301, 337
Brug, G.J., 95, 125, 129
Büchi, F.N., 358
Bürgers, A., 49
Butler–Volmer equation, 86, 285

C
Carlin, H.J., 426
Carman–Kozeny equation, 322, 326
Carnes, B., 291, 294
Catalyst-layer modeling, PEFC
 active phase volume fraction, 309
 agglomerate-type structure, 307–308
 impedance models
  electrochemical impedance spectroscopy (EIS), 317, 319
  equivalent-circuit of porous electrode, 317
 modeling equations
  CL flooding approaches, 314–315
  electrocatalyst and electrolyte interface, 311
  embedded agglomerate model, 311–313

ionomer, 310
  kinetic expressions, 310–311
  surface concentration, 313–314
  optimization analysis
    CL and GDL capillary properties, 316–317
    macrohomogeneous approach, 315–316
    two-phase and three-phase interface, 308
Catalyst (platinum) utilization
  cathode catalyst layer, 234
  effectiveness factor, 231, 232, 234
  Faradaic current density, 232, 233
  oxygen reduction reaction, 232
  parameterization, 232
Cathode catalyst layer
  agglomerates, 225
  capillary equilibrium, 231
  composition, 225, 226
  current-voltage curve, 230
  pore size distributions, 229
  rate of vaporization, 224
  sensitive dependencies, 229
  stability diagram, 231
  three state model, 228
  water balance modeling, 225
  Young–Laplace equation, 227
Cathode humidification temperature, 276–277
Cell-design strategies
  alternate cooling approaches, 364–365
  gas-flow direction
    CFD models, 358
    counterflow and coflow, 357–358
    flow orientation manipulation, 357
  interdigitated flow fields (IDFF)
    CFD models, 361
    gas-diffusion layer (GDL), 360–361
    mass-transport limitation reduction, 360
    pressure-drop calculation, 361
    water-transport plates, 363–364
  optimal cell hydration, 357
  reactant flow and cross flow orientation, 358–359

Chen, F.L., 367, 370
Cheng, K.T., 55
Chen, K.S., 354, 355
Chang, S.M., 373
Choe, S.Y., 371
Choi, P., 294
Chu, H.S., 349, 373
Client–server model, 29
Cold-start process
  automotive process, 376–377
  frozen state startup process
    bootstrap start, 384–386
    cell-level models, 388–392
    procedural strategies, 386–387
    stack-level models, 387–388
  shutdown and freezing
    cell-level models, 379–384
    stack-level models, 378–379
Constant phase element (CPE) model
  CPE behavior, 126–127
  disk electrodes, 127–130
  distribution function, 124–125
  double layer capacitance, 125–126
  faradaic reactions, 124
  fractal electrode model, 119–120
  Hull cell simulations, 127
  kinetic dispersion, 131–133
  porous electrode model, 94–95
Continuous porous model
  diffusion pores equation, 114–115
  impedance evaluation, 115
  polymer fuel cell, 117–118
  principle, 113
  solution theory, 115–117
Corey, A.T., 326
Corrosion
  carbon, 261–262
  catalyst support, 260
  PEMFC, 260–262
Costa, P., 362
Cyclic voltammetry
  electrochemical system, 435
  interaction potential, 440, 441
  Nernst approximation, 440, 441
  open-circuit voltage, 440
  potential sweep rate, 435–436

Index

Cylindrical pore electrode model
  de Levie impedance equation, 69–70
  equivalent electrical circuit, 73–74
  phasors ratio, 70–71
  principle, 68–69
  transmission line impedance equation, 72–73

D

Darcy's law, 300, 302, 362
Darling, R.M., 363
Datta, R., 294
de Brujin, F.A., 394
De Francesco, M., 387
de Levie impedance equation, 69–70
Desorption timescale, 153–154
Devan, S., 115
DeVidts, P., 312
Diffusive timescale, 164–165
Direct methanol fuel cell (DMFC) technology, 170
Disk electrodes, 127–130
Djilali, N., 291, 294, 316, 322, 345, 364
Douglas, M., 58
Drachman, R.J., 53
Drake, G.W.F., 49, 59, 61

E

EESS. *See* Electrochemical energy-storage system
Eides, M.I., 55
Eikema, K.S.E., 60
Eikerling, M., 178, 294, 315
Electrochemical cell interface (ECI), 504
Electrochemical energy-storage system (EESS)
  adaptive filter, 426–427
  composite power system, 420
  control system architectures, 424
  conventional lithium ion technology, 428
  electrochemical modeling
    constant power operation, 442–456
    cyclic voltammetry, 435–442
    equivalent circuit, 431–434
    equilibrium voltage, 428–429
  state estimators
    algorithm verification and validation, 502–511
    generalized weighted recursive least squares, 466–475
    method of least squares, 456–466
    regression analysis, 475–477
    variable forgetting factors, 493–502
  state of charge (SOC), 418, 419
  state of health (SOH) and state of power (SOP), 419
  weighted recursive least squares (WRLS), 425
Electrochemical impedance spectroscopy (EIS), 67, 317, 319
Electrochemical modeling
  constant power operation
    acetonitrile-based capacitor, 445
    Arrhenius relationship, 449, 453
    constant-power discharge and charge, 447–450
    Coulombic capacity, 452
    current and voltage histories, 443, 444
    Lambert W function, 442, 443
    lithium ion cell, 446
    NiMH module, 455
  cyclic voltammetry
    electrochemical system, 435
    Nernst approximation, 440, 441
    open-circuit voltage, 440
    potential sweep rate, 435–436
  equivalent circuit, 431–434
Electrochemical reactions
  constant phase element (CPE) model
    CPE behavior, 126–127
    disk electrodes, 127–130
    distribution function, 124–125
    double layer capacitance, 125–126
    faradaic reactions, 124
    fractal electrode model, 119–120
    Hull cell simulations, 127
    kinetic dispersion, 131–133
    porous electrode model, 94–95
  continuous porous model
    diffusion pores equation, 114–115
    impedancies evaluation, 115

polymer fuel cell, 117–118
principle, 113
solution theory, 115–117
cylindrical pore electrode model
de Levie impedance equation, 69–70
equivalent electrical circuit, 73–74
phasors ratio, 70–71
principle, 68–69
transmission line impedance equation, 72–73
fractal electrode model
CPE behavior, 119–120
distribution function, 123
faradaic reaction, 121–122
quasi-random surfaces, 120–121
von Koch line segments, 118–119
red-ox porous electrode
absence of dc current, 82–84
concentration and potential gradient, 105–110
gradient concentration, 95–105
pores distribution, 110–113
presence of dc current, 85–95
V-grooved pore electrodes
ac signal penetration length, 79–81
electronic resistivity, 79
model kinetics, 78–79
pore geometry, 74–75
pore shape and size, 75–78
Electrochemical system
computer engineering aspects, 27–30
constructing modeling systems, 27
data communication, 28
software introduction, 29
mathematical modeling
definition, 2
geometric and physical properties specification, 4–6
postprocessing and analysis, 14–15
solution method specification, 6–13
solution process, 13–14
software designing, 26–27
Electronic resistivity, 79
Elliott, J.A, 178, 187
Equivalent circuit
battery, 433, 434, 455, 457
regressed capacitance and resistance, 510
symmetric supercapacitor, 432, 455
Equivalent electrical circuit model, 73–74
Ergun equation, 326
Euler, J., 312

F

Faradaic impedances
flat electrodes, 102–103
polarization resistance, 101
semicircle formation, 104–105
Thiele modulus, 103–104
Faradaic reactions, 124
Faraday's law, 289, 304
Farhat, Z.N., 309
Ferreira, P.J., 256, 259
Fickian diffusion model, 147, 152–153
Fick's equation, 96–97
Fick's law, 309, 314
Fimrite, J., 291
Finite-difference method
approximtion values, 17
electrochemical setup, 2
linear boundary value problem, 16
Finite element methods
blending functions, 25–26
electrochemical setup, 2
electrostatic problem, 18
five-point sampling Laplace equation, 21–23
Hermite polynomials and B splines, 24–25
interpolating polynomials, 23–24
weighted-residual formulation, 19–21
Fractal electrode model
CPE behavior, 119–120
distribution function, 123
faradaic reaction, 121–122
quasi-random surfaces, 120–121
von Koch line segments, 118–119
Freund, D.E., 49
Friede, W., 367
Frost heave, 382–383
Frozen state startup process
bootstrap start, 384–386

Index

cell-level models
 1-D model, 389–390
 nonisothermal models, 390–392
 semi-empirical approach, 388–389
 procedural strategies, 386–387
 stack-level models, 387–388
Fuel starvation
 electrode potentials, 262, 263
 localized, 263–264
Fuller, T.F., 371
Fundamental governing equations
 conservation equations
  charge conservation, 289
  energy conservation, 289–290
  mass conservation, 288–289
  principal equation types, 288
 kinetics
  Butler–Volmer expression, 285–286
  electrochemical reaction, 285
  electrode overpotential, 286–287
  hydrogen-oxidation reaction (HOR), 287
  oxygen-reduction reaction (ORR), 287–288
  platinum metal electrode, 286
 thermodynamics, 284–285
  efficiency definition, 285
  thermoneutral potential, 284

G

Gas channels, liquid water
 droplet models and GDL/gas-channel interface
  boundary condition, 352
  contact-angle hysteresis, 355
  drag force, 354
  droplet-specific studies, 353
  droplet-stability diagrams, 355–356
  Reynolds number, 354–356
  surface-tension force, 353–354
 gas-channel analysis
  cathode inlet relative humidity effects, 349
  droplets effect on cell performance, 350
  dynamics, 352
  multiphase models, 348–349
  water movement, flow channel, 350–351
 liquidwater transport mechanisms, 347–348
 reactant starvation, 347
Gas-diffusion layers (GDL), water movement, 320–321
 macroscopic analysis
  anisotropic properties, 334–338
  compression, 338–343
  microporous layers, 328–331
  temperature-gradient (heat-pipe) effect, 331–334
  two-phase-flow parameter determination, 325–327
 microscopic treatments
  capillary-pressure-saturation curves, 323
  capillary-tree and channeling mechanism, 318, 321
  dominant water pathways, 321–322
  microscopic models, 322
  mixed-wettability system, 324
  relative permeability, 322–323
Gauss, C.F., 427
Ge, J.B., 340
Geometric specification, 4–5
Ghausi, M.S., 426
Gibbs free energy, 284, 385
Gibbs–Thomson equation, 381
Giordano, A.B., 426
Goldman, S.P., 49, 61
Gostick, J.T., 335
Grotch, H., 55
Guilin, H., 374
Guilminot, E, 257
Guo, Q.Z., 319
Guvelioglu, G.H., 344

H

Haddad, A., 367
Hagstrom, S.A., 49, 50
Hammer, B., 195
Hardware-in-the-loop (HWIL) system, 503, 504
Hartree–Fock method, 39
Haykin, S., 426, 427

Heat-pipe effect. *See* Temperature-gradient effect
Heat-transfer coefficient, 387–388
He, G.L., 357
Helium atom
  basic sets
    advantages, 49
    doubling, advantages, 46
    exponential scale factors variation, 47
    ground state, convergence study, 49
    principles, 47
    screened hydrogenic energy, 48
  calculation methods
    Bohr–Sommerfeld quantum theory, 44
    configuration interaction (CI) calculation, 44–45
    ground state energy, 44–45
  computational methods, 49–50
  coordinate system, 39
  correlated variational basis sets
    basic set members, 41
    Hylleraas–Undheim–McDonald theorem, 42–43
    Pekeris shell, 41
    Rayleigh-Schrödinger variational theorem, 42
    trial wave function, 44
  Hartree–Fock method, 39
  Schrödinger equation, 38, 40
Henry, K.S., 383
He, S., 383
HEV. *See* Hybrid electric vehicle
Hickner, M.A., 279
Higher temperature operation
  advantages and disadvantages, 394
  cathode layers (CLs), 396
  novel membrane synthesis, 394–396
  polybenzimidazole (PBI) system, 396–397
  procedures, 392–394
High precision atomic theory
  atomic units
    Bohr radius, 37
    correction factor, 38
    of energy, 37

  correction methods
    mass polarization, 52–53
    quantum electrodynamic, 55–59
    relativistic, 53–55
  helium energy levels
    P-state ionization energy, 61
    QED breakdown, 62
    S-state ionization energy, 60
  Kepler's laws of planetary motion, 34
  mass polarization
    center-of-mass (CM) frame, 52
    normal and specific isotope shift, 52
    perturbation approach, 52–53
  nonrelativistic helium atom
    basis sets principles, 47
    calculations, 44–46
    computational methods, 49–50
    coordinate system, 39
    correlated variational basis sets, 40–44
    doubling the basis set, 46
    exponential scale factors variation, 47
    ground state, convergence study, 49
    Hartree–Fock method, 39
    Schrödinger equation, 38
    screened hydrogenic energy, 48
    variational basis sets, 50
  nonrelativistic hydrogen atom
    classical mechanics, 35
    gravitational interaction energy, 34
    Hamiltonian approach, 35
    ideas and concepts, 34
    radial function $R_{nl}(\rho)$, 37
    Rydberg formula, 36
    Schrödinger's equation, 36
  quantum electrodynamic corrections
    Bethe logarithm, 56–57
    electron-electron QED, 57
    electron self energy, Feynman diagram, 55
  relativistic corrections
    Briet interactions, 53–54
    finite nuclear mass and recoil terms, 55
Hill, R.N., 49

Index

Himanen, O., 342
Hishinuma, Y., 390, 391
Hitz, C., 76
Horgorvorst, H., 60
Hottinen, T., 342
HPSP. *See* Hybrid powertrain simulation program
Huang, V.M.W., 127
Huang, W., 371
Hu, J.W., 397
Hull cell simulations, 127
Hwang, J.J., 290, 364
HWIL. *See* Hardware-in-the-loop
Hybrid electric vehicles (HEVs)
  electric-traction system, 421
  electrochemical cells, 428
  NiMH state estimator, 472
  propulsion system architecture, 420–421
  zero-emission-vehicle (ZEV) range, 430
Hybrid powertrain simulation program (HPSP), 505
Hydrogen atom
  center-of-mass plus relative coordinates, 35
  classical mechanics, 35
  gravitational interaction energy, 34
  ideas and concepts, 34
  radial function $R_{nl}(\rho)$, 37
  Rydberg formula, 36
  Schrödinger's equation, 36
Hylleraas, E.A., 40, 44
Hylleraas–Undheim–McDonald theorem, 42–43

I

Ihonen, J., 340
Inoue, G., 362
In situ visualization, PEFC
  gas-diffusion layers, 280–281
  imaging techniques, 278–279

J

Jianren, F., 374
Jiao, K., 350
Jorcin, J.B., 127
Ju, H., 345
Ju, H.C., 345

K

Kabir, P.K., 56, 57
Karan, K., 331
Kaviany, M., 299
Kazim, A., 361
Keiser, H., 75
Kelvin equation, 305
Kernel identification, 30
Khandelwal, M., 335
Klahn, B., 43
Knudsen diffusion coefficient, 154
Knudsen number, 153–154
Kornyshev, A.A., 185
Korobov, V., 58
Korobov, V.I., 49
Kreuer, K.D., 291
Krishna, R., 148
Kroll, N.M., 58
Kulhavý, R., 469

L

Lai, M.C., 350
Laker, K.R., 426
Lambert W function, 442, 443
Lasia, A., 76
Lattice-Boltzmann model, 322–323
Least square methods
  algorithm, 458, 464
  cell hysteresis voltage, 459
  NiMH battery, 457–459
  open-circuit voltage, 459, 460
  regression voltage, 461
  robustness, 465
Lee, C.I., 349
Lee, S.J., 397
Lennard–Jones (LJ) potential, 207, 212
Leverett J-fuction, 322
Levy, R., 426
Liquid-phase transport
  frost heave, 382–383
  Gibbs–Thomson equation, 381–382
  pure-diffusion model, 383–384
Lithium-ion cell
  algorithm convergence test, 485, 490
  discharge power test, 486, 487
  electrochemical parameters, 489, 491
  open-circuit potential, 480, 481, 488
  power capability projections, 492

recursive skewness analysis, 483
skewness, determinant, and voltage error, 489
12-V Panasonic HV1255 VRLA module, 480
weight factor, 484
Lithium, variational basic sets, 50
Litster, S., 280, 364
Liu, H.T., 361
Li, X.G., 360
Ljnug, L., 428, 469
Loch, J.P.G., 383
Lucatorto, T.B., 60

M

Macroscale (bulk) transport
  general formulations
    flux vector components, 146
    mass conservation, 145–146
    physical model assumptions, 147–148
  scale analysis
    diffusional model, 148
    diffusion timescale, 151
    large Peclet number, 150–151
    small Peclet number, 149–150
Macroscopic analysis, GDL
  anisotropic properties
    collective anisotropies, 338, 339
    electronic and thermal conductivity, 335
    in-plane and through-plane permeability, 335–338
    relative Knudsen diffusivity, 334
  GDL compression
    consequence of, 339
    contact resistances, 340
    current-density distribution, 342
    and gas channel expansion, 340–341
    modeling complexity, 342–343
    in situ state *vs.* ex situ state, 338
  microporous layer (MPL)
    advantages, 328
    half-cell and full-cell models, 331
    hydrophilic pore fraction, 330–331
    oxygen transfer limitations, 329
    PEFC performance, 328–329
    water pressure and saturation profiles, 329–330
  temperature-gradient effect
    liquid-saturation contours, 333
    mass-transport limitation, 333–334
    nonisothermal modeling, 331–332
    water and thermal management coupling, 332
  two-phase-flow parameter determination
    absolute permeability, 325–326
    Carman–Kozeny equation, 326
    effective permeability, 325
    Leverett J-function, 327
    relative permeability, 326–327
Manke, I., 280
Mao, L., 391
Marangos, J.P., 60
Markicevic, B., 322
Mass transport
  description and representation, 142–144
  PS gas sensor timescales, 163–164
  sensor properties, 144–145
Mathematical modeling
  computer modeling, 3
  geometric and physical properties specification, 4–6
  plate electrode geometrics, 4
  postprocessing and analysis, 14–15
  solution method specification, 6–13
    analytic solutions, 6
    finite-difference methods, 8–10
    finite-element methods, 10–11
    Galerkin method, 12–13
    Laplace equation, 6–7
    sampling theory, 7–8
    solution process, 13–14
Mathias, M., 261
McIlrath, T.J., 60
Membrane modeling, PEFC
  concentrated solution theory
    binary friction model, 291–292
    Schlögl's equation, 293
    transport equations, 292
    water content calculation, 292–293

membrane microstructure, 291
membrane pretreatment, 290
other transport through membrane, 297–298
water content and properties
constraint treatment, 295–296
membrane coefficient, 296–297
molecular-dynamic-type models, 295
water-uptake isotherm models, 293–295
Mench, M.M., 335, 383
Meng, H., 344, 353
Meyers, J.P., 115, 253, 254, 257, 364
Microscale transport, 161–163
Miller, R.D., 382
Modeling process, 2
Model reference adaptive system (MRAS), 424–425
Moore, J.B., 469
Moore, R.M., 388
Morgan III, J.D., 49
MRAS. *See* Model reference adaptive system
Mueller, F., 366, 372
Multiscale formulation
diffusive timescale, 164–165
mass transport timescales, 163–164
simple adsorption model, 165–166
Munroe, N.D.H., 397

N

Nam, J.H., 299
Nanoscale transport
anlytical solution
orthogonal series expansion solution, 159–161
steady-state problem, 158–159
nanopores
diffusion model, 154–155
simplified model, 155–156
transient properties, 157–158
nanopores continuum assumption
adsorption and desorption time scales, 153–154
molecular length and time scale, 152–153

Navier–Stokes equations, 300, 361
Nazarov, I., 295
Nernst equation, 284, 286, 440, 441
Newman, J., 277, 294, 295, 297, 312, 332, 345, 358, 371, 431
Neyerlin, K.C., 287
Nguyen, T.V., 361
Nonnenmacher, W., 312
Nørskov, J.K., 195
Nyikos, L., 119, 120

O

O'Brian, T.R., 60
Ohm's law, 292, 306, 309
Okada, T., 294
Ong, I.J., 431
Oszcipok, M., 388
Ota, K.I., 256

P

Pachucki, K., 51, 58, 61, 63, 64
Paddison, S.J., 178, 187
Pajkossy, T., 119, 120
Parallel hybrid propulsion system, 421
Park, J., 360
Pasaogullari, U., 328, 331, 337
Peclet number, 149–151
Pekeris, C.L., 45
Pekeris shell, 41
Peltier coefficient, 290
PEMFC. *See* Proton-exchange membrane fuel cells
Penetrability coefficient, 111
Peng, J., 397
Perry, M.L., 263, 265
Pesaran, A.A., 376, 378, 387
Petersen, M.K., 178
Phaes field model, 26–27
Pharaoh, J.G., 335, 336
Phasors ratio, 70–71
Pillar, S., 426
Plackett, R.L., 427
Plate electrode geometry, 4
Platinum nanoparticle catalyst
carbon-supported
cyclic voltammograms, 251
electrochemical oxidation, 250

chemical state
  Pourbaix diagrams, 251
  solubility, 252
dissolution
  equilibrium concentration vs.
    electrode potential, 254
  potential cycling, 256, 257
  particle growth, 257–260
Poiseulle's equation, 326
Polymer electrolyte fuel cells (PEFCs)
  basic methodology
    geometric dimensionality, 281–282
    macroscopic and microscopic
      models, 281
    pseudo-dimensional models, 282
  catalyst-layer modeling
    active phase volume fraction, 309
    agglomerate-type structure,
      307–308
    impedance models, 317–319
    modeling equations, 309–315
    optimization analyses, 315–317
    two-phase and three-phase
      interface, 308
  cell-design strategies
    alternate cooling approaches,
      364–365
    gas-flow direction, 357–359
    interdigitated flow fields, 360–363
    optimal cell hydration, 357
    water-transport plates, 363–364
  cold-start process
    automotive process, 376–377
    frozen state startup process,
      384–392
    shutdown and freezing, 377–384
  continuous porous model, 117–118
  electron transport, 306–307
  fundamental governing equations
    conservation equations, 288–290
    kinetics, 285–288
    thermodynamics, 284–285
  gas channels, liquid water, 347–348
    droplet models and GDL/gas-
      channel interface, 352–357
    gas-channel analyses, 348–352
    liquidwater transport mechanisms,
      347–348
    reactant starvation, 347
  higher temperature operation
    advantages and disadvantages, 394
    cathode layers (CLs), 396
    novel membrane synthesis,
      394–396
    polybenzimidazole (PBI) system,
      396–397
    procedures, 392–394
  hydrogen, 170, 171
  low-relative-humidity operation
    3-D velocity profiles, 345
    membrane-cathode interface,
      345–347
    reactant stream humidification,
      343–344
  macroscopic modeling, 277
  materials modeling
    complex process, 176
    length scales, 175, 176
    multi-scale phenomena, 176
  membrane modeling
    concentrated solution theory,
      291–293
    membrane microstructure, 291
    membrane pretreatment, 290
    other transport through membrane,
      297–298
    water content and properties,
      293–297
  model implementation and boundary
    conditions, 319–320
  multilayered design, 171, 172
  nonuniformities, 343
  PEM and catalyst layer
    agglomerates, 205
    Carbon–Nafion–Water–Solvent
      (CNWS), 208
    coarse-grained molecular dynamics
      (CG-MD), 205, 208, 211
    complex interactions, 204
    computational approach, 206
    Coulombic interaction, 207
    Derjaguin–Landau–Verwey–
      Overbeek (DLVO), 213

Index

hydrated Nafion membrane, 211, 212
interaction parameters, 210
Lennard–Jones (LJ) potential, 207, 212
microstructure and pore size distribution, 204
site–site radial distribution function, 209
solvent dielectric constant, 210
structural complexity, 206
structural formation process, 205
structure–performance relationship, 204
platinum nanoparticle electrocatalysis
active site model, 200, 202
adsorption energies, 194
catalyst poison, 197
chronoamperometric current transients, 200, 202
complex surface reaction mechanism, 196
heterogeneous surface model, 198
hydrogen reduction kinetics, 195
kinetic modeling, 199
kinetic Monte Carlo (kMC) simulations, 201
methanol electrooxidation, 195
orbital Free DFTcalculation, 196, 197
specific exchange current density, 193
spillover effect, 194
Tafel-plots, 201, 203
transient current, 199
polarization curve, 274–275
polymer electrolyte membrane (PEM), 172
proton transport, 182–193
typical 7-layer structure, 171–172
random heterogeneous media
composite porous catalyst layers, 213
fractal internal surface, 214, 215
membrane electrode assemblies (MEAs), 213
scales reconciling
catalyst utilization, 231–235
cathode catalyst layer, 223–231
water management, 219–223
shutdown and freezing, 377
cell-level models, 379–384
stack-level models, 378–379
in situ visualization of water, 278–281
startup from frozen state, 384–387
cell-level models, 388–392
stack-level models, 387–388
structure and water
block-copolymer systems, 217
electro-osmotic drag effect, 209
Joule heating, 217
mass transport phenomena, 217
membrane dehydration, 216
pore size distributions, 218
two phase models, 218
transient operation and load changes
single-phase-flow models, 367–372
time-constant analysis, 366
two-phase-flow models, 372–376
water-management strategies, 365
two-phase flow
gas-diffusion layer, 298–299
gas-phase transport, 300–303
liquid and gas phase coupling, 303–306
liquid-phase transport, 300
Polymer electrolyte membrane (PEM)
proton transport
activation energy, 183, 184
Car-Parinello molecular dynamics, 191
charge transfer theory, 185
conductivity, 183
Coulomb barrier, 186
empirical valence bond (EVB) approach, 184–185
formation energy, 189, 190
microscopic mechanism, 184
molecular modelling, 187
objectives, 192
Poisson–Boltzmann theory, 185
sulfonate ions, 185

water binding and molecular mechanisms, 186
Zundel-ion, 187, 188
structural evolution, 182, 183
typical 7-layer structure, 171–172
water management
diffusion models, 221, 222
electro-osmotic coupling, 219
Gibbs free energy, 223
hydraulic permeation model, 221, 222
molar flux, 220
pressure gradient, 220
proton conductivity, 219
proton current density, 220
Porous electrodes
continuous porous model
diffusion pores equation, 114–115
impedancies evaluation, 115
polymer fuel cell, 117–118
principle, 113
solution theory, 115–117
cylindrical pore electrode model, 67–74
definition, 67
red-ox and double layer capacitance
absence of dc current, 82–84
concentration and potential gradient, 105–110
pores distribution, 110–113
presence of concentration gradient, 95–105
presence of dc current, 85–95
V-grooved pore electrodes, 74–81
Porous silicon (PS) gas sensors
analytical solutions, 142
chemical sensors, 141–142
dissolution process, 139–140
macroscale (bulk) transport
general formulations, 145–148
scale analysis, 148–151
mass transport
description and representation, 142–144
sensor response, 144–145
microscale transport, 161–163
multiscale formulation, 163–166
nanoscale transport
analytical solution, 158–161
nanopores continuum assumption, 152–154
nanopores diffusion model, 154–155
nanopores simplified model, 155–156
nanopore transient response, 157–158
visible photoluminescence (PL), 140–141
Postprocessing and analysis modeling, 14–15
Press, W.H., 467
Promislow, K., 295, 299
Proportional-integral-differential (PID) schemes, 424–425
Proton-exchange membrane fuel cells (PEMFCs)
alloy effects
crystallinity and exchange current densities, 267
degradation rate, 268
phosphoric acid system, 267
stability, 268
corrosion
carbon, 261–262
catalyst support, 260
fuel starvation
electrode potentials, 262, 263
localized, 263–264
platinum nanoparticle catalyst
carbon-supported, 250–251
chemical state, 251–253
dissolution, 253–257
particle growth, 257–260
start/stop cycling, 264–265
temperature and relative humidity, 266
Proton transport (PT)
activation energy, 183, 184
Car-Parinello molecular dynamics, 191
charge transfer theory, 185
conductivity, 183
Coulomb barrier, 186

Index

empirical valence bond (EVB)
   approach, 184–185
formation energy, 189, 190
microscopic mechanism, 184
molecular modelling, 187
objectives, 192
Poisson–Boltzmann theory, 185
sulfonate ions, 185
water binding and molecular
   mechanisms, 186
Zundel-ion, 187, 188

Q
Quan, P., 350

R
Rand, D.A., 256
Rao, R.M., 368
Rayleigh-Schrödinger variational
   theorem, 42
Red-ox porous electrode
   absence of dc current
      intermediate length pores, 82–83
      semi-infinite pores, 83–84
      shallow pores, 83
      transfer resistance, 82
   concentration and potential gradient
      diffusion coefficients, 105–106
      electroreduction process, 107–108
      impedance complex, 109–110
      limitations, 106–107
   gradient concentration
      current density–potential relation,
         95–96
      faradaic and double layer
         impedances, 101–105
      Fick's equation, 96–97
      limitations, 98–99
      linearized current, 99–100
      Thiele modulus, 97
   pores distribution
      distribution functions, 111–112
      Fredholm integral equation,
         112–113
      transmission line ladder network,
         110–111
   presence of dc current
      Butler–Volmer equation, 86
      current density, 85–86
      electrode impedances, 88–90
      semi-infinite length pores, 87
      simulated impedances, 93–95
      skewed impedances, 90–92
      Tafel curves, 87–88
Reiser, C.A., 263, 265
Rempel, A.W., 383
Rengaswamy, R., 368
Reverse current mechanism, 265
Reynolds number, 354–356
Roen, L.M., 261
Rolston, S.L., 60
Rost, J.-M., 49
Roudgar, A., 178

S
Salpeter, E.E., 44, 56, 57
Sansonetti, C.J., 60
Santhanagopalan, S., 426
Sapirstein, J., 55
Scanlan, J.O., 426
Schiff, B., 45
Schlögl's equation, 293
Schröder's paradox, 296
Schrödinger's equation, 36, 38, 40
Schulz, V.P., 323
Schwartz, C., 49
Semi-infinite pores plot, 83–84
Shah, A.A., 312, 316, 374
Shallow pores plot, 83
Shan, Y.Y., 371, 372
Shelyuto, V.A., 55
Shutdown and freezing process
   cell-level models
      liquid-phase transport, 381–384
      vapor-phase transport, 379–381
   stack-level models, 378–379
Sims, J.S., 49, 50
Single-phase-flow models
   isothermal transient model, 367–370
   lumped model
      0-D, 1-D, 2-D models, 368
      3-D isothermal model, 368–370
      membrane hydration effects,
         367–368
   nonisothermal transient model,
      370–372
   steady-state performance, 370–372

Sinha, P.K., 323
Smith, K.A., 426
SOC. *See* State of charge
Söderström, T., 428, 469
SOH. *See* State of health
Solidification process, 27
Solid oxide fuel cells (SOFCs), 170
Solution method specification
 analytic solutions, 6
 finite-difference methods, 8–10
 finite-element methods, 10–11
 Galerkin method, 12–13
 Laplace equation, 6–7
 sampling theory, 7–8
Song, D.T., 316, 374
Song, H.K., 111
SOP. *See* State of power
Spohr, E., 178, 185
Springer, T.E., 117, 319, 389
Srinivasan, S., 312
State estimators
 algorithm verification and validation
  capacitor voltage, 506
  hardware-in-the-loop (HWIL) system, 503, 504
  maximum discharge power, 508
  test protocol, 509
  velocity *vs* time relationship, 509, 510
 generalized weighted recursive least squares
  algorithm, 466, 470
  instantaneous error, 466
  matrix system of equations, 469–470
  parameter models, 473–475
  weight factor, 468
 least square method
  algorithm, 458, 464
  cell hysteresis voltage, 459
  NiMH battery, 457–459
  open-circuit voltage, 459, 460
  regression voltage, 461
  robustness, 465
 lithium-ion cell
  algorithm convergence test, 485, 490
  discharge power test, 486, 487
  electrochemical parameters, 489, 491
  open-circuit potential, 480, 481, 488
  power capability projections, 492
  recursive skewness analysis, 483
  skewness, determinant, and voltage error, 489
  12-V Panasonic HV1255 VRLA module, 480
  weight factor, 484
 regression analysis
  determinant value, 475–476
  skewness, 476–477
 state of power (SOP)
  constant-voltage, 478–480
  maximum discharge power, 477
 variable forgetting factors
  high-power-density lithium ion battery, 493
  optimized values, 499
  power projections, 501
State of charge (SOC)
 experiment-theory comparison, 438, 439
 regressed combined and voltage-based, 465
 schematic representation, 418
State of health (SOH)
 definition, 501
 electrochemical parameters, 491
 schematic representation, 419
State of power (SOP)
 composite power system, 420
 constant-voltage, 478–480
 maximum discharge power, 477
Stearns, S.D, 426
Stefan–Maxwell equations, 291, 300–302
Stenger, H.G., 344
St-Pierre, J., 294, 358
Sucher, J., 56
Sundaresan, M., 388
Surface diffusion coefficient, 155

Index

T
Tafel curves, 87–89
Tate, E.D., 425
Temkin, A., 53
Temperature-gradient effect, 331–334
Thermoneutral potential, 284
Thiele modulus, 97, 102–103
Thomas-Alyea, K.E., 312
Three-phase electric-traction system, 421
Tiedemann, W., 312
Tobias, C.W., 312
Transfer resistance, 82
Transient operation, load changes
    single-phase-flow models, 367–372
    time-constant analysis, 366
    water-management strategies, 365
Transmission line impedance equation, 72–73
Tretter, S.A., 427
Two-phase-flow models
    cell uniform temperature, 372–374
    1-D and 3-D CFD model, 374–375
    gas-diffusion layer, 298–299
    gas-phase transport
        dusty-gas model, 302–303
        gas-phase volume fraction, 301
        Knudsen diffusion, 302
        Stefan–Maxwell equations, 300–301
    liquid and gas phase coupling
        capillary pressure, 303
        cathode GDL/CL interface, 304
        Kelvin equation, 305
        multiphase mixture model, 306
    liquid-phase transport, 300
    oxygen mole fraction, 375–376

U
Ubachs, W., 60
Udell, K.S., 323, 327

V
Vahidi, A., 494
Van Valkenburg, M.E., 426
Van Zee, J.W., 345, 368, 374
Vapor-phase transport, 379–381
Vassen, W., 60
V-grooved pore electrodes
    ac signal penetration length, 79–81
    electronic resistivity, 79
    model kinetics, 78–79
    pore geometry, 74–75
    pore shape and size, 75–78
Visible photoluminescence, 140–141
Vogel, H.J., 323
Von Koch line segments, fractal model, 118–119
Vorobev, A., 367
Voth, G.A., 178, 185

W
Walbran, S., 185
Wang, C.Y., 277, 323, 328, 332, 344, 345, 353, 368, 391, 396
Wang, G.Q., 308
Wang, L., 361
Wang, Q.P., 316
Wang, Y., 332, 345, 368
Warshel, A., 184
Weber, A.Z., 294, 295, 297, 299, 315, 332, 345, 358, 363, 373
Weighted recursive least squares (WRLS)
    algorithm, 466, 470
    application, 425–426
    instantaneous error, 466
    lead acid and lithium ion cell characteristics, 472
    matrix system equations, 469–470
    parameter models, 473–475
    parameter regression, 425
    statistics, 483
    step-by-step comparison, 428
    weight factor, 468, 484
Weighted-residual formulation, 19–21
Wen, J., 60
Wesselingh, J.A., 148
Westbrook, N., 60
White, R.E., 312, 319, 426
Widrow, B., 426
Wiegman, H.L.N., 425
Wiezell, K., 319
Wilkinson, D.P., 358
Williams, M.V., 335

Wintgen, D., 49
Wittenmark, B., 428
WRLS. *See* Weighted recursive least squares
Wyllie equation, 322

Y
Yamada, H., 361
Yan, W.M., 361, 368
Yan, Z.-C., 59
Yasuda, K., 257

Yelkhovsky, A., 58
Yi, J.S., 349
Young–Laplace equation, 227
Yu, H.M., 367
Yu, P.T., 263, 265

Z
Zhang, F.Y., 354, 357
Zhan, Z.G., 350
Ziegler, C., 367, 373
Zou, J., 361

Printed in the United States of America